Essentials and Examples of Applied Mathematics

Second Edition

Professor William W. Guo

Central Queensland University Australia

Pearson Originals is an imprint of Pearson Australia, a division of Pearson Australia Group Pty Ltd. It has been established to provide academics throughout Australia and New Zealand with fast and efficient access to the printing, warehousing and distribution services of Australia's leading educational publisher, ensuring a smooth supply channel to your campus bookseller.

For more information about the Pearson Originals service, contact the Editorial Department, Pearson Australia, 707 Collins Street, Melbourne, VIC 3008.

Project Management Team Leader: Jill Gillies
Production Manager: Katie Young
Courseware Associate: Jessica Darnell

Print ISBN 978-0-655-70362-4
eBook ISBN 978-0-655-70363-1

Printed and bound in Australia by The SOS Print + Media Group

Pearson Australia
707 Collins Street
MELBOURNE VIC 3008
www.pearson.com.au

Dr. William Guo is a professor in mathematics education and applied computing with Central Queensland University Australia since 2007. He is specialized in teaching applied mathematics for engineering, science and education students at all levels. Collaborated with Pearson Australia, he has published three original mathematics textbooks to facilitate effective learning and application of mathematics for his students in engineering, science and education studies since 2014. His teaching excellence has been recognized with multiple university teaching awards since 2012. He has chaired and/or delivered keynote speeches at many international conferences and presented Master Classes on mathematics in science for secondary students organized by local schools and colleges. He has initiated and served as the inaugural Editor in Chief for international journal *STEM Education* (STEME) with the American Institute of Mathematical Sciences (AIMS) since 2020. Professor Guo has been an Editorial Board Member for *MDPI Mathematics* since 2020 and the Sectional Editor in Chief for *Electronic Research Archive* (ERA) since 2022. He has authored or co-authored over one hundred papers in international journals and conference proceedings in applied computational intelligence, applied mathematics and data analysis, and STEM Education. His scholarship in learning and teaching was recognized nationally by Australian Research Council (ARC) as an Expert Peer Reviewer in STEM Education for the Excellence in Research of Australia (ERA) assessment.

Common Formulas and Rules

1. Algebraic rules

$$(a\pm b)^2 = a^2 \pm 2ab + b^2,\quad (a+b)(a-b) = a^2 - b^2,\quad (x+a)(x+b) = x^2 + (a+b)x + ab$$

$$(a+b)^3 = a^3 + 3a^2b + 3ab^2 + b^3,\quad (a-b)^3 = a^3 - 3a^2b + 3ab^2 - b^3$$

$$\frac{b}{a} \pm \frac{c}{a} = \frac{b \pm c}{a},\quad \frac{a}{b} \pm \frac{c}{d} = \frac{ad \pm bc}{bd},\quad \frac{a}{b}\cdot\frac{c}{d} = \frac{ac}{bd},\quad \frac{a}{b} \div \frac{c}{d} = \frac{ad}{bc},\quad \sqrt[n]{\frac{a}{b}} = \frac{\sqrt[n]{a}}{\sqrt[n]{b}}$$

$$ax^2 + bx + c = 0 \longleftarrow r_{1,2} = \frac{-b \pm \sqrt{b^2 - 4ac}}{2a} \qquad (a \neq 0)$$

$$\begin{cases} a_1x + b_1y = c_1 \\ a_2x + b_2y = c_2 \end{cases} \longleftarrow \begin{cases} x = \dfrac{b_2c_1 - b_1c_2}{a_1b_2 - a_2b_1} \\ y = \dfrac{a_1c_2 - a_2c_1}{a_1b_2 - a_2b_1} \end{cases}$$

$$\frac{1}{(x+a)(x+b)} = \frac{1}{b-a}\left(\frac{1}{x+a} - \frac{1}{x+b}\right),\quad \frac{cx+d}{(x+a)(x+b)} = \frac{d-ac}{b-a}\frac{1}{x+a} - \frac{d-bc}{b-a}\frac{1}{x+b}$$

2. Rules for exponents and logarithms

$$a^0 = 1,\quad a^{m+n} = a^m a^n,\quad a^{mn} = (a^m)^n = (a^n)^m,\quad a^{m-n} = \frac{a^m}{a^n},\quad a^{-n} = \frac{1}{a^n},\quad \left(\frac{x}{y}\right)^{-n} = \left(\frac{y}{x}\right)^n$$

$$a^m b^m = (ab)^m,\quad \frac{a^m}{b^m} = \left(\frac{a}{b}\right)^m = a^m b^{-m},\quad a^m b^{-n} = \frac{a^m}{b^n},\quad a^{\log_a b} = b,\quad e^{\ln b} = b,\quad x^{1/n} = \sqrt[n]{x}$$

$$\sqrt[n]{x^n} = (\sqrt[n]{x})^n = x,\quad x^{m/n} = \sqrt[n]{x^m} = (\sqrt[n]{x})^m,\quad \sqrt[n]{\sqrt[m]{x}} = \sqrt[m]{\sqrt[n]{x}} = \sqrt[mn]{x},\quad x^{-1/n} = \frac{1}{x^{1/n}} = \frac{1}{\sqrt[n]{x}}$$

$$(xy)^{1/n} = \sqrt[n]{x}\sqrt[n]{y} = \sqrt[n]{xy},\quad (xy)^{-1/n} = \frac{1}{\sqrt[n]{xy}} = \frac{1}{\sqrt[n]{x}\sqrt[n]{y}},\quad \left(\frac{x}{y}\right)^{1/n} = \sqrt[n]{\frac{x}{y}} = \frac{\sqrt[n]{x}}{\sqrt[n]{y}},\quad \left(\frac{x}{y}\right)^{-1/n} = \sqrt[n]{\frac{y}{x}}$$

$$\log_a 1 = 0,\quad \log_a a = 1,\quad \log_a a^m = m,\quad \log_a c = \frac{\log_b c}{\log_b a},\quad \log_a b = \frac{1}{\log_b a}$$

$$\log(ab) = \log a + \log b,\quad \log(ab)^m = m(\log a + \log b),\quad \log(a^m b^n) = m\log a + n\log b$$

$$\log\left(\frac{a}{b}\right) = \log a - \log b,\quad \log\left(\frac{a}{b}\right)^m = m(\log a - \log b),\quad \log\left(\frac{a^m}{b^n}\right) = m\log a - n\log b$$

3. Trigonometric functions and identities

$$\tan x = \frac{\sin x}{\cos x} = \frac{1}{\cot x}, \quad \cot x = \frac{\cos x}{\sin x} = \frac{1}{\tan x}, \quad \sec x = \frac{1}{\cos x}, \quad \csc x = \frac{1}{\sin x}$$

$$\sin^2 x + \cos^2 x = 1, \quad \tan x \cot x = 1, \quad \csc x \sin x = 1, \quad \cos x \sec x = 1$$

$$\sin(-x) = -\sin x, \quad \cos(-x) = \cos x, \quad \tan(-x) = -\tan x$$

$$\sin(90° - x) = \cos x, \quad \cos(90° - x) = \sin x, \quad \tan(90° - x) = \cot x$$

$$\sin(90° + x) = \cos x, \quad \cos(90° + x) = -\sin x, \quad \tan(90° + \alpha) = -\cot x$$

$$\sin(180° - x) = \sin x, \quad \cos(180° - x) = -\cos x, \quad \tan(180° - x) = -\tan x$$

$$\sin(180° + x) = -\sin x, \quad \cos(180° + x) = -\cos x, \quad \tan(180° + x) = \tan x$$

$$\sin(360° - x) = -\sin x, \quad \cos(360° - x) = \cos x, \quad \tan(360° - x) = -\tan x$$

$$\sin(360° + x) = \sin x, \quad \cos(360° + x) = \cos x, \quad \tan(360° + x) = \tan x$$

$$\sin(n \cdot 360° - x) = -\sin x, \quad \cos(n \cdot 360° - x) = \cos x, \quad \tan(n \cdot 360° - x) = -\tan x$$

$$\sin(n \cdot 360° + x) = \sin x, \quad \cos(n \cdot 360° + x) = \cos x, \quad \tan(n \cdot 360° + x) = \tan x$$

$$\sin 2x = 2\sin x \cos x, \quad \cos 2x = \cos^2 x - \sin^2 x = 2\cos^2 x - 1 = 1 - 2\sin^2 x, \quad \tan 2x = \frac{2\tan x}{1 - \tan^2 x}$$

$$\sin\frac{x}{2} = \pm\sqrt{\frac{1-\cos x}{2}}, \quad \cos\frac{x}{2} = \pm\sqrt{\frac{1+\cos x}{2}}, \quad \tan\frac{x}{2} = \pm\sqrt{\frac{1-\cos x}{1+\cos x}} = \frac{\sin x}{1+\cos x} = \frac{1-\cos x}{\sin x}$$

$$\sin(x - y) = \sin x \cos y - \sin y \cos x, \quad \cos(x - y) = \cos x \cos y + \sin x \sin y$$

$$\sin(x + y) = \sin x \cos y + \sin y \cos x, \quad \cos(x + y) = \cos x \cos y - \sin x \sin y$$

$$\tan(x + y) = \frac{\tan x + \tan y}{1 - \tan x \tan y}, \quad \tan(x - y) = \frac{\tan x - \tan y}{1 + \tan x \tan y}$$

$$\sin x + \sin y = 2\sin\frac{x+y}{2}\cos\frac{x-y}{2}, \quad \sin x - \sin y = 2\cos\frac{x+y}{2}\sin\frac{x-y}{2}$$

$$\cos x + \cos y = 2\cos\frac{x+y}{2}\cos\frac{x-y}{2}, \quad \cos x - \cos y = -2\sin\frac{x+y}{2}\sin\frac{x-y}{2}$$

$$\sin x \sin y = -\frac{1}{2}[\cos(x+y) - \cos(x-y)], \quad \sin x \cos y = \frac{1}{2}[\sin(x+y) + \sin(x-y)]$$

$$\cos x \cos y = \frac{1}{2}[\cos(x+y) + \cos(x-y)], \quad c^2 = a^2 + b^2 \text{ (Pythagorean theorem for right triangles)}$$

$$\frac{a}{\sin A} = \frac{b}{\sin B} = \frac{c}{\sin C} \Longleftrightarrow \frac{\sin A}{a} = \frac{\sin B}{b} = \frac{\sin C}{c} \text{ (law of sines)}$$

$$\left.\begin{aligned} a^2 &= b^2 + c^2 - 2bc\cos A \\ b^2 &= a^2 + c^2 - 2ac\cos B \\ c^2 &= a^2 + b^2 - 2ab\cos C \end{aligned}\right\} \Longleftrightarrow \begin{cases} \cos A = \dfrac{b^2 + c^2 - a^2}{2bc} \\ \cos B = \dfrac{a^2 + c^2 - b^2}{2ac} \\ \cos C = \dfrac{a^2 + b^2 - c^2}{2ab} \end{cases} \text{ (law of cosines)}$$

4. Properties of some geometric objects

Geometric object		Property
Rectangle	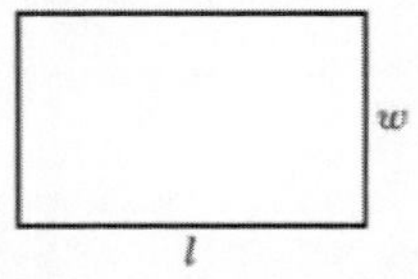	l = length, w = width Area: $A = lw$ Perimeter: $P = 2l + 2w = 2(l + w)$
Square	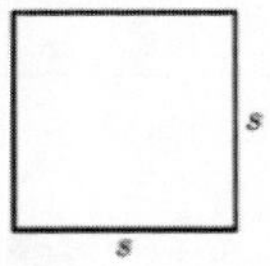	s = side Area: $A = s^2$ Perimeter: $P = 4s$
Right triangle	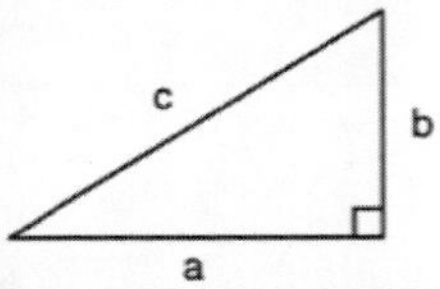	Area: $A = \frac{1}{2}ab$ Perimeter: $P = a + b + c$ Pythagorean theorem: $a^2 + b^2 = c^2$
Triangle	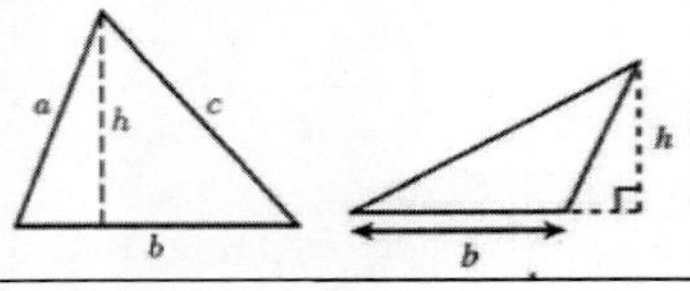	b = base, h = height Area: $A = \frac{1}{2}bh$ Perimeter: $P = a + b + c$
Equilateral triangle	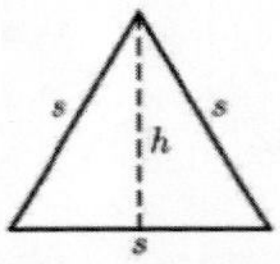	s = side, height: $h = \frac{\sqrt{3}}{2}s$ Area: $A = \frac{1}{2}sh = \frac{\sqrt{3}}{4}s^2$ Perimeter: $P = 3s$
Parallelogram	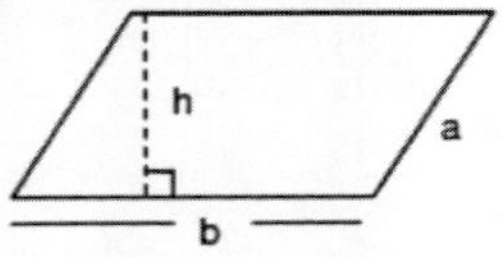	b = bases, h = height, a = sides Area: $A = bh$ Perimeter: $P = 2a + 2b = 2(a + b)$
Trapezoid	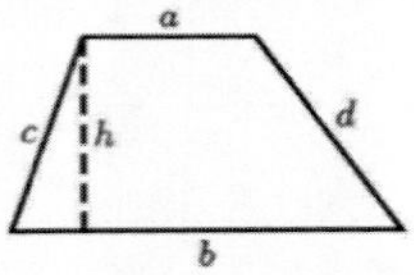	a,b = bases, h = height, c,d = sides Area: $A = \frac{1}{2}(a + b)h$ Perimeter: $P = a + b + c + d$
Circle	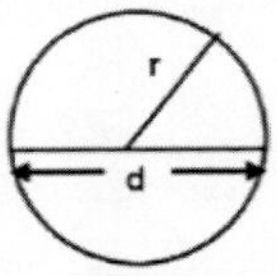	r = radius, d = diametre = $2r$ Area: $A = \pi r^2$ Circumference: $C = 2\pi r = \pi d$
Annulus	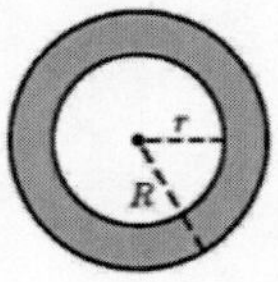	r = inner radius, R = outer radius Area: $A = \pi(R^2 - r^2)$
Ellipse	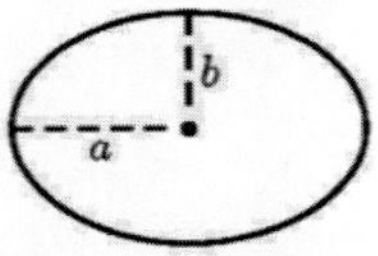	a = semimajor axis, b = semiminor axis Area: $A = \pi ab$

Rectangular solid		l = length, w = width , h = height Volume: $V = lwh$ Surface area: $S = 2lw + 2lh + 2wh$
Cube		s = side Volume: $V = s^3$ Surface area: $S = 6s^2$
Right circular cylinder		r = radius, h = height Volume: $V = \pi r^2 h$ Surface area: $S = 2\pi rh + 2\pi r^2$
Right circular cone		r = radius, h = height Volume: $V = \frac{1}{3}\pi r^2 h$ Surface area: $S = \pi r\sqrt{r^2 + h^2} + \pi r^2$
Frustum of cone		r = top radius, R = base radius, h = height, s = slant height Volume: $V = \frac{\pi}{3}(r^2 + rR + R^2)h$ Surface area: $S = \pi s(r + R) + \pi r^2 + \pi R^2$
Square pyramid		h = height, s = side Volume: $V = \frac{1}{3}s^2 h$ Surface area: $S = s(s + \sqrt{s^2 + 4h^2})$
Sphere		r = radius Volume: $V = \frac{4}{3}\pi r^3$ Surface area: $S = 4\pi r^2$

5. Limits

$$\lim_{x\to a}[f(x)\pm g(x)] = \lim_{x\to a} f(x) \pm \lim_{x\to a} g(x);\ \lim_{x\to a} f(x)g(x) = \lim_{x\to a} f(x)\cdot \lim_{x\to a} g(x)$$

$$\lim_{x\to a}\frac{f(x)}{g(x)} = \frac{\lim_{x\to a} f(x)}{\lim_{x\to a} g(x)} \text{ if } \lim_{x\to a} g(x)\neq 0;\ \lim_{x\to a}[kf(x)] = k\lim_{x\to a} f(x),\quad k\text{: constant}$$

$$\lim_{x\to a}[f(x)]^n = [\lim_{x\to a} f(x)]^n \ \&\ \lim_{x\to a}\sqrt[n]{f(x)} = \sqrt[n]{\lim_{x\to a} f(x)},\ n\text{: positive integer}$$

If $f(x)\le g(x)\le h(x)$ & $\lim_{x\to a} f(x) = \lim_{x\to a} h(x) = L,\ \lim_{x\to a} g(x) = L$

$$\lim_{x\to\infty}(1+\frac{1}{x})^x = e \rightleftharpoons \lim_{x\to 0}(1+x)^{\frac{1}{x}} = e \longrightarrow \lim_{x\to 0}\frac{e^x-1}{x} = 1$$

$$\lim_{x\to\infty}(1-\frac{1}{x})^x = \frac{1}{e} \rightleftharpoons \lim_{x\to 0}(1-x)^{\frac{1}{x}} = \frac{1}{e};\quad \lim_{x\to\infty}(1+\frac{a}{x})^x = e^a \rightleftharpoons \lim_{x\to 0}(1+ax)^{\frac{1}{x}} = e^a$$

$$\lim_{x\to 0}\frac{\sin x}{x} = 1 \rightleftharpoons \lim_{x\to 0}\frac{x}{\sin x} = 1;\quad \lim_{x\to 0}\frac{\sin ax}{x} = a \rightleftharpoons \lim_{x\to 0}\frac{x}{\sin ax} = \frac{1}{a}$$

$$\lim_{x\to 0}\frac{\tan x}{x} = 1 \rightleftharpoons \lim_{x\to 0}\frac{x}{\tan x} = 1$$

6. Rules for differentiation

$$\frac{dy}{dx} = y' = f'(x) = \lim_{\delta x\to 0}\frac{\delta y}{\delta x} = \lim_{\delta x\to 0}\frac{f(x+\delta x)-f(x)}{\delta x}$$

$(cu)' = cu'$ (c: constant)

$(au+bv)' = au'+bv'$ (a, b: constants)

$(uv)' = u'v+uv'$ (product rule)

$(\frac{u}{v})' = \frac{u'v-uv'}{v^2}$ (quotient rule)

$\frac{du}{dt} = \frac{du}{dx}\frac{dx}{dt}$ (chain rule)

$(\ln x)' = \frac{1}{x}$ (logarithmic rule)

$[\ln f(x)]' = \frac{f'(x)}{f(x)}$ (logarithmic rule)

$\frac{dy}{dx} = \frac{\frac{dy}{dt}}{\frac{dx}{dt}}$ (parametric rule)

7. Common derivatives

$f(x)$	$f'(x)$	$f(x)$	$f'(x)$
c (constant)	0	$\sin ax$	$a\cos ax$
x^r (r: real number)	rx^{r-1}	$\cos ax$	$-a\sin ax$
x ($r=1$)	$1 \equiv \frac{d(x)}{dx} = (x)'$	$\tan ax$	$a\sec^2 ax = \frac{a}{\cos^2 ax}$
$\frac{1}{x}$ ($r=-1$)	$\frac{-1}{x^2}$	$\cot ax$	$-a\csc^2 ax = \frac{-a}{\sin^2 ax}$
$\sqrt{x}$ ($r=\frac{1}{2}$)	$\frac{1}{2\sqrt{x}}$	$\sec ax$	$\frac{a\sin ax}{\cos^2 ax}$
e^{rx}	re^{rx}	$\csc ax$	$\frac{-a\cos ax}{\sin^2 ax}$
e^x ($r=1$)	e^x	$\arcsin x$	$\frac{1}{\sqrt{1-x^2}}$
e^{-x} ($r=-1$)	$-e^{-x}$	$\arccos x$	$\frac{-1}{\sqrt{1-x^2}}$
a^{rx} ($a>0$)	$ra^{rx}\ln a$	$\arctan x$	$\frac{1}{1+x^2}$
x^{ax} ($x>0$)	$ax^{ax}(1+\ln x)$	$\sinh ax$	$a\cosh ax$
$\ln x$	$\frac{1}{x}$	$\cosh ax$	$a\sinh ax$
$\log_a x$	$\frac{1}{x\ln a}$	$\tanh ax$	$\frac{a}{\cosh^2 ax}$

8. Rules for integration

$$\int cf(x)dx = c\int f(x)dx;\ \int[af(x)+bg(x)]dx = a\int f(x)dx + b\int g(x)dx \quad \text{(a,b,c: constants)}$$

$$\int f'(x)dx = \int d[f(x)] = f(x)+c;\ \int uv'dx = uv - \int u'vdx \quad \text{(integration by parts)}$$

$$\int f(x)dx = \int f(\varphi(t))\varphi'(t)dt \ (\text{let } x=\varphi(t));\ \int\frac{f'(x)dx}{f(x)} = \ln f(x)+c \quad \text{(logarithmic rule)}$$

$$\text{If } F(x) = \int f(x)dx,\ \int_a^b f(x)dx = F(x)\Big|_a^b = F(b)-F(a) \ (\textit{Newton} - \textit{Leibniz formula})$$

$$\int_a^b f(x)dx = -\int_b^a f(x)dx;\ \int_a^a f(x)dx = 0;\ f(x) = \frac{d}{dx}\int_a^x f(x)dx \quad (a<x)$$

$$\int_{-a}^a f(x)dx = \begin{cases} 2\int_0^a f(x)dx & \text{if } f(x) \text{ is an even function.} \\ 0 & \text{if } f(x) \text{ is an odd function.} \end{cases}$$

9. Common integrals

	$f(x)$	$f(x)dx = F'(x)dx$	$d[F(x)]$	$F(x) = \int f(x)dx = \int d[F(x)]$
1	0	0	$d[c]$	c (constant)
2	a (constant)	$ax^0 dx$	$d[ax]$	$ax + c$
3	$x^r\ (r \neq -1)$	$x^r dx$	$d[\frac{1}{r+1}x^{r+1}]$	$\frac{1}{r+1}x^{r+1} + c$
4	$\frac{1}{\sqrt{x}}\ (r = -\frac{1}{2})$	$x^{-\frac{1}{2}}dx$	$d[2\sqrt{x}]$	$2\sqrt{x} + c$
5	$\frac{1}{x^2}\ (r = -2)$	$x^{-2}dx$	$d[-\frac{1}{x}]$	$-\frac{1}{x} + c$
6	e^{rx}	$e^{rx}dx$	$d[\frac{1}{r}e^{rx}]$	$\frac{1}{r}e^{rx} + c$
7	e^x	$e^x dx$	$d[e^x]$	$e^x + c$
8	e^{-x}	$e^{-x}dx$	$d[-e^{-x}]$	$-e^{-x} + c$
9	$\frac{1}{x}\ (x > 0)$	$\frac{dx}{x}$	$d[\ln x]$	$\ln x + c$
10	$\sin ax$	$\sin axdx$	$d[-\frac{1}{a}\cos ax]$	$-\frac{1}{a}\cos ax + c$
11	$\cos ax$	$\cos axdx$	$d[\frac{1}{a}\sin ax]$	$\frac{1}{a}\sin ax + c$
12	$\sinh ax$	$\sinh axdx$	$d[\frac{1}{a}\cosh ax]$	$\frac{1}{a}\cosh ax + c$
13	$\cosh ax$	$\cosh axdx$	$d[\frac{1}{a}\sinh ax]$	$\frac{1}{a}\sinh ax + c$

10. Geometric applications of definite integration

Physical plane area:

$$A = \int_a^c [f(x) - g(x)]dx + \int_c^d [g(x) - f(x)]dx \leftarrow f(x) > g(x) \text{ in } [a,c]; f(x) < g(x) \text{ in } [c,b]$$

Volume of solids of revolution:

$$V = \pi\int_a^b y^2 dx = \pi\int_a^b [f(x)]^2 dx \text{ or } V = \pi\int_c^d x^2 dy = \pi\int_c^d [g(y)]^2 dy$$

Length of arc:

$$L = \int_a^b \sqrt{1 + (\frac{dy}{dx})^2}\, dx \text{ or } L = \int_a^b \sqrt{(\frac{dx}{dt})^2 + (\frac{dy}{dt})^2}\, dt$$

Surface area of solid of revolution:

$$S = 2\pi\int_a^b y\sqrt{1 + (\frac{dy}{dx})^2}\, dx \text{ or } L = 2\pi\int_a^b y\sqrt{(\frac{dx}{dt})^2 + (\frac{dy}{dt})^2}\, dt$$

11. Taylor and Maclaurin series

$$f(x)=f(a)+f'(a)(x-a)+f''(a)\frac{(x-a)^2}{2!}+...+f^{(n)}(a)\frac{(x-a)^n}{n!}+... \text{ Taylor series}$$

$$f(x)=f(0)+f'(0)x+f''(0)\frac{x^2}{2!}+...+f^{(n)}(0)\frac{x^n}{n!}+... \text{ Maclaurin series}$$

$$e^x=1+x+\frac{x^2}{2!}+\frac{x^3}{3!}...+\frac{x^n}{n!}+...=\sum_{n=0}^{\infty}\frac{x^n}{n!} \qquad -\infty<x<\infty$$

$$\sin x=x-\frac{x^3}{3!}+\frac{x^5}{5!}-\frac{x^7}{7!}...=\sum_{n=1}^{\infty}\frac{(-1)^{n-1}x^{2n-1}}{(2n-1)!} \qquad -\infty<x<\infty$$

$$\cos x=1-\frac{x^2}{2!}+\frac{x^4}{4!}-\frac{x^6}{6!}...=\sum_{n=0}^{\infty}\frac{(-1)^n x^{2n}}{(2n)!} \qquad -\infty<x<\infty$$

$$\frac{1}{1-x}=1+x+x^2+x^3...+x^n+...=\sum_{n=0}^{\infty}x^n \qquad -1<x<1$$

$$\frac{1}{1+x}=1-x+x^2-x^3...+x^n+...=\sum_{n=0}^{\infty}(-1)^n x^n \qquad -1<x<1$$

Preface (second edition)

The first edition of this textbook was published by Pearson in 2018. Since then it has been used by engineering students in their first-year studies and by education (and other) students in their first-year and second-year studies at Central Queensland University Australia. It has made a great impact on improving learning experience and outcomes of hundreds of students since 2018. This is evidenced by the overwhelming positive feedback from so many students who used this textbook from 2018 to 2020 when I coordinated and taught those mathematics courses. Here are some anonymous comments from our past students:

2018

- The textbook was definitely the best aspect of the unit. Most things were clearly explained, with lots of examples and diagrams. Very helpful.
- Textbook was great, had a lot of information that was needed for the assignments and weekly learning.
- Having a textbook tailored exactly to the unit was great.

2019

- The textbook although difficult to understand at first, becomes a key asset in succeeding at this course. It is like a dictionary for mathematics and also has many challenging exercises to practice and hone in your skills. This will be especially useful to refer back to as the remainder of my degree lacks further mathematic units. Thanks to William for creating the textbook!
- Good teacher and good textbook.
- Moodle site combined with the textbook and learning activities where easy to navigate. I loved being able to read the textbook, then just skip ahead in the recorded zooms to the parts I had trouble with.

2020

- Textbook had topics in a coherent order with course content that made study more flowing instead of jumping from cover to cover between exercises!
- The best aspect I have found in my study is the lecturer and the textbook.
- The best aspect of the course was the textbook which provides many worked examples.

Such overwhelming support from so many students made two of those mathematics courses the "CQUniversity Student Voice Commendation Award for Educators of the Year" in 2018 and 2019 respectively.

This new edition keeps the overall structure of the original version with more over 500 worked examples solved by the author by hand to support student's self-learning and understanding of applied mathematics with this book. The enhancement of the new edition is mainly in the calculus part. This allows students in all STEM disciplines to use it as either a textbook for multiple semesters or a reference book in applied mathematics from basics to elementary calculus. Solutions to all exercises are also included at the end of each chapter. The comprehensive coverage makes this book a general purpose textbook for many disciplines in universities over the world. This is a useful mathematics handbook for mathematics teachers in secondary schools as well.

My sincere appreciation goes to my students and colleagues at Central Queensland University for their encouragement and support all the time. Special thanks also go to the Customs Team at Pearson Australia for their support and assistance since 2013.

Professor William Guo
Central Queensland University, Rockhampton, Australia
July 2020

General Study Guide

This study guide was crafted by the author in 2014 and has been recommended to students in multiple mathematics courses since then. A large majority of students, particularly distance students, have found that this study guide has provided them with an effective and efficient way to approach, familiarise, comprehend, and eventually master a new topic in mathematics. You may find this is useful in guiding your study too.

1. Read through the designated sections of the textbook before a lecture. Particularly pay attention to the new terms and procedures of problem solving demonstrated in the examples.
2. Make a note or mark where you have any doubt. Then try to get the answer from the lecture or tutorial (either in class or through the videos that would be uploaded online after the class). If still unclear, ask your tutor or coordinator for further help.
3. First read and try to understand the solved examples in the text; then close the book and try to solve these examples the same way in the text. If you get the same outcome, congratulate yourself and move onto solving similar and new problems.
4. During your study, wherever you encounter a problem that requires knowledge and techniques you learnt before but you feel unclear, go back to review that part. It's very likely the same knowledge and techniques will be required again later.

Table of Contents

CHAPTER 1

1 Review of Basic Algebra

CHAPTER OBJECTIVES

- Review real numbers and arithmetic operations
- Review exponents, roots and logarithms with real numbers
- Review algebraic expressions and operations
- Review factorizing algebraic expressions
- Review fractions and operations
- Review equations and properties
- Review linear and quadratic equations and solutions
- Review systems of linear equations and solutions

Essential statements on basic algebra:

- Knowledge and techniques of basic algebra are the foundation of applied mathematics.
- It is vital to master the fundamental knowledge and techniques of basic algebra for a smooth progress in studying other mathematical units in the future.

Key topics:

- Real numbers and arithmetic operations
- Algebraic expressions and operations
- Factorizing algebraic expressions
- Fractions and operations
- Linear and quadratic equations
- Systems of linear equations

Flowchart of mathematical knowledge building

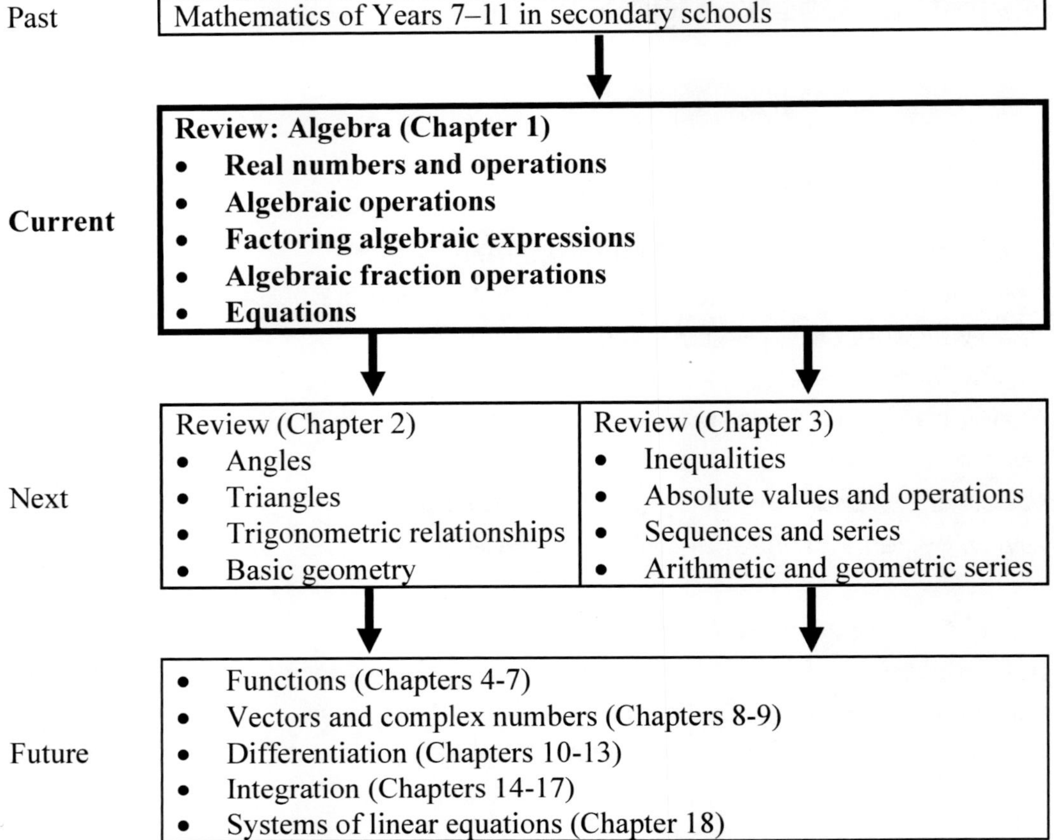

Chapter 1: Review of Basic Algebra

1.1 Numbers and Operations

1.1.1 Summary of real numbers and arithmetic operations

Numbers

Numbers can be classified into *real numbers* and *complex numbers*, and each can be further classified into some subclasses shown in Figure 1.1.

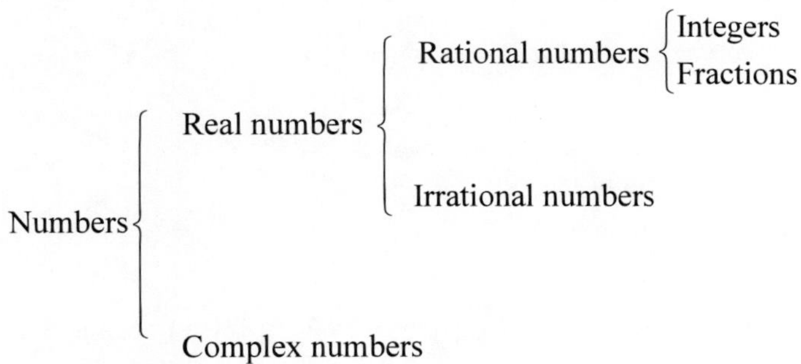

Figure 1.1: Classification of numbers

Our primary focus is the real numbers that are comprised of *rational numbers* and *irrational numbers*. The rational numbers include all integers and all fractions that can be written as a quotient of two integers, such as $\frac{3}{7}, \frac{-1}{13}$ and $\frac{117}{299}$. The rest of the real numbers, which are neither integers nor expressed as a quotient of two integers, forms the irrational numbers, such as $\sqrt{2}$, π, and e (the base of the natural logarithm). In the decimal form, irrational numbers are *nonterminating and nonrepeating decimal numbers*.

All rational and irrational numbers form the collection (or set) of the real numbers, which in theory form the real-number line, a continuous horizontal line, in which all real numbers are located in ascending order from negative infinity (left) to positive infinity (right) as shown in Figure 1.2. A few examples of rational numbers are shown in the real-number line in Figure 1.2. In theory, all irrational real numbers, such as $-\sqrt{3}$, $\sqrt{2}$ and π, can also be plotted in this line. Their positions in the line must be the approximate locations given the nature of irrationality of these numbers.

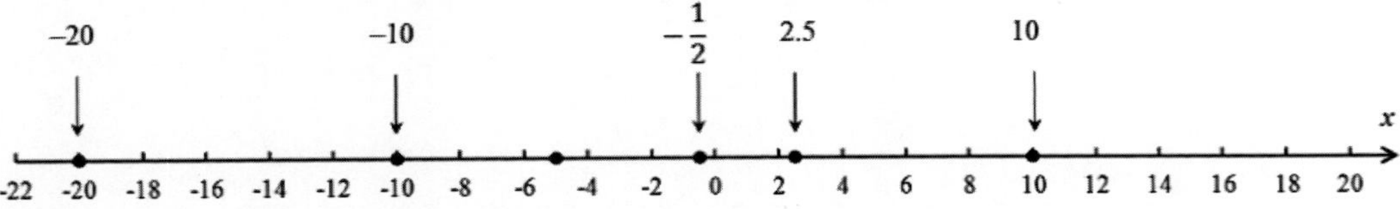

Figure 1.2: The real-number line

Arithmetic operations of real numbers

All real numbers follow the rules of arithmetic operations.

$$\begin{cases} a+b=b+a \quad (\textit{commutative rule of addition}) \\ \text{If } b=-a,\, a+(-a)=(-a)+a=0 \quad (\textit{inverse rule of addition}) \\ ab=ba \quad (\textit{commutative rule of multiplication}) \\ \text{If } b=\dfrac{1}{a},\, ab=a\dfrac{1}{a}=\dfrac{a}{a}=1 \quad (\textit{inverse rule of multiplication}) \end{cases} \tag{1.1}$$

For example, let $a = 2$ and $b = 3$, $a + b = 2 + 3 = 3 + 2 = 5$, $a + (–a) = 2 + (–2) = 2 – 2 = (–2) + 2 = 0$, $a + (–b) = 2 + (–3) = 2 – 3 = (–3) + 2 = –1$, $ab = 2×3 = 3×2 = 6$, $a\frac{1}{a}=\frac{2}{2}=1$.

$$\begin{cases} a+(b+c)=(a+b)+c \quad (\textit{associative rule of addition}) \\ a(bc)=(ab)c \quad (\textit{associative rule of multiplication}) \end{cases} \tag{1.2}$$

For example, let $a = 2$, $b = 3$ and $c = 5$, $a + (b + c) = 2 + (3 + 5) = (2 + 3) + 5 = 10$, and $a(bc) = 2×(3×5) = (2×3) ×5 = 30$.

$$a(b+c)=ab+ac=ba+ca=(c+b)a \quad (\textit{distributive rule}) \tag{1.3}$$

For example, let $a = 2$, $b = 3$ and $c = 5$, $a(b + c) = 2(3+5) = 2×3+2×5 = 3×2+5×2 = (3+5)2 = 16$.

$$\frac{a}{b}=a(\frac{1}{b})=(\frac{1}{b})a \quad (b\neq 0,\ \textit{commutative rule of division}) \tag{1.4}$$

For example, let $a = 2$ and $b = 3$, $\frac{a}{b}=2\times(\frac{1}{3})=(\frac{1}{3})\times 2=\frac{2}{3}$.

$$\frac{ab}{c}=a(\frac{b}{c})=(\frac{a}{c})b \quad (c\neq 0,\ \textit{associatative rule of division}) \tag{1.5}$$

For example, let $a = 2$, $b = 3$ and $c = 5$, $\frac{ab}{c}=\frac{2\times 3}{5}=2\times(\frac{3}{5})=(\frac{2}{5})\times 3=\frac{3\times 2}{5}=\frac{6}{5}$.

$$\begin{cases} \dfrac{a+b}{c}=\dfrac{1}{c}(a+b)=\dfrac{a}{c}+\dfrac{b}{c}=\dfrac{b}{c}+\dfrac{a}{c} \quad (c\neq 0,\ \textit{distributative rule}) \\ \dfrac{a+b}{c+d}=\dfrac{a}{c+d}+\dfrac{b}{c+d}=\dfrac{b}{c+d}+\dfrac{a}{c+d}\neq\dfrac{a}{c}+\dfrac{b}{d} \quad (c+d\neq 0,\ \textit{distributative rule}) \end{cases} \tag{1.6}$$

For example, let a = 2, b = 3, c = 1 and d = 5, $\frac{a+b}{c+d}=\frac{a}{c+d}+\frac{b}{c+d}=\frac{2}{1+5}+\frac{3}{1+5}=\frac{2}{6}+\frac{3}{6}=\frac{5}{6}$, which equals $\frac{a+b}{c+d}=\frac{2+3}{1+5}=\frac{5}{6}$, but is not equal to $\frac{a}{c}+\frac{b}{d}=\frac{2}{1}+\frac{3}{5}=2+\frac{3}{5}$.

Subtracting a positive number b from a can be regarded as the addition of a and the negative b, i.e., $a-b=a+(-b)$. Therefore, operational rules of number additions above can be easily applied to subtractive operations by replacing the number to be subtracted with the addition of the negative of that number. For instance, $2-3=2+(-3)=-1$, and $\frac{a-b}{c}=\frac{a+(-b)}{c}=\frac{2-3}{5}=\frac{-1}{5}$.

The sign "–" in front of a number can also be viewed as an action that *inverses the value* represented by this number. For example, if a = 2 (number a represents the value of 2), and b = –3 (number b represents the value of –3), then $-a = -(+2) = -2$, and $-b = -(-3) = +3$. Both values represented by a and b are inversed by the sign "–" in front of them. This nature of "–" in front of a number can be used to derive different arithmetic operations, some being listed in Table 1.1.

Table 1.1 Arithmetic rules and examples of numbers with the negative sign

Rule	*Example*
$a-b=a+(-b)$	$2-3=2+(-3)=-1$
$a-(-b)=a+b$	$2-(-3)=2+3=5$
$-a=(-1)a$, $-(-a)=a$	$-3=(-1)(3)$, $-(-3)=3$
$-(a+b)=-a-b$	$-(2+3)=-2-3=-5$
$-(a-b)=-a+b$	$-(2-3)=-2+3=1$
$-ab=(-a)b=a(-b)$	$-2\times3=(-2)\times3=2\times(-3)=-6$
$(-a)(-b)=ab$	$(-2)(-3)=2\times3=6$
$(-a)(-b)(-c)=-abc$	$(-2)(-3)(-5)=-2\times3\times5=-30$
$a(b-c)=ab-ac$	$2(3-5)=2\times3-2\times5=6-10=-4$
$-a(b+c)=-ab-ac$	$-2(3+5)=-2\times3-2\times5=-6-10=-16$
$a(-b-c)=-a(b+c)=-ab-ac$	$2(-3-5)=-2(3+5)=-6-10=-16$
$-a(-b-c)=-(-a)(b+c)=a(b+c)=ab+ac$	$-2(-3-5)=-(-2)(3+5)=2(3+5)=16$
$\frac{-a}{b}=\frac{a}{-b}=-\frac{a}{b}$, $\frac{-a}{-b}=\frac{a}{b}$	$\frac{-2}{3}=\frac{2}{-3}=-\frac{2}{3}$, $\frac{-2}{-3}=\frac{2}{3}$
$\frac{-ab}{c}=\frac{ab}{-c}=-\frac{ab}{c}$	$\frac{-2\times3}{5}=\frac{2\times3}{-5}=-\frac{6}{5}$
$\frac{a-b}{c}=\frac{a}{c}-\frac{b}{c}$, $\frac{a\pm b}{-c}=-\frac{a\pm b}{c}$	$\frac{2-3}{5}=\frac{2}{5}-\frac{3}{5}=-\frac{1}{5}$, $\frac{2+3}{-5}=-1$, $\frac{2-3}{-5}=-\frac{2-3}{5}=\frac{1}{5}$
$\frac{-(a+b)}{c+d}=\frac{a+b}{-(c+d)}=\frac{-a-b}{c+d}=-\frac{a+b}{c+d}$	$\frac{-(2+3)}{5+1}=\frac{2+3}{-6}=\frac{-2-3}{6}=-\frac{5}{6}$
$\frac{-(a-b)}{c+d}=-\frac{a-b}{c+d}=\frac{-a+b}{c+d}$	$\frac{-(2-3)}{5+1}=-\frac{2-3}{6}=\frac{-2+3}{6}=\frac{1}{6}$

Arithmetic operations with fractions

A fraction is expressed as a quotient of two numbers, for example, $a \div b$ or a/b. Number a is called the *numerator* (or *dividend*) and number b is the *denominator* (or *divisor*). If a/b is inversed to b/a, b becomes the numerator and a becomes the denominator. *Remember we assume all denominators are nonzero*. Following are the commonly used rules in arithmetic operations with fractions.

Addition and subtraction

- Addition and subtraction with two or more fractions are based on the *principle of the (least) common denominator – fractions with the same denominator can only be added together or subtracted from each other.*

The rules and examples of fraction additions and subtractions are shown in Table 1.2.

Table 1.2 Rules and examples of fraction additions and subtractions

Rule	*Example*
$\frac{a}{d} \pm \frac{b}{d} \pm \frac{c}{d} = \frac{a \pm b \pm c}{d}$	$\frac{1}{7} + \frac{5}{7} - \frac{2}{7} = \frac{1+5-2}{7} = \frac{4}{7}$
$\frac{a}{b} \pm \frac{c}{d} = \frac{a}{b}(\frac{d}{d}) \pm (\frac{b}{b})\frac{c}{d} = \frac{ad}{bd} \pm \frac{bc}{bd} = \frac{ad \pm bc}{bd}$	$\frac{2}{3} + \frac{1}{4} = \frac{2\times 4}{3\times 4} + \frac{1\times 3}{4\times 3} = \frac{8+3}{12} = \frac{11}{12}$
$a \pm \frac{b}{c} = \frac{a}{1}(\frac{c}{c}) \pm \frac{b}{c} = \frac{ac}{c} \pm \frac{b}{c} = \frac{ac \pm b}{c}$	$2 - \frac{2}{3} = \frac{3\times 2}{3} - \frac{2}{3} = \frac{6}{3} - \frac{2}{3} = \frac{6-2}{3} = \frac{4}{3} = 1\frac{1}{3}$

Multiplication and division

- Multiplication with two or more fractions is based on the *principle of parallel multiplication among all numerators and among all denominators separately*, e.g., $\frac{2}{3} \times \frac{1}{2} = \frac{2\times 1}{3\times 2} = \frac{2}{6} = \frac{1}{3}$.
- Dividing a fractional numerator by another fractional denominator is done by converting the division to the multiplication of the numerator with the inverse of the denominator, e.g., $\frac{2}{3} \div \frac{4}{5} = \frac{2}{3} \times \frac{5}{4} = \frac{2\times 5}{3\times 4} = \frac{10}{12} = \frac{5}{6}$.

Some rules and examples of fraction multiplications and divisions are shown in Table 1.3.

Table 1.3 Rules and examples of fraction multiplications and divisions

Rule	*Example*
$\frac{a}{c} \times \frac{b}{d} = \frac{ab}{cd}$, $\frac{a}{c} \times \frac{b}{d} \times \frac{e}{f} = \frac{abe}{cdf}$	$\frac{-\sqrt{2}}{2} \times \frac{1}{3} \times \frac{5}{-7} = \frac{(-\sqrt{2})\times 1\times 5}{2\times 3\times(-7)} = \frac{-5\sqrt{2}}{-42} = \frac{5\sqrt{2}}{42}$
$a \times \frac{b}{c} = \frac{a}{c} \times b = \frac{ab}{c}$, $a \times \frac{b}{a} = \frac{b}{a} \times a = b$	$3 \times \frac{2}{5} = \frac{6}{5}$, $\frac{5}{7} \times (-\sqrt{2}) = \frac{-5\sqrt{2}}{7}$, $\frac{2}{7} \times 7 = 7 \times \frac{2}{7} = 2$
$\frac{a}{b} \div \frac{c}{d} = \frac{a}{b} \times \frac{d}{c} = \frac{ad}{bc}$	$\frac{-\sqrt{2}}{2} \div \frac{5}{-7} = \frac{-\sqrt{2}}{2} \times \frac{-7}{5} = \frac{7\sqrt{2}}{10}$
$a \div \frac{b}{c} = \frac{a}{1} \times \frac{c}{b} = \frac{ac}{b}$, $\frac{a}{b} \div c = \frac{a}{b} \times \frac{1}{c} = \frac{a}{bc}$	$3 \div \frac{-3}{2} = 3 \times \frac{2}{-3} = \frac{6}{-3} = -2$, $\frac{2}{3} \div 5 = \frac{2}{3} \times \frac{1}{5} = \frac{2}{15}$

1.1.2 Summary of exponents and roots with real numbers

Exponents (Indices or power)

Let x be a real number (actually x can represent any number). If x multiplies itself by n times, it is written as x^n, i.e., $\underbrace{x \cdot x \cdot x \cdots x}_{n} = x^n$; x is called the *base* and n is called the *exponent* (or *index* or *power*) with integers only. This definition can be further expanded to form the basic rules of exponents as follows.

$$\begin{cases} x^n = \underbrace{x \cdot x \cdot x \cdots x}_{n} \\ x^{-n} = \dfrac{1}{x^n} = \dfrac{1}{\underbrace{x \cdot x \cdot x \cdots x}_{n}} \quad (x \neq 0) \\ \dfrac{1}{x^{-n}} = x^n = \underbrace{x \cdot x \cdot x \cdots x}_{n} \quad (x \neq 0) \\ x^0 = 1 \end{cases} \tag{1.7}$$

The rules of exponent operations and examples are shown in Table 1.4.

Table 1.4 Rules and examples of exponent operations on real numbers

Rule	*Example*
$x^n = \underbrace{x \cdot x \cdot x \cdots x}_{n}$, $x^0 = 1$	$(-3)^3 = (-3)\cdot(-3)\cdot(-3) = -27$, $(\frac{1}{5})^2 = (\frac{1}{5})\cdot(\frac{1}{5}) = \frac{1}{25}$, $3^0 = 1$, $(\frac{1}{5})^0 = 1$
$x^{-n} = \frac{1}{x^n}$, $\frac{1}{x^{-n}} = x^n$	$2^{-4} = \frac{1}{2^4} = \frac{1}{2\cdot2\cdot2\cdot2} = \frac{1}{16}$, $\frac{1}{5^{-2}} = 5^2 = 5\cdot5 = 25$
$x^{m+n} = x^{n+m} = x^m x^n$	$2^5 = 2^{2+3} = 2^2 2^3 = (2\cdot2)(2\cdot2\cdot2) = 4\times8 = 32$
$x^{m-n} = x^m x^{-n} = \frac{x^m}{x^n}$	$2^{2-3} = 2^{-1} = \frac{1}{2}$, or $2^{2-3} = \frac{2^2}{2^3} = \frac{1}{2}$, $2^{3-3} = 2^0 = 1$, or $2^{3-3} = \frac{2^3}{2^3} = 1$
$x^{mn} = (x^n)^m = (x^m)^n$	$2^{2\times3} = 2^6 = (2^3)^2 = 8^2 = (2^2)^3 = 4^3 = 64$
$x^{-mn} = \frac{1}{x^{mn}}$	$2^{-2\times3} = 2^{-6} = \frac{1}{2^6} = \frac{1}{64}$
$(xy)^n = x^n y^n$	$[(-2)\times3]^2 = (-6)^2 = 36$, or $[(-2)\times3]^2 = (-2)^2 3^2 = 4\times9 = 36$
$(\frac{x}{y})^n = \frac{x^n}{y^n} = x^n y^{-n}$	$(\frac{2}{3})^3 = 2^3 3^{-3} = \frac{2^3}{3^3} = \frac{8}{27}$, $(\frac{2}{-3})^3 = \frac{2^3}{(-3)^3} = \frac{8}{-27} = -\frac{8}{27}$
$(\frac{x}{y})^{-n} = x^{-n} y^n = (\frac{y}{x})^n$	$(\frac{2}{3})^{-3} = 2^{-3} 3^3 = (\frac{3}{2})^3 = \frac{3^3}{2^3} = \frac{27}{8}$, $(\frac{-1}{3})^{-2} = \frac{3^2}{(-1)^2} = \frac{9}{1} = 9$
$(x+y)^{-n} = \frac{1}{(x+y)^n}$	$(2+3)^{-2} = 5^{-2} = \frac{1}{5^2} = \frac{1}{25}$, or $(2+3)^{-2} = \frac{1}{(2+3)^2} = \frac{1}{5^2} = \frac{1}{25}$

It must be noted that:

- $(x+y)^n \neq x^n + y^n$, for example, $(2+3)^2 = 5^2 = 25$, but $2^2+3^2 = 4+9 = 13$;
- $\frac{1}{(x+y)^n} \neq \frac{1}{x^n+y^n}$, for example, $\frac{1}{(2+3)^2} = \frac{1}{5^2} = \frac{1}{25}$, but $\frac{1}{2^2+3^2} = \frac{1}{4+9} = \frac{1}{13}$.

Roots (Radicals)

The *root* (or *radical*) operation is the inverse of the exponent operation on a real number y. Let $x = y^n$ be the result of base y to exponent n; the root of x is to find the base y from $x = y^n$. This can be achieved by applying the fractional exponent $1/n$ to x, i.e., $x^{1/n} = (y^n)^{\frac{1}{n}} = y^{\frac{n}{n}} = y$. This is denoted as $x^{1/n} = \sqrt[n]{x}$. Considering the properties of exponents, for the outcome of a root operation to be a real number, root operations have the following basic property on real numbers:

$$\sqrt[n]{x} = \begin{cases} \text{a positive number if } x \text{ is positive for any real number } n. \\ 0 \text{ if } x = 0.\ (n > 0) \\ \text{a negative number if } x \text{ is negative AND } n \text{ must be odd numbers.} \end{cases} \quad (1.8)$$

The rules of root operations and some examples are shown in Table 1.5.

Table 1.5. Rules and examples of root operations on real numbers

Rule	*Example*
$x^{1/n} = \sqrt[n]{x}$	$8^{\frac{1}{3}} = \sqrt[3]{8},\ (-27)^{\frac{1}{3}} = \sqrt[3]{-27}$
$\sqrt[n]{x^n} = (\sqrt[n]{x})^n = x$	$\sqrt[3]{8} = \sqrt[3]{2^3} = 2^{3/3} = 2,\ \sqrt[3]{-27} = \sqrt[3]{(-3)^3} = -3$
$x^{-1/n} = \frac{1}{x^{1/n}} = \frac{1}{\sqrt[n]{x}}$	$2^{-1/4} = \frac{1}{2^{1/4}} = \frac{1}{\sqrt[4]{2}},\ (-125)^{-1/3} = \frac{1}{(-125)^{1/3}} = \frac{1}{\sqrt[3]{-125}} = \frac{1}{-5} = -\frac{1}{5}$
$x^{m/n} = \sqrt[n]{x^m} = (\sqrt[n]{x})^m$	$2^{2/3} = (\sqrt[3]{2})^2 = \sqrt[3]{2^2} = \sqrt[3]{4},\ (-3)^{3/5} = \sqrt[5]{(-3)^3} = \sqrt[5]{-27}$
$\sqrt[n]{\sqrt[m]{x}} = \sqrt[m]{\sqrt[n]{x}} = \sqrt[mn]{x}$	$\sqrt[6]{100} = \sqrt[2]{\sqrt[3]{100}} = \sqrt[3]{\sqrt[2]{100}} = \sqrt[3]{\sqrt[2]{10^2}} = \sqrt[3]{10}$
$(xy)^{1/n} = \sqrt[n]{x}\sqrt[n]{y} = \sqrt[n]{xy}$	$\sqrt[3]{24} = \sqrt[3]{8\times 3} = \sqrt[3]{8}\sqrt[3]{3} = \sqrt[3]{2^3}\sqrt[3]{3} = 2\sqrt[3]{3}$
$(xy)^{-1/n} = \frac{1}{\sqrt[n]{xy}} = \frac{1}{\sqrt[n]{x}\sqrt[n]{y}}$	$(32)^{-1/4} = \frac{1}{32^{1/4}} = \frac{1}{\sqrt[4]{32}} = \frac{1}{\sqrt[4]{16\times 2}} = \frac{1}{\sqrt[4]{16}\sqrt[4]{2}} = \frac{1}{\sqrt[4]{2^4}\sqrt[4]{2}} = \frac{1}{2\sqrt[4]{2}}$
$(\frac{x}{y})^{1/n} = \sqrt[n]{\frac{x}{y}} = \frac{\sqrt[n]{x}}{\sqrt[n]{y}}$	$(\frac{8}{27})^{1/3} = \sqrt[3]{\frac{8}{27}} = \frac{\sqrt[3]{8}}{\sqrt[3]{27}} = \frac{\sqrt[3]{2^3}}{\sqrt[3]{3^3}} = \frac{2}{3}$
$(\frac{x}{y})^{-1/n} = (\frac{y}{x})^{1/n} = \sqrt[n]{\frac{y}{x}} = \frac{\sqrt[n]{y}}{\sqrt[n]{x}}$	$(\frac{8}{27})^{-1/3} = (\frac{27}{8})^{1/3} = \frac{\sqrt[3]{27}}{\sqrt[3]{8}} = \frac{\sqrt[3]{3^3}}{\sqrt[3]{2^3}} = \frac{3}{2}$

Simplification of expressions with exponents and/or roots: examples

Example 1.1: Simplify the following expressions using the rules of real number operations:

a) $x^4x^{-3}x^{1/2}$, b) $a^5b^{-3}c^2a^{-2}b^{1/3}c^{-2}b^3$, c) $(\frac{1}{8})^{-3/2}(\frac{4}{81})^{-3/4}(\frac{2}{3})^3$, d) $(\frac{9}{64}x^3 3^{-3})^{1/2}$.

Solution

a) $x^4x^{-3}x^{1/2} = x^{4-3+\frac{1}{2}} = x^{1+\frac{1}{2}} = x^{\frac{3}{2}} = \sqrt[2]{x^3}$

b) $a^5b^{-3}c^2a^{-2}b^{1/3}c^{-2}b^3 = (a)^{5-2}(b)^{-3+3+\frac{1}{3}}(c)^{2-2} = a^3b^{1/3}c^0 = a^3\sqrt[3]{b}$

c) $(\frac{1}{8})^{-3/2}(\frac{4}{81})^{-3/4}(\frac{2}{3})^3 = (8)^{3/2}(\frac{81}{4})^{3/4}(\frac{2^3}{3^3}) = (2^3)^{3/2}(\frac{3^4}{2^2})^{3/4}(\frac{2^3}{3^3})$

$= 2^{9/2}\times 3^{4\times\frac{3}{4}}\times 2^{-2\times\frac{3}{4}}\times 2^3\times 3^{-3} = 2^{9/2}\times 3^3\times 2^{-\frac{3}{2}}\times 2^3\times 3^{-3} = 2^{\frac{9}{2}-\frac{3}{2}+3}3^{3-3} = 2^{\frac{6}{2}+3}3^0 = 2^{3+3} = 2^6 = 64$

d) $(\frac{9}{64}x^3 3^{-3})^{1/2} = (\frac{9}{64})^{1/2}(x^3)^{1/2}(3^{-3})^{1/2} = [(\frac{3}{8})^2]^{1/2}x^{3/2}3^{-3/2} = \frac{1}{8}x^{3/2}3^{1-3/2} = \frac{\sqrt{x^3}}{8\sqrt{3}} = \frac{1}{8}\sqrt{\frac{x^3}{3}}$

Example 1.2: Simplify the following expressions using the rules of real number operations:

a) $\frac{a-b}{ab}(\frac{1}{a}-\frac{1}{b})^{-2}$, b) $\sqrt[3]{x^4y^5}(\frac{y^2}{x^3})\div(\frac{y^{3/2}}{x^{2/3}})$, c) $9^{-2}\bullet 8^{-1}\bullet\sqrt{9}\bullet\sqrt[3]{8}\bullet 3^3\bullet 2^4/12$.

Solution

a) $\frac{a-b}{ab}(\frac{1}{a}-\frac{1}{b})^{-2} = \frac{a-b}{ab}(\frac{b}{ab}-\frac{a}{ab})^{-2} = \frac{a-b}{ab}(\frac{b-a}{ab})^{-2} = \frac{-(b-a)}{ab}(\frac{ab}{b-a})^2 = \frac{-ab}{b-a}$.

b) $\sqrt[3]{x^4y^5}(\frac{y^2}{x^3})\div(\frac{y^{3/2}}{x^{2/3}}) = x^{4/3}y^{5/3}(\frac{y^2}{x^3})(\frac{x^{2/3}}{y^{3/2}}) = x^{4/3}y^{5/3}y^2x^{-3}x^{2/3}y^{-3/2} = x^{\frac{4}{3}-3+\frac{2}{3}}y^{\frac{5}{3}+2-\frac{3}{2}}$

$= x^{\frac{6}{3}-3}y^{\frac{10}{6}+2-\frac{9}{6}} = x^{2-3}y^{\frac{1}{6}+2} = x^{-1}y^2y^{\frac{1}{6}} = x^{-1}y^2\sqrt[6]{y} = \frac{y^2\sqrt[6]{y}}{x}$.

c) $9^{-2}\bullet 8^{-1}\bullet\sqrt{9}\bullet\sqrt[3]{8}\bullet 3^3\bullet 2^4/12 = \frac{1}{9^2}\bullet\frac{1}{8}\bullet\sqrt{3^2}\bullet\sqrt[3]{2^3}\bullet 3^3\bullet 2^4\bullet\frac{1}{12} = \frac{1}{3^4}\bullet\frac{1}{2^3}\bullet 3\bullet 2\bullet 3^3\bullet 2^4\bullet\frac{1}{3\times 4}$

$= \frac{1}{3^4}\bullet\frac{1}{2^3}\bullet 3^4\bullet 2^5\bullet\frac{1}{3\times 2^2} = \frac{3^4}{3^5}\bullet\frac{2^5}{2^5} = \frac{1}{3}\bullet 1 = \frac{1}{3}$.

Rationalisation of irrational denominators

In fractions, if the denominator contains irrational numbers as roots, those roots are often converted to rational numbers by multiplying one or more common factors of similar roots to both the numerator and the denominator of the fraction. The chosen factor(s) should make the roots in the denominator become numbers alone or in integer exponents. The following examples demonstrate how to rationalise irrational denominators in fractions.

Example 1.3: Rationalise the following expressions using the rules of real number operations:

a) $\sqrt{\frac{5}{11}}$, b) $\frac{a+b}{\sqrt{a^3b\sqrt[3]{c^2}}}$, c) $\frac{x^2y^3}{(x-y)\sqrt{x^3y^5}}$.

Solution

a) $\sqrt{\frac{5}{11}}=\frac{\sqrt{5}}{\sqrt{11}}=\frac{\sqrt{11}\sqrt{5}}{\sqrt{11}\sqrt{11}}=\frac{\sqrt{11\times 5}}{11}=\frac{\sqrt{55}}{11}$

b) $\frac{a+b}{\sqrt{a^3b\sqrt[3]{c^2}}}=\frac{a+b}{a^{\frac{3}{2}}b^{\frac{1}{2}}c^{\frac{2}{3}}}=\frac{(a^{\frac{1}{2}}b^{\frac{1}{2}}c^{\frac{1}{3}})(a+b)}{(a^{\frac{1}{2}}b^{\frac{1}{2}}c^{\frac{1}{3}})a^{\frac{3}{2}}b^{\frac{1}{2}}c^{\frac{2}{3}}}=\frac{\sqrt{ab}\sqrt[3]{c}(a+b)}{a^{\frac{4}{2}}b^{\frac{2}{2}}c^{\frac{3}{3}}}=\frac{\sqrt{ab}\sqrt[3]{c}(a+b)}{a^2bc}$

c) $\frac{x^2y^3}{(x-y)\sqrt{x^3y^5}}=\frac{x^2y^3(\sqrt{xy})}{(x-y)\sqrt{x^3y^5}(\sqrt{xy})}=\frac{x^2y^3(\sqrt{xy})}{(x-y)\sqrt{x^4y^6}}=\frac{x^2y^3(\sqrt{xy})}{(x-y)x^2y^3}=\frac{\sqrt{xy}}{x-y}$

Transformation of negative exponents and roots to positive expressions

If any part in an expression of real numbers has a negative exponent or root, we normally convert such negative exponent or root to positive. This is largely based on the properties $x^{-n}=\frac{1}{x^n}$ and $x^n=\frac{1}{x^{-n}}$. This is another type of rationalisation. The examples below demonstrate how to convert expressions with negative exponent or root to its positive expression.

Example 1.4: Rationalise the following expressions using the rules of real number operations:

a) $\frac{a^2b^{-3}}{c^{-2}}$, b) $2^{-2}(5\sqrt{100}-\sqrt[3]{1000}+8\sqrt{2})8^{-1/3}$.

Solution

a) $\frac{a^2b^{-3}}{c^{-2}}=\frac{a^2c^2}{b^3}=\frac{(ac)^2}{b^3}$

b) $2^{-2}(5\sqrt{100}-\sqrt[3]{1000}+8\sqrt{2})8^{-1/3}=\frac{5\sqrt{100}-\sqrt[3]{1000}+8\sqrt{2}}{2^2 8^{1/3}}=\frac{5\sqrt{10^2}-\sqrt[3]{10^3}+8\sqrt{2}}{4\sqrt[3]{2^3}}$

$=\frac{5\times 10-10+8\sqrt{2}}{4\times 2}=\frac{40+8\sqrt{2}}{8}=\frac{40}{8}+\frac{8\sqrt{2}}{8}=5+\sqrt{2}.$

1.1.3 Summary of logarithmic operations with real numbers

Logarithm of real numbers

Root on a real number a is an inverse operation of the exponentiation that produced the real number a in order to find the base of the exponentiation. Logarithm on a positive real number a is another inverse operation of exponentiation that produced the real number a in order to find the index n of the given base b. For example, given $16 = 2^4$, it means 16 is the result of *four* self-multiplications of the base 2; in other words, 16 can be reduced to 1 by dividing 16 by 2 exactly 4 times (further dividing 1 by the base creates negative indices). This can be expressed as $\log_2 16 = \log_2 2^4 = 4$. In fact, any positive number a can be in the place of 16 and any positive number b can be the base. Hence, in mathematical terms, the logarithm of a positive number a to another positive number, base b, is a new real number c expressed as

$$\boxed{c = \log_b a \qquad (a > 0,\ b > 0\ \&\ b \neq 1)}. \tag{1.9}$$

Given a positive value a to a base b, its value of logarithm can be obtained by looking up a logarithmic table in the past, or using a calculator or other electronic devices, or computer software nowadays.

Special logarithms

The ***common logarithm*** is the logarithm to base 10, the inverse of the exponent $a = 10^c$,

$$\boxed{c = \log_{10} a = \log a. \quad (a > 0)} \tag{1.10}$$

The ***binary logarithm*** is the logarithm to base 2, the inverse of the exponent $a = 2^c$:

$$\boxed{c = \log_2 a. \ (a > 0)} \tag{1.11}$$

Example 1.5: Find the logarithm of the following real numbers by calculator and explain its meaning:

a) $\log 100$, b) $\log_3 81$, c) $\log_2 0.25$, d) $\log 50$, e) $\log_5 0.5$, f) $\log_3 30$

Solution

a) $\log 100 = \log 10^3 = 3$. Number 100 is three times of self-multiplication of base 10.

b) $\log_3 81 = \log_3 3^4 = 4$. Number 81 is four times of self-multiplication of base 3.

c) $\log_2 0.25 = -2$. Number 0.25 is two times of division of 1 by base 2.

d) $\log 50 \approx 1.6990$. Number 50 is about 1.7 times of self-multiplication of base 10.

e) $\log_5 0.5 \approx -0.4307$. Number 0.5 is about 0.43 times of division of 1 by base 5.

f) $\log_3 30 \approx 3.0959$. Number 30 is about 3.1 times of self-multiplication of base 3.

Logarithmic rules

The commonly used logarithmic rules are summarized in Table 1.6 with examples.

Table 1.6. Rules and examples of logarithmic operations on real numbers

$a>0$ and $a\neq 1$; $b>0$ and $b\neq 1$; $x>0$ and $y>0$		
	Rules	***Examples***
1	$\log_b 1=0$	$\log 1=0,\ \log_2 1=0,\ \ln 1=0,\ \log_{15.2} 1=0,\ \log_{0.5} 1=0$
2	$\log_b b=1$	$\log 10=1,\ \ln e=1,\ \log_{15.2} 15.2=1,\ \log_{0.5} 0.5=1$
3	$\log_b b^x=x$ (b^x can be exactly divided by b for x times.)	$\log 10^2=2,\ \log_2 2^{8.1}=8.1,\ \log_{0.5} 0.5^{\sqrt{2}}=\sqrt{2}$
4	$\log_b x^n=n\log_b x$ or $n\log_b x=\log_b x^n$	$\log x^2=2\log x,\ \log_2 4^x=x\log_2 4=x\log_2 2^2=2x$ $\frac{1}{2}\log_{0.5} 0.25=\log_{0.5}(0.25)^{\frac{1}{2}}=\log_{0.5}[(0.5)^2]^{\frac{1}{2}}=\log_{0.5}(0.5)^{\frac{2}{2}}=1$
5	$\log_b(xy)=\log_b x+\log_b y$ or $\log_b x+\log_b y=\log_b(xy)$	$\log(10\sqrt{10})=\log 10+\log\sqrt{10}=\log 10+\log 10^{\frac{1}{2}}=1+\frac{1}{2}=\frac{3}{2}$ $\log_{0.5} 0.25+\log_{0.5} 2=\log_{0.5}(0.25\times 2)=\log_{0.5} 0.5=1$
6	$\log_b(xy)^n=n(\log_b x+\log_b y)$ or $n(\log_b x+\log_b y)=\log_b(xy)^n$	$\log_2(4\times 2)^3=3(\log_2 4+\log_2 2)=3[(\log_2 2^2)+1]=3(2+1)=9$ $\frac{1}{3}(\log 3+\log 9)=\log(3\times 9)^{\frac{1}{3}}=\log[(3^3)^{\frac{1}{3}}]=\log(3)^{\frac{3}{3}}=\log 3$
7	$\log_b(\frac{x}{y})=\log_b x-\log_b y$ or $\log_b x-\log_b y=\log_b(\frac{x}{y})$	$\log_2(\frac{1}{2})=\log_2 1-\log_2 2=0-1=-1$ $\log 300-\log 3=\log\frac{300}{3}=\log 100=\log 10^2=2$
8	$\log_b(\frac{x}{y})^n=n(\log_b x-\log_b y)$ or $n(\log_b x-\log_b y)=\log_b(\frac{x}{y})^n$	$\log_2(\frac{1}{4})^3=3(\log_2 1-\log_2 4)=3[0-(\log_2 2^2)]=3(0-2)=-6$ $\frac{1}{2}(\log 50-\log 2)=\frac{1}{2}\log(\frac{50}{2})=\frac{1}{2}\log 25=\frac{1}{2}\log 5^2=\log 5$
9	$b^{\log_b x}=x$	$10^{\log 3}=3,\ 2^{\log_2 10}=10,\ 0.5^{\log_{0.5} 8.2}=8.2$
10	$\log_b x=\frac{\log_a x}{\log_a b}$ (base exchange)	$\log_{0.1} 10=\frac{\log 10}{\log 0.1}=\frac{1}{\log(\frac{1}{10})}=\frac{1}{\log 1-\log 10}=\frac{1}{0-1}=-1$

Exercises 1.1

1. Evaluate the following expressions:

a) $2+(-5)-(-4)$ b) $-(a+b)+(2a+3b)-(a-2b)$

c) $(x-2y)-(2x-y)+(3x+4y)$ d) $-3(2x-y)+2(-x-y)$

e) $3-\frac{3}{4}+\frac{1}{2}$ f) $\frac{1}{2a}-\frac{2}{b}$ g) $\frac{y}{x}+\frac{z}{y}+\frac{x}{z}$ h) $\frac{1}{ab}+\frac{2}{ac}-\frac{3}{bc}$

2. Evaluate the following expressions:

a) $\frac{2-x}{7}\times\frac{21}{x-2}+\frac{x-3}{5}\times\frac{20}{x-3}$ b) $3\div\frac{3}{4}\times\frac{1}{2}$ c) $ab(\frac{2}{a}-\frac{1}{b})$

d) $(\frac{y}{x}-\frac{z}{y}-\frac{x}{z})xyz$ e) $3abc(\frac{3}{ab}-\frac{2}{ac}-\frac{1}{bc})$ f) $\frac{ac}{b}\times\frac{ab}{c}\div\frac{a}{bc}$ g) $\frac{2z}{xy}\div\frac{z}{y}\times\frac{x}{y}$

3. Simplify the following expressions:

a) $(\frac{-3}{4})^2$ b) $(\frac{2}{3})^{-3}$ c) $x^{-4}x^2x^{-1/2}$ d) $a^{-4}b^3c^{-2}a^2b^{1/2}c^4b^{3/2}$

e) $(\frac{1}{9})^{2/3}(\frac{81}{16})^{-3/4}(\frac{3}{2})^3$ f) $(\frac{27}{64}x^{-3}3^6)^{1/3}$ g) $(1+2)^{-3}$ h) $[(-2)\times 3]^3$

i) $\frac{(ab)^2}{a-b}(\frac{1}{a}-\frac{1}{b})^2$ j) $\sqrt[4]{x^4y^6}(\frac{y^{3/2}}{x^{3/4}})\div(\frac{y^2}{x^3})$ k) $2^{-2}\bullet 3^2\bullet\sqrt{9}\bullet\sqrt[3]{8}\bullet 3^{-3}\bullet 2^4$

4. Rationalise or evaluate the following expressions:

a) $\sqrt{\frac{2-x}{7}}$ b) $\frac{a+b}{\sqrt[3]{a^3b}\sqrt{c+1}}$ c) $\frac{x^{1/2}y^{7/2}}{(x-y)\sqrt[2]{x^3y^5}}$

d) $(3\sqrt{100}+\sqrt[3]{1000}+8\sqrt{2})8^{-1/2}$ e) $\frac{a^{-2}c^4}{b^{-3}}$ f) $\sqrt[2]{5\sqrt[3]{125}}$

g) $\sqrt[3]{16\sqrt[2]{16}}$ h) $(256)^{-1/4}$ i) $(\frac{27}{64})^{-1/3}$

5. Find the logarithm of the following real numbers by logarithmic rules and verify by calculator:

a) $\log 100000$, b) $\log 500$, c) $\log 0.01$, d) $\log_2 1024$, e) $\log_5 5^{-3}$,

f) $\log_5 \sqrt[2]{5\sqrt[3]{125}}$, g) $\log_2(\frac{27}{64})^{-1/3}$, h) $\log_4 \sqrt[3]{16\sqrt[2]{16}}$, i) $\log_2 \frac{16^{1/2}4^{7/2}}{\sqrt[2]{4^3 16^5}}$

1.2 Algebraic Expressions and Operations

1.2.1 Summary of algebraic expressions and basic operations

Algebraic expressions

Real numbers can form explicit expressions in which the actual values of the numbers are given, such as $2^{-2}(5\sqrt{100}-\sqrt[3]{1000}+8\sqrt{2})8^{-1/3}$. This expression represents a certain value of $5+\sqrt{2}$ as seen in Example 1.4. If we replace some numbers in this expression by symbols, e.g., x, y and z for $2^{-2}(5\sqrt{x}-\sqrt[3]{y}+8\sqrt{2})z^{-1/3}$, such a mixed expression with known numbers and unknown symbols is called an implicit algebraic expression (or simply algebraic expression). Symbols x, y and z are called *variables* as they can be any number as allowed. For example, any number is allowed for both y and z ($\neq 0$), but only $x \geq 0$ ($x<0$ returns a complex number from $\sqrt{x}$!). Those constant values in front of the variable terms, such as 2^{-2}, 5, $8\sqrt{2}$, are called the *coefficients* of the variable terms. In particular, both 2^{-2} and $z^{-1/3}$ can also be called the *common factors* (or simply *factors*) of $(5\sqrt{x}-\sqrt[3]{y}+8\sqrt{2})$ as these three terms share both 2^{-2} and $z^{-1/3}$.

The greatest advantage of using an implicit algebraic expression instead of an explicit number expression is that an algebraic expression can represent many different results because the variables can be assigned any allowed values according to different circumstances. As we regard variables only representing real numbers in our current study, those rules summarised for real number operations in the previous section can also be applied to algebraic expressions.

Example 1.6: Simplify the following algebraic expressions:

a) $(-3x^2-2x+1)-(5x+3)+(8x^2+7x-5)$, b) $(5xy^2-2xy)-3y(x+y)$.

Solution

a) $(-3x^2-2x+1)-(5x+3)+(8x^2+7x-5)=-3x^2-2x+1-5x-3+8x^2+7x-5$
$=-3x^2+8x^2-2x-5x+7x+1-3-5=5x^2-7x+7x-7=5x^2-7.$

b) $(5xy^2-2xy)-3y(x+y)=5xy^2-2xy-3xy-3y^2=5xy^2-5xy-3y^2=y(5xy-5x-3y).$

Here y is the common factor.

Order of removing grouping brackets in combined algebraic expressions

A complicated algebraic expression may contain multiple groups of different operations separated by different pairs of brackets, such as $3\{x[(-3y^2+5\sqrt{10})-(2x+7)]+(3\sqrt[3]{10y^3}+8\sqrt{10}x)\}$. Normally operations at the same level are grouped using the same kind of brackets, braces { } at the outermost, parentheses ( ) at the innermost, and square brackets [ ] in between parentheses and braces. Removing brackets in any combinative algebraic expression follows the two principles below:

1) Removing the innermost brackets (parentheses) first, then the square brackets, and last the braces;
2) Within the same group at any level, exponents and roots going first, then multiplications and divisions, and lastly additions and subtractions.

Example 1.7: Expand and simplify the following algebraic expressions:

a) $3\{x[(-3y^2+5\sqrt{9})-(2x+7)]+(3\sqrt[3]{27y^3}+8\sqrt{4}x)\}$

b) $\{-\sqrt{xy}+[(5xy^2-2xy)/(xy)-5y(x+1)+2]\}\sqrt{xy}$

Solution

a) $3\{x[(-3y^2+5\sqrt{9})-(2x+7)]+(3\sqrt[3]{27y^3}+8\sqrt{4}x)\}=3\{x[-3y^2+5\times 3-2x-7]+(3\times 3y+8\times 2x)\}$
$=3\{-3xy^2+15x-2x^2-7x+9y+16x\}=3\{-3xy^2+24x-2x^2+9y\}=-9xy^2-6x^2+72x+27y$

b) $\{-\sqrt{xy}+[(5xy^2-2xy)/(xy)-5y(x+1)+2]\}\sqrt{xy}=\{-\sqrt{xy}+[(\frac{5xy^2}{xy}-\frac{2xy}{xy})-5xy-5y+2]\}\sqrt{xy}$
$=\{-\sqrt{xy}+[5y-2-5xy-5y+2]\}\sqrt{xy}=\{-\sqrt{xy}-5xy\}\sqrt{xy}=-\sqrt{xy}\sqrt{xy}-5xy\sqrt{xy}$
$=-xy-5xy\sqrt{xy}=-xy(1+5\sqrt{xy})$

1.2.2 Multiplication and division with algebraic expressions

Multiplication with algebraic expressions

Multiplication of two or more algebraic expressions largely follows the same rules of multiplications with real numbers summarised in the previous section. Regarding x as a variable and a, b, c and d as constants, some special multinomial multiplications are listed in Table 1.7.

Table 1.7 Special products of algebraic expressions

Product	*Remark*
$(ax+b)(cx+d)=acx^2+(ad+bc)x+bd$	Product of two general sums
$(x+a)(x+b)=x^2+(a+b)x+ab$	Product of two simple sums
$(x+a)^2=(x+a)(x+a)=x^2+2ax+a^2$	Square of a sum
$(x+a)(x-b)=x^2+(a-b)x-ab$	Product of a sum and a difference
$(x+a)(x-a)=x^2-a^2$	Product of the sum and the difference
$(x-a)(x-b)=x^2-(a+b)x+ab$	Product of two differences
$(x-a)^2=(x-a)(x-a)=x^2-2ax+a^2$	Square of a difference
$(x+a)^3=(x+a)(x+a)^2=x^3+3ax^2+3a^2x+a^3$	Cube of a sum
$(x-a)^3=(x-a)(x-a)^2=x^3-3ax^2+3a^2x-a^3$	Cube of a difference

Example 1.8: Expand and simplify the following algebraic expressions:

a) $(3x+5)(2x^2-3x+1)$, b) $(\sqrt{xy}+2)(\sqrt{xy}-2)$, c) $(3x+5)(2x+1)$,
d) $(x-3)^2-(x+2)^2$, e) $(x-2)^3$.

Solution

a) We regard the first sum as ONE number and then apply the distributive rule to this product repeatedly to obtain the result.

$$\underline{(3x+5)}(2x^2-3x+1)=\underline{(3x+5)}(2x^2)+\underline{(3x+5)}(-3x)+\underline{(3x+5)}=6x^3+10x^2-9x^2-15x+3x+5$$
$$=6x^3+x^2-12x+5$$

b) Use the product of the sum and the difference by regarding $\sqrt{xy}$ as the variable.

$$(\sqrt{xy}+2)(\sqrt{xy}-2)=(\sqrt{xy})^2-2^2=xy-4$$

c) Use the product of general sums or apply the distributive rule by regarding any sum as ONE number first.

$$(3x+5)(2x+1)=6x^2+(3+10)x+5=6x^2+13x+5$$

d) Firstly apply the product of the difference and the product of the sum respectively, and then simplify. The other way is to apply the product of the sum and the difference by regarding both as single numbers.

$$(x-3)^2-(x+2)^2=(x^2-6x+9)-(x^2+4x+4)=x^2-6x+9-x^2-4x-4=-10x+5$$
$$\underline{(x-3)}^2-\underline{(x+2)}^2=[\underline{(x-3)}+\underline{(x+2)}][\underline{(x-3)}-\underline{(x+2)}]=[2x-1][-5]=-10x+5$$

e) Use the cube of a difference.

$$(x-2)^3=x^3-3\times 2x^2+3\times 2^2x-2^3=x^3-6x^2+12x-8$$

Division with algebraic expressions

If an algebraic expression is divided by a combined term of constants and variables jointed together by multiplication and/or division, dividing can be carried out just like the real number division in the previous section, e.g., $\frac{a-b}{c}=\frac{a}{c}+\frac{-b}{c}$ and $\frac{a}{b}\div\frac{c}{d}=\frac{a}{b}\times\frac{d}{c}$.

Example 1.9: Find the results of algebraic divisions given below:

a) $(3x^3y^2 - 3xy^3 + 6x\sqrt{y})/(2x^2\sqrt{y})$, b) $(\sqrt{xy^3} + 2xy) \div \sqrt{\frac{x}{y}}$.

Solution

a) $$(3x^3y^2 - 3xy^3 + 6x\sqrt{y})/(2x^2\sqrt{y}) = \frac{3x^3y^2 - 3xy^3 + 6x\sqrt{y}}{2x^2\sqrt{y}} = \frac{3x^3y^2}{2x^2\sqrt{y}} - \frac{3xy^3}{2x^2\sqrt{y}} + \frac{6x\sqrt{y}}{2x^2\sqrt{y}}$$

$$= \frac{3}{2}x^{3-2}y^{2-\frac{1}{2}} - \frac{3}{2}x^{1-2}y^{3-\frac{1}{2}} + 3x^{1-2}y^{\frac{1}{2}-\frac{1}{2}} = \frac{3}{2}xy^{\frac{3}{2}} - \frac{3}{2}x^{-1}y^{\frac{5}{2}} + 3x^{-1}y^0 = \frac{3}{2}xy\sqrt{y} - \frac{3y^2\sqrt{y}}{2x} + \frac{3}{x}.$$

b) $$(\sqrt{xy^3} + 2xy) \div \sqrt{\frac{x}{y}} = (\sqrt{xy^3} + 2xy) \times \sqrt{\frac{y}{x}} = \frac{(\sqrt{xy^3} + 2xy)\sqrt{y}}{\sqrt{x}} = \frac{\sqrt{x}y^{\frac{3}{2}+\frac{1}{2}} + 2xy^{1+\frac{1}{2}}}{\sqrt{x}}$$

$$= \frac{\sqrt{x}y^{\frac{4}{2}}}{\sqrt{x}} + \frac{2xy^{\frac{3}{2}}}{\sqrt{x}} = y^2 + 2x^{1-\frac{1}{2}}y^{\frac{3}{2}} = y^2 + 2x^{\frac{1}{2}}y^{\frac{3}{2}} = y^2 + 2y\sqrt{xy} = y(y + 2\sqrt{xy}).$$

If both the numerator and denominator are multinomials, dividing is carried out by a method called *long division*. Long division only applies to situations where the highest exponent (order) of variables in the numerator is equal to or higher than that of variables in the denominator. The outcome of the division can be expressed as

$$\boxed{\frac{\text{devidend (or numerator)}}{\text{divisor (or denominator)}} = \text{quotient} + \frac{\text{remainder}}{\text{divisor}}}. \tag{1.12}$$

Example 1.10: Find the results of algebraic divisions below by long division:

a) $\dfrac{2x^3 - 7x^2 + 4x - 3}{x - 3}$, b) $\dfrac{6x^2 + 13x + 7}{2x + 1}$.

Solution

a) In long division, quotient can be variable or constant or both combined.

$$\begin{array}{rl}
 & \underline{2x^2 - x + 1} \quad \leftarrow \textit{quotient} \\
\textit{divisor} \rightarrow x - 3 \big) & 2x^3 - 7x^2 + 4x - 3 \leftarrow \textit{devidend} \\
 & \underline{2x^3 - 6x^2} \\
 & \quad -x^2 + 4x - 3 \\
 & \quad \underline{-x^2 + 3x} \\
 & \qquad\quad x - 3 \\
 & \qquad\quad \underline{x - 3} \\
\textit{remainder} \rightarrow & \qquad\quad 0
\end{array}$$

Therefore, $\dfrac{2x^3 - 7x^2 + 4x - 3}{x - 3} = 2x^2 - x + 1$.

b)

$$\begin{array}{r|l} & 3x+5 \\ \hline 2x+1 & 6x^2+13x+7 \\ & \underline{6x^2+3x} \\ & \quad 10x+7 \\ & \quad \underline{10x+5} \\ & \qquad 2 \end{array}$$

Therefore, $\dfrac{6x^2+13x+7}{2x+1} = 3x+5+\dfrac{2}{2x+1}$.

Exercises 1.2

1. Evaluate the following algebraic expressions:

   a) $(2x+1)(x-3)$ b) $(x+9)^2$ c) $(x-7)^2$ d) $(-x-2)(x-2)$

   e) $(x^2+2x+1)+(3x-5)-(2x^2-7x+5)$ f) $(x^2y+3xy)-3y(x-2y)$

2. Expand and simplify the following algebraic expressions:

   a) $x^2\{y[(3x^2+5\sqrt{4})-(2y+5)]-(\sqrt[3]{8x^3}+\sqrt{4x})\}$ b) $(\sqrt{xy}+2)(\sqrt{xy}-2)$

   c) $\{\sqrt{2xy}+[(xy^2+2xy)/\sqrt{2xy}+2x(y+3)-1]\}\sqrt{2xy}$ d) $(3x+5)(2x^2-3x+1)$

   e) $(3x+5)(2x+1)$ f) $(x-3)^2-(x-2)^2$ g) $(x-2)^3$

3. Find the results of algebraic divisions given below:

   a) $(x^3y^2+2xy^3-3y\sqrt{x})/(y\sqrt{x})$ b) $(\sqrt{x^3y^2}+\sqrt{2xy})\div\sqrt{\dfrac{y}{x}}$

   c) $(-\sqrt{xy^3}+2xy)(\sqrt{xy^3}+2xy)\div(xy)$

4. Find the results of algebraic divisions given below by long division:

   a) $\dfrac{2x^3+7x^2-4x+2}{2x-1}$ b) $\dfrac{4x^3+x^2-3x+2}{x-1}$ c) $\dfrac{4x^2-12x+5}{2x+1}$

   d) $\dfrac{2x^2-13x+1}{2-x}$

1.3 Factorising Algebraic Expressions

The concept of factorising algebraic expressions

Factorising an algebraic expression is the inverse of multiplication of algebraic expressions summarised in the previous section. Multiplication of two algebraic expressions is to expand and then simplify the product to individual terms, such as $(x+a)(x+b)=x^2+(a+b)x+ab$. Factorising an expression is to find the common factors whose mutual product can reproduce the expression, e.g., $x^2-ax=x(x-a)$, and $x^2+(a+b)x+ab=(x+a)(x+b)$.

Example 1.11: Factorise the following algebraic expressions:

a) x^4-x^2-4x+4, b) $2x^3-x^2-8x+4$, c) $(x-3)^3-x^2+9$, d) $\sqrt{2xy}-(2xy)^{\frac{3}{2}}$.

Solution

a) We can use $x^2-a^2=(x+a)(x-a)$ to factorise part of this expression.

$$x^4-x^2-4x+4=(x^4-x^2)-4(x-1)=x^2(x^2-1)-4(x-1)=x^2(x-1)(x+1)-4(x-1)$$
$$=(x-1)[x^2(x+1)-4]=(x-1)(x^3+x^2-4)$$

b) We first rearrange the terms and then take the common factors out.

$$2x^3-x^2-8x+4=(2x^3-8x)-(x^2-4)=2x\underline{(x^2-4)}-\underline{(x^2-4)}=(x^2-4)(2x-1)$$
$$=(x^2-2^2)(2x-1)=(x+2)(x-2)(2x-1)$$

c) Identify one common factor and take it out; then factorise the remaining terms.

$$(x-3)^3-x^2+9=(x-3)^3-(x^2-9)=(x-3)^3-(x^2-3^2)=(x-3)^3-(x-3)(x+3)$$
$$=\underline{(x-3)}[(x-3)^2-(x+3)]=(x-3)[(x^2-6x+9)-x-3)]=(x-3)(x^2-7x+6)$$
$$=(x-3)(x-1)(x-6).$$

d) $\sqrt{2xy}-(2xy)^{\frac{3}{2}}=\sqrt{2xy}-2xy\sqrt{2xy}=\sqrt{2xy}(1-2xy).$

Factorising trinomials

Trinomials refer to the algebraic expressions with three terms joined together by additions and/or subtractions, such as $x^2+(a+b)x+ab$ and $x^2+2xy+y^2$. Not all trinomials can be factorised to factors with real numbers. Some certain types of trinomials can be factorised using the *rectangular decomposition method* illustrated below. The two vertical products form the two ends of the trinomial; the sum of the two cross products forms the middle term of the trinomial; the two horizontal sums become the common factors for the trinomial. Note all coefficients can be either positive or negative.

$$
\begin{array}{cccccccl}
ax & & & + & & & b & \leftarrow \textit{factor } (ax+b) \\
\uparrow & \nwarrow & & & & \nearrow & \uparrow & \\
\times & & & \times & & & \times & \\
\downarrow & \swarrow & & & & \searrow & \downarrow & \\
cx & & \downarrow & + & \downarrow & & d & \leftarrow \textit{factor } (cx+d) \\
acx^2 & & +bcx & & +adx & & +bd & \\
\multicolumn{7}{l}{\underline{acx^2+(ad+bc)x+bd=(ax+b)(cx+d)}} &
\end{array}
$$

Figure 1.3: Rectangular decomposition for factoring trinomials

Example 1.12: Factorise the following algebraic expressions:

a) x^2-4x+3, b) $3x^2+10x+3$, c) $2x^2+7xy+3y^2$.

Solution

a) We use the rectangular decomposition method to factorise this expression.

$$
\begin{array}{cccccccl}
x & & & + & & & -3 & \leftarrow \textit{factor } (x-3) \\
\uparrow & \nwarrow & & & & \nearrow & \uparrow & \\
\times & & & \times & & & \times & \\
\downarrow & \swarrow & & & & \searrow & \downarrow & \\
x & & \downarrow & + & \downarrow & & -1 & \leftarrow \textit{factor } (x-1) \\
x^2 & & -3x & & -x & & +3 &
\end{array}
$$

Thus, $x^2-4x+3=(x-3)(x-1)$.

b) Use the rectangular decomposition method to factorise this expression.

$$\begin{array}{cccccccl}
3x & & & + & & & 1 & \leftarrow \textit{factor } (3x+1) \\
\uparrow & \nwarrow & & & & \nearrow & \uparrow & \\
\times & & & \times & & & \times & \\
\downarrow & \swarrow & & & & \searrow & \downarrow & \\
x & \downarrow & & + & \downarrow & & 3 & \leftarrow \textit{factor } (x+3) \\
3x^2 & +x & & & +9x & & +3 &
\end{array}$$

Thus, $3x^2+10x+3=(3x+1)(x+3)$.

c) Use the rectangular decomposition method to factorise this expression.

$$\begin{array}{cccccccl}
x & & & + & & & 3y & \leftarrow \textit{factor } (x+3y) \\
\uparrow & \nwarrow & & & & \nearrow & \uparrow & \\
\times & & & \times & & & \times & \\
\downarrow & \swarrow & & & & \searrow & \downarrow & \\
2x & \downarrow & & + & \downarrow & & y & \leftarrow \textit{factor } (2x+y) \\
2x^2 & +6xy & & & +xy & & +3y^2 &
\end{array}$$

Thus, $2x^2+7xy+3y^2=(x+3y)(2x+y)$.

Exercises 1.3

1. Factorise the following algebraic expressions:

   a) $x^2+8xy+16y^2$
   b) $(x-2)^2-x^2+4$
   c) $(\sqrt{x}-\sqrt{y})(\sqrt{x}+\sqrt{y})+2\sqrt{xy}+2y$
   d) x^3-4x^2+x-4
   e) $x^4-2x^3+x^2-x+1$

2. Factorise the following algebraic expressions using the rectangular decomposition method:

   a) $x^2+4x-21$
   b) $x^2-7x-18$
   c) $4x^2+5x+1$
   d) $5x^2-6x-8$
   e) $3x^2-5xy+2y^2$

1.4 Algebraic Fraction Operations

The principles of fraction operations with real numbers are applicable to algebraic fractions. These are demonstrated by the following examples.

Simplifying algebraic fractions

Simplifying algebraic fractions is achieved through firstly identifying the common factor(s) in both the numerator and the denominator and then cancelling out the same common factor(s) in both the numerator and the denominator together to obtain the simplified fraction or sometimes a standalone expression.

Example 1.13: Simplify the following algebraic fractions:

a) $\dfrac{x^2-4x+3}{x^2-x-6}$, b) $\dfrac{x^2-2x-3}{x^3-3x^2-2x+6}$, c) $\dfrac{2x^2+7xy+3y^2}{3x^2+11xy+6y^2}$.

Solution

a) $\dfrac{x^2-4x+3}{x^2-x-6}=\dfrac{(x-1)\cancel{(x-3)}}{(x+2)\cancel{(x-3)}}=\dfrac{x-1}{x+2}\quad (x\neq -2,3)$

b) $\dfrac{x^2-2x-3}{x^3-3x^2-2x+6}=\dfrac{(x+1)(x-3)}{(x^3-3x^2)-(2x-6)}=\dfrac{(x+1)(x-3)}{x^2(x-3)-2(x-3)}$

$=\dfrac{(x+1)\cancel{(x-3)}}{(x^2-2)\cancel{(x-3)}}=\dfrac{x+1}{x^2-2}\quad (x\neq \pm\sqrt{2},3)$

c) $\dfrac{2x^2+7xy+3y^2}{3x^2+11xy+6y^2}=\dfrac{\cancel{(x+3y)}(2x+y)}{\cancel{(x+3y)}(3x+2y)}=\dfrac{2x+y}{3x+2y}\quad (x+3y\neq 0, 3x+2y\neq 0)$

Multiplication with algebraic fractions

Fraction multiplication begins with identifying any common factor(s) in all involved fractions. Simplification should go first by cancelling out the identified same common factors, followed by parallel multiplications with the remaining terms in the numerator and the denominator separately.

Example 1.14: Evaluate the following fraction multiplications:

a) $\dfrac{4x}{2x+6}\bullet\dfrac{x-1}{x-5}$, b) $\dfrac{x^2-4x+3}{x^2+4x+3}\bullet\dfrac{x+1}{x-3}$, c) $\dfrac{x^2+2xy-3y^2}{x^2-xy-2y^2}\bullet\dfrac{x+y}{x^2+4xy-5y^2}$.

Solution

a) $\dfrac{4x}{2x+6}\bullet\dfrac{x-1}{x-5}=\dfrac{4x(x-1)}{2(x+3)(x-5)}=\dfrac{2x(x-1)}{(x+3)(x-5)}=\dfrac{2x^2-2x}{x^2-2x-15}\quad (x\neq -3,5)$

b) $\frac{x^2-4x+3}{x^2+4x+3}\bullet\frac{x+1}{x-3}=\frac{(x-1)(x-3)}{(x+1)(x+3)}\bullet\frac{x+1}{x-3}=\frac{(x-1)\cancel{(x-3)}\cancel{(x+1)}}{(x+3)\cancel{(x+1)}\cancel{(x-3)}}=\frac{x-1}{x+3}$ $\quad(x\neq -1,\pm 3)$

c) $\frac{x^2+2xy-3y^2}{x^2-xy-2y^2}\bullet\frac{x+y}{x^2+4xy-5y^2}=\frac{(x-y)(x+3y)}{(x+y)(x-2y)}\bullet\frac{x+y}{(x+5y)(x-y)}$

$=\frac{\cancel{(x-y)}(x+3y)\cancel{(x+y)}}{\cancel{(x+y)}(x-2y)(x+5y)\cancel{(x-y)}}=\frac{x+3y}{(x-2y)(x+5y)}=\frac{x+3y}{x^2+3xy-10y^2}$ $\quad(x\neq \pm y,2y,-5y)$

Division with algebraic fractions

Fraction division begins with converting the division to multiplication by inversing the fraction divisor. The remaining process is the same as the fraction multiplication.

Example 1.15: Evaluate the following fraction divisions:

a) $\frac{2x+1}{x+3}\div\frac{-2x-1}{x-2}$, b) $\frac{x^2-4x+3}{x^2+4x+3}\div\frac{x-1}{x+3}$, c) $\frac{x^2-4xy+3y^2}{x^2-xy-2y^2}\div\frac{x-y}{x+y}$.

Solution

a) $\frac{2x+1}{x+3}\div\frac{-2x-1}{x-2}=\frac{2x+1}{x+3}\bullet\frac{x-2}{-2x-1}=\frac{\cancel{2x+1}}{x+3}\bullet\frac{x-2}{-\cancel{(2x+1)}}=-\frac{x-2}{x+3}$ $\quad(x\neq -\frac{1}{2},-3)$

b) $\frac{x^2-4x+3}{x^2+4x+3}\div\frac{x-1}{x+3}=\frac{(x-1)(x-3)}{(x+1)(x+3)}\bullet\frac{x+3}{x-1}=\frac{\cancel{(x-1)}(x-3)\cancel{(x+3)}}{\cancel{(x+3)}(x+1)\cancel{(x-1)}}=\frac{x-3}{x+1}$ $\quad(x\neq \pm 1,-3)$

c) $\frac{x^2-4xy+3y^2}{x^2-xy-2y^2}\div\frac{x-y}{x+y}=\frac{\cancel{(x-y)}(x-3y)}{\cancel{(x+y)}(x-2y)}\bullet\frac{\cancel{x+y}}{\cancel{x-y}}=\frac{x-3y}{x-2y}$ $\quad(x\neq \pm y,2y)$

Rationalising algebraic denominators

Like real number fractions, if the algebraic denominator contains irrational terms as roots, those roots may be possibly removed by multiplying one or more common factors of similar roots to both the numerator and denominator. Formula $(x+y)(x-y)=x^2-y^2$ is commonly used in rationalising denominators.

Example 1.16: Rationalise the following fractions:

a) $\frac{2x}{x+\sqrt{3}}$, b) $\frac{\sqrt{y}-\sqrt{x}}{\sqrt{y}+\sqrt{x}}$, c) $\frac{x^2-3}{(x-\sqrt{3})(y+\sqrt{3})}$.

Solution

a) $\frac{2x}{x+\sqrt{3}}=\frac{2x}{x+\sqrt{3}}\bullet\frac{x-\sqrt{3}}{x-\sqrt{3}}=\frac{2x(x-\sqrt{3})}{x^2-(\sqrt{3})^2}=\frac{2x^2-2\sqrt{3}x}{x^2-3}$ $\quad(x\neq \pm\sqrt{3})$

b) $\frac{\sqrt{y}-\sqrt{x}}{\sqrt{y}+\sqrt{x}}=\frac{\sqrt{y}-\sqrt{x}}{\sqrt{y}+\sqrt{x}}\bullet\frac{\sqrt{y}-\sqrt{x}}{\sqrt{y}-\sqrt{x}}=\frac{(\sqrt{y}-\sqrt{x})^2}{(\sqrt{y})^2-(\sqrt{x})^2}=\frac{x-2\sqrt{xy}+y}{y-x} \quad (x,y\geq 0, x\neq y)$

c) $\frac{x^2-3}{(x-\sqrt{3})(y+\sqrt{3})}=\frac{x^2-3}{(x-\sqrt{3})(y+\sqrt{3})}\bullet\frac{(x+\sqrt{3})(y-\sqrt{3})}{(x+\sqrt{3})(y-\sqrt{3})}$

$=\frac{(x^2-3)(xy+\sqrt{3}y-\sqrt{3}x-3)}{(x-\sqrt{3})(x+\sqrt{3})(y+\sqrt{3})(y-\sqrt{3})}=\frac{(x^2-3)[xy-\sqrt{3}(x-y)-3]}{[(x^2-(\sqrt{3})^2][y^2-(\sqrt{3})^2]}$

$=\frac{\cancel{(x^2-3)}[xy-\sqrt{3}(x-y)-3]}{\cancel{(x^2-3)}(y^2-3)}=\frac{xy-\sqrt{3}(x-y)-3}{y^2-3} \quad (x,y\neq\pm\sqrt{3})$

Addition and subtraction with algebraic fractions

Similar to real number additions and subtractions, algebraic fraction additions and subtractions follow the same _principle of the (least) common denominator – fractions with the same denominator can only be added together or subtracted from each other_.

Example 1.17: Evaluate the following algebraic fraction additions and/or subtractions:

a) $\frac{5}{3x+2}+\frac{2x-5}{3x+2}-\frac{x+2}{3x+2}$, b) $\frac{1}{x}+\frac{2}{y}-\frac{3}{z}$, c) $\frac{3x}{x-2}+\frac{x+2}{x+5}$,

d) $1+\frac{1}{x-1}-\frac{1}{x+1}$, e) $\frac{x^2+2x-3}{x^2-5x+4}-\frac{x^2+5x+6}{x^2+2x}$.

Solution

a) $\frac{5}{3x+2}+\frac{2x-5}{3x+2}-\frac{x+2}{3x+2}=\frac{5+(2x-5)-(x+2)}{3x+2}=\frac{5+2x-5-x-2}{3x+2}=\frac{x-2}{3x+2} \quad (x\neq-\frac{2}{3})$

b) $\frac{1}{x}+\frac{2}{y}-\frac{3}{z}=\frac{(yz)}{x(yz)}+\frac{2(xz)}{y(xz)}-\frac{3(xy)}{(xy)z}=\frac{yz+2xz-3xy}{xyz}=\frac{yz-xy+2xz-2xy}{xyz}$

$=\frac{y(z-x)+2x(z-y)}{xyz} \quad (x,y,z\neq 0)$

c) $\frac{3x}{x-2}+\frac{x+2}{x+5}=\frac{3x(x+5)}{(x-2)(x+5)}+\frac{(x-2)(x+2)}{(x-2)(x+5)}=\frac{3x^2+15x}{(x-2)(x+5)}+\frac{x^2-4}{(x-2)(x+5)}$

$=\frac{3x^2+15x+x^2-4}{(x-2)(x+5)}=\frac{4x^2+15x-4}{(x-2)(x+5)}=\frac{4x^2+15x-4}{x^2+3x-10} \quad (x\neq-5,2)$

d) $1+\frac{1}{x-1}-\frac{1}{x+1}=1+\frac{x+1}{(x-1)(x+1)}-\frac{x-1}{(x-1)(x+1)}=\frac{x^2-1}{x^2-1}+\frac{x+1-x+1}{x^2-1}=\frac{x^2+1}{x^2-1} \quad (x\neq\pm1)$

e) $\frac{x^2+2x-3}{x^2-5x+4}-\frac{x^2+5x+6}{x^2+2x}=\frac{(x-1)(x+3)}{(x-1)(x-4)}-\frac{(x+2)(x+3)}{x(x+2)}=\frac{x+3}{x-4}-\frac{x+3}{x}$

$=(x+3)(\frac{1}{x-4}-\frac{1}{x})=(x+3)[\frac{x}{x(x-4)}-\frac{(x-4)}{x(x-4)}]=(x+3)[\frac{x-x+4}{x(x-4)}]=\frac{4(x+3)}{x(x-4)} \quad (x\neq-2,0,1,4)$

Exercises 1.4

1. Simplify the following algebraic fractions:

a) $\dfrac{x^2+4x-12}{x^2+x-6}$ b) $\dfrac{2x^2-5x+2}{x^3-2x^2+2x-4}$ c) $\dfrac{4x^2+7xy+3y^2}{x^2-5xy-6y^2}$

2. Evaluate the following fraction multiplications:

a) $\dfrac{4x-2}{x+1}\bullet\dfrac{x-1}{2x-1}$ b) $\dfrac{x^2-4x+4}{x^2-x-2}\bullet\dfrac{x+4}{x-2}$ c) $\dfrac{x^2-2xy-3y^2}{x^2+xy-2y^2}\bullet\dfrac{x-y}{x^2-4xy-5y^2}$

d) $\dfrac{-2x-3}{x-1}\div\dfrac{2x^2+3x}{2x+1}$ e) $\dfrac{x^2-4}{x^2+4x}\div\dfrac{x-2}{x+4}$ f) $\dfrac{x^2-5xy+6y^2}{x^2-4xy-5y^2}\div\dfrac{x-2y}{x+y}$

3. Rationalise the following fractions:

a) $\dfrac{\sqrt{xy}}{\sqrt{x}+\sqrt{y}}$ b) $\dfrac{\sqrt{y-1}-\sqrt{x+1}}{\sqrt{y-1}+\sqrt{x+1}}$ c) $\dfrac{\sqrt{x}-\sqrt{3y}}{\sqrt{x}(y+\sqrt{3})}$

4. Evaluate the following fraction additions and/or subtractions:

a) $1+\dfrac{2x+5}{3x-2}+\dfrac{x+2}{3x-2}$ b) $xyx(\dfrac{1}{x}+\dfrac{2}{y}+\dfrac{3}{z})$ c) $1+\dfrac{3}{x-2}+\dfrac{1}{x+2}$

d) $\dfrac{x^2-2x-3}{x^2+2x+1}-\dfrac{x^2+5x-6}{x^2-1}$

1.5 Equations

1.5.1 Equations and general properties

Equations

If two algebraic expressions can produce the same result, the two algebraic expressions form an equation linked together by the **equal sign** "=". Since both sides equal to each other, exchanging two expressions on sides does not change the nature of the equation. Some examples of equations are shown in Table 1.8:

Table 1.8 Examples of equations

Equation	*Remark*
$x-5=-2$ $2x+3y=x-2y$ $2z=3x+5y$	Linear equation with one unknown variable (x) Linear equation with two unknown variables (x, y) Linear equation with three unknown variables (x, y, z)
$y=\frac{2}{3x}$ or $\frac{1}{2x}=10$	Reciprocal equations
$\frac{x-3}{x+1}=\frac{2}{3x+1}$ or $\frac{3}{x-2}=\frac{1-x}{2x^2+1}$	Fractional equations
$2x^2-3=x^2+2x+5$ or $x^2-3x-2=0$	Quadratic equations (2nd-order of one unknown, x)
$x^3+2x^2-3x+5=3x^3-2x$	Cubic equation (3rd-order of one unknown, x)
$x^n+\cdots-2x+3=0$	nth polynomial equation (nth-order of one unknown, x)
$2+\sqrt{3x-2}=\sqrt{x+1}$ or $\sqrt{x^2-1}=8$	Radical equations

The value that makes an equation remain true (or balanced) is called the *solution* or *root* to that equation. For example, obviously $x = 2$ is the only solution to $x + 1 = 3$ because it makes the statement of this equation true. Some equations may have more than one solution, which will be seen in some later sections and chapters.

General operations for equations

An equation can be reorganised in different forms of *equivalent equations* by manipulating both sides of the equation simultaneously as long as such manipulations do not violate the fundamental rules. Some commonly used manipulations are summarised as follows.

Addition and subtraction

Adding to or subtracting from both sides of an equation the same term transfers the equation to an equivalent equation. This adding or subtracting operation equals that *moving any term in an equation from one side to the other side changes the sign of that moving term.* For instance,

$$2x^2-3=x^2+2x+5 \to 2x^2-3-(x^2)=x^2+2x+5-(x^2) \to x^2-3=2x+5$$

$$2x^2-3=\underline{x^2}+2x+5 \to 2x^2-3\underline{-x^2}=2x+5 \to x^2-3=2x+5$$

$$x^2\underline{-3}=\underline{2x}+5 \to x^2\underline{-2x}=5+\underline{3} \to x^2-2x=\underline{8} \to x^2-2x-8=0$$

Multiplication and division

Multiplying and/or dividing both sides of an equation with the same *nonzero* factor transfers the equation to an equivalent equation. This multiplying and/or dividing operation implies:

1) The common factors shared by both sides of the equation can be cancelled out;
2) A fraction equation can be converted to an equivalent normal equation by multiplying both sides with the product of all denominators;
3) Inversing both sides of a fraction equation comes with an equivalent fraction equation.

Some examples are shown below.

$$2x^2-8=x^2+2x+4 \rightarrow \frac{2(x-2)(x+2)}{\underline{x+2}}=\frac{(x+2)^2}{\underline{x+2}} \rightarrow 2x-4=x+2 \quad (x\neq -2)$$

$$2-x^2=-2x^2+x-3 \rightarrow (-1)(2-x^2)=(-1)(-2x^2+x-3) \rightarrow x^2-2=2x^2-x+3$$

$$\frac{2x}{x-3}=\frac{x+2}{x-2} \rightarrow \underline{(x-3)(x-2)}\frac{2x}{x-3}=\frac{x+2}{x-2}\underline{(x-3)(x-2)} \rightarrow 2x(x-2)=(x+2)(x-3) \quad (x\neq 2,3)$$

$$\frac{2x}{x-3}=\frac{x+2}{x-2} \rightarrow \frac{x-3}{2x}=\frac{x-2}{x+2} \quad (x\neq 0,\pm 2,3)$$

Exponent and radical operations

Applying the same exponent to both sides of an equation transfers the equation to an equivalent equation. Applying the same radical to both sides of an equation transfers the equation to an equivalent equation if meeting the following conditions:

1) Only positive results are allowed to come from any even radical operation;
2) Both positive and negative results are allowed to come from any odd radical operation.

Some examples are shown below.

$$\sqrt{x^2+4}-3=x-2 \rightarrow \sqrt{x^2+4}=x+1 \rightarrow (\sqrt{x^2+4})^2=(x+1)^2 \rightarrow x^2+4=x^2+2x+1$$

$$x^2=2y-3 \rightarrow \sqrt{x^2}=\sqrt{2y-3} \rightarrow x=\pm\sqrt{2y-3} \quad (2y-3\geq 0)$$

$$x^3=\frac{y+3}{z} \rightarrow \sqrt[3]{x^3}=\sqrt[3]{\frac{y+3}{z}} \rightarrow x=\sqrt[3]{\frac{y+3}{z}} \quad (z\neq 0)$$

1.5.2 Solving linear and quadratic equations

As many equations can be simplified to either linear or quadratic equations, here we focus on solving linear and quadratic equations.

Linear equations

Any single variable linear equation can be reorganised as the standard form shown below:

$$ax+b=0 \tag{1.13}$$

where a and b are known constants. Its solution is

$$x=\frac{-b}{a} \quad (a \neq 0). \tag{1.14}$$

Example 1.18: Solve the following questions:

a) $x-5=-2$, b) $\frac{3x+1}{x-3}=1$, c) $\frac{5}{x+\sqrt{3}}=2$.

Solution

a) $x-5=-2 \longrightarrow x-3=0 \longrightarrow x=3$

b) $\frac{3x+1}{x-3}=1 \longrightarrow (x-3)\frac{3x+1}{x-3}=1(x-3) \longrightarrow 3x+1=x-3 \longrightarrow 2x=-4 \longrightarrow x=-2$

c) $\frac{5}{x+\sqrt{3}}=2 \longrightarrow 2(x+\sqrt{3})=\frac{5}{x+\sqrt{3}}(x+\sqrt{3}) \longrightarrow 2x+2\sqrt{3}=5 \longrightarrow 2x=5-2\sqrt{3}$

$x=2.5-\sqrt{3}$

Quadratic equations

Any quadratic equation can be reorganised as the standard form shown below:

$$\boxed{ax^2+bx+c=0} \tag{1.15}$$

where a, b and c are known constants. In theory, any quadratic equation can be factorised as

$(x-r_1)(x-r_2)=0$, which gives

$x-r_1=0$ and $x-r_2=0$ or $x=r_1$ and $x=r_2$.

Both r_1 and r_2 are the TWO roots or solutions to this quadratic equation, and can be explicitly expressed as

$$r_{1,2} = \frac{-b \pm \sqrt{b^2 - 4ac}}{2a} \text{ or } r_1 = \frac{-b + \sqrt{b^2 - 4ac}}{2a} \text{ and } r_2 = \frac{-b - \sqrt{b^2 - 4ac}}{2a} \quad (a \neq 0) \qquad (1.16)$$

If $b^2 - 4ac > 0 \Rightarrow r_1$ and r_2 are two distinct real numbers;
If $b^2 - 4ac = 0 \Rightarrow r_1$ and r_2 are two repeated real numbers;
If $b^2 - 4ac < 0 \Rightarrow r_1$ and r_2 are two conjugate complex numbers.

Although formula (1.16) is the generic formula to find the solutions to a standard quadratic equation, finding complete common factors (if exist) for a quadratic equation by factorisation is always preferred in solving quadratic equations.

Example 1.19: Solve the following questions:

a) $x^2 + 2x - 3 = 0$, b) $\frac{3x}{x+2} - \frac{x}{x+1} = 0$, c) $\sqrt{2x^2 - 7} - 3 = x - 2$, d) $x^2 + 4x + 4 = 0$.

Solution

a) By factoring the equation

$$x^2 + 2x - 3 = 0 \longrightarrow (x-1)(x+3) = 0 \longrightarrow \begin{cases} x - 1 = 0 \\ x + 3 = 0 \end{cases} \longrightarrow \begin{cases} x_1 = 1 \\ x_2 = -3 \end{cases}$$

By using formula (1.16)

$$r_{1,2} = \frac{-2 \pm \sqrt{2^2 - 4 \cdot 1 \cdot (-3)}}{2 \cdot 1} = \frac{-2 \pm \sqrt{4+12}}{2} = \frac{-2 \pm \sqrt{16}}{2} = \frac{-2 \pm 4}{2} = -1 \pm 2 \longrightarrow \begin{cases} x_1 = r_1 = -1 + 2 = 1 \\ x_2 = r_2 = -1 - 2 = -3 \end{cases}$$

b) By rearranging the equation

$$\frac{3x}{x+2} - \frac{x}{x+1} = 0 \longrightarrow \frac{3x}{x+2} = \frac{x}{x+1} \longrightarrow \underline{(x+2)(x+1)\frac{3x}{x+2} = \frac{x}{x+1}(x+2)(x+1)}$$

$$3x(x+1) = x(x+2) \longrightarrow 3x^2 + 3x = x^2 + 2x \longrightarrow 2x^2 + x = 0 \longrightarrow x(2x+1) = 0$$

$$\begin{cases} x = 0 \\ 2x + 1 = 0 \end{cases} \longrightarrow \begin{cases} x = 0 \\ 2x = -1 \end{cases} \longrightarrow \begin{cases} x_1 = 0 \\ x_2 = -\frac{1}{2} \end{cases}$$

c) By rationalising and then factorizing the equation

$$\sqrt{2x^2 - 7} - 3 = x - 2 \longrightarrow \sqrt{2x^2 - 7} = x + 1 \longrightarrow (\sqrt{2x^2 - 7})^2 = (x+1)^2$$

$$2x^2 - 7 = x^2 + 2x + 1 \longrightarrow x^2 - 2x - 8 = 0 \longrightarrow (x+2)(x-4) = 0 \longrightarrow \begin{cases} x_1 = -2 \\ x_2 = 4 \end{cases}$$

d) By factorizing the equation

$$x^2+4x+4=0 \longrightarrow (x+2)^2=0 \longrightarrow \begin{cases} x+2=0 \\ x+2=0 \end{cases} \longrightarrow \begin{cases} x_1=-2 \\ x_2=-2 \end{cases} \text{ or } x_1=x_2=-2.$$

By using formula (1.16)

$$r_{1,2}=\frac{-4\pm\sqrt{4^2-4\bullet1\bullet4}}{2\bullet1}=\frac{-4\pm\sqrt{16-16}}{2}=\frac{-4}{2}=-2 \longrightarrow \begin{cases} x_1=r_1=-2 \\ x_2=r_2=-2 \end{cases} \text{ or } x_1=x_2=-2.$$

1.5.3 Systems of linear equations

All the equations presented by far have only one unknown number (x) in the equations. If two unknown numbers, x and y, are contained in two interrelated linear equations, these two interrelated linear equations form *a system of linear equations* which can be generalised as

$$\boxed{\begin{cases} a_1x+b_1y=c_1 \\ a_2x+b_2y=c_2 \end{cases}} \tag{1.17}$$

where a_i, b_i, and c_i (i = 1 or 2) are known constants. This definition can be extended to a system of three, four and more interrelated linear equations with three, four and more unknowns respectively. The solution to system (1.17) is

$$\boxed{\begin{cases} x=\dfrac{b_2c_1-b_1c_2}{a_1b_2-a_2b_1} \\ y=\dfrac{a_1c_2-a_2c_1}{a_1b_2-a_2b_1} \end{cases}} \tag{1.18}$$

which can be obtained by using either *substitution* or *elimination* demonstrated as follows.

Substitution

The method of substitution takes four steps to solve the system of two linear equations (1.17):

1) Isolate one unknown (e.g., y) from the first (or second) linear equation in the system:

$$a_1x+b_1y=c_1 \longrightarrow b_1y=c_1-a_1x \longrightarrow y=\frac{1}{b_1}(c_1-a_1x)$$

2) Substitute this isolated unknown y into the second (or the first) linear equation in the system to make it as an equivalent equation of one unknown (x in this case);

$$a_2x+b_2y=c_2 \longrightarrow a_2x+\frac{b_2}{b_1}(c_1-a_1x)=c_2 \longrightarrow a_2x+\frac{b_2c_1-a_1b_2x}{b_1}=c_2 \longleftarrow \times b_1 \text{ both sides}$$

$$a\,b_1x+b_2c_1-a_1b_2x=b_1c_2 \longrightarrow (a_2b_1-a_1b_2)x=b_1c_2-b_2c_1$$

3) Solve the equivalent linear equation to find x;

$$x=\frac{b_1c_2-b_2c_1}{a_2b_1-a_1b_2}=\frac{-(b_2c_1-b_1c_2)}{-(a_1b_2-a_2b_1)}=\frac{b_2c_1-b_1c_2}{a_1b_2-a_2b_1}$$

4) Substitute x found in Step 3 back to the y expression isolated in Step 1 to determine y.

$$\begin{aligned} y&=\frac{1}{b_1}(c_1-a_1x)=\frac{1}{b_1}(c_1-a_1\frac{b_2c_1-b_1c_2}{a_1b_2-a_2b_1})=\frac{1}{b_1}[\frac{c_1(a_1b_2-a_2b_1)}{a_1b_2-a_2b_1}-\frac{a_1(b_2c_1-b_1c_2)}{a_1b_2-a_2b_1}] \\ &=\frac{1}{b_1}[\frac{a_1b_2c_1-a_2b_1c_1}{a_1b_2-a_2b_1}-\frac{a_1b_2c_1-a_1b_1c_2}{a_1b_2-a_2b_1}]=\frac{1}{b_1}[\frac{\cancel{a_1b_2c_1}-a_2b_1c_1-\cancel{a_1b_2c_1}+a_1b_1c_2}{a_1b_2-a_2b_1}] \\ &=\frac{1}{b_1}[\frac{b_1(a_1c_2-a_2c_1)}{a_1b_2-a_2b_1}]=\frac{a_1c_2-a_2c_1}{a_1b_2-a_2b_1}. \end{aligned}$$

Note you can also follow the same process to isolated x first. Both will produce the same result.

Elimination

The method of elimination takes three steps to solve the system of two linear equations (1.17):

1) Multiply b_2 to both sides of the first linear equation and multiply b_1 to both sides of the second linear equation in system (1.17). This makes the y-term in both equations same.

$$\begin{cases} a_1x+b_1y=c_1 \\ a_2x+b_2y=c_2 \end{cases} \xrightarrow[\times b_1]{\times b_2} \begin{cases} a_1b_2x+b_1b_2y=b_2c_1 \\ a_2b_1x+b_1b_2y=b_1c_2 \end{cases}$$

2) Subtract the second equation from the first equation on both sides to eliminate the y-terms; then solve the resultant equation to find x;

$$\begin{array}{l} a_1b_2x+b_1b_2y=b_2c_1 \\ a_2b_1x+b_1b_2y=b_1c_2 \\ \quad - \\ \hline a_1b_2x-a_2b_1x=b_2c_1-b_1c_2 \end{array}$$

$$(a_1b_2-a_2b_1)x=b_2c_1-b_1c_2 \longrightarrow x=\frac{b_2c_1-b_1c_2}{a_1b_2-a_2b_1}$$

3) Substitute x into either linear equation to obtain value for y.

Example 1.20: Solve the following system of linear equations:

$$\begin{cases} -2x+3y=8 \\ x+2y=3 \end{cases}$$

Solution

Apply substitution to this system as x can be easily isolated from the second equations:

$$x+2y=3 \longrightarrow x=3-2y.$$

Substitute $x=3-2y$ into the first equation:

$$-2x+3y=8 \longrightarrow -2(3-2y)+3y=8 \longrightarrow -6+4y+3y=8 \longrightarrow 7y=14 \longrightarrow y=2.$$

Substitute $y = 2$ into $x=3-2y$ to determine the x value: $x=3-2y=3-2\times 2=-1.$

Therefore, the solution is $x = -1$ and $y = 2$.

Example 1.21: Solve the following system of linear equations:

$$\begin{cases} -2x+3y=2 \\ 6x-5y=2 \end{cases}$$

Solution

Apply elimination to this system to cancel out x in both equations:

$$\begin{cases} -2x+3y=2 \\ 6x-5y=2 \end{cases} \xrightarrow[\rightarrow]{\times 3} \begin{cases} -6x+9y=6 \\ 6x-5y=2 \end{cases} \xrightarrow[+]{\text{add both sides}} 4y=8 \longrightarrow y=2.$$

Substitute $y = 2$ into the first equation;

$$-2x+3y=2 \longrightarrow 2x=3y-2 \longrightarrow x=\frac{3y-2}{2}=\frac{3\times 2-2}{2}=\frac{4}{2}=2.$$

Therefore, the solution is $x = 2$ and $y = 2$.

Example 1.22: Solve the following system of linear equations:

$$\begin{cases} 3x+4y=2 \\ 4x+5y=3 \end{cases}$$

Solution

Use formula (1.18) to solve this system of equations given $a_1 = 3$, $b_1 = 4$, $c_1 = 2$, $a_2 = 4$, $b_2 = 5$ and $c_2 = 3$:

$$\begin{cases} x = \dfrac{b_2c_1 - b_1c_2}{a_1b_2 - a_2b_1} = \dfrac{5\times 2 - 4\times 3}{3\times 5 - 4\times 4} = \dfrac{10-12}{15-16} = \dfrac{-2}{-1} = 2 \\ y = \dfrac{a_1c_2 - a_2c_1}{a_1b_2 - a_2b_1} = \dfrac{3\times 3 - 4\times 2}{3\times 5 - 4\times 4} = \dfrac{9-8}{15-16} = \dfrac{1}{-1} = -1 \end{cases}$$

Therefore, the solution is $x = 2$ and $y = -1$.

Example 1.23: A mathematics course is offered to both education students and engineering students. Suppose there are totally 99 students enrolled in the course. The number of engineering students is 3 short of the double of the education students. How many education students and engineering students are enrolled in this course respectively?

Solution

Let x and y be the numbers of education and engineering students enrolled respectively. The total number is 99, i.e., $x + y = 99$. The number of engineering students is the double number of education students less 3, i.e., $y = 2x - 3$. These two linear equations form the following system:

$$\begin{cases} x+y=99 \\ y=2x-3 \end{cases} \longrightarrow \begin{cases} x+y=99 \\ 2x-y=3 \end{cases}$$

Apply elimination to this system as the y-terms in both equations are the same but with opposite signs.

$$\begin{cases} x+y=99 \\ 2x-y=3 \end{cases} \xrightarrow{+} 3x=102 \longrightarrow x=34.$$

Substitute $x = 34$ into the second equation (or the first one) to find y:

$$y = 2x-3 = 2\times 34 - 3 = 68 - 3 = 65.$$

Therefore, there are 34 education students and 65 engineering students in this course.

Exercises 1.5

1. Solve the following equations:

   a) $x+3=-7$

   b) $\dfrac{x-1}{3x-5}=1$

   c) $\dfrac{5}{\sqrt{x}+\sqrt{3}}=\sqrt{x}-\sqrt{3}$

   d) $\sqrt{x^2-2}-1=x-2$

   e) $x^2-100=0$

   f) $x^3+81=0$

2. Solve the following equations:

   a) $x^2+3x-10=0$

   b) $\dfrac{2x-1}{x-1}-\dfrac{x-2}{x+1}=0$

   c) $\sqrt{2x^2-1}-3=x-1$

   d) $2x^2-4x-3=0$

   e) $y^2+3y+1=0$

   f) $5t^2+8t+1=0$

3. Solve the following systems of linear equations using both substitution and elimination:

   a) $\begin{cases} x-3y=8 \\ x-2y=3 \end{cases}$

   b) $\begin{cases} 4x-3y=1 \\ 2x+y=3 \end{cases}$

   c) $\begin{cases} -x+7y=1 \\ 3x-16y=2 \end{cases}$

   d) $\begin{cases} 5x-2y=3 \\ 2x+y=3 \end{cases}$

   e) $\begin{cases} a+9b=1 \\ 2a-3b=2 \end{cases}$

4. Two salt solutions, 21% and 5% of salt respectively, are available. To make 1000 ml of salt solution with 11% salt using these two solutions, what amounts of solutions from each of these two solutions must be mixed together?

Chapter 1: Exercises Answers

Exercises 1.1

1. a) 1; b) $4b$; c) $2x+3y$; d) $-8x+y$; e) 11/4; f) $\dfrac{b-4a}{2ab}$;

   g) $\dfrac{y^2z+xz^2+x^2y}{xyz}$; h) $\dfrac{c+2b-3a}{abc}$

2. a) 1; b) 2; c) $2b-a$; d) $y^2z-xz^2-x^2y$; e) $-3a-6b+9c$; f) abc; g) $\dfrac{2}{y}$

3. a) 9/16; b) 27/8; c) $x^{-\frac{5}{2}}=\dfrac{1}{x^{5/2}}$; d) $\dfrac{b^5c^2}{a^2}=b^5(\dfrac{c}{a})^2$; e) $(\dfrac{1}{81})^{1/3}$; f) $\dfrac{27}{4x}$

   g) 1/27; h) –216; i) $-(b-a)=a-b$; j) $x^{\frac{13}{4}}y$; k) 8

4. a) $\dfrac{\sqrt{14-7x}}{7}$; b) $\dfrac{\sqrt[3]{b^2}(a+b)\sqrt{c+1}}{ab(c+1)}$; c) $\dfrac{y}{x(x-y)}$; d) $10\sqrt{2}+4$;

   e) $\dfrac{b^3c^4}{a^2}$; f) 5; g) 4; h) 1/4; i) 4/3

5. a) 5, b) 2.6990, c) –2, d) 10, e) –3, f) 1, g) 0.4150,
   h) 1, i) –4

Exercises 1.2

1. a) $2x^2-5x-3$; b) $x^2+18x+81$; c) $x^2-14x+49$;
   d) $-x^2+4$; e) $-x^2+12x-9$; f) x^2y+6y^2

2. a) $3x^4y-2x^2y^2+5x^2y-4x^3$; b) $xy-4$; c) $4xy+xy^2+(2xy+6x-1)\sqrt{2xy}$;
   d) $6x^3+x^2-12x+5$; e) $6x^2+13x+5$; f) $-2x+5$; g) $x^3-6x^2+12x-8$

3. a) $yx^2\sqrt{x}+2\sqrt{x}y^2-3$; b) $x(\sqrt{2}+x\sqrt{y})$; c) $(4x-y)y$

4. a) $x^2+4x+\dfrac{2}{2x-1}$; b) $4x^2+5x+2+\dfrac{4}{x-1}$; c) $2x-7+\dfrac{12}{2x+1}$; d) $-2x+9+\dfrac{17}{x-2}$

Exercises 1.3

1. a) $(x+4y)^2$; b) $-4(x-2)$; c) $(\sqrt{x}+\sqrt{y})^2$; d) $(x-4)(x^2+1)$; e) $(x-1)(x^3-x^2-1)$

2. a) $(x-3)(x+7)$; b) $(x-9)(x+2)$; c) $(x+1)(4x+1)$; d) $(x-2)(5x+4)$;
   e) $(x-y)(3x-2y)$

Exercises 1.4

1. a) $\dfrac{x+6}{x+3}$; b) $\dfrac{2x-1}{x^2+2}$; c) $\dfrac{4x+3y}{x-6y}$

2. a) $\frac{2(x-1)}{x+1}$; b) $\frac{x+4}{x+1}$; c) $\frac{x-3y}{(x+2y)(x-5y)}$; d) $\frac{-2x-1}{x(x-1)}$; e) $\frac{x+2}{x}$; f) $\frac{x-3y}{x-5y}$

3. a) $\frac{x\sqrt{y}-y\sqrt{x}}{x-y}$; b) $\frac{x+y-2\sqrt{(x+1)(y-1)}}{y-x-2}$; c) $\frac{(x-\sqrt{3xy})(y-\sqrt{3})}{x(y^2-3)}$

4. a) $\frac{6x+5}{3x-2}$; b) $\frac{x(yz+2xz+3xy)}{z}$; c) $\frac{x^2+4x}{x^2-4}=\frac{x(x+4)}{x^2-4}$; d) $\frac{-9}{x+1}$

Exercises 1.5

1. a) -10; b) 2; c) 8; d) 3/2; e) ± 10; f) $-3\sqrt[3]{3}$

2. a) $2,-5$; b) $-2\pm\sqrt{7}$; c) $5,-1$; d) $\frac{2\pm\sqrt{10}}{2}$; e) $\frac{-3\pm\sqrt{5}}{2}$; f) $\frac{-4\pm\sqrt{11}}{5}$

3. a) $\begin{cases} x=-7 \\ y=-5 \end{cases}$; b) $\begin{cases} x=1 \\ y=1 \end{cases}$; c) $\begin{cases} x=6 \\ y=1 \end{cases}$; d) $\begin{cases} x=1 \\ y=1 \end{cases}$; e) $\begin{cases} a=1 \\ b=0 \end{cases}$

4. 21% – 375 ml; 5% – 625 ml

CHAPTER 2

2 Review of Triangles and Trigonometry

CHAPTER OBJECTIVES

- Review plane angles, triangles and their general properties
- Review right triangles and trigonometric functions of acute angles
- Review general trigonometric functions and identities
- Review oblique triangles and laws of sines and cosines
- Refresh key features of basic geometric shapes

Essential statements on triangles and trigonometry:

- Triangles and properties are the key to understand trigonometric functions and identities.
- It is vital to master trigonometric functions and identities for solving various scientific and engineering problems.

Key topics:

- Plane angles and arcs
- Triangles and properties
- Trigonometric functions and identities
- Law of sines and law of cosines
- Key features of basic geometric shapes

Flowchart of mathematical knowledge building

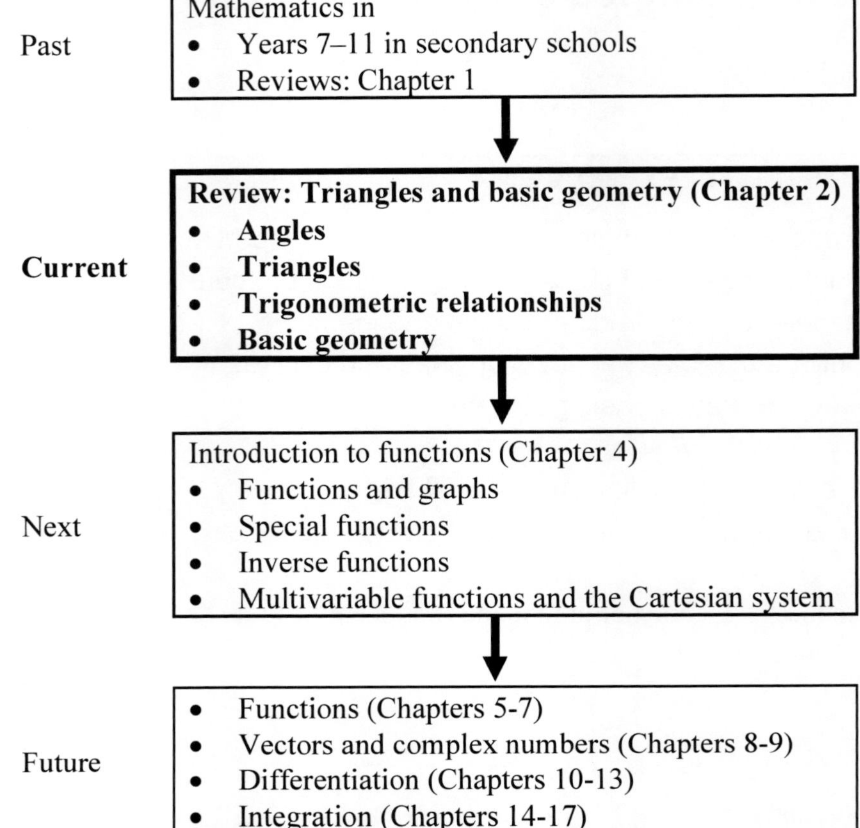

Chapter 2: Review of Triangles and Trigonometry

2.1 Plane Angles and General Properties of Triangles

2.1.1 Plane angles

The plane angles

A directed straight line is called a *ray*. In Figure 2.1a, $\overrightarrow{OA}$ and $\overrightarrow{OB}$ are two rays originated from point *O* and pass points *A* and *B* respectively. These two rays originated from the same point *O*, or in other words, intersected at point *O*, form an angle $\angle O$ *or* $\angle AOB$ at point *O*. Point *O* is called the *vertex* of angle $\angle O$; $\overrightarrow{OA}$ and $\overrightarrow{OB}$ are called the *sides* of angle $\angle O$. Since both $\overrightarrow{OA}$ and $\overrightarrow{OB}$ are on the same plane, this angle is called the *plane angle*. This differs from the *space angle* where for instance another ray $\overrightarrow{OC}$ also originated from point *O* is normal to both $\overrightarrow{OA}$ and $\overrightarrow{OB}$, i.e., $\overrightarrow{OC}$ is not lying on the plane formed by $\overrightarrow{OA}$ and $\overrightarrow{OB}$. As all angles discussed in this chapter are assumed to be on the same plane, a plane angle is simply shortened as an angle in this book.

An angle can also be regarded as the angular difference created by rotating ONE ray, for instance ***r***, about the origin *O* anticlockwise from its *initial angular position* ($\theta_0 = 0$) to any of other angular positions θ_i (Figure 2.1b). Understandably if the ray ***r*** rotates about the origin *O* clockwise from its initial angular position ($\theta_0 = 0$) to other angular positions, it creates *negative* angles.

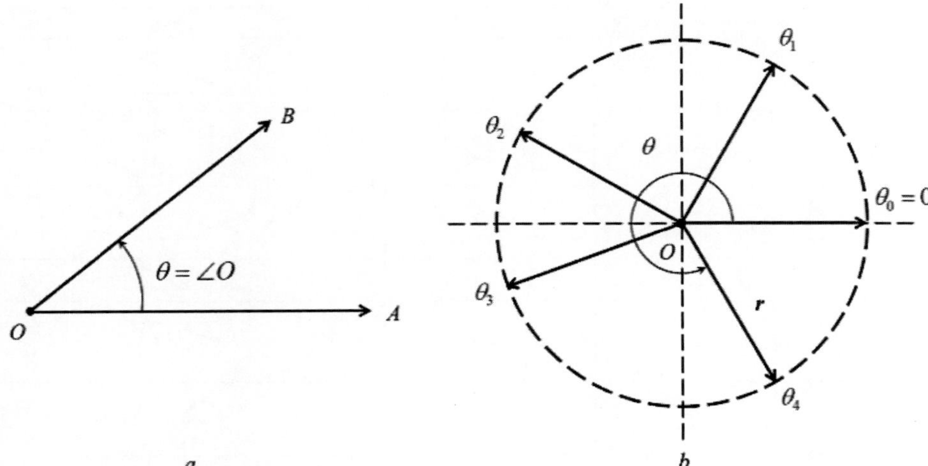

Figure 2.1: The plane angle (a) and the change of angle by rotating a ray (b)

Measuring angles by radians and degrees

Angles are measured by two commonly used scales: radians (rad) and degrees (°). Since a full rotation (or a revolution) is of an angle of 2π radians or 360° (degrees), hence

$$2\pi = 360^\circ \longrightarrow \pi = 180^\circ \longrightarrow \frac{\pi}{2} = 90^\circ.$$

These mappings can be generalised by the following conversions between radians and degrees:

$$\boxed{\begin{aligned} &1\text{ radian} = \frac{180^\circ}{\pi} \approx \frac{180^\circ}{3.14159} \approx 57.296^\circ \\ &1\text{ degree} = \frac{\pi}{180^\circ} \approx \frac{3.14159}{180^\circ} \approx 0.017453\text{ rad.} \end{aligned}} \qquad (2.1)$$

This set offers direct conversion of any angle in one scale to the other. Some commonly used angles in both scales are listed in Table 2.1. Angles are also classified into several types according to the size of angle, which is summarised in Table 2.2.

Table 2.1: Commonly used angles in radians and degrees

Radian	0	$\frac{\pi}{6}$	$\frac{\pi}{4}$	$\frac{\pi}{3}$	$\frac{\pi}{2}$	$\frac{2\pi}{3}$	$\frac{3\pi}{4}$	$\frac{5\pi}{6}$	π	$\frac{7\pi}{6}$	$\frac{5\pi}{4}$	$\frac{4\pi}{3}$	$\frac{3\pi}{2}$	$\frac{7\pi}{4}$	2π
Degree	0°	30°	45°	60°	90°	120°	135°	150°	180°	210°	225°	240°	270°	315°	360°

Table 2.2: Types of angles

Type	Range (degree)	Range (radian)	Diagram
Acute	$0^\circ < \theta < 90^\circ$	$0 < \theta < \frac{\pi}{2}$	
Right	$\theta = 90^\circ$	$\theta = \frac{\pi}{2}$	
Obtuse	$90^\circ < \theta < 180^\circ$	$\frac{\pi}{2} < \theta < \pi$	
Straight	$\theta = 180^\circ$	$\theta = \pi$	
Reflect	$180^\circ < \theta < 360^\circ$	$\pi < \theta < 2\pi$	
Full turn	360°	2π	

Example 2.1: Convert –230° and 333.25° to radians respectively.

Solution

$$-230^\circ = -230^\circ \times \frac{\pi}{180^\circ} = \frac{-230^\circ\pi}{180^\circ} \approx -1.27778\pi \approx -1.27778 \times 3.14159 \approx -4.01426 \text{ rad}$$

$$333.25^\circ = 333.25^\circ \times \frac{\pi}{180^\circ} = \frac{333.25^\circ\pi}{180^\circ} \approx 1.85139\pi \approx 1.85139 \times 3.14159 \approx 5.81631 \text{ rad}$$

Example 2.2: Convert $\frac{\pi}{10}$ and 1.517 in radians to degrees respectively.

Solution

$$\frac{\pi}{10} = \frac{\pi}{10} \times \frac{180°}{\pi} = \frac{180°}{10} = 18°$$

$$1.517 = 1.517 \times \frac{180°}{\pi} \approx 1.517 \times 57.296° \approx 86.918°.$$

The degree scale has two more subunits: *minutes* (') and *seconds* ("). Conversions among degrees, minutes and seconds are as follows:

$$\begin{array}{|lcl|} 1'\text{(minute)} = 60''\text{(seconds)} & \rightleftharpoons & 1''\text{(second)} = (\frac{1}{60})'\text{(minutes)}; \\ 1°\text{(degree)} = 60'\text{(minutes)} & \rightleftharpoons & 1'\text{(minute)} = (\frac{1}{60})°\text{(degrees)}; \\ 1°\text{(degree)} = 3600'\text{(seconds)} & \rightleftharpoons & 1''\text{(second)} = (\frac{1}{3600})°\text{(degrees)}. \end{array} \quad (2.2)$$

Example 2.3: Convert 57.296°, –230.13°, 1.517 rad to the form of *degree-minute-second.*

Solution

We use set (2.2) to convert the given angles to the form of *degree-minute-second* as follows.

$$\begin{aligned} 57.296° &= 57° + 0.296° = 57° + 0.296 \times 60' = 57° + 17.76' = 57°17' + 0.76' \\ &= 57°17' + 0.76 \times 60'' = 57°17' + 45.6'' = 57°17'45.6'' \\ -230.13° &= -230° - 0.13° = -230° - 0.13 \times 60' = -230° - 7.8' = -230°7' - 0.8' \\ &= -230°7' - 0.8 \times 60'' = -230°7' - 48'' = -230°7'48'' \\ 1.517 \text{ rad} &= 1.517 \times \frac{180°}{\pi} \approx 1.517 \times 57.296° \approx 86.918° = 86° + 0.918 \times 60' = 86° + 55.08' \\ &= 86°55' + 0.08 \times 60'' = 86°55' + 4.8'' = 86°55'4.8''. \end{aligned}$$

Example 2.4: Convert 57°35′27″ and –48°50′45″ in *degree-minute-second* to radians.

Solution

We use set (2.2) to convert the given angles in the form of *degree-minute-second* to angles in radians.

$$57°35'27'' = 57° + 35' + 27'' = 57° + (\frac{35}{60})° + (\frac{27}{3600})° \approx 57° + 0.5833° + 0.0075° = 57.5908°$$

$$= 57.5908° \times \frac{\pi}{180°} \approx 0.31995 \times 3.14159 \approx 1.00515$$

$$-48°50'44'' = -48° - 50' - 44'' = -48° - (\frac{50}{60})° - (\frac{44}{3600})° \approx -48° - 0.8333° - 0.0122° = -48.8455°$$

$$= -48.8455° \times \frac{\pi}{180°} \approx -0.27136 \times 3.14159 \approx -0.8525.$$

Arc length and area of a sector

On a circle of radius r, an *arc* is a segment of the circumference of the circle that is confined within a central angle of θ in radians. The corresponding section of the circle confined within the central angle θ is called a *sector* of the circle (Figure 2.2). The length (s) of the arc and the area (A) of the sector on a circle of radius r can be determined by

$$\boxed{\begin{aligned} s &= r\theta \\ A &= \frac{1}{2}r^2\theta \end{aligned}} \qquad (2.3)$$

The central angle must be in radians for calculating both the arc length and sector area. If an angle is given in degree or in degree-minute-second, it must be converted to the radian before applying set (2.3).

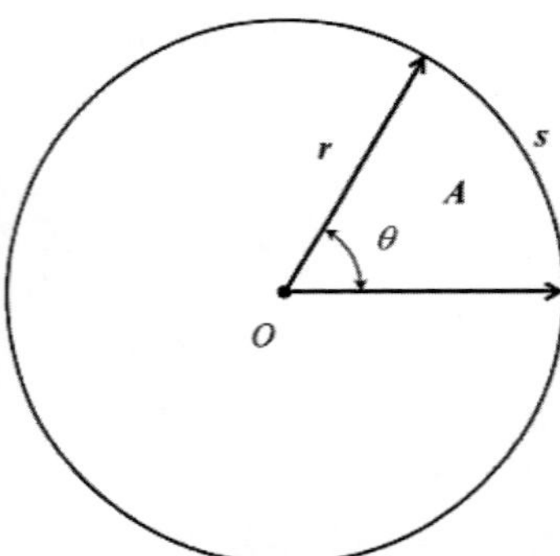

Figure 2.2: Arc as a segment of the circumference of a circle and sector as a section of a circle

Example 2.5: For a circle of radius 10 cm, find the length of the arc and the area of the sector confined by a central angle of 65°.

Solution

We convert 65° to radians and then use set (2.3) to calculate the arc length and sector area required.

$$65° = 65° \times \frac{\pi}{180°} \approx 0.36111 \times 3.14159 \approx 1.13446 \text{ rad}$$

$$s = r\theta \approx 10 \times 1.13446 = 11.3446 \text{ cm}, \qquad A = \frac{1}{2}r^2\theta \approx \frac{1}{2} \times 10^2 \times 1.13446 = 56.723 \text{ cm}^2.$$

Example 2.6: The area of a sector with a central angle of 0.2π is 31.4159 cm². What is the radius of this circle?

Solution

$$A = \frac{1}{2}r^2\theta \longrightarrow 2A = r^2\theta \longrightarrow r^2 = \frac{2A}{\theta} \longrightarrow r = \sqrt{\frac{2A}{\theta}} \approx \sqrt{\frac{2\times 31.4159}{0.2\times 3.14159}} = \sqrt{10\times 10} = 10\text{ cm.}$$

Example 2.7: If a section on a circle has an arc length of 20 cm and a sector area of 45 cm², find the radius of this circle and the central angle of this sector.

Solution

Since $\frac{A}{s} = \frac{\frac{1}{2}r^2\theta}{r\theta} = \frac{1}{2}r$, thus $r = \frac{2A}{s} = \frac{2\times 45}{20} = 4.5\text{ cm.}$

Since $s = r\theta$, thus $\theta = \frac{s}{r} = \frac{20}{4.5} \approx 4.44444\text{ rad} = 4.44444\times\frac{180°}{\pi} \approx 254.6478°.$

Relationships between two associated angles

Complementary angles

If the sum of two adjacent angles (α and β) sharing the same vertex is 90°, these two angles are complementary to each other, i.e., $\alpha + \beta = 90°$ or $\alpha = 90° - \beta$ (Figure 2.3a).

Supplementary angles

If the sum of two adjacent angles (α and β) sharing the same vertex is 180°, these two angles are supplementary to each other, i.e., $\alpha + \beta = 180°$ or $\alpha = 180° - \beta$ (Figure 2.3b).

Conjugate angles

If the sum of two adjacent angles (α and β) sharing the same vertex is 360°, these two angles are called conjugate angles, i.e., $\alpha + \beta = 360°$ or $\alpha = 360° - \beta$ (Figure 2.3c).

Vertically opposite angles

Two intersecting lines form four angles at the point of intersection. Any two opposite angles at the point of intersection have the same measure (Figure 2.3d).

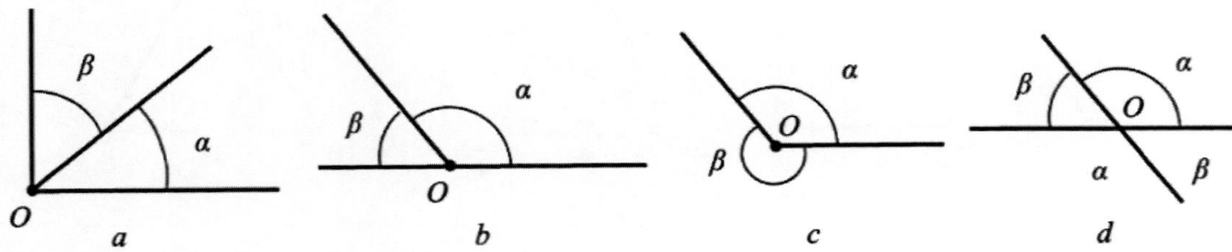

Figure 2.3: Complementary (a), supplementary (b), conjugate (c), vertically opposite (d) angles

Example 2.8: Given α = 76°, find the corresponding complementary, supplementary and conjugate angles respectively.

Solution

Complementary angle: $\beta = 90° - \alpha = 90° - 76° = 14°$

Supplementary angle: $\beta = 180° - \alpha = 180° - 76° = 104°$

Conjugate angle: $\beta = 360° - \alpha = 360° - 76° = 284°$.

2.1.2 Triangles and general properties

Types of triangles

Triangles are categorised by their angles as *acute*, *obtuse*, and *right* triangles. In an acute triangle, all three angles are less than 90° (Figure 2.4a). If any angle in a triangle is greater than 90°, it is an obtuse triangle (Figure 2.4b). If a triangle has one angle of 90°, it is called a right triangle (Figure 2.4c).

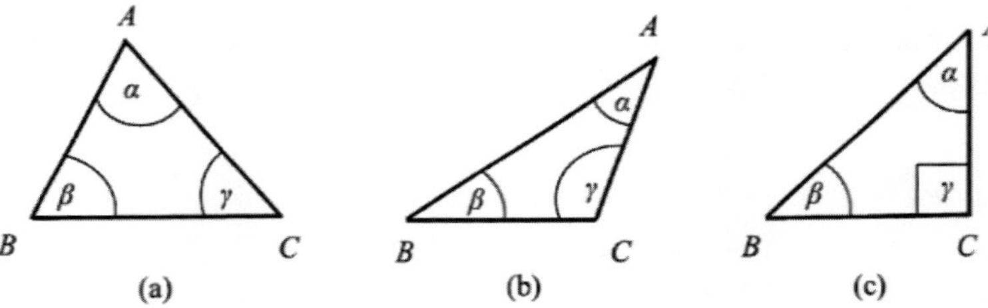

Figure 2.4: Acute triangle (a), obtuse triangle (b), and right triangle (c)

Triangles can also be categorised by their sides as *scalene*, *isosceles*, and *equilateral* triangles. Any triangle with three different sides is called a scalene triangle (Figure 2.5a). If two sides in a triangle are equal, it is an isosceles triangle (Figure 2.5b). Any triangle with three equal sides is called an equilateral triangle (Figure 2.5c).

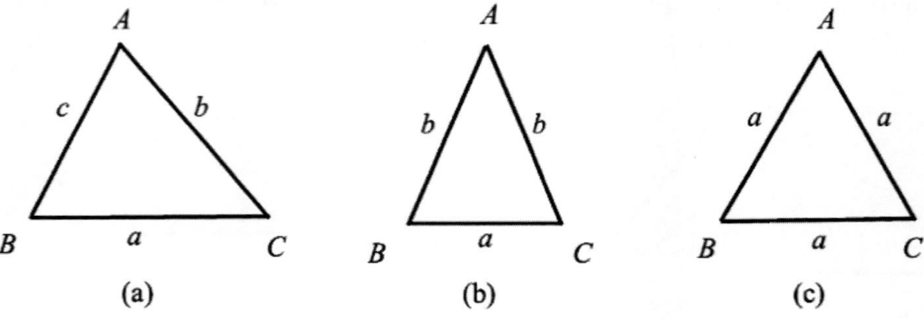

Figure 2.5: Scalene triangle (a), isosceles triangle (b), and equilateral triangle (c)

Main properties of triangles

- The sum of all three interior angles of any triangle is 180°, i.e., $\alpha + \beta + \gamma = 180°$.
- Angles lying opposite the equal sides are equal too and vice versa.
- All three angles of an equilateral triangle are equal to 60°.
- In any triangle, if one side is extended, its exterior angle equals the sum of the two opposite interior angles, i.e., $\gamma' = \beta + \alpha$ (Figure 2.6a).
- The sum of the three exterior angles of any triangle is 360°, i.e., $\alpha' + \beta' + \gamma' = 360°$ (Figure 2.6b).
- Any side of a triangle is less than the sum of the other two sides and greater than their difference.
- The side opposite to the largest angle in a triangle is the greatest side and vice versa.

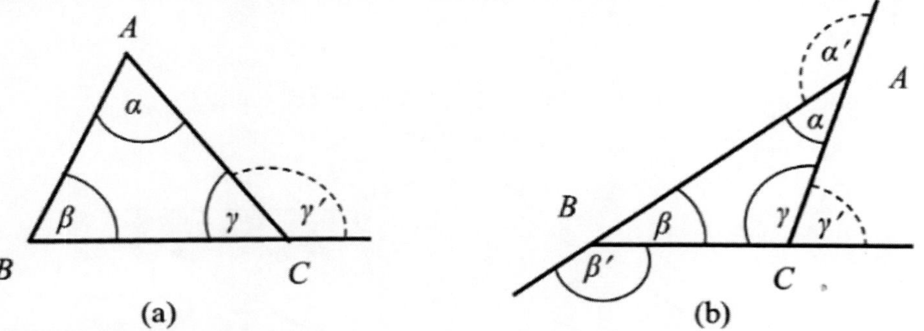

Figure 2.6: An exterior angle of a triangle (a), and three exterior angles of a triangle (b)

Example 2.9: Referring to Figure 2.6a, prove if one side of a triangle is extended, its exterior angle equals the sum of the two opposite interior angles, i.e., $\gamma' = \alpha + \beta$.

Solution

Since the two angles γ and γ' are supplementary to each other, thus $\gamma + \gamma' = 180°$. Also the sum of the three interior angles in any triangle is 180°, i.e., $\alpha + \beta + \gamma = 180°$, therefore

$$\gamma + \gamma' = \alpha + \beta + \gamma, \text{ or } \gamma' = \alpha + \beta.$$

Example 2.10: Referring to Figure 2.6b, prove that the sum of the three exterior angles of any triangle is 360°, i.e., $\alpha' + \beta' + \gamma' = 360°$.

Solution

Since each pair of the interior and exterior angles are supplementary to each other, thus

$$\alpha + \alpha' = 180°$$
$$\beta + \beta' = 180°$$
$$\gamma + \gamma' = 180°$$
$$+ \overline{\qquad\qquad\qquad\qquad}$$
$$(\alpha + \beta + \gamma) + (\alpha' + \beta' + \gamma') = 540°$$

As the sum of the three interior angles in any triangle is 180°, i.e., $\alpha + \beta + \gamma = 180°$, therefore

$$\alpha' + \beta' + \gamma' = 540° - (\alpha + \beta + \gamma) = 540° - 180° = 360°.$$

Congruence of triangles

Two triangles ΔABC and $\Delta A'B'C'$ are *congruent* if they have identical size and shape, i.e., their corresponding angles and sides are equal, denoted as $\Delta ABC \cong \Delta A'B'C'$. Two triangles are congruent (Figure 2.7)

- If a pair of corresponding angles and the included side are equal (ASA for angle-side-angle), or
- If a pair of corresponding sides and the included angle are equal (SAS), or
- If all their corresponding sides are equal (SSS).

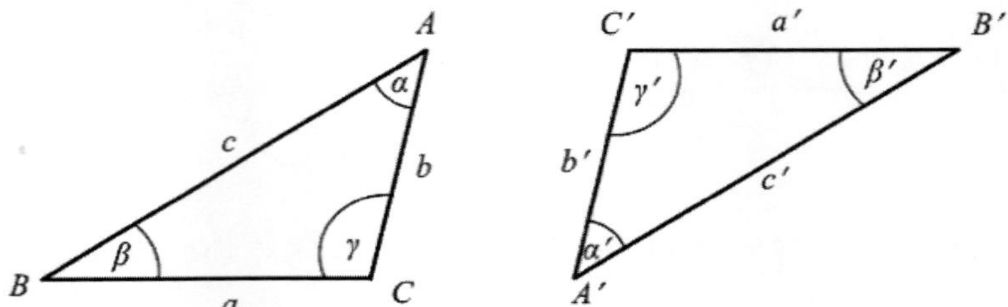

Figure 2.7: Congruent triangles ΔABC (left) and $\Delta A'B'C'$ (right)

Similarities of triangles

Two triangles ΔABC and $\Delta A'B'C'$ are *similar* if they are identical in shape but different in scale, i.e., their corresponding angles are equal but sides are proportional (Figure 2.8), denoted as $\Delta ABC \sim \Delta A'B'C'$.

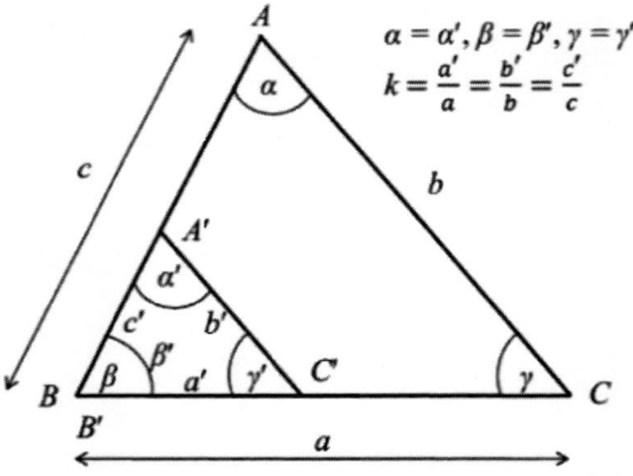

Figure 2.8: Similar triangles ΔABC and $\Delta A'B'C'$

Two triangles are similar

- If all their corresponding angles are equal, or
- If all their corresponding sides are proportional, or
- If an angle of ΔABC is congruent to one angle of $\Delta A'B'C'$ and the sides that include those angles are proportional.

Example 2.11: The three angles of ΔABC are the same to those of ΔDEF. However, the longest side of ΔABC is 20 cm whereas the longest side of ΔDEF is 5 cm. Evaluate

(a) Whether ΔABC is similar to ΔDEF?
(b) What is the proportion of ΔDEF with respect to ΔABC if they are similar?

Solution

(a) Since the three angles of both ΔABC and ΔDEF are the same, these two triangles are similar.
(b) As ΔABC and ΔDEF are similar, the ratio of any corresponding pair of sides gives the proportion of these two triangles. The two longest sides produce

$$k = \frac{\text{longest side in } \Delta DEF}{\text{longest side in } \Delta ABC} = \frac{5}{20} = \frac{1}{4}.$$

Hence ΔDEF is a quarter of ΔABC in scale, or ΔABC is four times larger than ΔDEF.

Area of a triangle

Area of a triangle is determined by

$$\boxed{A = \frac{1}{2}(base)(height) = \frac{1}{2}bh \quad \text{(any triangle)}} \tag{2.4}$$

The base (b) and height (h) of a triangle are illustrated in Figure 2.9.

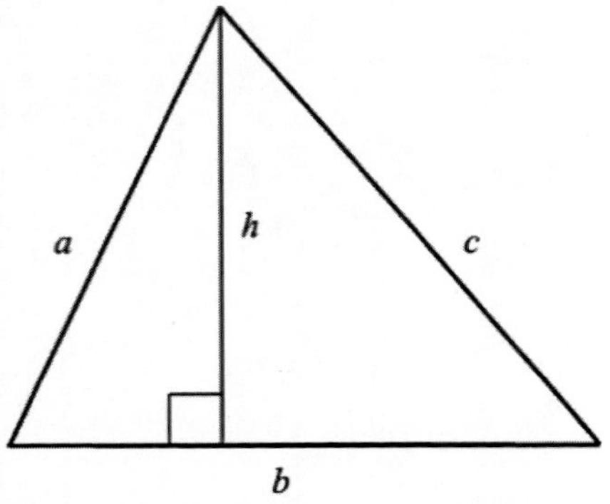

Figure 2.9: The base (b) and height (h) of a triangle

2.1.3 Right triangles and trigonometric functions of acute angles

The Pythagorean theorem for right triangles

In any right triangle, the side opposite to the right angle is called the *hypotenuse* whereas the other two sides that form the right angle are called *sides* or *legs* (Figure 2.10a). Referring to angle α, the side opposite to angle α is called the *opposite side* (of α) whereas the other side is called the *adjacent side* (of α) shown in Figure 2.10b. On the other hand, by referring to angle β, the side opposite to angle β becomes the opposite side (of β) whereas the other side becomes the adjacent side (of β) shown in Figure 2.10c.

The Pythagorean theorem for any right triangle is stated as: the square of hypotenuse c is the sum of the squares of the two legs or sides a and b, i.e.,

$$\boxed{c^2 = a^2 + b^2.} \tag{2.5}$$

This can be reorganised alternatively as

$$\begin{cases} c^2 = a^2 + b^2 \\ a^2 = c^2 - b^2 \\ b^2 = c^2 - a^2 \end{cases} \Longleftrightarrow \begin{cases} c = \sqrt{a^2 + b^2} \\ a = \sqrt{c^2 - b^2} \\ b = \sqrt{c^2 - a^2} \end{cases}.$$

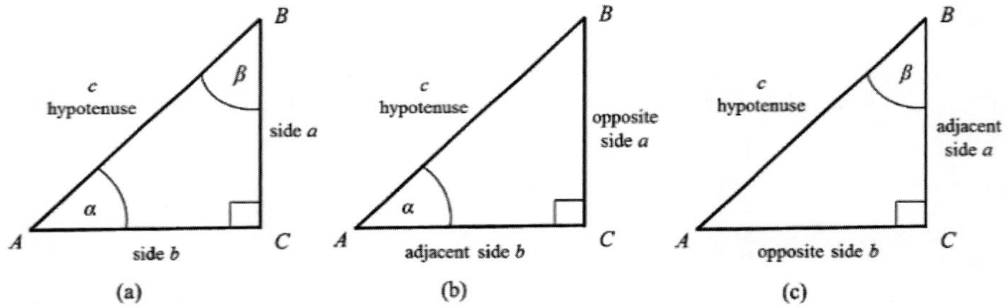

Figure 2.10: Right triangle (a); right triangle referring to α (b); right triangle referring to β (c)

Example 2.12: Refer to Figure 2.10a; find the length of the third side in the following right triangles:

(a) Given $a = 4$ cm and $b = 3$ cm;
(b) Given $a = 4$ cm and $c = 5$ cm;
(c) Given $b = 8$ cm and $c = 10$ cm.

Solution

a) $c = \sqrt{a^2 + b^2} = \sqrt{4^2 + 3^2} = \sqrt{16 + 9} = \sqrt{25} = 5$ cm.

b) $b = \sqrt{c^2 - a^2} = \sqrt{5^2 - 4^2} = \sqrt{25 - 16} = \sqrt{9} = 3$ cm.

c) $a = \sqrt{c^2 - b^2} = \sqrt{10^2 - 8^2} = \sqrt{100 - 64} = \sqrt{36} = 6$ cm.

Example 2.13: Referring to Figure 2.9, given $a = 5$ cm, $c = 4\sqrt{2}$ cm, and $h = 4$ cm, find the area of this scalene triangle.

Solution

The triangle in Figure 2.9 is divided into two right triangles. The base of the left-side right triangle is

$$b_L = \sqrt{a^2 - h^2} = \sqrt{5^2 - 4^2} = \sqrt{25 - 16} = \sqrt{9} = 3 \text{ cm}.$$

The base of the right-side right triangle is

$$b_R = \sqrt{c^2 - h^2} = \sqrt{(4\sqrt{2})^2 - 4^2} = \sqrt{32 - 16} = \sqrt{16} = 4 \text{ cm}.$$

By formula (2.4), the area of this scalene triangle is

$$A = \frac{1}{2}bh = \frac{1}{2}(b_L + b_R)h = \frac{1}{2}(3+4)\times 4 = \frac{7}{2}\times 4 = 7\times 2 = 14 \text{ cm}^2.$$

Trigonometric functions of acute angles

In any right triangle, the ratios of its three sides define six trigonometric functions with respect to any of the two acute angles α or β (Figure 2.10a). Referring to angle α shown in Figure 2.10b, the six trigonometric functions, known as sine (sin), cosine (cos), tangent (tan), cotangent (cot), secant (sec), and cosecant (csc), are defined as

$$\sin\alpha = \frac{\text{opposite side}}{\text{hypotenuse}} = \frac{a}{c}; \qquad (2.6)$$

$$\cos\alpha = \frac{\text{adjacent side}}{\text{hypotenuse}} = \frac{b}{c}; \qquad (2.7)$$

$$\tan\alpha = \frac{\text{opposite side}}{\text{adjacent side}} = \frac{a}{b} = \frac{\sin\alpha}{\cos\alpha}; \qquad (2.8)$$

$$\sec\alpha = \frac{\text{hypotenuse}}{\text{adjacent side}} = \frac{c}{b} = \frac{1}{\cos\alpha}; \qquad (2.9)$$

$$\csc\alpha = \frac{\text{hypotenuse}}{\text{opposite side}} = \frac{c}{a} = \frac{1}{\sin\alpha}; \qquad (2.10)$$

$$\cot\alpha = \frac{\text{adjacent side}}{\text{opposite side}} = \frac{b}{a} = \frac{1}{\tan\alpha} = \frac{\cos\alpha}{\sin\alpha}. \qquad (2.11)$$

Although the six trigonometric functions are defined, only sine and cosine functions are called the primary functions. This is because other four functions can be derived from either or both of sine and cosine functions. Some values of special acute angles for sine, cosine, and tangent functions are given in Table 2.3. The corresponding values for cosecant, secant and cotangent functions are simply the reciprocal of those for sine, cosine, and tangent functions respectively.

Table 2.3: Values of special acute angles for sine, cosine, and tangent functions

θ	$\sin\theta$	$\cos\theta$	$\tan\theta$
0°	0	1	0
30°	$\frac{1}{2}$	$\frac{\sqrt{3}}{2}$	$\frac{\sqrt{3}}{3}$
45°	$\frac{\sqrt{2}}{2}$	$\frac{\sqrt{2}}{2}$	1
60°	$\frac{\sqrt{3}}{2}$	$\frac{1}{2}$	$\sqrt{3}$
90°	1	0	∞

Through these trigonometric functions, another way to find the measure of an angle is to use the *inverse trigonometric functions*, commonly the inverse sine, cosine and tangent functions:

$$\begin{aligned} \alpha &= \sin^{-1}(\frac{a}{c}) \\ \alpha &= \cos^{-1}(\frac{b}{c}) \, . \\ \alpha &= \tan^{-1}(\frac{a}{b}) \end{aligned} \tag{2.12}$$

The measure of an angle from formula (2.12) used to be determined by looking up the trigonometric tables in the past. Nowadays any scientific calculator can accurately return the value of an acute angle in either degree or radian depending on the setting of your calculator.

The similar process of defining the six trigonometric functions with respect to angle α can be applied to angle β by referring to Figure 2.10c. For example,

$$\sin\beta = \frac{\text{opposite side of } \beta}{\text{hypotenuse}} = \frac{b}{c} = \cos\alpha.$$

As angles α and β are complementary to each other, i.e., $\alpha + \beta = 90°$ or $\beta = 90° - \alpha$, thus

$$\sin\beta = \sin(90° - \alpha) = \cos\alpha. \tag{2.13}$$

Similarly,

$$\cos\beta = \cos(90° - \alpha) = \sin\alpha; \quad (2.14)$$

$$\tan\beta = \tan(90° - \alpha) = \cot\alpha; \quad (2.15)$$

$$\cot\beta = \cot(90° - \alpha) = \tan\alpha; \quad (2.16)$$

$$\sec\beta = \sec(90° - \alpha) = \csc\alpha; \quad (2.17)$$

$$\csc\beta = \csc(90° - \alpha) = \sec\alpha. \quad (2.18)$$

The Pythagorean theorem in trigonometric form is expressed as

$$\begin{cases} \sin^2\alpha + \cos^2\alpha = 1 \\ \sin^2\alpha = 1 - \cos^2\alpha \\ \cos^2\alpha = 1 - \sin^2\alpha \end{cases} \Longleftrightarrow \begin{cases} \sqrt{\sin^2\alpha + \cos^2\alpha} = 1 \\ \sin\alpha = \sqrt{1 - \cos^2\alpha} \\ \cos\alpha = \sqrt{1 - \sin^2\alpha} \end{cases}. \quad (2.19)$$

This trigonometric form of Pythagorean theorem shows that *an angle exists independently, regardless of the sides that form the angle*.

Example 2.14: Prove the trigonometric form of Pythagorean theorem $\sin^2\alpha + \cos^2\alpha = 1$.

Solution

The Pythagorean theorem $c^2 = a^2 + b^2$ can be reorganised as:

$$c^2 = a^2 + b^2 \xrightarrow{\text{both sides diveded by } c^2} \frac{a^2}{c^2} + \frac{b^2}{c^2} = 1 \longrightarrow (\frac{a}{c})^2 + (\frac{b}{c})^2 = 1.$$

As $\sin\alpha = \frac{a}{c}$ and $\cos\alpha = \frac{b}{c}$, substitute them into above formula to prove

$$\sin^2\alpha + \cos^2\alpha = 1.$$

Example 2.15: Referring to the right triangle in Figure 2.10b, given a = 3 cm and $b = \sqrt{3}$ cm, find the values for the hypotenuse, sine, cosine, tangent, and angles α and β respectively.

Solution

All these values can be obtained by choosing appropriate formulae from (2.5-19) and looking up the mappings in Table 2.3 as follows.

$$c = \sqrt{a^2 + b^2} = \sqrt{3^2 + (\sqrt{3})^2} = \sqrt{9+3} = \sqrt{12} = \sqrt{4\times 3} = \sqrt{4}\sqrt{3} = 2\sqrt{3} \text{ cm}$$

$$\sin\alpha = \frac{a}{c} = \frac{3}{2\sqrt{3}} = \frac{3(\sqrt{3})}{2\sqrt{3}(\sqrt{3})} = \frac{3\sqrt{3}}{2\times 3} = \frac{\sqrt{3}}{2}, \quad \cos\alpha = \frac{b}{c} = \frac{\sqrt{3}}{2\sqrt{3}} = \frac{1}{2}, \quad \tan\alpha = \frac{a}{b} = \frac{3}{\sqrt{3}} = \sqrt{3}$$

$$\alpha = \sin^{-1}(\frac{\sqrt{3}}{2}) = \cos^{-1}(\frac{1}{2}) = \tan^{-1}(\sqrt{3}) = 60°, \quad \beta = 90° - \alpha = 90° - 60° = 30°.$$

Alternatively, these values can be obtained by using other formulae as follows.

$$\tan\alpha = \frac{a}{b} = \frac{3}{\sqrt{3}} = \sqrt{3} \longrightarrow \alpha = \tan^{-1}(\sqrt{3}) = 60°$$

$$\sin\alpha = \sin 60° = \frac{\sqrt{3}}{2} \longrightarrow \cos\alpha = \sqrt{1-\sin^2\alpha} = \sqrt{1-(\frac{\sqrt{3}}{2})^2} = \sqrt{\frac{4}{4}-\frac{3}{4}} = \sqrt{\frac{1}{4}} = \frac{1}{2}$$

$$\cos\alpha = \frac{b}{c} \longrightarrow c = \frac{b}{\cos\alpha} = \frac{\sqrt{3}}{1/2} = 2\sqrt{3}\text{ cm}, \quad \sin\beta = \cos\alpha = \frac{1}{2} \longrightarrow \beta = \sin^{-1}(\frac{1}{2}) = 30°.$$

Solving right triangles

The process of solving unknown elements of any right triangle if given two known elements is summarised in Table 2.4.

Table 2.4: Summary of solving right angles

Known elements	Solving unknown elements	Methods
a and b	c, α and β	$c = \sqrt{a^2+b^2}$, $\alpha = \tan^{-1}\frac{a}{b}$, $\beta = 90° - \alpha$
a and c	b, α and β	$b = \sqrt{c^2-a^2}$, $\alpha = \tan^{-1}\frac{a}{b}$, $\beta = 90° - \alpha$
b and c	a, α and β	$a = \sqrt{c^2-b^2}$, $\alpha = \tan^{-1}\frac{a}{b}$, $\beta = 90° - \alpha$
a and α	b, c and β	$b = \frac{a}{\tan\alpha}$, $c = \frac{a}{\sin\alpha}$, $\beta = 90° - \alpha$
a and β	b, c and α	$\alpha = 90° - \beta$, $b = \frac{a}{\tan\alpha}$, $c = \frac{a}{\sin\alpha}$
b and α	a, c and β	$a = b\tan\alpha$, $c = \frac{b}{\cos\alpha}$, $\beta = 90° - \alpha$
b and β	a, c and α	$\alpha = 90° - \beta$, $a = b\tan\alpha$, $c = \frac{b}{\cos\alpha}$
c and α	a, b and β	$a = c\sin\alpha$, $b = c\cos\alpha$, $\beta = 90° - \alpha$
c and β	a, b and α	$\alpha = 90° - \beta$, $a = c\sin\alpha$, $b = c\cos\alpha$

Example 2.16: Find the height and area of an equilateral triangle with side a shown in Figure 2.11.

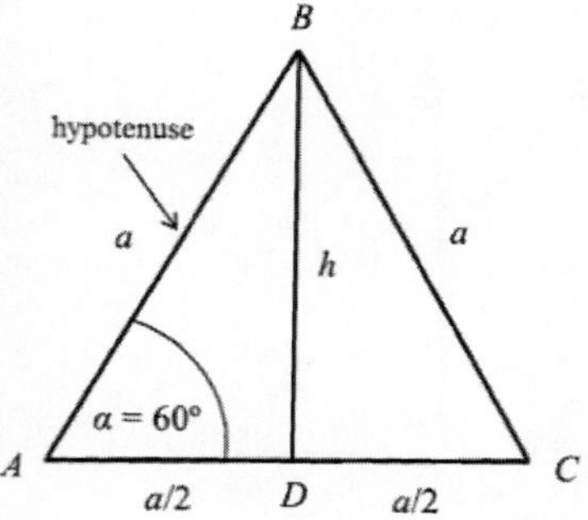

Figure 2.11: Equilateral triangle as two identical right triangles with side a

Solution

As shown in Figure 2.11, the equilateral triangle can be divided into two identical right triangles. The two angles with the baseline have 60° each. The height is the opposite side to either of these two angles.

$$h = \sqrt{a^2 - (\frac{a}{2})^2} = \sqrt{\frac{4a^2}{4} - \frac{a^2}{4}} = \sqrt{\frac{3a^2}{4}} = \frac{\sqrt{3}}{2}a, \text{ or } h = a\sin 60^\circ = \frac{\sqrt{3}}{2}a$$

$$A = \frac{1}{2}bh = \frac{1}{2}a \times \frac{\sqrt{3}}{2}a = \frac{\sqrt{3}}{4}a^2.$$

2.1.4 Applications of right triangles

Example 2.17: Referring to the figure below, if a building 60 m high casts a shadow of 70 m on the ground, what angle above the horizon is the Sun?

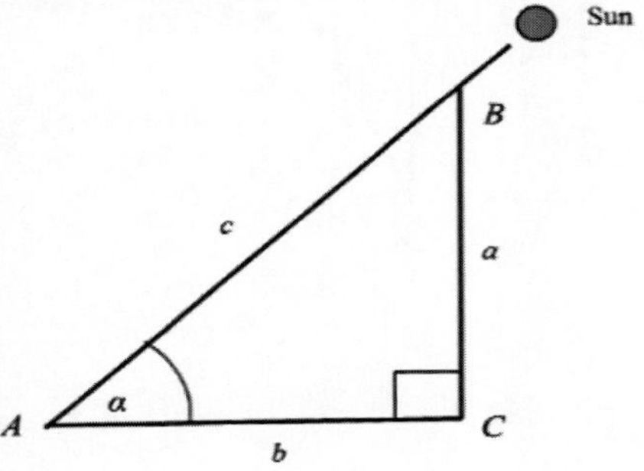

Solution

Since the building is 60 m high or a = 60 m and its shadow is 70 m long or b = 70 m, the angle of the Sun above the horizon can be determined using the inverse of tangent function as follows

$$\alpha = \tan^{-1}(\frac{a}{b}) = \tan^{-1}(\frac{60}{70}) \approx \tan^{-1}(0.8671) \approx 40.6^\circ.$$

Example 2.18: Refer to the figure below for estimating the width of a river. C and A are two points on the same bank of the river. Points A, B and C form a right triangle. Given $b = 55$ m and $\alpha = 38°$, find the width of this river.

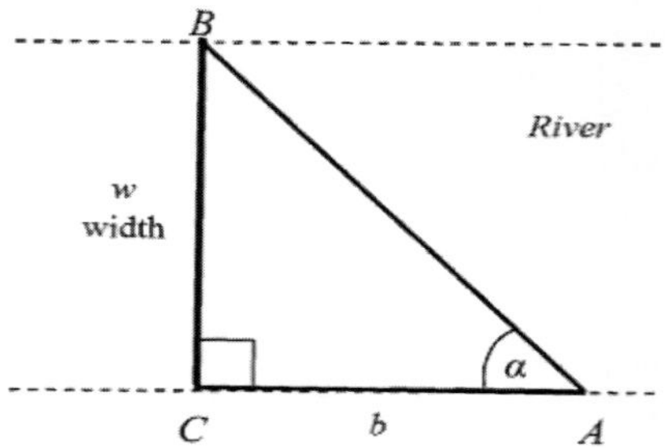

Solution

The width of this river can be estimated using the tangent function as below:

$$\tan\alpha = \frac{w}{b} \longrightarrow w = b\tan\alpha = 55 \times \tan 38° \approx 55 \times 0.7813 = 42.9715 \approx 43 \text{ m.}$$

Example 2.19: Refer to the figure below. Ross wants to swim across a river 100 m wide directly towards the other bank with a constant ground speed of $V_s = 2$ m/s during the journey. The current (water) of the river has a ground speed of $V_w = 1$ m/s. If Ross begins swimming at point A, how far will Ross be drifted away from point C on the other bank (the opposite of A) when touching on the other bank at point B? How long will be the total journey in the water? Which direction will Ross swim towards actually?

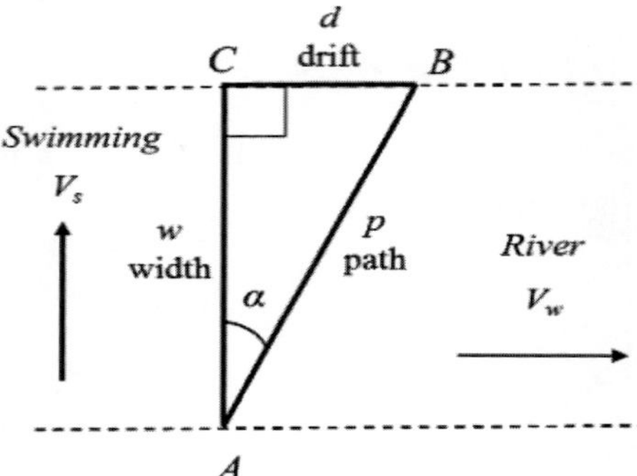

Solution

The time Ross swims across the river is $t = \frac{w}{V_s} = \frac{100}{2} = 50$ s. During this time, Ross will be carried by the current downstream, or drifted away from point C by $d = V_w t = 1 \times 50 = 50$ m.

The total journey will be $p = \sqrt{d^2 + w^2} = \sqrt{50^2 + 100^2} = \sqrt{2500 + 10000} = \sqrt{12500} \approx 111.8$ m.

The actual direction Ross will swim is about $\alpha = \tan^{-1}(\frac{d}{w}) = \tan^{-1}(\frac{50}{100}) = \tan^{-1}(0.5) \approx 26.6°$.

Example 2.20: Refer to the figure below. A ladder leans against the side of a house 5.5 m high. The foot is restricted to be 2 metres due to ground obstacles. What should be the minimum length of the ladder? Given a ladder 6.5 m long, how long the ladder will be extended beyond the edge of the top side of the house?

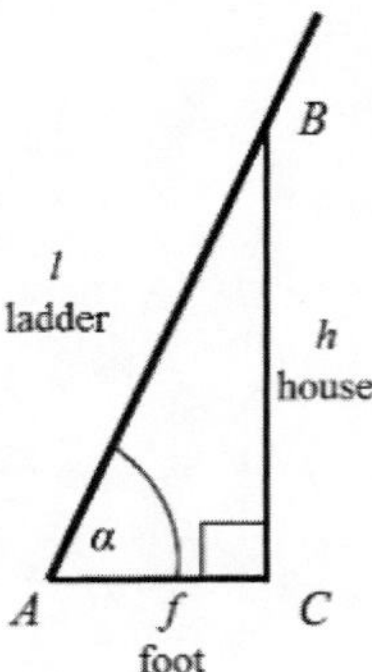

Solution

The minimum length of the ladder is

$$l = \sqrt{f^2 + h^2} = \sqrt{2^2 + 5.5^2} = \sqrt{34.25} \approx 5.85\,\text{m}.$$

A ladder 6.5 m long meets this minimum requirement. If this ladder is used, an extra length of 0.65 m (= 6.5 – 5.85) will extend beyond the edge of the top side of the house.

Example 2.21: Refer to the figure below. A small maintain (hill) is beside and above a lake. To estimate the height of this hill, measurements at A and D on the ground are taken. The angle above the horizon from A to the point B on top of the hill is α = 45°. The angle above the horizon from D to B is δ = 30°. The distance between A and D is d = 50 m. Find the height of this hill.

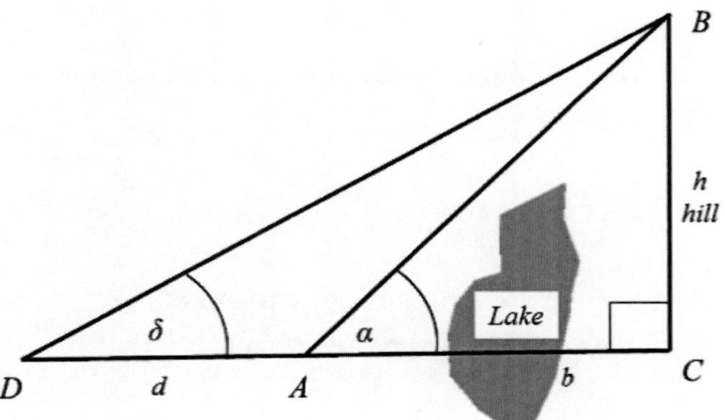

Solution

Assuming the distance between A and C is b, for right triangle ABC, the tangent gives

$$\tan\alpha = \frac{h}{b} \longrightarrow h = b\tan\alpha = b\tan 45° = b \times 1 = b.$$

For right triangle *DBC*, the tangent gives

$$\tan\delta = \frac{h}{b+d} \longrightarrow h = (b+d)\tan\delta = (b+50)\tan 30° = (b+50)\times\frac{\sqrt{3}}{3}$$

$$b+50 = \frac{3}{\sqrt{3}}h \longrightarrow b = \sqrt{3}h - 50.$$

Substitute $b = h$ obtained from right triangle *ABC*, into above equation:

$$h = \sqrt{3}h - 50 \longrightarrow \sqrt{3}h - h = 50 \longrightarrow (\sqrt{3}-1)h = 50$$

$$h = \frac{50}{\sqrt{3}-1} \approx \frac{50}{0.7321} \approx 68.3\text{ m}.$$

Exercises 2.1

1. Convert the following angles in degrees to radians.
   a) 115°
   b) –30°
   c) 330°

2. Convert the following angles in radians to degrees.
   a) $\frac{\pi}{5}$
   b) $-\frac{3\pi}{7}$
   c) 2.5

3. For a circle of radius 20 cm, find the length of the arc and the area of the sector confined by a central angle of 45°.

4. The area of a sector with a central angle of 0.3π is 63 cm^2. What is the radius of this circle?

5. If a section on a circle has the arc length of 30 cm and the sector area of 150 cm^2, find the radius of this circle and the central angle of this sector.

6. Find the corresponding complementary, supplementary and conjugate angles of following angles respectively:
   a) 65°
   b) 15°
   c) 80°

7. Regarded as standard right triangles, find the length of the third side of the followings:

   a) Given $a = 4$ cm and $b = 5$ cm
   b) Given $a = 5$ cm and $c = 15$ cm
   c) Given $b = 60$ cm and $c = 100$ cm

8. Refer to the figure below. Given $c = 18$ cm, $a = 6\sqrt{3}$ cm, and $h = 9$ cm, find the area of this scalene triangle.

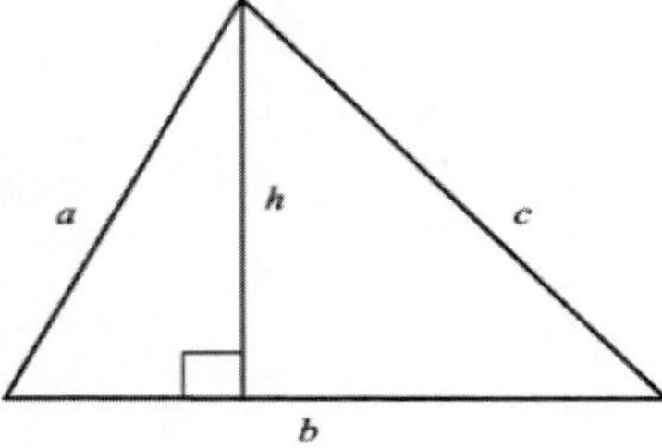

9. Referring to the right triangle below, given $a = 3$ cm and $b = \sqrt{3}$ cm, find the values for the hypotenuse, sine, cosine, tangent, and angle β respectively.

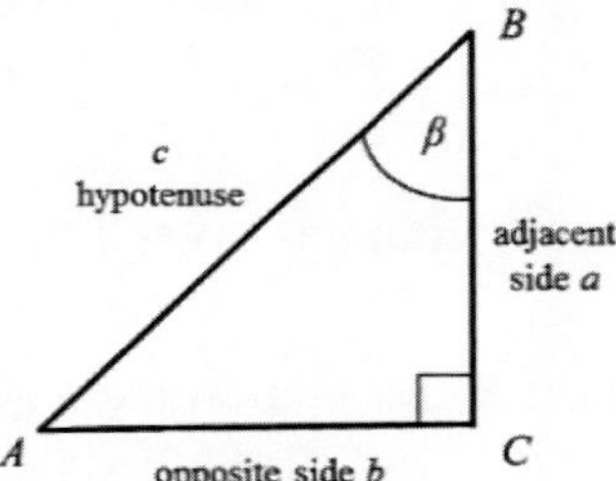

10. Refer to the figure below for estimating the width of a river. C and A are two points on the same bank of the river. Points A, B and C form a right triangle. Given $b = 66$ m and $\alpha = 60°$, find the width of this river.

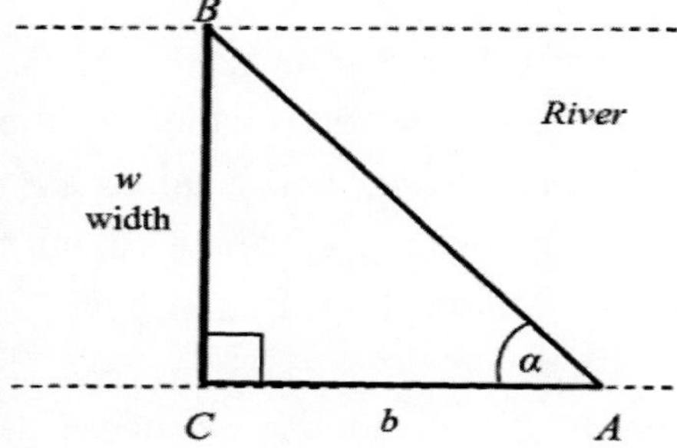

11. Refer to the figure in Example 2.21. A hill is beside and above a lake. To estimate the height of this hill, measurements at A and D on the ground are taken. The angle above the horizon from A to the point B on top of the hill is $\alpha = 60°$. The angle above the horizon from D to B is $\delta = 45°$. The distance between A and D is $d = 40$ m. Find the height of this hill.

2.2 General Trigonometric Functions and Identities

2.2.1 Trigonometric functions of general angles

Coordinates on a plane

Although the trigonometric functions are introduced using a right triangle for acute angles, these functions are applicable to any angle in a full turn of 360°. This is realised by placing a generic right triangle onto a plane of *x*-*y* coordinates shown in Figure 2.12a.

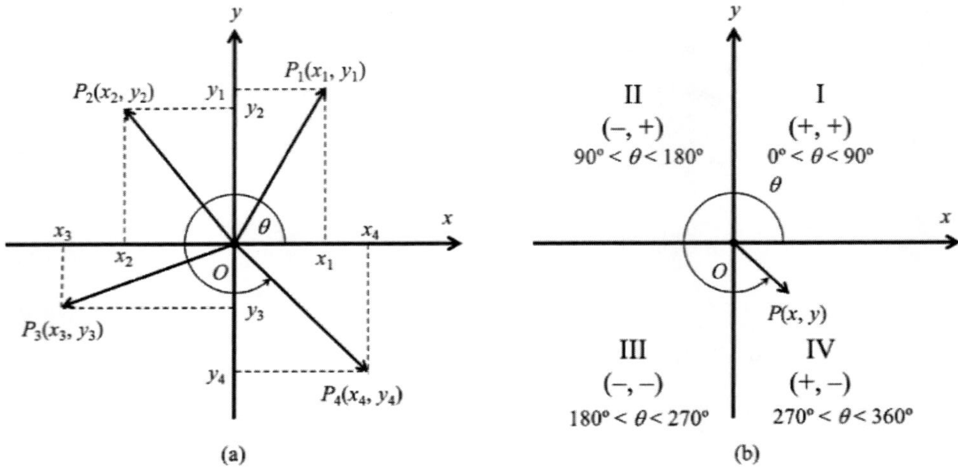

Figure 2.12: A plane of *x*-*y* coordinates (a) and the quadrants of coordinates (b)

Every pair of an *x*-coordinate (*x*) and a *y*-coordinate (*y*) is placed as a point *P*(*x*, *y*) on this plane, such as $P_1(x_1, y_1)$, $P_2(x_2, y_2)$, $P_3(x_3, y_3)$, and $P_4(x_4, y_4)$ in Figure 2.12a. The distance from the origin *O*(0, 0) to *P*(*x*, *y*) is defined by

$$r = \overrightarrow{OP} = \sqrt{x^2 + y^2} \text{ or } r^2 = x^2 + y^2. \tag{2.20}$$

Thus in Figure 2.12a $r_1 = \overrightarrow{OP_1} = \sqrt{x_1^2 + y_1^2}, \ldots, r_4 = \overrightarrow{OP_4} = \sqrt{x_4^2 + y_4^2}$.

Since both *x* and *y* can take positive and negative values independently, the four *x*-*y* combinations divide the plane into four *quadrants*, called quadrants I, II, III, and IV counter-clockwise (Figure 2.12b). In quadrant I, both *x* and *y* values are *positive* whereas both *x* and *y* values are *negative* in quadrant III, opposite to quadrant I about the origin. In quadrant II, *x* is *negative* but *y* is *positive*. Oppositely to quadrant II, *x* is *positive* but *y* is *negative* in quadrant IV.

Any generic point *P*(*x*, *y*) on the plane can also be regarded as a directed segment of line $\overrightarrow{OP}$ with length $r = \overrightarrow{OP} = \sqrt{x^2 + y^2}$ rotated with a generic angle θ about the origin counter-clockwise (Figure 2.12b). In terms of angle θ, quadrant I is within $0° < \theta < 90°$; quadrant II is within $90° < \theta < 180°$; quadrant III is within $180° < \theta < 270°$; quadrant IV is within $270° < \theta < 360°$.

Trigonometric functions defined by the coordinates on a plane

The trigonometric functions introduced using a right triangle for acute angles are applicable to any angle in a full turn of 360°. This is realised by placing a generic right triangle onto a plane of x-y coordinates in any of the four quadrants shown in Figure 2.13. For the right triangle OAP in any quadrant, the six trigonometric functions are defined by the same set of formulae below.

$$
\begin{aligned}
\sin\theta &= \frac{\text{opposite side}}{\text{hypotenuse}} = \frac{y}{r} = \frac{1}{\csc\theta} \\
\cos\theta &= \frac{\text{adjacent side}}{\text{hypotenuse}} = \frac{x}{r} = \frac{1}{\sec\theta} \\
\tan\theta &= \frac{\text{opposite side}}{\text{adjacent side}} = \frac{y}{x} = \frac{1}{\cot\theta} \\
\sec\theta &= \frac{\text{hypotenuse}}{\text{adjacent side}} = \frac{r}{x} = \frac{1}{\cos\theta} \\
\csc\theta &= \frac{\text{hypotenuse}}{\text{opposite side}} = \frac{r}{y} = \frac{1}{\sin\theta} \\
\cot\theta &= \frac{\text{adjacent side}}{\text{opposite side}} = \frac{x}{y} = \frac{1}{\tan\theta}
\end{aligned}
\qquad (2.21)
$$

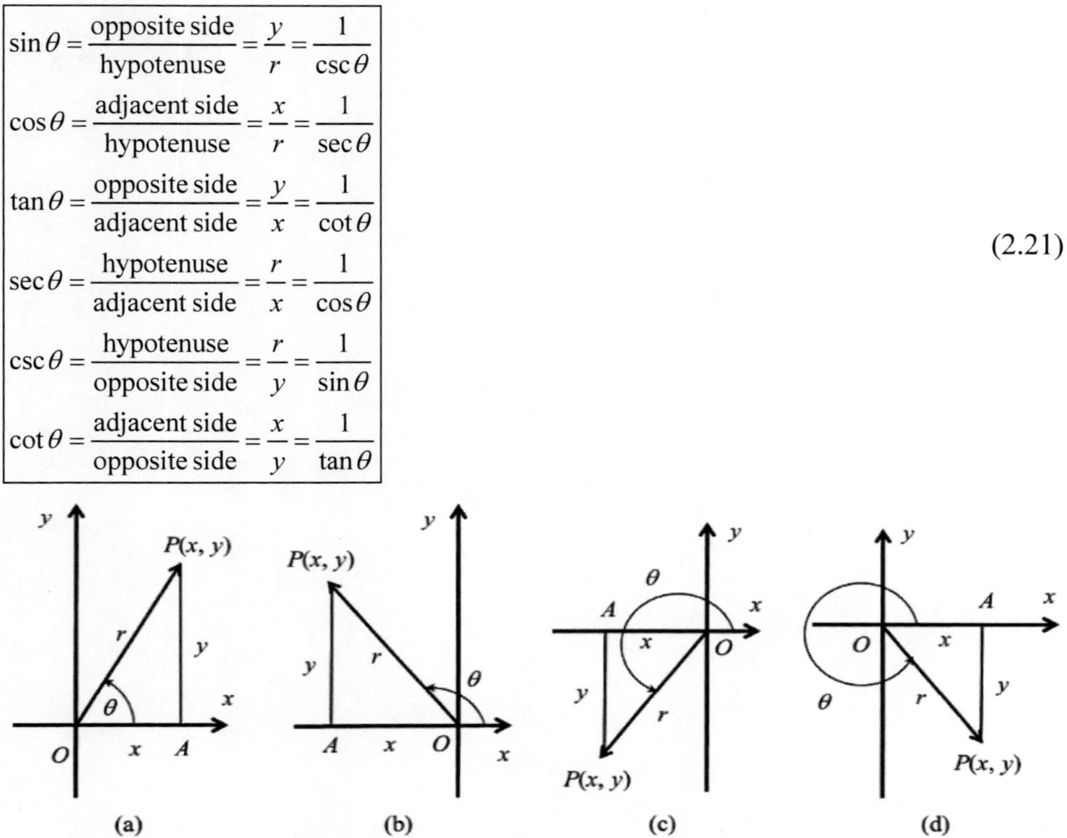

Figure 2.13: A generic right triangle in quadrants I (a), II (b), III (c) and IV (d) on the x-y plane

Since $r = \sqrt{x^2 + y^2}$ is always positive, the sign of any of the six trigonometric functions in a quadrant is dependent on the sign of x or y or the ratio of x and y in that quadrant. Referring to Figure 2.12b, the signs of these functions in the four quadrants are summarised in Table 2.5.

Table 2.5: Summary of signs of trigonometric functions in different quadrants

Quadrant	$\sin\theta$ ($\csc\theta$)	$\cos\theta$ ($\sec\theta$)	$\tan\theta$ ($\cot\theta$)
I (0° < θ < 90°)	*positive* ***y/r*** → +	*positive* ***x/r*** → +	***y/x*** (+/+) → +
II (90° < θ < 180°)	*positive* ***y/r*** → +	*negative* ***x/r*** → –	***y/x*** (+/–) → –
III (180° < θ < 270°)	*negative* ***y/r*** → –	*negative* ***x/r*** → –	***y/x*** (–/–) → +
IV (270° < θ < 360°)	*negative* ***y/r*** → –	*positive* ***x/r*** → +	***y/x*** (–/+) → –

Example 2.22: Given four points on the *x-y* plane at $A(1, 3)$, $B(-4, 3)$, $C(-4, -3)$ and $D(2, -2)$, find the values of the six trigonometric functions at each of these points respectively. Observe if the signs of these results are consistent with the status shown in Table 2.5.

Solution

Both x and y at point $A(1, 3)$ are positive so this point is in quadrant I.

$$r = \sqrt{x^2 + y^2} = \sqrt{1^2 + 3^2} = \sqrt{1+9} = \sqrt{10}$$

$$\sin\alpha = \frac{y}{r} = \frac{3}{\sqrt{10}} = \frac{3\sqrt{10}}{\sqrt{10}\sqrt{10}} = \frac{3\sqrt{10}}{10}, \qquad \cos\alpha = \frac{x}{r} = \frac{1}{\sqrt{10}} = \frac{\sqrt{10}}{\sqrt{10}\sqrt{10}} = \frac{\sqrt{10}}{10}$$

$$\tan\alpha = \frac{y}{x} = \frac{3}{1} = 3, \qquad \cot\alpha = \frac{1}{\tan\alpha} = \frac{1}{3}$$

$$\sec\alpha = \frac{r}{x} = \frac{\sqrt{10}}{1} = \sqrt{10}, \qquad \csc\alpha = \frac{r}{y} = \frac{\sqrt{10}}{3}.$$

All these values are positive as point A is in quadrant I. This is consistent with the status in Table 2.5.

At point $B(-4, 3)$, x is negative and y is positive. Thus point B is in quadrant II.

$$r = \sqrt{(-4)^2 + 3^2} = \sqrt{16+9} = \sqrt{25} = 5$$

$$\sin\beta = \frac{y}{r} = \frac{3}{5}, \qquad \csc\beta = \frac{1}{\sin\beta} = \frac{5}{3},$$

$$\cos\beta = \frac{x}{r} = \frac{-4}{5}, \qquad \sec\beta = \frac{1}{\cos\beta} = -\frac{5}{4}$$

$$\tan\beta = \frac{y}{x} = \frac{3}{-4} = -\frac{3}{4}, \qquad \cot\beta = \frac{1}{\tan\beta} = -\frac{4}{3}.$$

As point B is in quadrant II, by Table 2.5, it is true that both sine and cosecant functions have positive values whereas the other four functions produce negative values.

At point $C(-4, -3)$, both x and y are negative; thus point C is in quadrant III.

$$r = \sqrt{(-4)^2 + (-3)^2} = \sqrt{16+9} = \sqrt{25} = 5$$

$$\sin\gamma = \frac{y}{r} = \frac{-3}{5}, \qquad \csc\gamma = \frac{1}{\sin\gamma} = -\frac{5}{3},$$

$$\cos\gamma = \frac{x}{r} = \frac{-4}{5}, \qquad \sec\gamma = \frac{1}{\cos\gamma} = -\frac{5}{4}$$

$$\tan\gamma = \frac{y}{x} = \frac{-3}{-4} = \frac{3}{4}, \qquad \cot\gamma = \frac{1}{\tan\gamma} = \frac{4}{3}.$$

As point C is in quadrant III, by Table 2.5, it is true that only tangent and cotangent functions have positive values whereas the other four functions produce negative values.

At point $D(2, -2)$, x is positive and y is negative; hence point D is in quadrant IV.

$$r = \sqrt{2^2 + (-2)^2} = \sqrt{4+4} = \sqrt{2\times 4} = 2\sqrt{2}$$

$$\sin\delta = \frac{y}{r} = \frac{-2}{2\sqrt{2}} = \frac{-1}{\sqrt{2}} = -\frac{\sqrt{2}}{2}, \qquad \csc\delta = \frac{1}{\sin\delta} = -\sqrt{2},$$

$$\cos\delta = \frac{x}{r} = \frac{2}{2\sqrt{2}} = \frac{1}{\sqrt{2}} = \frac{\sqrt{2}}{2}, \qquad \sec\delta = \frac{1}{\cos\delta} = \sqrt{2}$$

$$\tan\delta = \frac{-2}{2} = -1, \qquad \cot\delta = \frac{1}{\tan\delta} = -1.$$

As point D is in quadrant IV, by Table 2.5, it is true that only cosine and secant functions have positive values whereas the other four functions produce negative values.

Example 2.23: Given $\cos\theta = -0.5$, find $\sin\theta$ and the quadrants where θ can be placed.

Solution

$$\sin\theta = \pm\sqrt{1-\cos^2\theta} = \pm\sqrt{1-(-0.5)^2} = \pm\sqrt{1-0.25} = \pm\sqrt{0.75} = \pm\sqrt{\frac{3}{4}} = \pm\frac{\sqrt{3}}{2}.$$

If $\sin\theta = \frac{\sqrt{3}}{2}$, sine function is positive and cosine function is negative. By Table 2.5, θ must be in quadrant II. This is the case of using calculators to find the angle ($\theta = 120°$). If $\sin\theta = -\frac{\sqrt{3}}{2}$, both sine and cosine functions are negative. By Table 2.5, θ must be in quadrant III, which cannot be produced by calculators ($\theta = 240°$).

Reduction to trigonometric functions of positive acute angles from other angles

Negative angles

If rotation about the origin is clockwise, the angles become negative (Figure 2.14). Because the quadrants of the x-y plane remain the same, rotation clockwise means the lower half ($y < 0$) of this plane is the reflection of its upper half ($y > 0$) about the x-axis, and vice versa. This indicates that the two mirrored points (P and P') about the x-axis in quadrants I and IV are mutually reflective (Figure 2.14a,d); the same is also true for the pair in quadrants II and III (Figure 2.14b,c). These facts lead to the following formulae of translating trigonometric functions of negative angles to that of positive angles.

$$\boxed{\begin{array}{ll} \sin(-\theta) = -\sin\theta, & \csc(-\theta) = -\csc\theta \\ \cos(-\theta) = \cos\theta, & \sec(-\theta) = \sec\theta \\ \tan(-\theta) = -\tan\theta, & \cot(-\theta) = -\cot\theta \end{array}} \qquad (2.22)$$

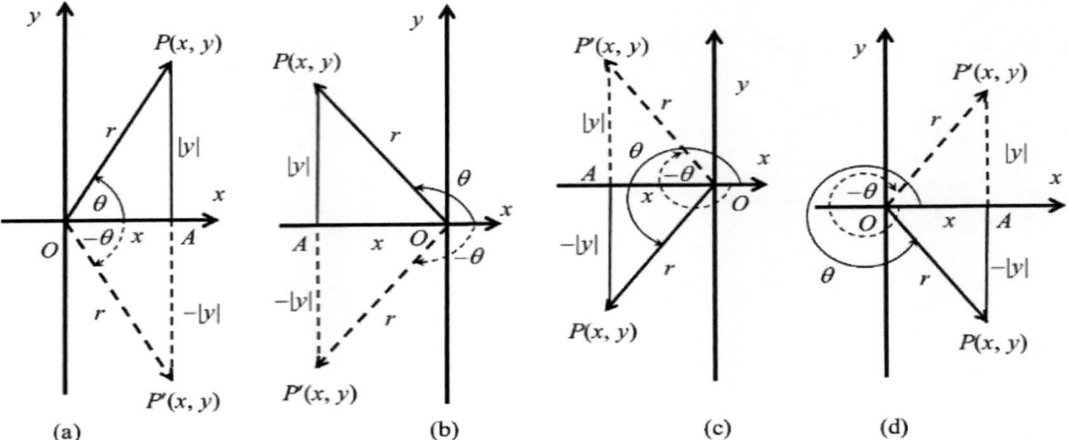

Figure 2.14: Reflective right triangles about the x-axis in quadrants I (a), II (b), III (c) and IV (d)

Example 2.24: Given $\theta = -30^\circ$, find its values for the six trigonometric functions respectively.

Solution

By formulae (2.22), these can be determined as follows:

$$\sin(-30^\circ) = -\sin 30^\circ = -\frac{1}{2}, \qquad \csc(-30^\circ) = -\csc 30^\circ = -2$$

$$\cos(-30^\circ) = \cos 30^\circ = \frac{\sqrt{3}}{2}, \qquad \sec(-30^\circ) = \sec 30^\circ = \frac{2}{\sqrt{3}} = \frac{2\sqrt{3}}{3}$$

$$\tan(-30^\circ) = -\tan 30^\circ = -\frac{\sqrt{3}}{3}, \qquad \cot(-30^\circ) = -\cot 30^\circ = -\frac{3}{\sqrt{3}} = -\sqrt{3}.$$

Extended complementary, supplementary, conjugate and coterminal angles

Similar analogies to the reduction of negative angles to positive angles for trigonometric functions can be applied to translating complementary, supplementary and conjugate angles to positive acute angles for trigonometric functions. These formulae are shown as follows.

- Extended complementary angles

$$\boxed{\begin{aligned}\sin(90^\circ - \theta) &= \cos\theta\\ \cos(90^\circ - \theta) &= \sin\theta\\ \tan(90^\circ - \theta) &= \cot\theta\\ \cot(90^\circ - \theta) &= \tan\theta\\ \sec(90^\circ - \theta) &= \csc\theta\\ \csc(90^\circ - \theta) &= \sec\theta\end{aligned}} \quad \text{and} \quad \boxed{\begin{aligned}\sin(90^\circ + \theta) &= \cos\theta\\ \cos(90^\circ + \theta) &= -\sin\theta\\ \tan(90^\circ + \theta) &= -\cot\theta\\ \cot(90^\circ + \theta) &= -\tan\theta\\ \sec(90^\circ + \theta) &= -\csc\theta\\ \csc(90^\circ + \theta) &= \sec\theta\end{aligned}} \qquad (2.23)$$

- Extended supplementary angles

$$\begin{array}{|l|}\hline \sin(180° - \theta) = \sin\theta \\ \cos(180° - \theta) = -\cos\theta \\ \tan(180° - \theta) = -\tan\theta \\ \cot(180° - \theta) = -\cot\theta \\ \sec(180° - \theta) = -\sec\theta \\ \csc(180° - \theta) = \csc\theta \\ \hline \end{array} \quad \text{and} \quad \begin{array}{|l|}\hline \sin(180° + \theta) = -\sin\theta \\ \cos(180° + \theta) = -\cos\theta \\ \tan(180° + \theta) = \tan\theta \\ \cot(180° + \theta) = \cot\theta \\ \sec(180° + \theta) = -\sec\theta \\ \csc(180° + \theta) = -\csc\theta \\ \hline \end{array} \qquad (2.24)$$

- Extended conjugate angles

$$\begin{array}{|l|}\hline \sin(360° - \theta) = -\sin\theta \\ \cos(360° - \theta) = \cos\theta \\ \tan(360° - \theta) = -\tan\theta \\ \cot(360° - \theta) = -\cot\theta \\ \sec(360° - \theta) = \sec\theta \\ \csc(360° - \theta) = -\csc\theta \\ \hline \end{array} \quad \text{and} \quad \begin{array}{|l|}\hline \sin(360° + \theta) = \sin\theta \\ \cos(360° + \theta) = \cos\theta \\ \tan(360° + \theta) = \tan\theta \\ \cot(360° + \theta) = \cot\theta \\ \sec(360° + \theta) = \sec\theta \\ \csc(360° + \theta) = \csc\theta \\ \hline \end{array} \qquad (2.25)$$

- Coterminal angles

Coterminal angles are those angles that reoccur in every full turn, i.e., $\alpha = n\times360° + \theta$.

$$\boxed{\left\{\begin{array}{l} \sin(n\bullet360° + \theta) = \sin\theta \\ \cos(n\bullet360° + \theta) = \cos\theta \\ \tan(n\bullet360° + \theta) = \tan\theta \\ \cot(n\bullet360° + \theta) = \cot\theta \\ \sec(n\bullet360° + \theta) = \sec\theta \\ \csc(n\bullet360° + \theta) = \csc\theta \end{array}\right. \quad (n = 1, 2, ...)} \qquad (2.26)$$

Example 2.25: Given $\alpha = 120°$, $\beta = 225°$ and $\theta = -330°$, use two different methods to find $\sin\alpha$, $\cos\beta$, and $\tan\theta$.

Solution

We can use one or more relationships from formulae (2.22-2.25) to obtain the results.

$$\sin\alpha = \sin 120° = \sin(180° - 60°) = \sin 60° = \sqrt{3}/2, \text{ or}$$
$$\sin\alpha = \sin 120° = \sin(90° + 30°) = \cos 30° = \sqrt{3}/2$$
$$\cos\beta = \cos 225° = \cos(180° + 45°) = -\cos 45° = -\sqrt{2}/2, \text{ or}$$
$$\cos\beta = \cos 225° = \cos(360° - 135°) = \cos 135° = \cos(180° - 45°) = -\cos 45° = -\sqrt{2}/2$$

$\tan\theta = \tan(-330°) = -\tan 330° = -[\tan(360° - 30°)] = -(-\tan 30°) = \sqrt{3}/3$, or

$\tan\theta = -\tan 330° = -\tan(180° + 150°) = -\tan 150° = -[\tan(180° - 30°)] = -(-\tan 30°) = \sqrt{3}/3$.

Example 2.26: Given $\alpha = 690°$ and $\beta = 900°$, find sin α and cosβ.

Solution

These can be determined using formulae (2.27) as below:

$$\sin\alpha = \sin 690° = \sin(720° - 30°) = \sin[2\times 360° + (-30°)] = \sin(-30°) = -\sin 30° = -\frac{1}{2}$$

$$\cos\beta = \cos 900° = \cos(720° + 180°) = \cos(2\times 360° + 180°) = \cos 180° = -1.$$

Values of special angles for trigonometric functions

Using one formula or combinations of above reduction formulae, values of special angles for the six trigonometric functions can be determined by the positive acute angles (Table 2.6).

Table 2.6: Values of special angles for trigonometric functions

θ	$\sin\theta$	$\cos\theta$	$\tan\theta$	$\cot\theta$	$\sec\theta$	$\csc\theta$
0°	0	1	0	$\pm\infty$	1	$\pm\infty$
30°	$1/2$	$\sqrt{3}/2$	$\sqrt{3}/3$	$\sqrt{3}$	$2\sqrt{3}/3$	2
45°	$\sqrt{2}/2$	$\sqrt{2}/2$	1	1	$\sqrt{2}$	$\sqrt{2}$
60°	$\sqrt{3}/2$	$1/2$	$\sqrt{3}$	$\sqrt{3}/3$	2	$2\sqrt{3}/3$
90°	1	0	$\pm\infty$	0	$\pm\infty$	1
120°	$\sqrt{3}/2$	$-1/2$	$-\sqrt{3}$	$-\sqrt{3}/3$	−2	$2\sqrt{3}/3$
135°	$\sqrt{2}/2$	$-\sqrt{2}/2$	−1	−1	$-\sqrt{2}$	$\sqrt{2}$
150°	$1/2$	$-\sqrt{3}/2$	$-\sqrt{3}/3$	$-\sqrt{3}$	$-2\sqrt{3}/3$	2
180°	0	−1	0	$\pm\infty$	−1	$\pm\infty$
210°	$-1/2$	$-\sqrt{3}/2$	$\sqrt{3}/3$	$\sqrt{3}$	$-2\sqrt{3}/3$	−2
225°	$-\sqrt{2}/2$	$-\sqrt{2}/2$	1	1	$-\sqrt{2}$	$-\sqrt{2}$
240°	$-\sqrt{3}/2$	$-1/2$	$\sqrt{3}$	$\sqrt{3}/3$	−2	$-2\sqrt{3}/3$
270°	−1	0	$\pm\infty$	0	$\pm\infty$	−1
300°	$-\sqrt{3}/2$	$1/2$	$-\sqrt{3}$	$-\sqrt{3}/3$	2	$-2\sqrt{3}/3$
315°	$-\sqrt{2}/2$	$\sqrt{2}/2$	−1	−1	$\sqrt{2}$	$-\sqrt{2}$
330°	$-1/2$	$\sqrt{3}/2$	$-\sqrt{3}/3$	$-\sqrt{3}$	$2\sqrt{3}/3$	−2
360°	0	1	0	$\pm\infty$	1	$\pm\infty$

Example 2.27: Given $\alpha = -405°$, find its values for the six trigonometric functions respectively.

Solution

These can be determined using formulae in (2.22-2.26) and the known values in Table 2.6.

$$\sin\alpha = \sin(-405°) = -\sin 405° = -\sin(360° + 45°) = -\sin 45° = -\frac{\sqrt{2}}{2}$$

$$\cos\alpha = \cos(-405°) = \cos 405° = \cos(360° + 45°) = \cos 45° = \frac{\sqrt{2}}{2}$$

$$\tan\alpha = \tan(-405°) = -\tan 405° = -\tan(360° + 45°) = -\tan 45° = -1$$

$$\cot\alpha = \frac{1}{\tan\alpha} = \frac{1}{\tan(-405°)} = -1$$

$$\sec\alpha = \frac{1}{\cos\alpha} = \frac{1}{\cos(-405°)} = \frac{2}{\sqrt{2}} = \sqrt{2}$$

$$\csc\alpha = \frac{1}{\sin\alpha} = \frac{1}{\sin(-405°)} = -\frac{2}{\sqrt{2}} = -\sqrt{2}.$$

Determining angles by inverse trigonometric functions

In theory, the inverse function of any trigonometric function can be used to determine the measure of any angle associated with that function shown in formulae (2.27). However, as coterminal angles and some extended supplementary and conjugate angles can produce the same value for a chosen trigonometric function, the angle determined by the inverse trigonometric function is not unique if no constraints are provided.

Normally, the angles resolved from an inverse trigonometric function within ONE full turn of 360° are called the *principal angles*. Any inverse trigonometric function will have two possible principal angles since each trigonometric function will have the same value twice within a full turn of 360°. Referring to Table 2.6, for example, both sin30° and sin150° share the same value of 0.5 whereas both tan45° and tan225° produce the same value of 1. If angle θ (either positive or negative) is the result of an inverse function resolved by a **calculator**, the two principal angles can be found from θ according to the nature of the trigonometric function involved. Table 2.7 presents a summary of determining the two principal angles from θ for sine, cosine and tangent functions.

$$\begin{array}{|ll|}
\theta = \sin^{-1} z & 0 \le |z| \le 1 \\
\theta = \cos^{-1} z & 0 \le |z| \le 1 \\
\theta = \tan^{-1} z & 0 \le |z| < \infty \\
\theta = \cot^{-1} z & 0 \le |z| < \infty \\
\theta = \sec^{-1} z & 1 \le |z| < \infty \\
\theta = \csc^{-1} z & 1 \le |z| < \infty
\end{array}. \qquad (2.27)$$

Table 2.7: Summary of principal angles of inverse sine, cosine and tangent functions
(θ is the angle obtained by calculator)

$\theta = \sin^{-1}z$	Principal angles	$\theta = \cos^{-1}z$	Principal angles	$\theta = \tan^{-1}z$	Principal angles
$0° < \theta < 90°$	θ & $180° - \theta$	$0° < \theta < 180°$	θ & $360° - \theta$	$0° < \theta < 90°$	θ & $180° + \theta$
$-90° < \theta < 0°$	$180° + \lvert\theta\rvert$ & $360° - \lvert\theta\rvert$			$-90° < \theta < 0°$	$180° - \lvert\theta\rvert$ & $360° - \lvert\theta\rvert$

Example 2.28: Given $\sin\alpha = -0.2588$, $\cos\beta = 0.6428$ and $\tan\gamma = -5.6713$, find α that is in quadrant III, β in any possible quadrant, and γ that is in quadrant II respectively.

Solution

These can be determined using formulae (2.27) and Table 2.7 as follows.

$$\theta = \sin^{-1}(-0.2588) \approx -15°$$

Since α is in quadrant III, by Table 2.7, the solution is $\alpha = 180° + |\theta| = 180° + 15° = 195°$.

$$\theta = \cos^{-1} 0.6428 \approx 50°$$

Since β is in any possible quadrant, by Table 2.7, the solution is $\beta = \theta = 50°$ and $\beta = 360° - \theta = 360° - 50° = 310°$.

$$\theta = \tan^{-1}(-5.6713) \approx -80°$$

Since γ is in quadrant II, by Table 2.7, the solution is $\gamma = 180° - |\theta| = 180° - 80° = 100°$.

2.2.2 Trigonometric identities and relationships

Trigonometric identities

From definitions of trigonometric functions and the Pythagorean theorem, the following relationships that always stand true for all trigonometric values of angles are called *trigonometric identities*.

$$\begin{array}{l}
\tan\theta = \dfrac{\sin\theta}{\cos\theta} \rightleftarrows \sin\theta = \cos\theta\tan\theta \\
\cot\theta = \dfrac{\cos\theta}{\sin\theta} \rightleftarrows \cos\theta = \sin\theta\cot\theta \text{ and } \tan\theta = \dfrac{1}{\cot\theta} \rightleftarrows \tan\theta\cot\theta = 1 \\
\sec\theta = \dfrac{1}{\cos\theta} \rightleftarrows \cos\theta\sec\theta = 1 \text{ and } \csc\theta = \dfrac{1}{\sin\theta} \rightleftarrows \sin\theta\csc\theta = 1 \\
\sin^2\theta + \cos^2\theta = 1 \rightleftarrows 1 + \tan^2\theta = \sec^2\theta \rightleftarrows 1 + \cot^2\theta = \csc^2\theta
\end{array} \qquad (2.28)$$

Trigonometric identities can be used to prove the equality or inequality of two different trigonometric expressions, and simplify an expression involving multiple trigonometric terms.

Example 2.29: Verify the equality of the following trigonometric expressions:

a) $\dfrac{2\sin^2\theta-\cos^2\theta}{\sin\theta\cos\theta}=2\tan\theta-\cot\theta$

b) $\dfrac{1+\cos\theta}{\sin\theta}=\dfrac{\sin\theta}{1-\cos\theta}$

c) $\sqrt{\dfrac{\sec\theta+\tan\theta}{\sec\theta-\tan\theta}}=\sec\theta+\tan\theta$

Solution

a) $\underline{\dfrac{2\sin^2\theta-\cos^2\theta}{\sin\theta\cos\theta}}=\dfrac{2\sin^2\theta}{\sin\theta\cos\theta}-\dfrac{\cos^2\theta}{\sin\theta\cos\theta}=\dfrac{2\sin\theta}{\cos\theta}-\dfrac{\cos\theta}{\sin\theta}=\underline{2\tan\theta-\cot\theta}$

b) $\underline{\dfrac{1+\cos\theta}{\sin\theta}}=\dfrac{1+\cos\theta}{\sin\theta}\bullet\dfrac{1-\cos\theta}{1-\cos\theta}=\dfrac{1^2-\cos^2\theta}{\sin\theta(1-\cos\theta)}=\dfrac{\sin^2\theta}{\sin\theta(1-\cos\theta)}=\underline{\dfrac{\sin\theta}{1-\cos\theta}}$

c) $\underline{\sqrt{\dfrac{\sec\theta+\tan\theta}{\sec\theta-\tan\theta}}}=\sqrt{\dfrac{\sec\theta+\tan\theta}{\sec\theta-\tan\theta}}\bullet\sqrt{\dfrac{\sec\theta+\tan\theta}{\sec\theta+\tan\theta}}=\sqrt{\dfrac{(\sec\theta+\tan\theta)^2}{\sec^2\theta-\tan^2\theta}}=\dfrac{\sec\theta+\tan\theta}{1}=\underline{\sec\theta+\tan\theta}$

Example 2.30: Simplify the following trigonometric expressions:

a) $\cos^2\theta(1+\tan^2\theta)$

b) $\tan\theta-\dfrac{\cos\theta}{1-\sin\theta}$

Solution

a) $\cos^2\theta(1+\tan^2\theta)=\cos^2\theta\sec^2\theta=\cos^2\theta\dfrac{1}{\cos^2\theta}=1$

b) $\tan\theta-\dfrac{\cos\theta}{1-\sin\theta}=\dfrac{\sin\theta}{\cos\theta}-\dfrac{\cos\theta}{1-\sin\theta}=\dfrac{\sin\theta}{\cos\theta}\bullet\dfrac{1-\sin\theta}{1-\sin\theta}-\dfrac{\cos\theta}{\cos\theta}\bullet\dfrac{\cos\theta}{1-\sin\theta}=\dfrac{\sin\theta(1-\sin\theta)-\cos^2\theta}{\cos\theta(1-\sin\theta)}$

$=\dfrac{\sin\theta-\sin^2\theta-\cos^2\theta}{\cos\theta(1-\sin\theta)}=\dfrac{\sin\theta-(\sin^2\theta+\cos^2\theta)}{\cos\theta(1-\sin\theta)}=\dfrac{\sin\theta-1}{\cos\theta(1-\sin\theta)}=\dfrac{-(1-\sin\theta)}{\cos\theta(1-\sin\theta)}=-\sec\theta.$

Trigonometric relationships of two angles

Addition formulae

$$\begin{aligned}
&\sin(\alpha+\beta)=\sin\alpha\cos\beta+\cos\alpha\sin\beta\\
&\cos(\alpha+\beta)=\cos\alpha\cos\beta-\sin\alpha\sin\beta\\
&\tan(\alpha+\beta)=\frac{\tan\alpha+\tan\beta}{1-\tan\alpha\tan\beta}
\end{aligned} \tag{2.29}$$

Double-angle formulae ($\alpha=\beta$ for addition formulae)

$$\begin{aligned}
&\sin 2\alpha=2\sin\alpha\cos\alpha\\
&\cos 2\alpha=\cos^2\alpha-\sin^2\alpha=1-2\sin^2\alpha=2\cos^2\alpha-1\\
&\tan 2\alpha=\frac{2\tan\alpha}{1-\tan^2\alpha}
\end{aligned} \tag{2.30}$$

Subtraction formulae

$$\begin{aligned}
&\sin(\alpha-\beta)=\sin\alpha\cos\beta-\cos\alpha\sin\beta\\
&\cos(\alpha-\beta)=\cos\alpha\cos\beta+\sin\alpha\sin\beta\\
&\tan(\alpha-\beta)=\frac{\tan\alpha-\tan\beta}{1+\tan\alpha\tan\beta}
\end{aligned} \tag{2.31}$$

Half-angle formulae

$$\begin{aligned}
&\sin\frac{\alpha}{2}=\pm\sqrt{\frac{1-\cos\alpha}{2}}\\
&\cos\frac{\alpha}{2}=\pm\sqrt{\frac{1+\cos\alpha}{2}}\\
&\tan\frac{\alpha}{2}=\pm\sqrt{\frac{1-\cos\alpha}{1+\cos\alpha}}=\frac{\sin\alpha}{1+\cos\alpha}=\frac{1-\cos\alpha}{\sin\alpha}
\end{aligned} \tag{2.32}$$

Products of sines and cosines

$$\begin{aligned}
&\sin\alpha\sin\beta=-\frac{1}{2}[\cos(\alpha+\beta)-\cos(\alpha-\beta)]\\
&\cos\alpha\cos\beta=\frac{1}{2}[\cos(\alpha+\beta)+\cos(\alpha-\beta)]\\
&\sin\alpha\cos\beta=\frac{1}{2}[\sin(\alpha+\beta)+\sin(\alpha-\beta)]\\
&\cos\alpha\sin\beta=\frac{1}{2}[\sin(\alpha+\beta)-\sin(\alpha-\beta)]
\end{aligned} \tag{2.33}$$

Sum and difference of sines and cosines

$$\begin{aligned}
\sin\alpha + \sin\beta &= 2\sin(\frac{\alpha+\beta}{2})\cos(\frac{\alpha-\beta}{2}) \\
\sin\alpha - \sin\beta &= 2\cos(\frac{\alpha+\beta}{2})\sin(\frac{\alpha-\beta}{2}) \\
\cos\alpha + \cos\beta &= 2\cos(\frac{\alpha+\beta}{2})\cos(\frac{\alpha-\beta}{2}) \\
\cos\alpha - \cos\beta &= 2\sin(\frac{\alpha+\beta}{2})\sin(\frac{\alpha-\beta}{2})
\end{aligned} \tag{2.34}$$

Example 2.31: Find the exact values for sin75°, cos75° and tan75°.

Solution

Use the addition formulae (2.29).

$$\sin 75° = \sin(30° + 45°) = \sin 30° \cos 45° + \cos 30° \sin 45° = \frac{1}{2}\bullet\frac{\sqrt{2}}{2} + \frac{\sqrt{3}}{2}\bullet\frac{\sqrt{2}}{2}$$

$$= \frac{\sqrt{2} + \sqrt{2}\sqrt{3}}{4} = \frac{\sqrt{2}(1+\sqrt{3})}{4}$$

$$\cos 75° = \cos(30° + 45°) = \cos 30° \cos 45° - \sin 30° \sin 45° = \frac{\sqrt{3}}{2}\bullet\frac{\sqrt{2}}{2} - \frac{1}{2}\bullet\frac{\sqrt{2}}{2}$$

$$= \frac{\sqrt{2}\sqrt{3} - \sqrt{2}}{4} = \frac{\sqrt{2}(\sqrt{3}-1)}{4}$$

$$\tan 75° = \tan(30° + 45°) = \frac{\tan 30° + \tan 45°}{1 - \tan 30° \tan 45°} = \frac{\frac{\sqrt{3}}{3} + 1}{1 - \frac{\sqrt{3}}{3}\bullet 1} = \frac{\frac{\sqrt{3}}{3} + \frac{3}{3}}{\frac{3}{3} - \frac{\sqrt{3}}{3}} = \frac{\frac{3+\sqrt{3}}{3}}{\frac{3-\sqrt{3}}{3}}$$

$$= \frac{3+\sqrt{3}}{3-\sqrt{3}} = \frac{(3+\sqrt{3})(3+\sqrt{3})}{(3-\sqrt{3})(3+\sqrt{3})} = \frac{3^2 + 6\sqrt{3} + (\sqrt{3})^2}{3^2 - (\sqrt{3})^2} = \frac{9 + 6\sqrt{3} + 3}{9-3} = \frac{12 + 6\sqrt{3}}{6} = 2 + \sqrt{3}.$$

Example 2.32: Find the exact values for sin15°, cos15° and tan15°.

Solution

Use the subtraction formulae (2.31).

$$\sin 15° = \sin(45° - 30°) = \sin 45° \cos 30° - \cos 45° \sin 30° = \frac{\sqrt{2}}{2}\bullet\frac{\sqrt{3}}{2} - \frac{\sqrt{2}}{2}\bullet\frac{1}{2} = \frac{\sqrt{2}\sqrt{3} - \sqrt{2}}{4} = \frac{\sqrt{2}(\sqrt{3}-1)}{4}$$

$$\cos 15° = \cos(45° - 30°) = \cos 45° \cos 30° + \sin 45° \sin 30° = \frac{\sqrt{2}}{2}\bullet\frac{\sqrt{3}}{2} + \frac{\sqrt{2}}{2}\bullet\frac{1}{2} = \frac{\sqrt{2}(\sqrt{3}+1)}{4}$$

$$\tan 15° = \frac{\sin 15°}{\cos 15°} = \frac{\frac{\sqrt{2}(\sqrt{3}-1)}{4}}{\frac{\sqrt{2}(\sqrt{3}+1)}{4}} = \frac{\sqrt{3}-1}{\sqrt{3}+1} = \frac{(\sqrt{3}-1)(\sqrt{3}-1)}{(\sqrt{3}+1)(\sqrt{3}-1)} = \frac{(\sqrt{3})^2 - 2\sqrt{3} + 1}{(\sqrt{3})^2 - 1} = \frac{4 - 2\sqrt{3}}{2} = 2 - \sqrt{3}$$

Example 2.33: Use sin20° = 0.3420 to evaluate sin10°, cos10°, tan10°, sin40°, cos40° and tan40° respectively.

Solution

The double-angle formulae (2.30) can be used to evaluate sin40°, cos40° and tan40° and the half-angle formula (2.32) are useful for sin10°, cos10°, and tan10°.

$$\cos 20° = \sqrt{1-\sin^2 20°} = \sqrt{1-0.342^2} \approx \sqrt{0.8830} \approx 0.9397$$
$$\sin 40° = \sin(2\times 20°) = 2\sin 20°\cos 20° = 2\times 0.3420\times 0.9397 \approx 0.6428$$
$$\cos 40° = \cos(2\times 20°) = \cos^2 20° - \sin^2 20° = 0.9397^2 - 0.3420^2 \approx 0.7661$$
$$\tan 40° = \frac{\sin 40°}{\cos 40°} = \frac{0.6428}{0.7661} \approx 0.8391$$
$$\sin 10° = \sin\frac{20°}{2} = \sqrt{\frac{1-\cos 20°}{2}} = \sqrt{\frac{1-0.9397}{2}} \approx 0.1736$$
$$\cos 10° = \sqrt{1-\sin^2 10°} = \sqrt{1-0.1736^2} \approx 0.9848 \longrightarrow \tan 10° = \frac{\sin 10°}{\cos 10°} = \frac{0.1736}{0.9848} \approx 0.1763$$

Example 2.34: If $\alpha + \beta + \gamma = 180°$, prove $\sin 2\alpha + \sin 2\beta + \sin 2\gamma = 4\sin\alpha\cdot\sin\beta\cdot\sin\gamma$.

Solution

$$\because \quad \cos 2\theta = \cos^2\theta - \sin^2\theta = 1-2\sin^2\theta \rightleftarrows 2\sin^2\theta = 1-\cos 2\theta \quad \text{and}$$
$$\alpha+\beta+\gamma = 180° \longrightarrow \gamma = 180° - (\alpha+\beta) \longrightarrow \alpha+\beta = 180° - \gamma$$
$$\therefore \quad \sin 2\gamma = \sin\{2[180° - (\alpha+\beta)]\} = \sin[360° - 2(\alpha+\beta)]$$
$$= -\sin(2\alpha+2\beta) = -\sin 2\alpha\cos 2\beta - \cos 2\alpha \sin 2\beta$$
$$\underline{\sin 2\alpha + \sin 2\beta + \sin 2\gamma} = \sin 2\alpha + \sin 2\beta - \sin 2\alpha\cos 2\beta - \cos 2\alpha\sin 2\beta$$
$$= \sin 2\alpha(1-\cos 2\beta) + \sin 2\beta(1-\cos 2\alpha) = \sin 2\alpha\cdot 2\sin^2\beta + \sin 2\beta\cdot 2\sin^2\alpha$$
$$= 2\sin 2\alpha\sin^2\beta + 2\sin 2\beta\sin^2\alpha = 4\sin\alpha\cos\alpha\sin^2\beta + 4\sin\beta\cos\beta\sin^2\alpha$$
$$= 4\sin\alpha\sin\beta(\cos\alpha\sin\beta + \cos\beta\sin\alpha) = 4\sin\alpha\sin\beta(\sin\alpha\cos\beta + \cos\alpha\sin\beta)$$
$$= 4\sin\alpha\sin\beta\sin(\alpha+\beta) = 4\sin\alpha\sin\beta\sin(180°-\gamma) = \underline{4\sin\alpha\sin\beta\sin\gamma}$$

Example 2.35: Use the addition and subtraction formulae (2.29) & (2.31) to derive the product formulae: $\sin\alpha\cos\beta = \frac{1}{2}[\sin(\alpha+\beta)+\sin(\alpha-\beta)]$ and $\cos\alpha\sin\beta = \frac{1}{2}[\sin(\alpha+\beta)-\sin(\alpha-\beta)]$.

Solution

$$\because \quad \sin(\alpha+\beta) = \sin\alpha\cos\beta + \cos\alpha\sin\beta$$
$$\sin(\alpha-\beta) = \sin\alpha\cos\beta - \cos\alpha\sin\beta$$
$$+ \quad \overline{}$$
$$\sin(\alpha+\beta) + \sin(\alpha-\beta) = 2\sin\alpha\cos\beta$$

$\therefore \quad \sin\alpha\cos\beta = \frac{1}{2}[\sin(\alpha+\beta)+\sin(\alpha-\beta)].$

$$\because \quad \sin(\alpha+\beta) = \sin\alpha\cos\beta + \cos\alpha\sin\beta$$
$$\sin(\alpha-\beta) = \sin\alpha\cos\beta - \cos\alpha\sin\beta$$
$$- \quad \overline{}$$
$$\sin(\alpha+\beta) - \sin(\alpha-\beta) = 2\cos\alpha\sin\beta$$

$\therefore \quad \cos\alpha\sin\beta = \frac{1}{2}[\sin(\alpha+\beta)-\sin(\alpha-\beta)].$

Example 2.36: Derive $\sin\alpha - \sin\beta = 2\cos(\frac{\alpha+\beta}{2})\sin(\frac{\alpha-\beta}{2})$.

Solution

Let $A = \alpha+\beta$ and $B = \alpha-\beta$, then

$$A+B = 2\alpha \longrightarrow \alpha = \frac{A+B}{2} \text{ and } A-B = 2\beta \longrightarrow \beta = \frac{A-B}{2}.$$

Substitute α and β into $\cos\alpha\sin\beta = \frac{1}{2}[\sin(\alpha+\beta)-\sin(\alpha-\beta)]$:

$$\cos\frac{A+B}{2}\sin\frac{A-B}{2} = \frac{1}{2}[\sin(\frac{A+B}{2}+\frac{A-B}{2}) - \sin(\frac{A+B}{2}-\frac{A-B}{2})]$$

$$2\cos\frac{A+B}{2}\sin\frac{A-B}{2} = \sin\frac{A+B+A-B}{2} - \sin\frac{A+B-A+B}{2}$$

$$\sin\frac{2A}{2} - \sin\frac{2B}{2} = 2\cos\frac{A+B}{2}\sin\frac{A-B}{2} \longrightarrow \sin A - \sin B = 2\cos\frac{A+B}{2}\sin\frac{A-B}{2}$$

Let $A = \alpha$ and $B = \beta$. This proves

$$\sin\alpha - \sin\beta = 2\cos(\frac{\alpha+\beta}{2})\sin(\frac{\alpha-\beta}{2}).$$

Exercises 2.2

1. Given four points on the x-y plane at $A(3, 2)$, $B(-2, 4)$, $C(-3, -3)$ and $D(2, -3)$, find the values of the six trigonometric functions at each of these points respectively.

2. Given $\sin\theta = -0.5$, find $\cos\theta$ and the quadrants where θ can be placed.

3. Given $\theta = -60°$, find its values for the six trigonometric functions respectively.

4. Given $\alpha = 135°$, $\beta = -240°$ and $\theta = 315°$, use two different methods to find $\sin\alpha$, $\cos\beta$, and $\tan\theta$.

5. Given $\alpha = 750°$ and $\beta = 540°$, find $\sin\alpha$ and $\cos\beta$.

6. Given $\alpha = -390°$, find its values for the six trigonometric functions respectively.

7. Given $\sin\alpha = 0.6428$, $\cos\beta = -0.2588$ and $\tan\gamma = 5.6713$, find α that is in quadrant II, β in any possible quadrant, and γ that is in quadrant III respectively.

8. Find the exact values for sin22.5°, cos22.5°, and tan22.55° respectively.

9. Use sin25° = 0.4226 to evaluate sin50°, cos50° and tan50° respectively.

10. Use the addition and subtraction formulae (2.29) & (2.31) to derive the product formulae: $\sin\alpha\sin\beta = \frac{1}{2}[\cos(\alpha-\beta) - \cos(\alpha+\beta)]$ and $\cos\alpha\cos\beta = \frac{1}{2}[\cos(\alpha-\beta) + \cos(\alpha+\beta)]$.

2.3 Oblique Triangles and Laws of Sines and Cosines

2.3.1 Oblique triangles and laws of sines and cosines

Oblique triangles

Any triangle that does not have a right angle is called an *oblique triangle*. Oblique triangles have either three acute angles (Figure 2.15a) or one obtuse angle and two acute angles (Figure 2.15b). By convention, the three angles in an oblique triangle are denoted by their vertex letters A, B and C and the lengths of the corresponding opposite sides are denoted by a, b and c respectively (Figure 2.15).

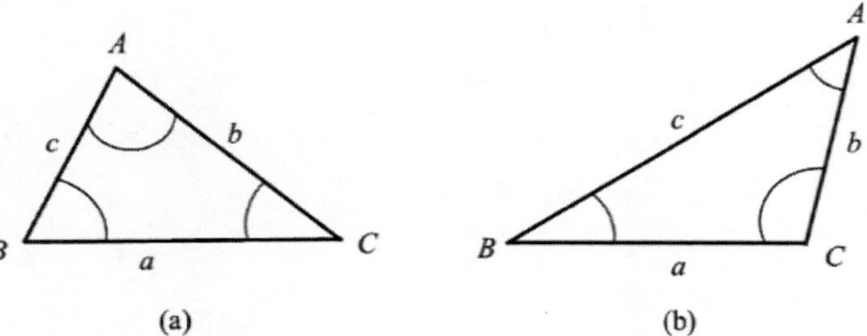

Figure 2.15: Oblique triangles: three acute angles (a); one obtuse angle and two acute angles (b)

Law of sines

In triangle ABC, the ratio of a side to the sine of the opposite angle (vice versa) is a constant:

$$\frac{a}{\sin A} = \frac{b}{\sin B} = \frac{c}{\sin C} \longleftrightarrow \frac{\sin A}{a} = \frac{\sin B}{b} = \frac{\sin C}{c} \tag{2.35}$$

Law of cosines

In any triangle ABC, the square of any side equals the sum of the squares of the other two sides reduced by twice the product of these two sides and the cosine of the included angle, i.e.,

$$\begin{aligned} a^2 &= b^2 + c^2 - 2bc\cos A \\ b^2 &= a^2 + c^2 - 2ac\cos B \\ c^2 &= a^2 + b^2 - 2ab\cos C \end{aligned} \tag{2.36}$$

Alternatively, law of cosines can be expressed as

$$\begin{aligned} \cos A &= \frac{b^2 + c^2 - a^2}{2bc} \\ \cos B &= \frac{a^2 + c^2 - b^2}{2ac} \\ \cos C &= \frac{a^2 + b^2 - c^2}{2ab} \end{aligned} \tag{2.37}$$

2.3.2 Solving oblique triangles

Given three known values mixed with angle(s) and side(s) or three sides of any oblique triangle, other unknown elements can be obtained using either the *law of sines* or the *law of cosines* or the combination.

Given three sides

- Use law of cosines (2.37) to find two (or three) angles;
- Use $A + B + C = 180°$ to find the third angle.

Given two sides and the included angle

- Use law of cosines (2.36) to find the third side;
- Use law of sines (2.35) or law of cosine (2.37) to find one opposite angle;
- Use $A + B + C = 180°$ to find the third angle.

Given two sides and one opposite angle

- Use law of sines (2.35) to find the other opposite angle;
- Use $A + B + C = 180°$ to find the third angle;
- Use law of sines (2.35) or law of cosines (2.36) to find the third side.

Given two angles and one side

- Use $A + B + C = 180°$ to find the third angle;
- Use law of sines (2.35) to find the other sides.

Example 2.37: Solve the triangle ABC, given $a = 25$, $b = 26.4$ and $c = 44.8$.

Solution

This is the case where three sides are known.

$$\cos A = \frac{b^2 + c^2 - a^2}{2bc} = \frac{26.4^2 + 44.8^2 - 25^2}{2 \times 26.4 \times 44.8} = \frac{2079}{2365.44} \approx 0.8789$$

$$A = \cos^{-1}(0.8789) \approx 28.49°$$

$$\cos B = \frac{a^2 + c^2 - b^2}{2ac} = \frac{25^2 + 44.8^2 - 26.4^2}{2 \times 25 \times 44.8} = \frac{1935.08}{2240} \approx 0.8639$$

$$B = \cos^{-1}(0.8639) \approx 30.24°$$

$$C = 180° - A - B = 180° - 28.49° - 30.24° = 121.27°.$$

Example 2.38: Solve the triangle *ABC*, given $a = 625$, $b = 291$ and $C = 44°$.

Solution

This is the case where two sides and the included angle are known.

$$c^2 = a^2 + b^2 - 2ab\cos C$$
$$c = \sqrt{a^2 + b^2 - 2ab\cos C} = \sqrt{625^2 + 291^2 - 2\times 625\times 291\times \cos 44°} \approx \sqrt{213646} \approx 462$$
$$\frac{\sin B}{b} = \frac{\sin C}{c} \longrightarrow \sin B = \frac{b\sin C}{c}$$
$$\sin B = \frac{b\sin C}{c} = \frac{291\times \sin 44°}{462} \approx \frac{291\times 0.6947}{462} \approx 0.4376$$
$$B = \sin^{-1}(0.4376) \approx 25.95° \approx 26°$$
$$A = 180° - B - C = 180° - 26° - 44° = 110°.$$

Example 2.39: Solve the triangle *ABC*, given $a = 5$, $c = 12$ and $C = 137°$.

Solution

This is the case where two sides and one opposite angle are known.

$$\frac{\sin A}{a} = \frac{\sin C}{c} \longrightarrow \sin A = \frac{a\sin C}{c}$$
$$\sin A = \frac{a\sin C}{c} = \frac{5\times \sin 137°}{12} \approx \frac{5\times 0.6820}{12} \approx 0.2842$$
$$A = \sin^{-1}(0.2842) \approx 16.5°$$
$$B = 180° - A - C = 180° - 16.5° - 137° = 26.5°.$$
$$\frac{b}{\sin B} = \frac{c}{\sin C} \longrightarrow b = \frac{c\sin B}{\sin C} = \frac{12\times \sin 26.5°}{\sin 137°} \approx \frac{12\times 0.4462}{0.6820} \approx 7.85$$

Example 2.40: Solve the triangle *ABC*, given $c = 20.5$ and $A = 47°$ and $B = 58°$

Solution

This is the case where two angles and the included side are known.

$$C = 180° - A - B = 180° - 47° - 58° = 75°$$
$$\frac{a}{\sin A} = \frac{b}{\sin B} = \frac{c}{\sin C} \longrightarrow a = \frac{c\sin A}{\sin C} \text{ and } b = \frac{c\sin B}{\sin C}$$
$$a = \frac{c\sin A}{\sin C} = \frac{20.5\times \sin 47°}{\sin 75°} \approx \frac{20.5\times 0.7314}{0.9659} \approx 15.5$$
$$b = \frac{c\sin B}{\sin C} = \frac{20.5\times \sin 58°}{\sin 75°} \approx \frac{20.5\times 0.8480}{0.9659} \approx 18.0$$

Example 2.41: Solve the triangle ABC, given $b = 35$ and $B = 32°$ and $C = 106°$

Solution

This is the case where two angles and one opposite side are known.

$$A = 180° - C - B = 180° - 106° - 32° = 42°$$

$$\frac{a}{\sin A} = \frac{b}{\sin B} = \frac{c}{\sin C} \longrightarrow a = \frac{b\sin A}{\sin B} \text{ and } c = \frac{b\sin C}{\sin B}$$

$$a = \frac{b\sin A}{\sin B} = \frac{35 \times \sin 42°}{\sin 32°} \approx \frac{35 \times 0.6691}{0.5299} \approx 44.2$$

$$c = \frac{b\sin C}{\sin B} = \frac{35 \times \sin 106°}{\sin 32°} \approx \frac{35 \times 0.9613}{0.5299} \approx 63.5$$

Example 2.42: Refer to the figure below. Two adjacent sides of a parallelogram are 35.7 and 48.2 cm respectively. The acute angle between them is 72.2°. Find the lengths of the two diagonals.

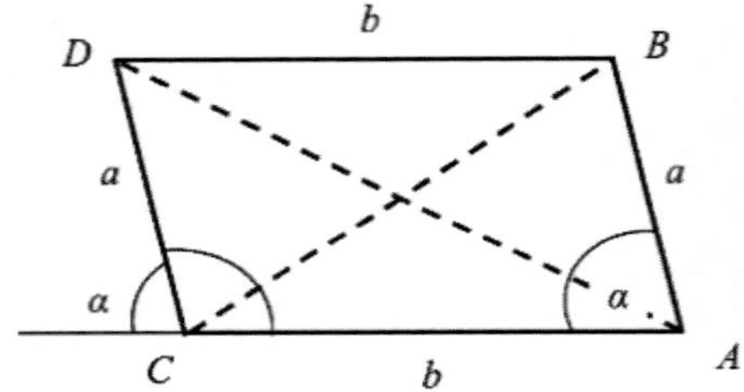

Solution

For triangle ABC, the diagonal BC can be obtained by using law of cosines directly.

$$BC^2 = a^2 + b^2 - 2ab\cos\alpha$$

$$BC = \sqrt{a^2 + b^2 - 2ab\cos\alpha} = \sqrt{35.7^2 + 48.2^2 - 2 \times 35.7 \times 48.2 \times \cos 72.2°} \approx \sqrt{2545.69} \approx 50.45 \text{ cm}$$

For triangle ACD, the diagonal AD can be obtained by using law of cosines with the obtuse angle C.

$$C = 180° - A = 180° - \alpha = 180° - 72.2° = 107.8°$$

$$AD^2 = a^2 + b^2 - 2ab\cos C$$

$$AD = \sqrt{a^2 + b^2 - 2ab\cos C} = \sqrt{35.7^2 + 48.2^2 - 2 \times 35.7 \times 48.2 \times \cos 107.8°} \approx \sqrt{4649.77} \approx 68.2 \text{ cm}$$

Exercises 2.3

1. Given $a = 30$, $b = 40$ and $c = 60$, solve this triangle.

2. Given $a = 45$, $b = 25$ and $c = 55$, solve this triangle.

3. Given $a = 62$, $b = 29$ and $C = 65°$, solve this triangle.

4. Given $a = 25$, $b = 45$ and $C = 145°$, solve this triangle.

5. Given $a = 55$, $c = 100$ and $C = 120°$, solve this triangle.

6. Given $a = 15$, $c = 32$ and $C = 150°$, solve this triangle.

7. Given $c = 75$ and $A = 55°$ and $B = 50°$, solve this triangle.

8. Given $c = 24$ and $A = 28°$ and $B = 72°$, solve this triangle.

9. Given $b = 75$ and $B = 50°$ and $C = 95°$, solve this triangle.

10. Refer to the figure below. Two adjacent sides of a parallelogram are 57 and 82 cm respectively. The acute angle between them is 69°. Find the lengths of the two diagonals.

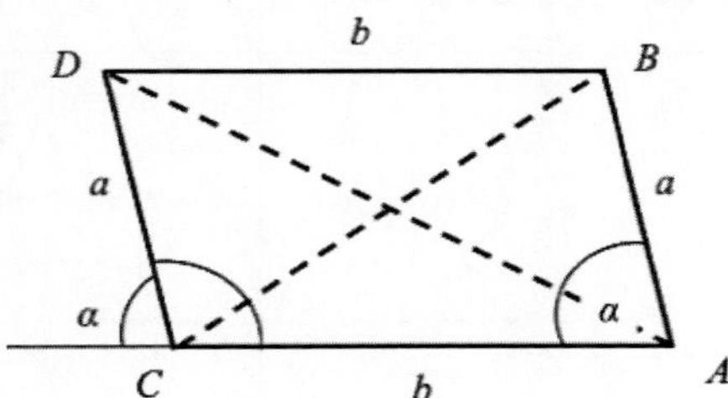

11. An antenna mast of 55 m high is placed on a sloping ground with an angle of 14°. The two cables stabilizing the mast make angles with the mast 43° and 36° respectively (See figure below). Find the length of the two cables.

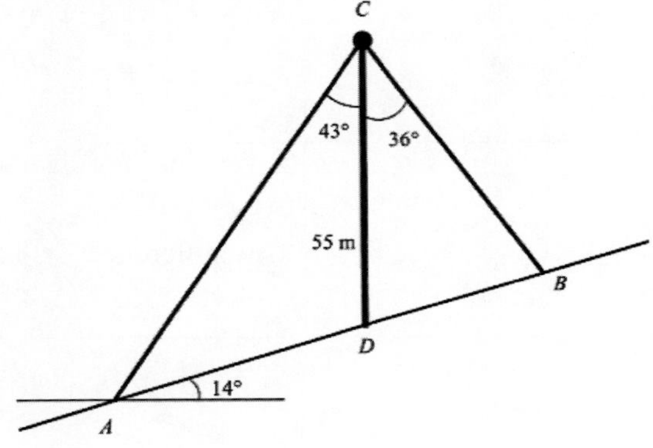

2.4 Summary of Basic Geometry

This section summaries properties of some commonly encountered geometric objects listed in Table 2.8. These properties are widely used in science, technology, and engineering.

Table 2.8 Properties of some geometric objects

Geometric object		*Property*
Rectangle	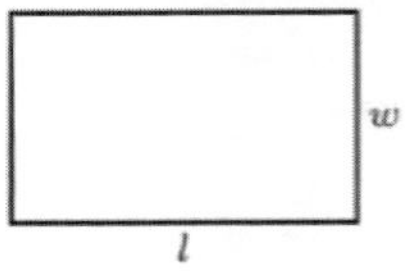	l = length, w = width Area: $A = lw$ Perimeter: $P = 2l + 2w = 2(l + w)$
Square	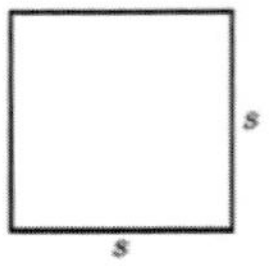	s = side Area: $A = s^2$ Perimeter: $P = 4s$
Right triangle	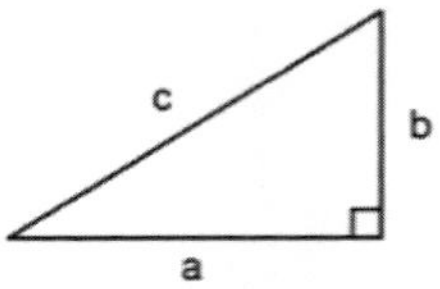	Area: $A = \frac{1}{2}ab$ Perimeter: $P = a + b + c$ Pythagorean theorem: $a^2 + b^2 = c^2$
Triangle	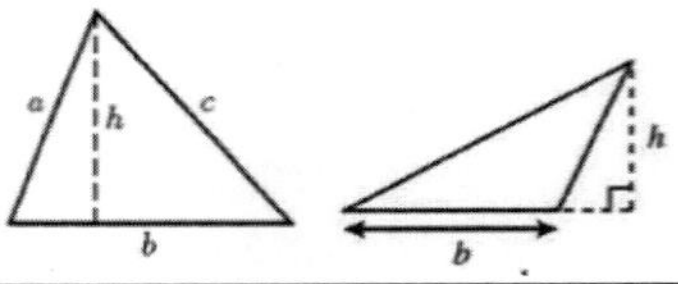	b = base, h = height Area: $A = \frac{1}{2}bh$ Perimeter: $P = a + b + c$
Equilateral triangle	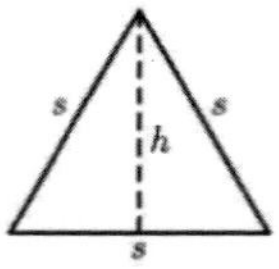	s = side, height: $h = \frac{\sqrt{3}}{2}s$ Area: $A = \frac{1}{2}sh = \frac{\sqrt{3}}{4}s^2$ Perimeter: $P = 3s$
Parallelogram	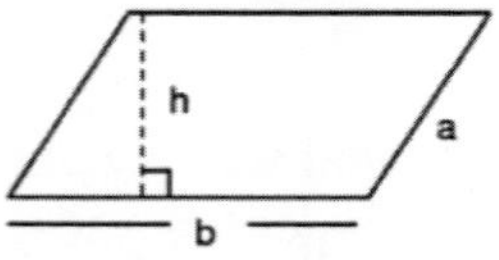	b = bases, h = height, a = sides Area: $A = bh$ Perimeter: $P = 2a + 2b = 2(a + b)$
Trapezoid	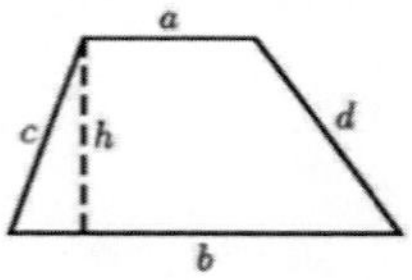	a,b = bases, h = height, c,d = sides Area: $A = \frac{1}{2}(a + b)h$ Perimeter: $P = a + b + c + d$
Circle	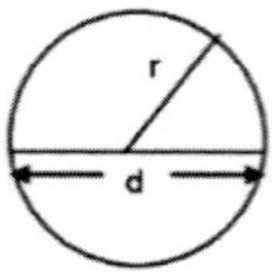	r = radius, d = diameter = $2r$ Area: $A = \pi r^2$ Circumference: $C = 2\pi r = \pi d$

Annulus	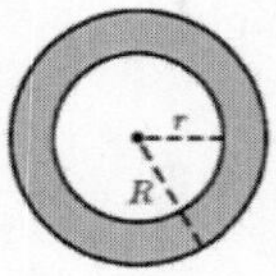	r = inner radius, R = outer radius Area: $A = \pi(R^2 - r^2)$
Ellipse	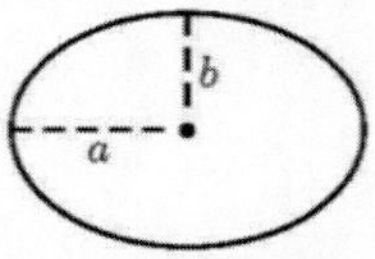	a = semimajor axis, b = semiminor axis Area: $A = \pi ab$
Rectangular solid	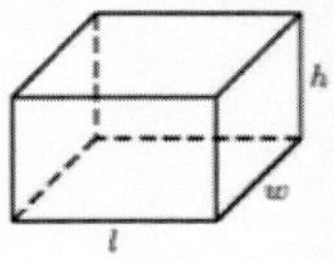	l = length, w = width , h = height Volume: $V = lwh$ Surface area: $S = 2lw + 2lh + 2wh$
Cube	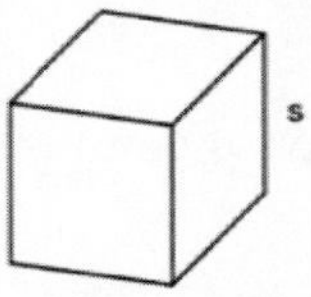	s = side Volume: $V = s^3$ Surface area: $S = 6s^2$
Right circular cylinder	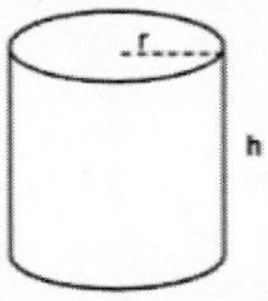	r = radius, h = height Volume: $V = \pi r^2 h$ Surface area: $S = 2\pi rh + 2\pi r^2$
Right circular cone	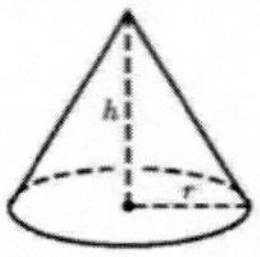	r = radius, h = height Volume: $V = \frac{1}{3}\pi r^2 h$ Surface area: $S = \pi r\sqrt{r^2 + h^2} + \pi r^2$
Frustum of cone	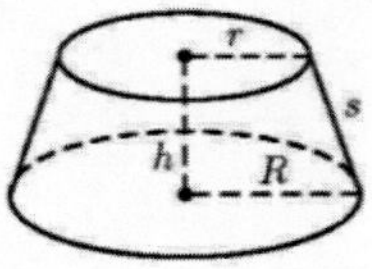	r = top radius, R = base radius, h = height, s = slant height Volume: $V = \frac{\pi}{3}(r^2 + rR + R^2)h$ Surface area: $S = \pi s(r + R) + \pi r^2 + \pi R^2$
Square pyramid	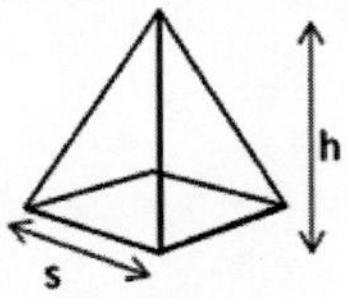	h = height, s = side Volume: $V = \frac{1}{3}s^2 h$ Surface area: $S = s(s + \sqrt{s^2 + 4h^2})$
Sphere	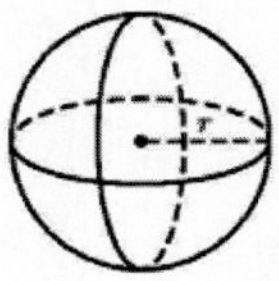	r = radius Volume: $V = \frac{4}{3}\pi r^3$ Surface area: $S = 4\pi r^2$

Example 2.43: A right circular cone has radius of 1 metre, and height of 2 metres. Taking π = 3.14, what is its surface area?

Solution

Given $r = 1$ metre and $h = 2$ metres, from Table 2.8, the surface area of this right circular cone is

$$S = \pi r\sqrt{r^2+h^2} + \pi r^2 = \pi r(\sqrt{r^2+h^2}+r) = 3.14\times 1(\sqrt{1^2+2^2}+1)$$
$$= 3.14(1+\sqrt{5}) \approx 3.14\times 3.236 \approx 10.161\ (m^2).$$

Example 2.44: The top radius of a frustum of cone is the same as its height of 2 metres. If its volume is 31.4 cubic metres, by taking $\pi = 3.14$, what is its base radius?

Solution

Given $r = h = 2$ metres, from Table 2.8, the volume of this frustum of cone is

$$V = \frac{\pi}{3}(r^2+rR+R^2)h \longrightarrow r^2+rR+R^2 = \frac{3V}{\pi h} \longrightarrow R^2+2R+4 = \frac{3\times 31.4}{3.14\times 2}$$
$$R^2+2R+4 = 15 \longrightarrow R^2+2R-11 = 0$$

Use formula (1.13) to solve this quadratic equation:

$$R = \frac{-b\pm\sqrt{b^2-4ac}}{2a} = \frac{-2\pm\sqrt{2^2+4\times 11}}{2} = \frac{-2\pm\sqrt{48}}{2} = \frac{-2\pm 4\sqrt{3}}{2} = -1\pm 2\sqrt{3}$$

Since only the positive value is meaningful for radius, the solution is

$$R = -1+2\sqrt{3} \approx -1+2\times 1.732 = 2.464\ (m).$$

Therefore, the base radius is about 2.464 m.

Exercises 2.4

1. The bottom and top sides of a symmetric trapezoid are 5 and 3 metres respectively. Its height is 2 metres. What is its area and what is its perimeter?

2. An open right circular cone has a radius of 3 metres, and a height of 3 metres. Taking $\pi = 3.14$, what is its surface area?

3. A right circular cone has a radius of 2 metres, and a height of 3 metres. Taking $\pi = 3.14$, what is its surface area?

4. The side of a square pyramid is 2 metres and its height is 3 metres. What is its volume?

5. The volume of a sphere is 100 cubic metres. By taking $\pi = 3.14$, what is its radius?

6. The bottom radius of a frustum of cone is 3 metres and its height is 2 metres. If its volume is 50 cubic metres, by taking $\pi = 3.14$, what is its top radius?

Chapter 2: Exercises Answers

Exercises 2.1

1. a) 2.007 *rad*; b) –0.5236 *rad*; c) 5.7596 *rad*
2. a) 36°; b) –77°08′34″; c) 143°14′22″
3. $\theta \approx 0.7854\text{ rad},\ s \approx 15.708\text{ cm},\ A \approx 157.08\text{ cm}^2$
4. $r \approx 11.56\text{ cm}$
5. $r = 10\text{ cm},\ \theta = 3\text{ rad} \approx 171.8875°$
6. a) Complementary angle: 25°, Supplementary angle: 115°, Conjugate angle: 295°
   b) Complementary angle: 75°, Supplementary angle: 165°, Conjugate angle: 345°
   c) Complementary angle: 10°, Supplementary angle: 100°, Conjugate angle: 280°
7. a) $c = \sqrt{41}\text{ cm}$, b) $b = 10\sqrt{2}\text{ cm}$, c) $a = 80\text{ cm}$
8. $A = 54\sqrt{3}\text{ cm}^2$
9. $c = 2\sqrt{3}\text{ cm},\ \sin\beta = \frac{1}{2},\ \cos\beta = \frac{\sqrt{3}}{2},\ \tan\beta = \frac{\sqrt{3}}{3},\ \beta = 30°$
10. $w = 114.312\text{ m}$
11. $h \approx 94.6\text{ m}$

Exercises 2.2

1.

$A(3, 2)$: $r = \sqrt{13},\ \sin\alpha = \frac{2\sqrt{13}}{13},\ \cos\alpha = \frac{3\sqrt{13}}{13},\ \tan\alpha = \frac{2}{3},\ \cot\alpha = \frac{3}{2},\ \sec\alpha = \frac{\sqrt{13}}{3},\ \csc\alpha = \frac{\sqrt{13}}{2}$

$B(-2, 4)$: $r = 2\sqrt{5},\ \sin\beta = \frac{2}{\sqrt{5}},\ \csc\beta = \frac{\sqrt{5}}{2},\ \cos\beta = \frac{-1}{\sqrt{5}},\ \sec\beta = -\sqrt{5},\ \tan\beta = -2,\ \cot\beta = -\frac{1}{2}$

$C(-3, -3)$: $r = 3\sqrt{2},\ \sin\gamma = -\frac{\sqrt{2}}{2},\ \csc\gamma = -\sqrt{2},\ \cos\gamma = -\frac{\sqrt{2}}{2},\ \sec\gamma = -\sqrt{2},\ \tan\gamma = 1,\ \cot\gamma = 1$

$D(2, -3)$:

$r = \sqrt{13},\ \sin\delta = -\frac{3\sqrt{13}}{13},\ \csc\delta = -\frac{\sqrt{13}}{3},\ \cos\delta = \frac{2\sqrt{13}}{13},\ \sec\delta = \frac{\sqrt{13}}{2},\ \tan\delta = -\frac{3}{2},\ \cot\delta = -\frac{2}{3}$

2. $\cos\theta = \pm\frac{\sqrt{3}}{2}$, *θ in Quadrant IV or Quadrant III.*

3. $\sin\theta = -\frac{\sqrt{3}}{2},\ \csc\theta = -\frac{2\sqrt{3}}{3},\ \cos\theta = \frac{1}{2},\ \sec\theta = 2,\ \tan\theta = -\sqrt{3},\ \cot\theta = -\frac{\sqrt{3}}{3}.$

4. $\sin 135° = \sin(180° - 45°) = \sin 45° = \frac{\sqrt{2}}{2}$, *or* $\sin 135° = \sin(90° + 45°) = \cos 45° = \frac{\sqrt{2}}{2}$;

$\cos(-240°) = \cos 240° = \cos(180° + 60°) = -\cos 60° = -\frac{1}{2}$, *or*

$\cos(-240°) = \cos 240° = \cos(360° - 120°) = \cos 120° = \cos(90° + 30°) = -\sin 30° = -\frac{1}{2}$;

$\tan 315° = \tan(360° - 45°) = -\tan 45° = -1,$ *or*
$\tan 315° = \tan(180° + 135°) = \tan 135° = \tan(90° + 45°) = -\cot 45° = -1.$

5. $\sin 750° = \sin(2{\bullet}360° + 30°) = \sin 30° = \dfrac{1}{2}, \quad \cos 540° = \cos(360° + 180°) = \cos 180° = -1.$

6. $\sin(-390°) = -\sin 390° = -\sin(360° + 30°) = -\sin 30° = -\dfrac{1}{2}, \quad \csc(-390°) = \dfrac{1}{\sin(-390°)} = -2,$

$\cos(-390°) = \cos 390° = \cos(360° + 30°) = \cos 30° = \dfrac{\sqrt{3}}{2}, \quad \sec(-390°) = \dfrac{1}{\cos(-390°)} = \dfrac{2}{\sqrt{3}} = \dfrac{2\sqrt{3}}{3},$

$\tan(-390°) = -\tan 390° = -\tan(360° + 30°) = -\tan 30° = -\dfrac{\sqrt{3}}{3}, \quad \cot(-390°) = \dfrac{1}{\tan(-390°)} = -\sqrt{3}.$

7. $\sin\alpha = 0.6428,\ Q\,II\ \alpha = 140°;\ \cos\beta = -0.2588,\ \beta = \begin{cases} 105°\ Q\,II \\ 255°\ Q\,III \end{cases};\ \tan\gamma = 5.6713, Q\,III\ \gamma = 260°$

8. $\sin 22.5° = \dfrac{\sqrt{2-\sqrt{2}}}{2}, \quad \cos 22.5° = \dfrac{\sqrt{2+\sqrt{2}}}{2}, \quad \tan 22.5° = \sqrt{2} - 1$

9. $\sin 50° = \sin(2{\bullet}25°) = 2\sin 25°\cos 25° = 2\sin 25°\sqrt{1-\sin^2 25°} = 2\times 0.4226\sqrt{1-0.4226^2} \approx 0.7660$
$\cos 50° = \sqrt{1-\sin^2 50°} = \sqrt{1-0.7660^2} = 0.6428; \quad \tan 50° = \dfrac{\sin 50°}{\cos 50°} = \dfrac{0.7660}{0.6428} \approx 1.1917$

Exercises 2.3

1. $A \approx 26.39°,\ B \approx 36.33°,\ C = 117.28°.$
2. $A \approx 53.78°,\ B \approx 26.63°,\ C = 99.59°.$
3. $c \approx 56.26,\ A \approx 87.19°,\ B = 27.81°.$
4. $c \approx 67.03,\ A \approx 12.35°,\ B = 22.65°.$
5. $A \approx 28.44°,\ B = 31.56°,\ b \approx 60.45$
6. $A \approx 13.56°,\ B = 16.44°,\ b \approx 18.11$
7. $C = 75°,\ a \approx 63.60,\ b \approx 59.48$
8. $C = 80°,\ a \approx 11.44,\ b \approx 23.18$
9. $A = 35°,\ a \approx 56.16,\ c \approx 97.53$
10. $BC \approx 81.38$ cm, $AD \approx 115.42$ cm
11. $AC \approx 97.98$ m, $BC \approx 57.56$ m

Exercises 2.4

1. $A = 8\ m^2,\ P = 8 + 2\sqrt{5}\ m$; 2. $S \approx 39.97\ m^2$; 3. $S \approx 35.20\ m^2$; 4. $V = 4\ m^3$; 5. $r \approx 2.88\ m$;
6. $r \approx 2.655\ m$

CHAPTER 3

3 Inequalities and Sequences

CHAPTER OBJECTIVES

- Review inequalities and operational rules
- Review absolute values, equations and inequalities
- Review fundamentals of sequences and series
- Review arithmetic and geometric sequences and series

Essential statements on inequalities, sequences and series:

- Ranges or domains are the key factor for inequalities.
- Sum of arithmetic or geometric series is very useful in understanding definite integration.

Key topics:

- Inequalities and ranges or domains
- Absolute values and absolute inequalities
- Arithmetic sequences and series
- Geometric sequences and series

Flowchart of mathematical knowledge building

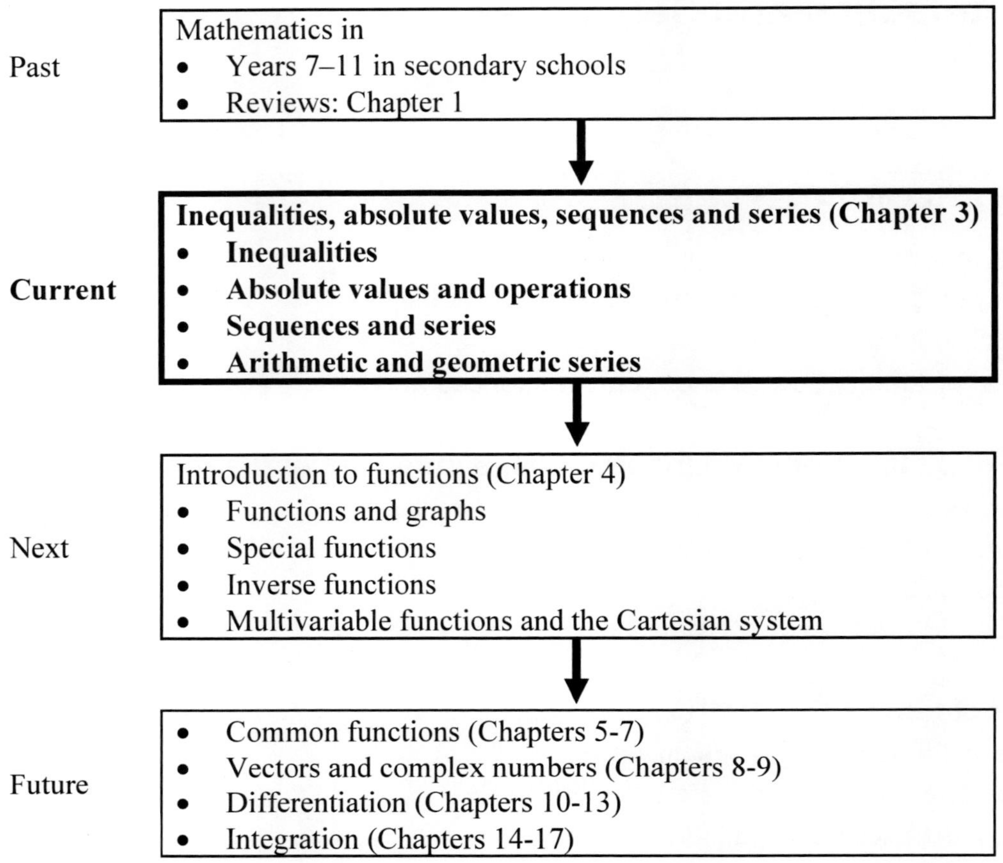

Past	Mathematics in • Years 7–11 in secondary schools • Reviews: Chapter 1
Current	**Inequalities, absolute values, sequences and series (Chapter 3)** • **Inequalities** • **Absolute values and operations** • **Sequences and series** • **Arithmetic and geometric series**
Next	Introduction to functions (Chapter 4) • Functions and graphs • Special functions • Inverse functions • Multivariable functions and the Cartesian system
Future	• Common functions (Chapters 5-7) • Vectors and complex numbers (Chapters 8-9) • Differentiation (Chapters 10-13) • Integration (Chapters 14-17)

Chapter 3: Inequalities and Sequences

3.1 Inequalities

3.1.1 Linear inequalities

Inequalities

For linear equation $x-4=0$, its solution is $x=4$, which is represented as a solid point in a real-number line (or x-axis) in Figure 3.1. Any value other than x = 4 in the real-number line does not satisfy this linear equation, or $x-4\neq 0$. Similarly $x-2=0$ represents only *one point* x = 2 whereas $x-2\neq 0$ represents infinitely many points except x = 2. Therefore, $x-2\neq 0$ is the *inequalities* complementing equation $x-2=0$. Look at the real-number line in Figure 3.1, $x-2\neq 0$ can be further split into two regions of $x-2<0$ (on the left of 2) and $x-2>0$ (on the right of 2). These two formulas represent all inequalities smaller than x = 2 and all inequalities larger than x = 2 respectively. Thus these two can not only serve the same purpose together as $x-2\neq 0$ does, but also explicitly point to a region where the inequalities are specifically referred, for example, $x-2>0$.

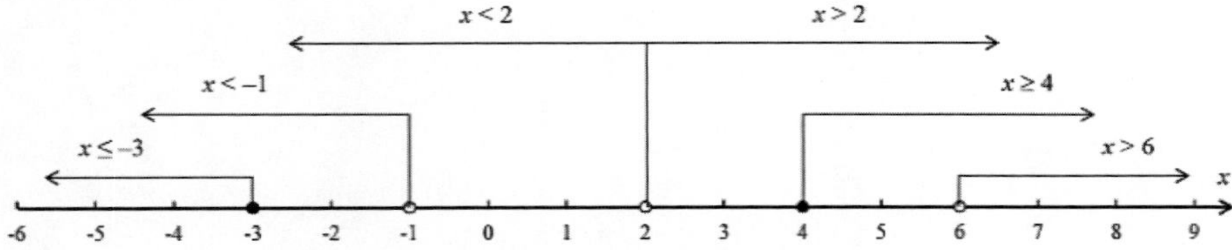

Figure 3.1: A real-number line (or x-axis)

In general, an inequality is defined as one quantity is *greater than* (>), or *less than* (<), or *greater than or equal to* (≥), or *less than or equal to* (≤), another quantity. These are expressed as

$$a>b \text{ or } c<d \text{ or } a\geq b \text{ or } c\leq d. \quad (3.1)$$

Let z = 6, then $x > z$ or x > 6 represents the infinite section on the right of 6 in the real-number line with an unfilled circle at x = 6 indicating this point is excluded (Figure 3.1). Let y = 4, then $x \geq y$ or $x \geq 4$ represents the infinite section on the right of 4 in the real-number line with a solid circle at x = 4 indicating this point is included (Figure 3.1). Similarly let v = –1, then $x < v$ or $x < -1$ represents the infinite section on the left of –1 in the real-number line with an unfilled circle at x = –1 indicating this point is excluded (Figure 3.1). Let u = –3, then $x \leq u$ or $x \leq -3$ represents the infinite section on the left of –3 in the real-number line with a solid circle at x = –3 indicating this point is included (Figure 3.1).

Properties of inequalities

1) Adding to or subtracting from both sides of an inequality the same number does not change the status of the original inequality, i.e.,

   If $a > b$, then $a \pm c > b \pm c$.

   For example, 10 > 6, then 10 – 3 > 6 – 3 and 10 + 3 > 6 + 3.

2) Multiplying or dividing both sides of an inequality by the same ***positive*** number does not change the status of the original inequality, i.e.,

   If $a < b$, then $ac < bc$ and $\frac{a}{c} < \frac{b}{c}$ $(c > 0)$.

   For example, 3 < 9 so 3×9 < 9×9 and 3/3 < 9/3.

3) Multiplying or dividing both sides of an inequality by the same ***negative*** number inverses the status of the original inequality, i.e.,

   If $a < b$, then $ac > bc$ and $\frac{a}{c} > \frac{b}{c}$ $(c < 0)$.

   For example, 3 < 9 so 3×(–9) > 9×(–9) or –27 > –81, and 3/(–3) > 9/(–3) or –1 > –3.

4) If both sides of an inequality are either all positive or all negative, the reciprocal action on both sides of the inequality inverses the status of the original inequality, i.e.,

   If $0 < a < b$, or $a < b < 0$, then $\frac{1}{a} > \frac{1}{b}$.

   For example, 3 < 9 so 1/3 > 1/9; –9 < –3 so –1/9 > –1/3.

5) If both sides of an inequality are all positive, the action of applying a positive exponent or a positive radical on both sides of the inequality does not change the status of the original inequality, i.e.,

   If $0 < a < b$ and $n > 0$, then $a^n < b^n$ or $\sqrt[n]{a} < \sqrt[n]{b}$.

   For example, 4 < 9 so $4^2 < 9^2$ and $\sqrt{4} < \sqrt{9}$.

6) Either side of an inequality can be replaced by another number (or term) if such number (term) keeps the status of the original inequality unchanged, i.e.,

   If $a < b \leq c$, then $a < b$ and $a < c$.

   For example, 4 < 6 < 9 so 4 < 6 and 4 < 9.

Linear inequalities

In inequalities discussed above, if either or both sides of an inequality have linear algebraic expressions, the inequality becomes a linear inequality. For example, $x-2 \geq 3x+1$, $x-2<0$, $x-2>0$, and $-2x+3 \leq x-1$ are linear inequalities. In mathematical terms, linear inequalities can be expressed as

$$ax+b \geq 0 \text{ or } ax+b>0 \text{ or } ax+b \leq 0 \text{ or } ax+b<0 \quad (a \neq 0). \tag{3.2}$$

We can solve a linear inequality using the properties of inequalities and algebraic rules reviewed in Chapter 1. Unlike the solution to a linear equation that is a solid point in a real-number line, the solution to an inequality is an infinite region or limited segment in a real-number line including or excluding the *boundary point* depending on the nature of the inequality. The boundary point is the solution point to the linear equation derived by replacing the "inequality sign" by "the equal sign" in the linear inequality. The boundary point is a solid dot in the real-number line if the inequality sign includes 'equal to'; otherwise an unfilled circle in the real-number line.

Example 3.1: Solve $-2x+3<x-3$. Plot the solution in the real-number line.

Solution

$-2x+3<x-3$	
$(-x)-2x+3+(-3)<(-x)+x-3+(-3)$	Property 1, same as exchanging sides by changing signs
$-3x<-6$	
$(-1)(-3x)>(-1)(-6)$	Property 3, multiplying both sides by –1
$3x>6$	
$x>2$	Property 2, dividing both sides by +3.

The solution is the region from 2 exclusively towards positive infinity in the real-number line in Figure 3.2, or in the range (2, ∞).

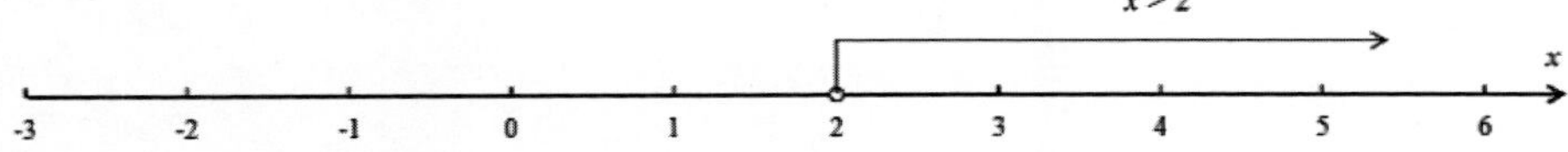

Figure 3.2: The real-number line of solution to Example 3.1

Example 3.2: Solve $x-2 \le 7-2x$. Plot the solution in the real-number line.

Solution

$x-2 \le 7-2x$	
$(2x)+x-2+(2) \le (2x)+7-2x+(2)$	Property 1, same as exchanging sides by changing signs
$3x \le 9$	
$x \le 3$	Property 2, dividing both sides by +3.

The solution is the region from 3 inclusively towards negative infinity in the real-number line in Figure 3.3, or in the range (–∞, 3]; the square bracket means 3 being included.

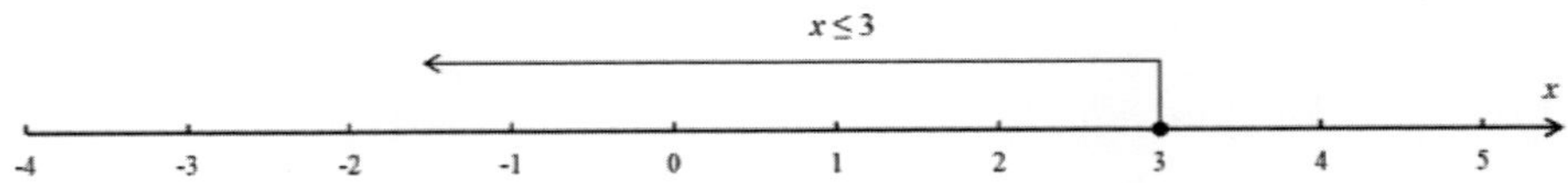

Figure 3.3: The real-number line of solution to Example 3.2

Example 3.3: Solve $x-1 \le 3-x$ and $-2x+3 < x+6$ together to determine the common region that meets both inequalities. Plot the common region in the real-number line.

Solution

$x-1 \le 3-x$	$-2x+3 < x+6$
$x-1+(x)+(1) \le 3-x+(x)+(1)$	$-2x+3+(-x)+(-3) < x+6+(-x)+(-3)$
$2x \le 4$	$-3x < 3 \xrightarrow{\times(-1)} 3x > -3$
$x \le 2$	$x > -1$

The solution is the common region shared by both inequalities, i.e., from –1 exclusive to 2 inclusive in the real-number line in Figure 3.4. or in the range (–1, 2].

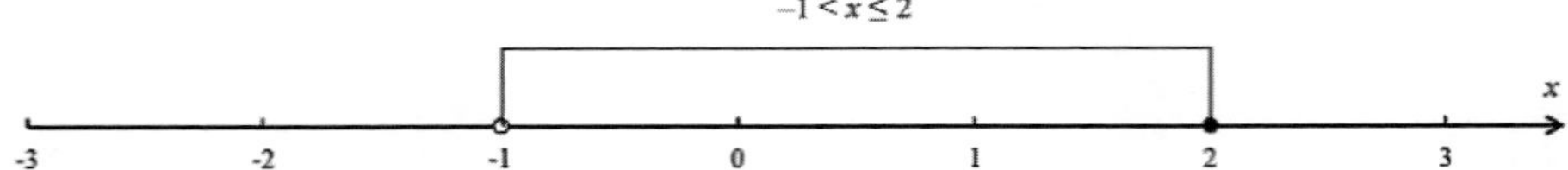

Figure 3.4: The real-number line of solution to Example 3.3

Example 3.4: Evaluate if $2x-1 \ge 5+x$ and $x-3 < -5$ have a common region shared by both inequalities. Plot the common region in the real-number line.

Solution

$2x-1 \ge 5+x$	$x-3 < -5$
$2x-x \ge 5+1$	$x < -5+3$
$x \ge 6$	$x < -2$

The two ranges are [6, ∞) and (–∞, –2) respectively, which have no common region shared by both inequalities. The regions are separated by a section from –1 to 6 in the real-number line in Figure 3.5.

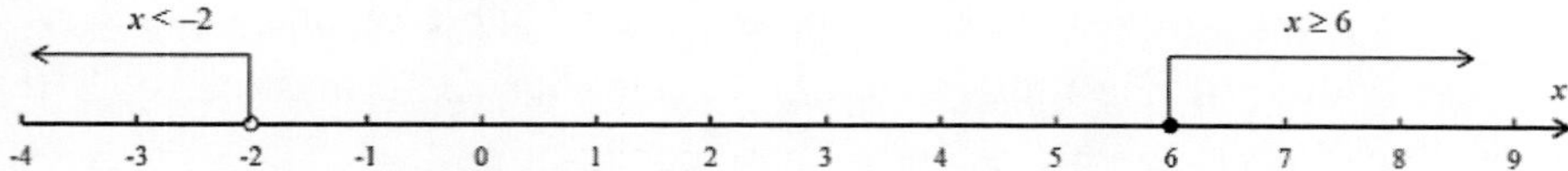

Figure 3.5: The real-number line of solution to Example 3.4

3.1.2 Applications of inequalities

Example 3.5: A publisher publishes textbooks for school and university students. The cost is \$35 per copy and the sale price is \$55 per book. For any book published, the total of fixed costs (expenses occurred during producing the book regardless of number of copies printed) is \$4000.

1) How many copies must be sold for the publisher to earn a profit?
2) If only 100 copies are expected to be sold, what sale price should be set to earn a profit?

Solution

1) Let x be the minimum number of copies required to be sold.

$profit = total\ revenue - total\ cost$

To earn a profit, $profit = total\ revenue - total\ cost > 0$ or $total\ revenue > total\ cost$

$total\ revenue$: $55x$

$total\ cost$: $35x + 4000$

Therefore, we have the following inequality:

$55x > 35x + 4000$, or $55x - (35x + 4000) > 0$.

$20x - 4000 > 0 \longrightarrow 20x > 4000 \longrightarrow x > 200.$

It requires at least 201 copies to be sold to earn a profit from publishing such a textbook.

2) Let y be the new sale price per copy.

$total\ revenue$ $100y$

$total\ cost$ $100 \times 35 + 4000 = 7500$

$100y - 7500 > 0 \longrightarrow y > 75.$

It means to earn a profit from publishing such a textbook, the sale price must be higher than \$75 per copy if only 100 copies to be sold.

Example 3.6: An organising team had a total budget of \$20000 for a contest. It was decided to allocate no more than one-third of the total budget to the prize pool and the rest to cover the expenditures of other activities. The amount in the prize pool should be awarded to 1 first-prize, 2 second-prizes, and 3 third-prizes. The first prize is the double of one second-prize and one second-prize is the double of one third-prize. Any prize must be a multiple of \$100. Find the highest possible allocations to the prizes under these constraints.

Solution

Let x be the amount of money for one 3rd-prize. One 2nd-prize should be $2x$ and the 1st-prize should be $2(2x) = 4x$.

$$\textit{total prizes} \leq \frac{1}{3}(\textit{total fund}) \qquad \underbrace{3\times x}_{3rd} + \underbrace{2\times(2x)}_{2nd} + \underbrace{1\times 4x}_{1st} \leq \frac{20000}{3}$$

Therefore, we have the following inequality:

$$11x \leq \frac{20000}{3} \longrightarrow x \leq \frac{20000}{33} = 606.06.$$

Since any prize must be a multiple of \$100, the highest allowed prizes should be: \$600 for a 3rd-prize, \$1200 for a 2nd-prize and \$2400 for the top prize.

Example 3.7: A small company used to hire a truck for transporting materials and products whenever required. The leasing contract is yearly based and the company has to pay \$600 per month, plus an extra cost of \$0.3 per km. If the company is to buy a truck, the fixed annual expense is \$8200 and other running cost is around \$0.25 per km. How many kilometres should the truck be driven in a year to make buying a truck cheaper than hiring a truck?

Solution

Let x be the driven kilometres per year. To make buying a truck cheaper than hiring, it must meet

$$\textit{total cost of buying} < \textit{total cost of hiring}$$
$$\textit{total cost of buying} = 0.25x + 8200$$
$$\textit{total cost of hiring} = 0.3x + 12\times 600 = 0.3x + 7200.$$

Therefore, we have the following inequality:

$$0.25x + 8200 < 0.3x + 7200 \longrightarrow -0.05x < -1000 \longrightarrow x > \frac{1000}{0.05} = 20000\ km.$$

Thus, buying a truck is cheaper than hiring a truck if driven for more than 20,000 km per year.

Example 3.8: A company produces square bricks. For the same-sized square bricks, if the error in area of the squared surface among the individual bricks must be within ±1%, what range of error in each side of the square should be allowed?

Solution

Let S be the side of the square and Δs be the error in the sides of the square. The area of the squared surface of a brick should be $(S \pm \Delta s)^2$. As the error in the surface area of the brick must be within ±1%, i.e., between $0.99S^2$ and $1.01S^2$, we should derive the following set of inequalities:

$$(S+\Delta s)^2 < 1.01S^2$$
$$(S-\Delta s)^2 > 0.99S^2.$$

These can be solved by applying square root to both sides of these inequalities as follows.

$$(S+\Delta s)^2 < 1.01S^2 \longrightarrow S+\Delta s < \sqrt{1.01}S \longrightarrow \Delta s < (\sqrt{1.01}-1)S = 0.00499S$$
$$(S-\Delta s)^2 > 0.99S^2 \longrightarrow S-\Delta s > \sqrt{0.99}S \longrightarrow \Delta s < (1-\sqrt{0.99})S = 0.00501S.$$

The two inequalities give similar error values for the side, approximately $0.005S$. This equals $\pm\frac{0.005S}{S}\% = \pm 0.5\%$, i.e., the error in each side of the square should be within (–0.5%, 0.5%).

Exercises 3.1

1. Solve the following inequalities and plot the solution in the real-number line.

   a) $2x+3>x-1$
   b) $-2x-5<x+3$
   c) $3x+2\le 10-x$
   d) $4x-7\ge x+3$
   e) $3x-1\ge 7-x$ and $-x+2<x+4$
   f) $4x-1\ge x-7$ and $2x-5<3$
   g) $3x-2<x-8$ and $x-1>2$

2. Some institutions have such grading policy that requires no more than 20% of the students in High Distinction (HD), no more than 30% of the students in Distinction (D). Suppose a unit has 200 students enrolled in a semester. Find the highest possible number of students who could be graded in either HD or D under such policy.

3. A coffee shop serves fast coffee in a shopping centre. The cost is \$2 per cup and the sale price is \$4 per cup. The total of fixed costs daily (daily wage, rental, materials etc.) is \$2000. How many cups of coffee must be sold daily for the shop to earn a profit?

4. To achieve High Distinction, a student will need reaching a total of 85 out of 100 marks or higher (85% or higher). Suppose a unit has 40 marks (40%) for assignments and 60 marks (60%) for the final exam. A student has achieved 32 out of 40 from all assignments. If the student wants to achieve a HD, what is the minimum percentage this student must achieve in the final exam?

3.2 Absolute Values, Equations and Inequalities

3.2.1 Absolute values

The concept of absolute values

In the real-number line, points 5 and –5 are equal in distance from 0 but on the opposite sides. If we only want to know the distance or quantity of a real number x from a reference point (e.g., the origin in a real-number line) regardless of its token, we use absolute operation on x to get it, denoted as $|x|$. As it is self-explanatory, $|x|$ is always positive. Therefore, if x is zero or positive, $|x|$ becomes itself; if x is originally negative, $|x|$ becomes the negative of itself, summarised as

$$\boxed{|x| = \begin{cases} x & \text{if } x \geq 0 \\ -x & \text{if } x < 0 \end{cases}}. \tag{3.3}$$

For example, if $x = 5$, $|x| = x = 5 > 0$; if $x = -5$, $|x| = -x = -(-5) = 5 > 0$, shown in Figure 3.6.

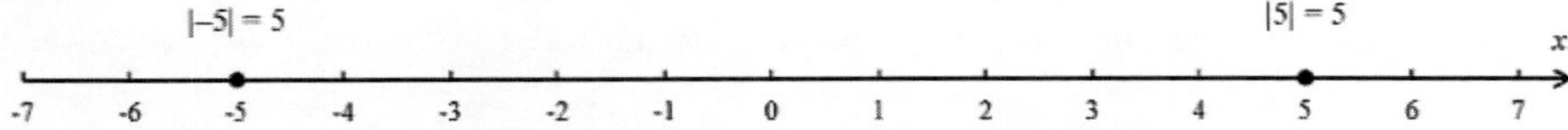

Figure 3.6: Absolute values in the real-number line

Properties of absolute values

Let x and y be real numbers. Absolute values have the following properties:

1. $|xy| = |x||y| = |y||x|$ (3.4)
2. $\left|\dfrac{x}{y}\right| = \dfrac{|x|}{|y|} \quad (y \neq 0)$ (3.5)
3. $|x - y| = |y - x|$ (3.6)
4. $-|x| \leq x \leq |x|$ (3.7)
5. $|x + y| = |y + x| \leq |x| + |y|$ (3.8)

Example 3.9: Given $x = 4$ and $y = -3$, evaluate $|x/y|$ and $|x - y|$; prove $|x + y| = |y + x| < |x| + |y|$ and $-|y| \leq y \leq |y|$ respectively.

Solution

$\left|\dfrac{4}{-3}\right| = \left|-\dfrac{4}{3}\right| = \dfrac{4}{3}$; $\dfrac{|4|}{|-3|} = \dfrac{|4|}{|3|} = \dfrac{4}{3}$, thus $\left|\dfrac{x}{y}\right| = \dfrac{|x|}{|y|} = \dfrac{4}{3}$

$|4 - (-3)| = |4 + 3| = 7$; $|-3 - 4| = |-7| = 7$, thus $|x - y| = |y - x|$

$|4 + (-3)| = |4 - 3| = |-3 + 4| = 1$; $|4| + |-3| = 4 + 3 = 7$, thus $|x + y| = 1 < |x| + |y| = 7$

$|y| = |-3| = 3$, $-|y| = -3$, thus $-|y| \leq y \leq |y|$ stands for $-3 \leq -3 \leq 3$, rationalised to $-3 = -3 < 3$.

3.2.2 Absolute-value equations

An absolute-value equation is a statement of equality between the two sides of an expression involving at least one absolute-value term, such as $|x+3| = 3x-2$ and $|-2x+1|-2=0$. Solving absolute-value equations takes the following two steps:

1) Replacing the absolute term using the rule of formula (3.3) by *a set of two normal equations*, one to deal with the case where the inner term is positive and the other to deal with the case where the inner term is negative;
2) Solving the two normal equations together to determine the possible solutions.

Example 3.10: Solve $|2x+1| = 5$.

Solution

The given absolute-value equation $|2x+1| = 5$ can be replaced by the two linear equations as follows:

$2x+1=5$ and $2x+1=-5$

For $2x+1=5 \longrightarrow 2x=4 \longrightarrow x=2$,

For $2x+1=-5 \longrightarrow 2x=-6 \longrightarrow x=-3$.

Thus, both $x = 2$ and $x = -3$ are the solutions to $|2x+1| = 5$.

Example 3.11: Solve $|-3x+2| = x-6$.

Solution

The absolute term $|-3x+2|$ cannot be negative, which means $x - 6 \geq 0$, or $x \geq 6$. Therefore, any solution must meet $x \geq 6$.

The absolute-value equation $|-3x+2| = x-6$ can be replaced by the two linear equations:

$-3x+2 = x-6$ and $-3x+2 = -(x-6)$

For $-3x+2 = x-6 \longrightarrow -4x = -8 \longrightarrow x = 2$,

For $-3x+2 = -(x-6) \longrightarrow -3x+2 = -x+6 \longrightarrow -2x = 4 \longrightarrow x = -2$.

However, both $x = 2$ and $x = -2$ do not meet $x \geq 6$. Hence $|-3x+2| = x-6$ has no solution.

Example 3.12: Solve $|2x-3|=|x+3|$.

Solution

This absolute-value equation involves two absolute terms. We first use absolute property (3.5) to reorganise these two absolute-value terms as one absolute-value fraction as follows:

$$|2x-3|=|x+3| \longrightarrow \frac{|2x-3|}{|x+3|}=1 \longrightarrow \left|\frac{2x-3}{x+3}\right|=1$$

This reorganised absolute-value equation can be replaced by the following two equations:

$$\frac{2x-3}{x+3}=1 \text{ or } 2x-3=x+3 \text{ and } \frac{2x-3}{x+3}=-1 \text{ or } 2x-3=-(x+3)$$

For $2x-3=x+3 \longrightarrow x=6$,

For $2x-3=-(x+3) \longrightarrow 2x-3=-x-3 \longrightarrow 3x=0 \longrightarrow x=0$.

Thus, both $x=0$ and $x=6$ are the solutions to $|2x-3|=|x+3|$.

Note that $\left|\frac{2x-3}{x+3}\right|=1$ is equivalent to $2x-3=x+3$ and $2x-3=-(x+3)$, just like the right side as $x+3$ without the absolute sign.

3.2.3 Absolute-value inequalities

Similar to an absolute-value equation that has two solutions in normal circumstances, an absolute-value inequality usually has two boundary points that either define a commonly shared segment or indicate two infinite regions in the real-number line. The possible solutions of absolute-value inequalities are summarised in Table 3.1.

Table 3.1 Possible solutions of absolute-value inequalities

Inequality ($p > 0$)	*Solution*
$\lvert a\rvert < p$	$-p < a < p$
$\lvert a\rvert \le p$	$-p \le a \le p$
$\lvert a\rvert > p$	$a < -p$ or $a > p$
$\lvert a\rvert \ge p$	$a \le -p$ or $a \ge p$

Example 3.13: Solve $\left|\frac{1}{2}x-1\right|\le 2$.

Solution

In $|a|\le p$, let $a=\frac{1}{2}x-1$ and $p = 2$, then the solution is

$$-2\le\frac{1}{2}x-1\le 2\longrightarrow -1\le\frac{1}{2}x\le 3\longrightarrow -2\le x\le 6.$$

Thus, the inclusive range [–2, 6] is the solution to $\left|\frac{1}{2}x-1\right|\le 2$, which is shown in Figure 3.7.

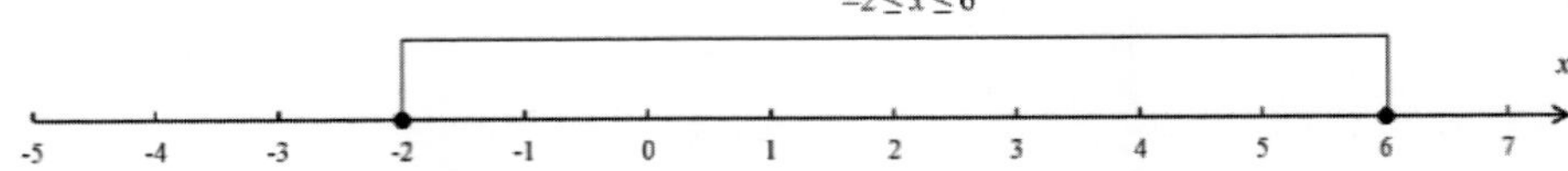

Figure 3.7: The real-number line of solution to Example 3.13

Example 3.14: Solve $|-2x+3|>5$.

Solution

$$-2x+3<-5 \text{ or } -2x+3>5$$

For $-2x+3<-5\longrightarrow 2x-3>5\longrightarrow 2x>8\longrightarrow x>4$

For $-2x+3>5\longrightarrow 2x-3<-5\longrightarrow 2x<-2\longrightarrow x<-1$.

Thus, either $x > 4$ or $x < -1$ is the solution to $|-2x+3|>5$, which is shown in Figure 3.8.

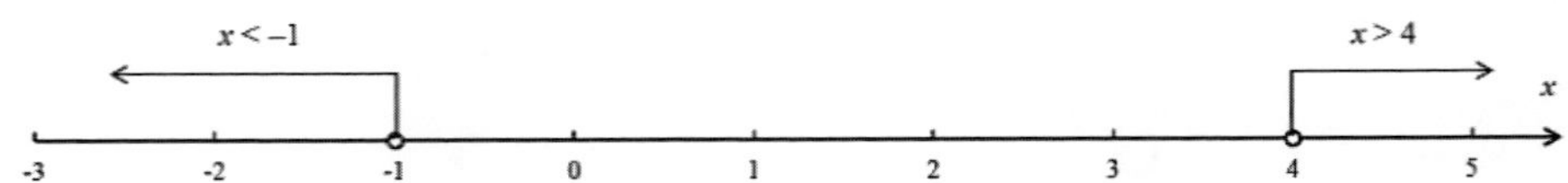

Figure 3.8: The real-number line of solution to Example 3.14

Example 3.15: Solve $|2x-3|\ge|x+3|$.

Solution

Similar to the first step in Example 3.12, the inequality can be reorganised as

$$|2x-3|\ge|x+3|\longrightarrow\frac{|2x-3|}{|x+3|}\ge 1\longrightarrow\left|\frac{2x-3}{x+3}\right|\ge 1$$

This can be solved by

$$\{\frac{2x-3}{x+3}\ge 1\longrightarrow 2x-3\ge x+3\} \text{ or } \{\frac{2x-3}{x+3}\le -1\longrightarrow 2x-3\le-(x+3)\}$$

For $2x-3 \geq x+3 \longrightarrow x \geq 6$

For $2x-3 \leq -(x+3) \longrightarrow 2x-3 \leq -x-3 \longrightarrow 3x \leq 0 \longrightarrow x \leq 0.$

Thus, either $x \leq 0$ or $x \geq 6$ is the solution to $|2x-3| \geq |x+3|$, which is shown in Figure 3.9.

Note that $\left|\frac{2x-3}{x+3}\right| \geq 1$ is equivalent to $2x-3 \geq x+3$ and $2x-3 \leq -(x+3)$, just like the right side as $x+3$ without the absolute sign.

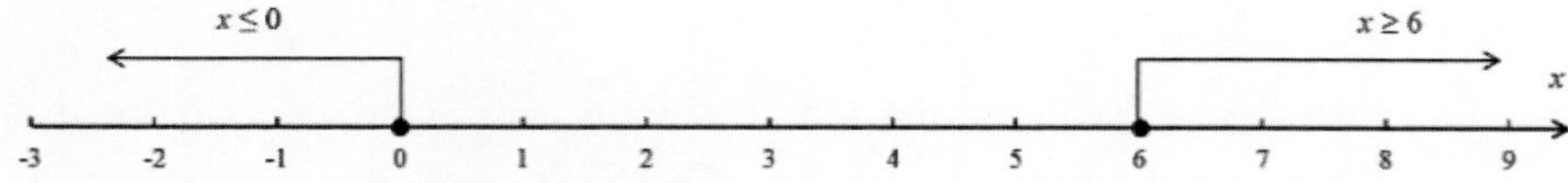

Figure 3.9: The real-number line of solution to Example 3.15

Exercises 3.2

1. Given $x = -2$ and $y = 3$, evaluate $|xy|$, $|x/y|$, $|x + y|$ and $|x - y|$.

2. Solve the following abstract equations:

   a) $|x-2|=7$
   b) $|3x+1|=x-9$
   c) $|-2x-1|=7-x$
   d) $|x+2|=|2x-1|$

3. Solve the following abstract inequalities:

   a) $|x-1|\leq 3$
   b) $|-x+2|>3$
   c) $|2x-1|\geq x+2$
   d) $|3x-1|\leq|x+5|$

3.3 Sequences and Series

3.3.1 Concepts of sequences and series

The concept of sequences

A sequence is a set of numbers (or terms in general) organised in the natural order indexed by continuous positive integers, i.e.,

$$\begin{aligned} &t_1,\ t_2,\ t_3,\cdots t_n && \text{(finite sequence)} \\ &t_1,\ t_2,\ t_3,\cdots t_n,\cdots && \text{(infinite sequence)} \end{aligned} \tag{3.9}$$

where the first sequence has n terms and such is a finite sequence; the second is an infinite sequence as it has infinitely many terms.

A sequence can begin with any positive integer index. Furthermore, a part of consecutive terms from a longer sequence T_n is called a subsequence of T_n. Some examples of (sub)sequences are shown below:

$$\begin{aligned} &2,\ 4,\ 6,\cdots 2n,\cdots && \text{(infinite sequence of even numbers)} \\ &2,\ 4,\ 6,\ 8,\ 10 && \text{(finite sequence or subsequence of even number sequence)} \\ &1,\ \frac{1}{2},\ \frac{1}{3},\cdots\frac{1}{n},\cdots && \text{(infinite sequence)} \end{aligned}$$

The general term of a sequence

In the infinite sequence of even numbers above, any individual term can be represented by $T_n = 2n$ where n takes the same number as the index, i.e., n = 1, 2, 3, …. Such expression is called the general term of a sequence. Similarly, the third sequence above has the general term $T_n = \frac{1}{n}$. Using the general term, a sequence can be expressed alternatively as

$$\begin{aligned} &[T_n]\quad (n = 1,2,\ldots 5) && \Leftarrow\ \text{finite sequence} \\ &[T_n]\quad (n = 1,2,\ldots) && \Leftarrow\ \text{infinite sequence} \end{aligned} \tag{3.10}$$

We can use the general term of a sequence to 1) explicitly find a specific term in the sequence by its index, 2) reconstruct the full sequence, and 3) extract a subsequence from a full sequence.

Example 3.16: The general term of an infinite sequence is

$$T_n = \frac{1}{n^2} \quad (n = 1, 2, ...).$$

a) Find the third and tenth terms of this sequence;
b) Express this sequence in the form of its natural order;
c) Express the subsequence from the third term to the sixth term inclusively in the form of its natural order.

Solution

a) Substitute $n = 3$ and $n = 10$ into the general term to get

$$T_3 = \frac{1}{3^2} = \frac{1}{9} \quad and \quad T_{10} = \frac{1}{10^2} = \frac{1}{100}$$

b) The natural order of this sequence is

$$\frac{1}{1^2}, \frac{1}{2^2}, \frac{1}{3^2}, ..., \frac{1}{n^2}, ... \quad or \quad 1, \frac{1}{4}, \frac{1}{9}, ..., \frac{1}{n^2}, ...$$

c) The natural order of the subsequence of this sequence from $n = 3$ to $n = 6$ is

$$\frac{1}{3^2}, \frac{1}{4^2}, \frac{1}{5^2}, \frac{1}{6^2} \quad or \quad \frac{1}{9}, \frac{1}{16}, \frac{1}{25}, \frac{1}{36}$$

The concept of series

A series is the sum of all terms of a sequence or subsequence, i.e,

$$\begin{aligned} S_n &= \sum_{i=1}^{n} T_i \quad (n: \text{a positive integer}) && \Leftarrow \text{finite sequence} \\ S &= \sum_{i=1}^{\infty} T_i && \Leftarrow \text{infinite sequence} \end{aligned} \qquad (3.11)$$

where the symbol Σ represents the 'sum' of the indicated terms from the bottom index (beginning term) to the top index (ending term). It is self-explanatory that the series of a finite sequence has a certain value. In mathematical terms, series of a finite sequence is always *convergent* to a certain value. However, series (or the sum) of an infinite sequence is either *convergent* or *divergent* depending on whether a certain value for the sum of the infinite sequence exists or not.

Example 3.17: Given $T_n = \frac{1}{n}$ $(n = 1, 2, ...)$

a) Find the sum of the first 4 terms of this sequence;
b) Express the sum of subsequence from the 5th to 20th terms by means of the series form.

Solution

a) The sum of the first 4 terms is

$$S_4 = T_1 + T_2 + T_3 + T_4 = 1 + \frac{1}{2} + \frac{1}{3} + \frac{1}{4} = \frac{12}{12} + \frac{6}{12} + \frac{4}{12} + \frac{3}{12} = \frac{25}{12} = 2\frac{1}{12}$$

b) The sum from the 5th term to the 20th term of this sequence is

$$S[5,20] = \sum_{i=5}^{20} T_i = \sum_{i=5}^{20} \frac{1}{n}.$$

3.3.2 Arithmetic sequences

An arithmetic sequence, also known as arithmetic progression, is a set of terms in which the difference (d) between any two consecutive terms ($a_n - a_{n-1}$) is the same constant, i.e., $d = a_n - a_{n-1}$. In other words, in an arithmetic sequence, the next term (a_n) is the sum of the previous term a_{n-1} and a constant d, i.e., $a_n = a_{n-1} + d$. Some examples of arithmetic sequences are as follows:

$$2, 4, 6, 8, ... \quad (d = 4 - 2 = 8 - 6 = 2 \; or \; 8 = 6 + 2, 6 = 4 + 2, 4 = 2 + 2)$$
$$109, 100, 91, ..., -8 \quad (d = 100 - 109 = 91 - 100 = -9 \; or \; 91 = 100 - 9, \; 100 = 109 - 9)$$

Arithmetic sequences have the following properties summarised in Table 3.2.

Table 3.2 Properties of arithmetic sequences

Property	*Remark*
$a_n - a_{n-1} = d$	The same difference is between any two consecutive terms.
$a_n = a_{n-1} + d$	
$a_n = a_1 + (n-1)d$	a_1 is the beginning term; a_n is the ending term.
$S_n = \frac{n}{2}(a_1 + a_n)$	
$S_n = \frac{n}{2}[2a_1 + (n-1)d]$	

Example 3.18: Given $a_n = n\sqrt{2}\ (n \geq 1)$,

a) Find the difference of this arithmetic sequence;
b) Find the tenth term of this arithmetic sequence;
c) Find the sum from the 4th term to the 8th term of this sequence;
d) Find the sum of the first 10 terms of this sequence.

Solution

a) Since $d = a_n - a_{n-1}$,

$$a_n = n\sqrt{2},\quad a_{n-1} = (n-1)\sqrt{2},\quad d = a_n - a_{n-1} = n\sqrt{2} - (n-1)\sqrt{2} = n\sqrt{2} - n\sqrt{2} + \sqrt{2} = \sqrt{2}$$

b) Let $n = 10$ in $a_n = n\sqrt{2}$.

$$a_{10} = 10\sqrt{2}$$

c) There are 5 terms from the 4th term to 8th term, i.e., $n = 5$.

$$S_n = S[4,8] = \frac{n}{2}(a_4 + a_8) = \frac{5}{2}(4\sqrt{2} + 8\sqrt{2}) = \frac{5}{2} \times 12\sqrt{2} = 30\sqrt{2}$$

d) The sum of the first 10 terms is

$$S_{10} = \frac{n}{2}[(2a_1 + (n-1)d] = \frac{10}{2}[(2 \times \sqrt{2} + (10-1) \times \sqrt{2})] = 5(2\sqrt{2} + 9\sqrt{2}) = 5 \times 11\sqrt{2} = 55\sqrt{2}.$$

Example 3.19: Given the first two terms of an arithmetic sequence as $a_1 = 40$ and $a_2 = 37$, find the sum of the first 8 terms of this arithmetic sequence.

Solution

$$d = a_2 - a_1 = 37 - 40 = -3$$

$$S_8 = \frac{n}{2}[(2a_1 + (n-1)d] = \frac{8}{2}[(2 \times 40 + (8-1) \times (-3)] = 4(80 - 21) = 4 \times 59 = 236.$$

Example 3.20: A new employee is offered an annual salary of \$42,000 with an annual raise of \$2000 from second year. Find the total amount earned in 10 years.

Solution

$$d = 2000,\quad a_1 = 42000,\quad n = 10$$

$$S_{10} = \frac{n}{2}[(2a_1 + (n-1)d] = \frac{10}{2}[(2 \times 42000 + (10-1) \times (2000)] = 5(84000 + 18000) = 510000.$$

The total amount earned in 10 years is \$510,000.

Example 3.21: The central section of a theatre has more than ten rows of seats. The first row has 5 seats and the last row has 16 seats. From the second row, each row has one more seat than its immediate front row has. Find the total number of seats in this section.

Solution

$d=1,\ \ a_1=5,\ \ a_n=16.\ \ As\ \ a_n=a_1+(n-1)d,\ \ 16=5+(n-1)\times 1 \longrightarrow 11=n-1 \longrightarrow n=12$

$S_{12}=\frac{n}{2}(a_1+a_n)=\frac{12}{2}(5+16)=6\times 21=126.$

The total number of seats in the central section is 126.

Example 3.22: An arithmetic sequence has ten terms. Its second term is 3 and the tenth term is 107. Find the common difference and the first term of this sequence.

Solution

$\because a_n=a_1+(n-1)d,\ \ a_2=3,\ a_{10}=107$

$\therefore \begin{cases} 3=a_1+(2-1)d \\ 107=a_1+(10-1)d \end{cases} \longrightarrow \begin{cases} a_1+d=3 \\ a_1+9d=107 \end{cases} \xrightarrow[(2)-(1)]{} 8d=104$

$d=13,\ \ a_1=a_2-d=3-13=-10$

3.3.3 Geometric sequences

A geometric sequence, also known as geometric progression, is a set of terms in which the next term is the previous term multiplied by a constant factor (r) known as the *common factor*, i.e., $a_n = ra_{n-1}$ or $r = a_n / a_{n-1}$. Some examples of geometric sequences are as follows:

$2,4,8,16,...\ \ (r=4/2=8/4=16/8=2\ or\ 16=2\times 8, 8=2\times 4, 4=2\times 2)$

$64,32,16,...,1\ \ (r=32/64=16/32=\frac{1}{2}\ or\ 32=64\times\frac{1}{2}, 16=32\times\frac{1}{2}, 1=2\times\frac{1}{2})$

Geometric sequences have the following properties summarised in Table 3.3.

Table 3.3 Properties of geometric sequences

Property	*Remark*
$a_n=ra_{n-1}$	The next term is the previous term multiplied by a common factor.
$r=\frac{a_n}{a_{n-1}}$	
$a_n=a_1r^{n-1}$	a_1 is the beginning term; a_n is the ending term.
$S_n=\frac{a_1(1-r^n)}{1-r}=\frac{a_1-ra_n}{1-r}$	
$S=S_\infty=\frac{a_1}{1-r}\ \ (\lvert r\rvert<1)$	

Example 3.23: The first two terms of a geometric sequence are $a_1 = 1$ and $a_2 = \frac{1}{10}$.

a) Find the common factor of this geometric sequence;
b) Find the fifth term of this sequence;
c) Find the sum of the first 3 terms of this sequence;
d) Find the sum of this sequence assuming it goes to infinity.

Solution

a) $r = a_2 / a_1 = \frac{1}{10} / 1 = \frac{1}{10}$

b) Let $n = 5$.

$$a_5 = a_1 r^{5-1} = 1 \times (\frac{1}{10})^4 = \frac{1}{10000}$$

c) $S_3 = \frac{a_1(1-r^3)}{1-r} = \frac{1-(\frac{1}{10})^3}{1-\frac{1}{10}} = \frac{\frac{1000}{1000} - \frac{1}{1000}}{\frac{10}{10} - \frac{1}{10}} = \frac{\frac{999}{1000}}{\frac{9}{10}} = \frac{111}{100}$

d) $S = \frac{a_1}{1-r} = \frac{1}{1-\frac{1}{10}} = \frac{1}{\frac{10}{10} - \frac{1}{10}} = \frac{1}{\frac{9}{10}} = \frac{10}{9}$.

Example 3.24: The first swing of a pendulum is 16 cm. From the second swing, each swing is 75% of the previous swing. Find the total distance moved in four swings.

Solution

$$r = 0.75 = \frac{3}{4}, \quad a_1 = 16, \quad n = 4$$

$$S_4 = \frac{a_1(1-r^4)}{1-r} = 16 \times \frac{1-(\frac{3}{4})^4}{1-\frac{3}{4}} = 16 \times \frac{\frac{256}{256} - \frac{81}{256}}{\frac{4}{4} - \frac{3}{4}} = 16 \times \frac{\frac{175}{256}}{\frac{1}{4}} = 16 \times \frac{175}{64} = \frac{175}{4} = 43.75\ cm.$$

The total distance travelled in four swings is 43.75 cm.

Example 3.25: The price of a house is currently \$300000. It is expected the house price to increase by 6% each year. Find the house price in 5 years and 10 years respectively.

Solution

$$r = 1.06, \quad a_1 = 300000$$
$$a_5 = a_1 r^{5-1} = 300000(1.06)^4 \approx 378743$$
$$a_{10} = a_1 r^{10-1} = 300000(1.06)^9 \approx 506844.$$

The house price in 5 years and 10 years will be around \$378,743 and \$506,844 respectively.

Example 3.26: A ball dropped from a height of 10 m rebounds to two-thirds of its height on each bounce. Find the total distance travelled by the ball before it comes to rest.

Solution

$$r = \frac{2}{3}, \quad a_1 = 10$$
$$S = \frac{a_1}{1-r} = \frac{10}{1-\frac{2}{3}} = \frac{10}{\frac{1}{3}} = 30.$$

Since the distance travelled during each bounce is the double of the height the ball bounces (up & down), except the first bounce that was only a drop down of 10 metres, the total distance travelled by the ball until it stops is

$$D = 2S - 10 = 2 \times 30 - 10 = 60 - 10 = 50 \ (metres).$$

Example 3.27: The common factor and the first term of an infinite geometric sequence are the same. The sum of this series is 2. Find the common factor (hence the first term) of this infinite geometric series.

Solution

$$\because r = a_1, \ S_\infty = \frac{a_1}{1-r} = 2$$
$$\therefore \frac{a_1}{1-r} = 2 \longrightarrow \frac{r}{1-r} = 2 \longrightarrow r = 2(1-r) \longrightarrow r = 2 - 2r \longrightarrow 3r = 2$$
$$r = a_1 = \frac{2}{3}$$

Exercises 3.3

1. Given the fourth and fifth terms of an arithmetic sequence as $a_4 = 40$ and $a_5 = 45$, find the sum of the first 8 terms of this arithmetic sequence.

2. Some second-hand books are given away for free. Ten books are allowed for the first guest, nine books for the second, eight books for the third, and so on until the last book to be picked by the tenth guest. Find the total number of books to give away.

3. The general term of an infinite sequence is

   $T_n = \frac{(-1)^n}{n} \quad (n = 1, 2, \ldots)$.

   a) Find the third and tenth terms of this sequence;
   b) Express this sequence in the form of its natural order;
   c) Express the subsequence from the third term to the sixth term inclusively in the form of its natural order;
   d) Find the sum of the first 4 terms of this sequence;
   e) Express the sum of the subsequence from the 5th term to the 20th term by means of the series form.

4. Given $a_n = 2n \quad (n \geq 1)$,

   a) Find the difference of this arithmetic sequence;
   b) Find the third term of this arithmetic sequence;
   c) Find the sum from the 5th term to the 10th term of this sequence;
   d) Find the sum of the first 10 terms of this sequence.

5. Given $a_n = -\frac{n}{10} \quad (n \geq 1)$,

   a) Find the difference of this arithmetic sequence;
   b) Find the tenth term of this arithmetic sequence;
   c) Find the sum from the 5th term to the 8th term of this sequence;
   d) Find the sum of the first 12 terms of this sequence.

6. The first two terms of a geometric sequence are $a_1 = 1$ and $a_2 = \frac{1}{3}$.

   a) Find the common factor of this geometric sequence;
   b) Find the fourth term of this sequence;
   c) Find the sum the first 4 terms of this sequence;
   d) Find the sum of this sequence assuming it goes to infinity.

7. The first swing of a pendulum is 10 cm. From the second swing, each swing is 50% of the previous swing. Find the total distance travelled in five swings.

8. The price of a house is currently \$500000. It is expected the house price to decrease by 1% each year. Find the house price after 10 years.

9. A farmer planted mango trees in a block of triangular land shown in the diagram below. The first row has one tree, and the subsequent row has two more trees than its previous row. The last row has 21 trees. Find the total number of trees in this block of land.

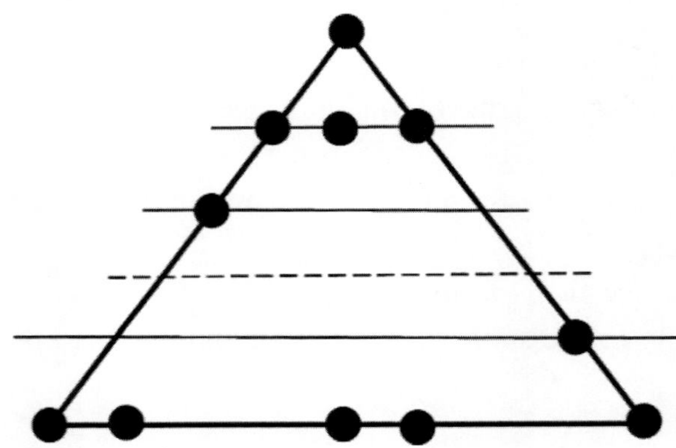

10. A car slows down every second. In the first second, it moved 40 meters, and then moved 75% of the distance of the previous second. Find the total distance travelled in 5 seconds.

Chapter 3: Exercises Answers

Exercises 3.1

1. a) $x > -4$; b) $x > -\frac{8}{3}$; c) $x \le 2$; d) $x \ge \frac{10}{3}$; e) $x \ge 2$; f) $-2 \le x < 4$;

   g) *No solution*
2. $x \le 40$ and $y \le 60$
3. $x > 1000$
4. $x \ge 87.5\%$

Exercises 3.2

1. $|xy| = 6$, $|x/y| = 2/3$, $|x + y| = 1$, $|x - y| = 5$
2. a) $x = 9$ or $x = -5$; b) *no solution*; c) $x = -8$ *and* $x = 2$; d) $x = 3$ or $x = -\frac{1}{3}$
3. a) $-2 \le x \le 4$; b) $x < -1$ or $x > 5$; c) $x \le -1/3$, or $x \ge 3$; d) $-1 \le x \le 3$.

Exercises 3.3

1. $S_8 = 340$.
2. $d = -1$ and $S_{10} = 55$.
3. a) $T_3 = -\frac{1}{3}, T_{10} = \frac{1}{10}$; b) $-1, \frac{1}{2}, -\frac{1}{3}, \ldots, \frac{(-1)^n}{n}, \ldots$; c) $-\frac{1}{3}, \frac{1}{4}, -\frac{1}{5}, \frac{1}{6}$; d) $S_4 = -\frac{7}{12}$;

   e) $S[5,20] = \sum_{i=5}^{20} T_i = \sum_{i=5}^{20} \frac{(-1)^n}{n}$
4. a) $d = 2$, b) $a_3 = 6$, c) $S[5,10] = 90$, d) $S_{10} = 110$
5. a) $d = -\frac{1}{10}$; b) $a_{10} = -1$; c) $S[5,8] = \frac{-13}{5}$; d) $S_{12} = \frac{-39}{5}$
6. a) $r = \frac{1}{3}$; b) $a_4 = \frac{1}{27}$; c) $S_4 = \frac{40}{27}$; d) $S = \frac{3}{2}$.
7. $S_5 = 19.375$ cm
8. $a_{11} \approx 452191(\$)$
9. 121 trees
10. 122.03 m

CHAPTER 4

4 Functions and Graphs

CHAPTER OBJECTIVES

- Introduce the concept of functions
- Use graphs to visually present functions
- Explain general properties of functions
- Study special functions and inverse functions
- Introduce multivariable functions and graphs in the Cartesian system
- Calculate the distance between two points in the Cartesian system

Essential statements on functions and graphs:

- Functions are the foundation of calculus.
- The Cartesian system is the mostly used coordinating system for function analysis and operations.
- Special functions and properties help understand and simplify combinative functions.
- 2D graphs are lines or curves whereas 3D graphs are planes or surfaces.

Key topics:

- Functions and properties
- Translations and reflections of functions
- Symmetric functions
- Implicit and parametric functions
- Inverse functions
- The Cartesian system
- The distance between two points in the Cartesian system
- 2D graphs and 3D graphs

Flowchart of mathematical knowledge building

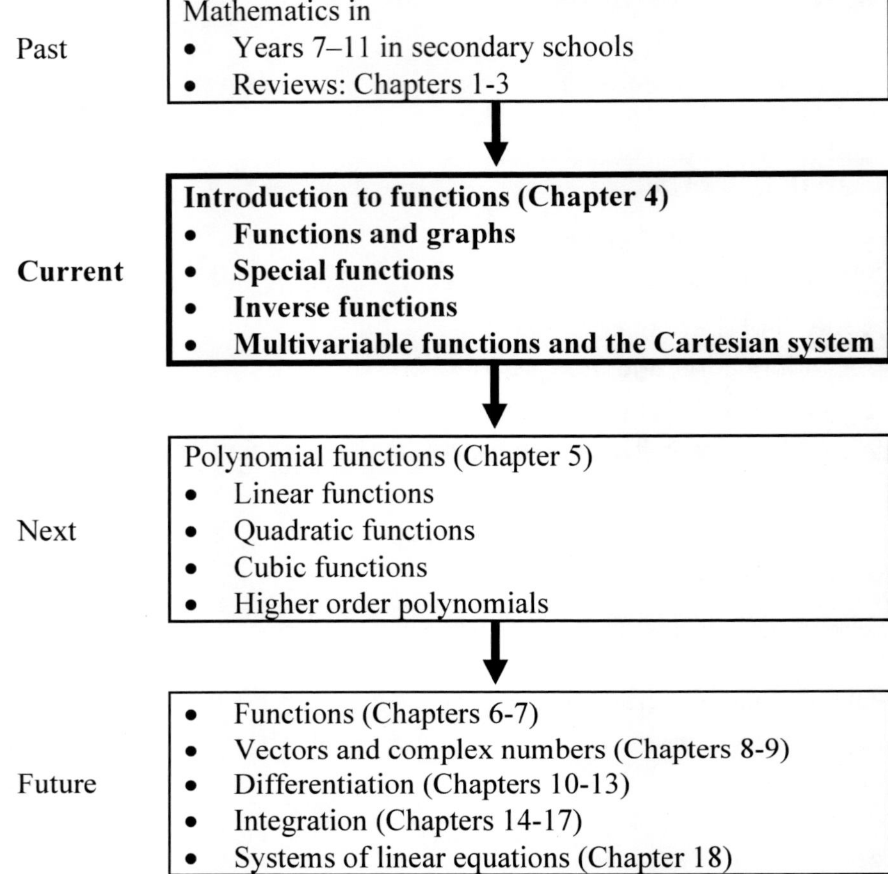

Chapter 4: Functions and Graphs

4.1 Introduction to Functions and Graphs

4.1.1 The concept of functions and graphic presentations

The perimeter (P) and area (A) of a square are defined by its size (s) as $P = 4s$ and $A = s^2$ respectively. Given a value to size s, values for the perimeter and area are then determined through both relationships. For example, given $s = 2$ meters, this then uniquely defines $P = 8$ meters and $A = 4$ m^2. In other words, the perimeter P and area A of a square depend on the size s of the square. Regarding the size s as a parameter that can change freely in its natural range ($s > 0$), or called *independent variable*, the perimeter P and area A of any square can change accordingly, thus P and A being called *dependent variables*. In general, *the relationship between one or more independent variables and a dependent variable is called the function between the variables*, such as $P = 4s$ and $A = s^2$. The paired values of s–P and s–A can be displayed as a plot of variable s versus dependent variable P or A shown in Figure 4.1.

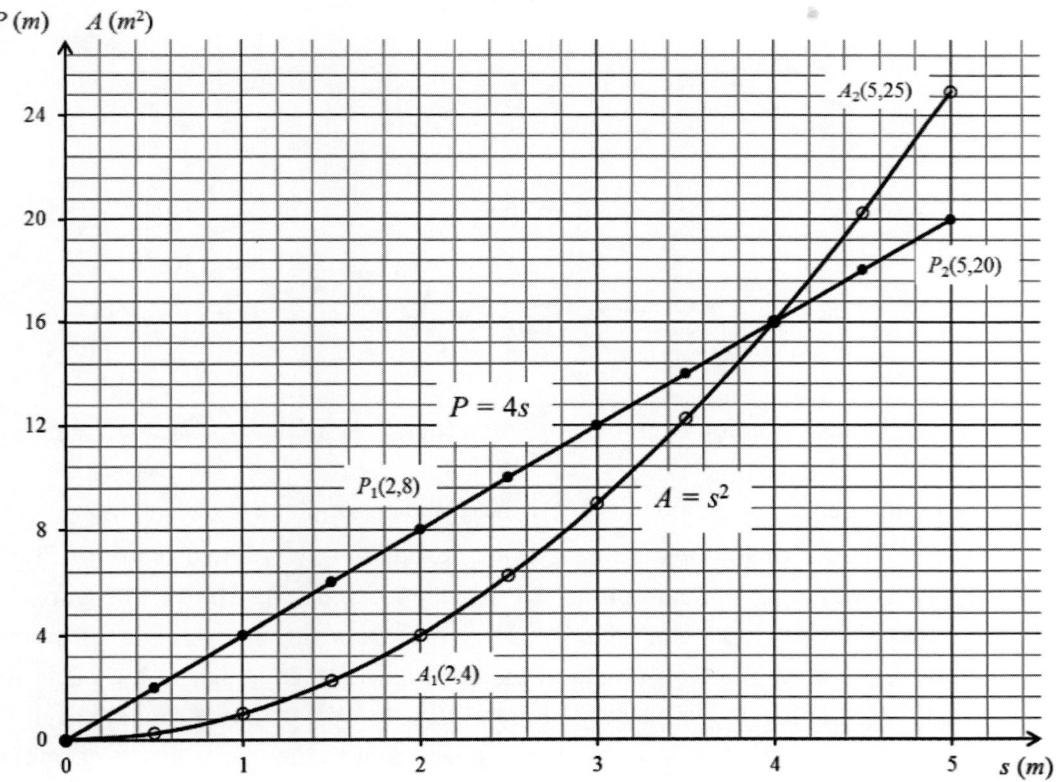

Figure 4.1: A graphic presentation of relationships $P = 4s$ and $A = s^2$

The horizontal axis represents all possible values for size s whereas the vertical axis represents all possible corresponding values for P or A. Function $P = 4s$ forms a straight line in the s–P plot and each s–P pair is a point in this P-line, e.g., $P_1(2, 8)$ represents point P_1 located at $s = 2$ and $P = 8$ in the P-line, and $P_2(5, 20)$ represents point P_2 located at $s = 5$ and $P = 20$ in the P-line. Function $A = s^2$ forms a curve concaving up in the s–A plot and each s–A pair is a point in this A-curve. For example, $A_1(2, 4)$ represents point A_1 located at $s = 2$ and $A = 4$ in the A-curve, and $A_2(5, 25)$ represents point A_2 located at $s = 5$ and $A = 25$ in the A-curve. The curves presented in Figure 4.1 are called the graphs of these two functions.

In generic terms, for functions with only one independent variable (called the elementary functions), the independent variable (or simply as variable) is often represented by x and its corresponding dependent variable is represented by y. The relationship or function between x and y is expressed generically as

$$y = f(x) \tag{4.1}$$

where $f(\ )$ refers to any *abstract function*. A real relationship becomes a substantive case of this abstract function. For example, let $y = P$ and $x = s$, the perimeter of a square becomes $y = f(x) = 4x$, which is equivalent to $P = 4s$. Similarly, let $y = A$ and $x = s$, the area of a square is $y = f(x) = x^2$, which is equivalent to $A = s^2$. Therefore, this abstract notion $y = f(x)$ is widely used for generic elementary function in mathematics.

A function is also regarded as a transformation that projects each of the possible values in the whole range of x (called the x-domain) to the corresponding values in the whole range of y (called the y-domain), which is illustrated in Figure 4.2.

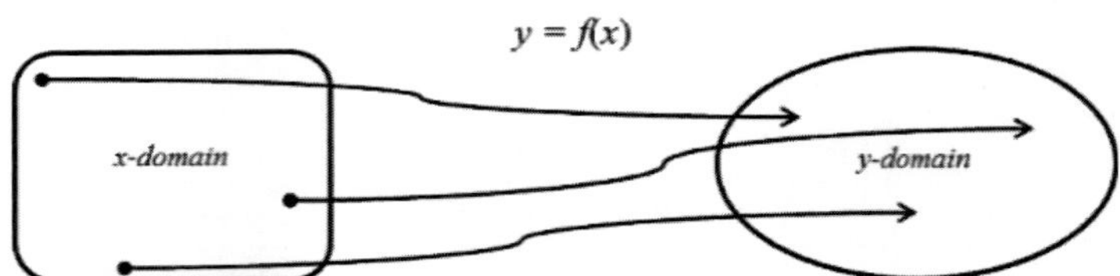

Figure 4.2: Transformation from the x-domain to the y-domain by function $y = f(x)$

By adopting the generic function $y = f(x)$, the graph of an elementary function (simply as function) can be plotted as pairs of x–y on an xy-plane, called a two-dimensional (or 2D) plane of the Cartesian coordinating system defined by the '*right-hand rule*' shown in Figure 4.3. As long as the rotation order 'X→Y' or 'Y→Z' or 'Z→X' from the index finger to the middle finger is followed, the thumb always points to the third axis perpendicular to the plane formed by the index and middle finger. For example, Figure 4.3a shows a potential rotation of the index finger to the middle finger, or Y→Z so the thumb must point to the x-axis that is normal to the yz-plane. In Figure 4.3b, the potential rotation indicates X→Y, so the thumb points to the z-axis that is normal to the xy-plane. Similarly Figure 4.3c is another variation of the Cartesian coordinating system.

In a 2D xy-plane, function $y = f(x)$ is plotted using the pairs of x–y, which can be pre-calculated in a tabular format for the convenience of plotting. The following examples show how to create a graph for a given function.

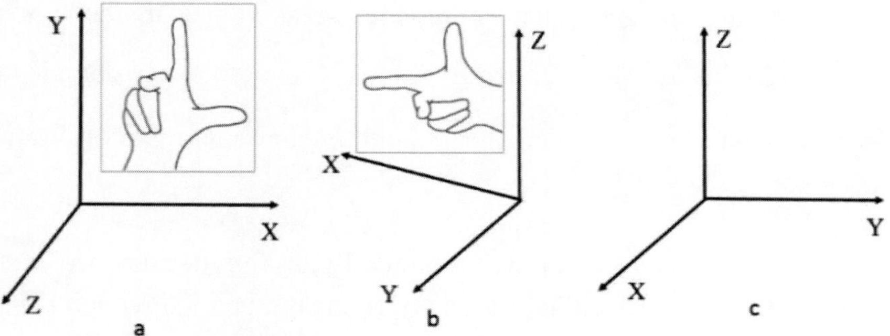

Figure 4.3: Variations of the Cartesian coordinating system defined by the 'right-hand rule'

Example 4.1: Given $y = \frac{1}{2}x^2 - x + 2$, plot this function as a graph in range [–4, 4] for x.

Solution

We first calculate the data pairs between x and y and organise them in the table below. These data pairs can be plotted on to the xy-plane one by one shown as the dots in Figure 4.4.

x	-4	-3.5	-3	-2.5	-2	-1.5	-1	-0.5	0	0.5	1	1.5	2	2.5	3	3.5	4
y	14	11.625	9.5	7.625	6	4.625	4.5	2.625	2	1.625	1.5	1.625	2	2.625	4.5	4.625	6

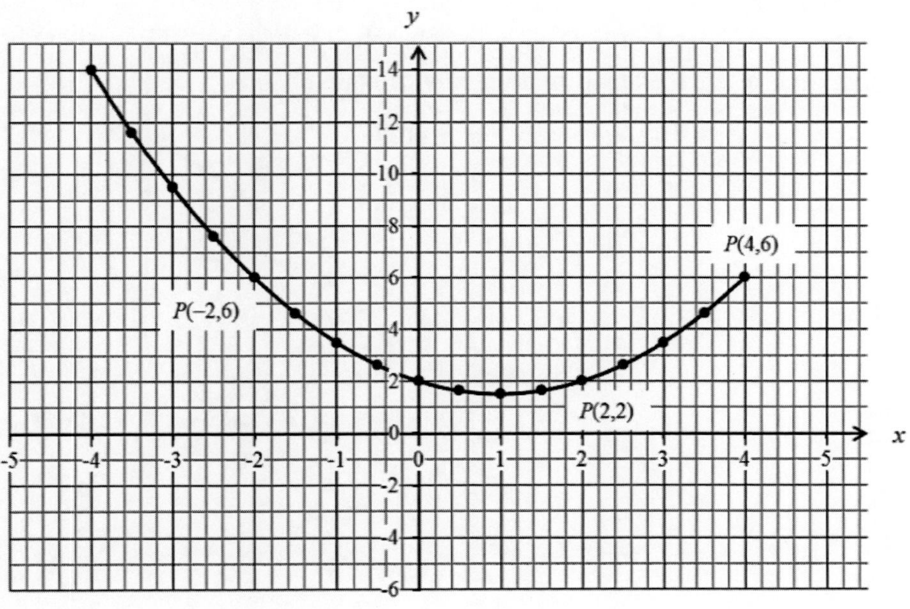

Figure 4.4: Graph of function $y = \frac{1}{2}x^2 - x + 2$ in Example 4.1

The solid line joining all these dots is the graph of function $y=\frac{1}{2}x^2-x+2$. The curve is continuous, which means that the infinitely many data pairs in this curve all belong to this function. If we only look at the real-number line of x in the data table above, values in the x-domain are linearly organised. By transforming the linear values in the x-domain through function $y=\frac{1}{2}x^2-x+2$, in the real-number line of y, values in the y-domain are no longer linearly organised. In other words, this function is a nonlinear transformation from the x-domain to the y-domain.

Example 4.2: The same temperature can be measured by different scales, for example, Celsius (°C) is widely used as the international unit in many countries, and Fahrenheit (°F) is commonly used in USA. The relationship between these two units is $t_F=1.8t_C+32$ where t_C is the Celsius degree and t_F is the Fahrenheit degree. Plot this function as a graph in the range [–40, 40] for t_C.

Solution

Let $x = t_C$ and $y = t_F$. Using an interval of 5 °C, we can calculate the data pairs between x and y and organise them in the table below:

x (t_C)	-40	-35	-30	-25	-20	-15	-10	-5	0	5	10	15	20	25	30	35	40
y (t_F)	-40	-31	-22	-13	-4	5	14	23	32	41	50	59	68	77	86	95	104

These data pairs can be plotted on to the xy-plane one by one shown as the dots in Figure 4.5. The solid line joining all these dots is the graph of function $t_F=1.8t_C+32$. This is a continuous straight line, which implies that the infinitely many data pairs in this line all belong to this function and both the Celsius degree t_C and the Fahrenheit degree t_F are linearly related. In other words, this function is a linear transformation between t_C and t_F.

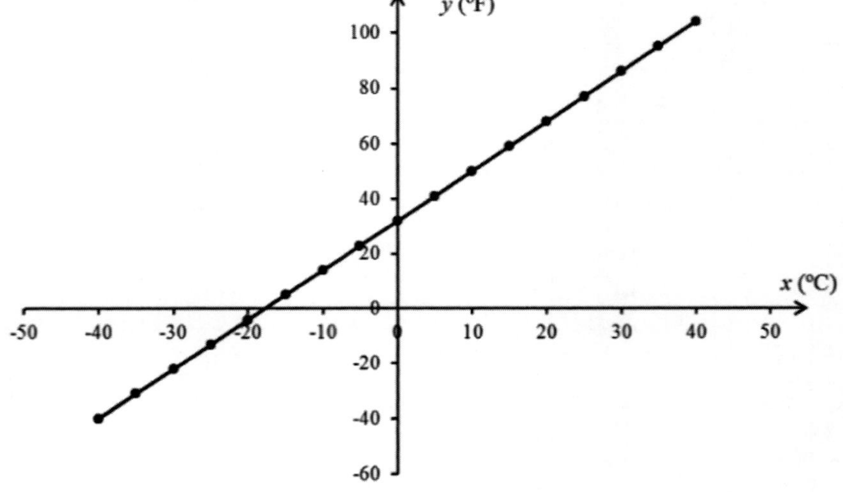

Figure 4.5: Graph of function $y=1.8x+32$ in Example 4.2

4.1.2 General properties of functions

The domain of a function

For function $y = f(x)$, it may be meaningful only for a certain range of x values. For example, if we express the area of a square $A = s^2$ as $y = x^2$, this function is only meaningful for any positive value of x, or size of the square. In this case, we state that function $y = x^2$ is defined in domain $x > 0$ or $(0, \infty)$. If we accept the concept of 'zero area', the domain is then adjusted to $x \geq 0$, or $[0, \infty)$. The domain of a function can be determined by solving the equation or inequality of the function constrained by any condition(s) of the dependent variable y. For example, as the area of a square CANNOT be negative, or $y \geq 0$, thus we have an inequality $x^2 \geq 0$. Solving this inequality gives $x \geq 0$.

Example 4.3: Determine the domain for $y = \sqrt{24 - 4x}$.

Solution

For y to be a real number, $24 - 4x$ must be zero or positive. This means

$$24 - 4x \geq 0 \longrightarrow -4x \geq -24 \longrightarrow x \leq 6.$$

Therefore, the domain for $y = \sqrt{24 - 4x}$ is $x \leq 6$ or $(-\infty, 6]$.

Example 4.4: Determine the domain for $y = \dfrac{x+1}{x-1}$.

Solution

Any value of x will make y be a real number, except $x - 1 = 0$ or $x = 1$ that makes the denominator zero. This means

$$x - 1 \neq 0 \longrightarrow x \neq 1.$$

Therefore, the domain for $y = \dfrac{x+1}{x-1}$ is $x \neq 1$ or $(-\infty, 1)$ and $(1, \infty)$.

Example 4.5: The distance between Brisbane and Rockhampton is about 600 km. If you drive a vehicle from Rockhampton to Brisbane to deliver goods in no more than ten hours, what is the range (domain) of the average speed for your travel? Assume the speed limit is 100 km/h.

Solution

Let y be the distance travelled and x be the average speed. At any time $t > 0$, the distance travelled is $y = xt$ and the average speed is $x = \dfrac{y}{t}$. Here are the two extreme cases:

- The slowest speed is the case of spending 10 hours to travel 600 km, i.e., $x \geq 600/10 = 60$ km/h.
- The fastest speed must be the allowed speed limit of 100 km/h, i.e., $x \leq 100$ km/h. Travelling with this fastest speed, it only takes 6 hours to go down from Rockhampton to Brisbane.

Therefore, the range (domain) of the average speed for this travel is $60\,(\text{km/h}) \leq x \leq 100\,(\text{km/h})$.

The combination of functions

Different functions that share the same independent variable may be combined together to form a new function of the same independent variable. The domain of the combined function needs to be reassessed since the new function may have a different domain from those of the original functions. The common combinations of functions are summarized in Table 4.1. Note these can be extended to combinations of three or more functions too.

Table 4.1 List of the common combinations of functions

Original function	*Combined function*
$y = f(x)$	$y = h(x) = af(x)$ (a : constant)
$y = f(x)$	$y = h(x) = (a+b)f(x) = af(x) + bf(x)$ (a, b : constants)
$y = f(x),\ y = g(x)$	$y = h(x) = a[f(x) \pm g(x)] = af(x) \pm ag(x)$ (a : constant)
$y = f(x),\ y = g(x)$	$y = h(x) = af(x) \pm bg(x)$ (a, b : constants)
$y = f(x),\ y = g(x)$	$y = h(x) = f(x) \times g(x)$
$y = f(x),\ y = g(x)$	$y = h(x) = \dfrac{f(x)}{g(x)}$ $g(x) \neq 0$

Example 4.6: Given $f(x) = x^2$ and $g(x) = \sqrt{24 - 4x}$, determine the domain for $2f(x) + 3g(x)$.

Solution

The new combined function *h*(*x*) is

$$h(x) = 2f(x) + 3g(x) = 2x^2 + 3\sqrt{24 - 4x}\,.$$

For *h*(*x*) to be a real number, 24 – 4*x* must be zero or positive. This means

$$24 - 4x \geq 0 \longrightarrow -4x \geq -24 \longrightarrow x \leq 6.$$

Therefore, the domain for the combined function *h*(*x*) is $x \leq 6$, or (–∞, 6], the same as that for $g(x) = \sqrt{24 - 4x}$.

Example 4.7: Given $f(x)=-2\sqrt{24-4x}$ and $g(x)=\sqrt{24-4x}$, find the domain for $f(x)\times g(x)$.

Solution

The new combined function $h(x)$ is

$$h(x)=f(x)\times g(x)=(-2\sqrt{24-4x})(\sqrt{24-4x})=-2(\sqrt{24-4x})^2=-2(24-4x)=8x-48.$$

Known from Example 4.6, both $f(x)$ and $g(x)$ are defined in $(-\infty, 6]$; however, $h(x)$ is defined in the whole range where x can be any real number.

Example 4.8: Given $f(x)=\sqrt{1-x}$ and $g(x)=\sqrt{1+x}$, what is the domain for $\frac{f(x)}{g(x)}$?

Solution

The new combined function $h(x)$ is

$$h(x)=\frac{\sqrt{1-x}}{\sqrt{1+x}}=\sqrt{\frac{1-x}{1+x}}.$$

For $h(x)$ to be a real number, $\frac{1-x}{1+x}\geq 0$ and $1+x\neq 0$ must be met. This means

$$x\neq -1, \text{and } \frac{1-x}{1+x}\geq 0 \longrightarrow \text{either } (1-x\geq 0 \text{ and } 1+x>0) \text{ or } (1-x\leq 0 \text{ and } 1+x<0).$$

For $(1-x\geq 0 \text{ and } 1+x>0)$: $x\leq 1$ and $x>-1$. The common range for both to share is $-1<x\leq 1$.

For $(1-x\leq 0 \text{ and } 1+x<0)$: $x\geq 1$ and $x<-1$. There is no common range shared by $x\geq 1$ and $x<-1$.

Therefore, the domain for the combined function $h(x)$ is $-1<x\leq 1$, or $(-1, 1]$, which is different from either $x\leq 1$ for $f(x)$ or $x\geq -1$ for $g(x)$.

The composition of functions

We know function $y = g(x)$ is a transformation from the x-domain to the y-domain. If another function $z = f(y)$ uses y as its independent variable, by substituting $y = g(x)$ into $f(y)$, function $z = f(y)$ becomes a *composition of functions* as

$$z=f(y)=f(g(x)). \tag{4.2}$$

This composition establishes a nested relationship between the x-domain and the z-domain through the y-domain, which can be illustrated in Figure 4.6.

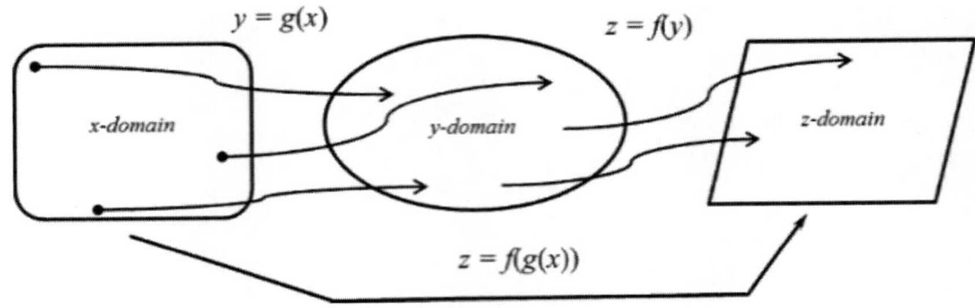

Figure 4.6: Transformation by the composition of functions $z = f(y) = f(g(x))$

Example 4.9: Given $z = f(y) = y^2$ and $y = g(x) = \sqrt{24 - 4x}$, determine function $h(x) = f(g(x))$.

Solution

The compositional function $h(x)$ is

$$h(x) = z = f(y) = y^2 = (g(x))^2 = (\sqrt{24 - 4x})^2 = 24 - 4x \text{ or}$$
$$h(x) = f(g(x)) = (g(x))^2 = (\sqrt{24 - 4x})^2 = 24 - 4x.$$

As $z = h(y) = y^2 \geq 0$, $24 - 4x \geq 0$ or $x \leq 6$. Thus function $h(x)$ is defined in $(-\infty, 6]$. If considering $h(x)$ as an independent function, its range is the whole x-domain, or $(-\infty, \infty)$.

Example 4.10: Given $f(x) = \sqrt{24 - 4x}$ and $g(x) = x^2$, determine $f(g(x))$.

Solution

The compositional function is

$$f(g(x)) = \sqrt{24 - 4g(x)} = \sqrt{24 - 4x^2} = 2\sqrt{6 - x^2}.$$

The compositional function is defined in the range of $6 - x^2 \geq 0$, i.e., $-\sqrt{6} \leq x \leq \sqrt{6}$.

Example 4.11: Given $f(x) = \sqrt{1 - x}$ and $g(x) = \sqrt{1 + x}$, determine function $h(x) = g(f(x)^2)$.

Solution

The compositional function $h(x)$ is

$$h(x) = g(f(x)^2) = \sqrt{1 + f(x)^2} = \sqrt{1 + (\sqrt{1 - x})^2} = \sqrt{1 + 1 - x} = \sqrt{2 - x}.$$

As a new function, this compositional function is defined in the range of $2 - x \geq 0$, i.e., $x \leq 2$, which is different from the domains for both $f(x)$ and $g(x)$.

4.1.3 Translations and reflections of functions

The translation of a function

Given a function $y = f(x)$, some simple operations with $f(x)$ will result in a new function $g(x)$ whose graph (of course values) can be obtained by moving $f(x)$ around. This is called the translation of function $f(x)$ to another function $g(x)$. The commonly encountered translations of functions are summarised in Table 4.2. Note that $g(x)$ and $f(x)$ may be defined in different domains.

Table 4.2 Translations of function $y = f(x)$ (constant $c > 0$)

New function	Graphic transition of the original function $y = f(x)$
$g(x) = f(x) + c$	Shift $y = f(x)$ upwards by c units
$g(x) = f(x) - c$	Shift $y = f(x)$ downwards by c units
$g(x) = f(x - c)$	Shift $y = f(x)$ to right by c units
$g(x) = f(x + c)$	Shift $y = f(x)$ to left by c units
$g(x) = cf(x) \quad (c > 1)$	Stretch $y = f(x)$ vertically away from the x-axis by a factor of c
$g(x) = cf(x) \quad (c < 1)$	Shrink $y = f(x)$ vertically towards the x-axis by a factor of c

Example 4.12: Given function $y = f(x) = x^2$, find the following translated functions and plot their graphs.

a) $g(x) = f(x) - 4$;
b) $g(x) = f(x + 2)$;
c) $g(x) = 2f(x)$;
d) $g(x) = \frac{1}{2}f(x)$.

<u>**Solution**</u>

a) $g(x) = f(x) - 4 = x^2 - 4$.

b) $g(x) = f(x+2) = (x+2)^2 = x^2 + 4x + 4$.

c) $g(x) = 2f(x) = 2x^2$.

d) $g(x) = \frac{1}{2}f(x) = \frac{1}{2}x^2 = 0.5x^2$.

The original and the translational functions are plotted in Figure 4.7.

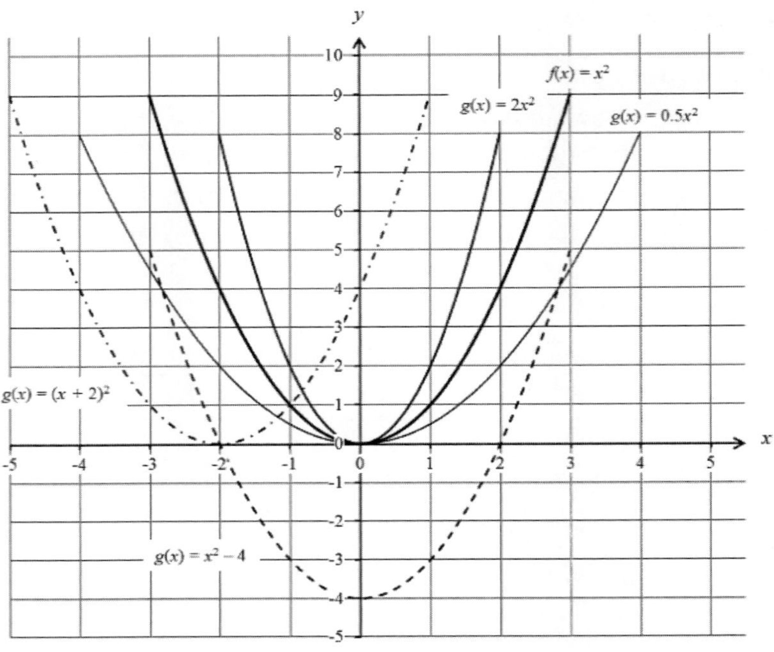

Figure 4.7: Graphs of the original and translational functions in Example 4.12

The reflection of a function

Given a function $y = f(x)$, flipping it about the x-axis or the y-axis would create a new function $g(x)$ that is the symmetry of $f(x)$ about the x-axis or the y-axis. This is called the reflection of function $f(x)$ about the x-axis or the y-axis, which is summarised in Table 4.3. Note that the reflected function $g(x)$ and the original function $f(x)$ may be defined in different domains.

Table 4.3 Symmetry and reflections of functions $y = f(x)$

Symmetric function	Graphic transition of the original function $y = f(x)$
$g(x) = -f(x)$	Reflect $y = f(x)$ about the x-axis, or symmetric about the x-axis
$g(x) = f(-x)$ if exists.	Reflect $y = f(x)$ about the y-axis, or symmetric about the y-axis

Example 4.13: Given function $f(x) = 0.5x^3 - 3x + 5$, find the following functions by reflecting $f(x)$ and plot their graphs.

a) $g(x) = -f(x)$, b) $g(x) = f(-x)$.

Solution

a) $g(x) = -f(x) = -(0.5x^3 - 3x + 5) = -0.5x^3 + 3x - 5.$

b) $g(x) = f(-x) = 0.5(-x)^3 - 3(-x) + 5 = -0.5x^3 + 3x + 5.$

The original function $f(x)$ is symmetric with $g(x) = -f(x)$ about the x-axis; function $g(x) = f(-x)$ is symmetric with the original $f(x)$ about the y-axis (Figure 4.8).

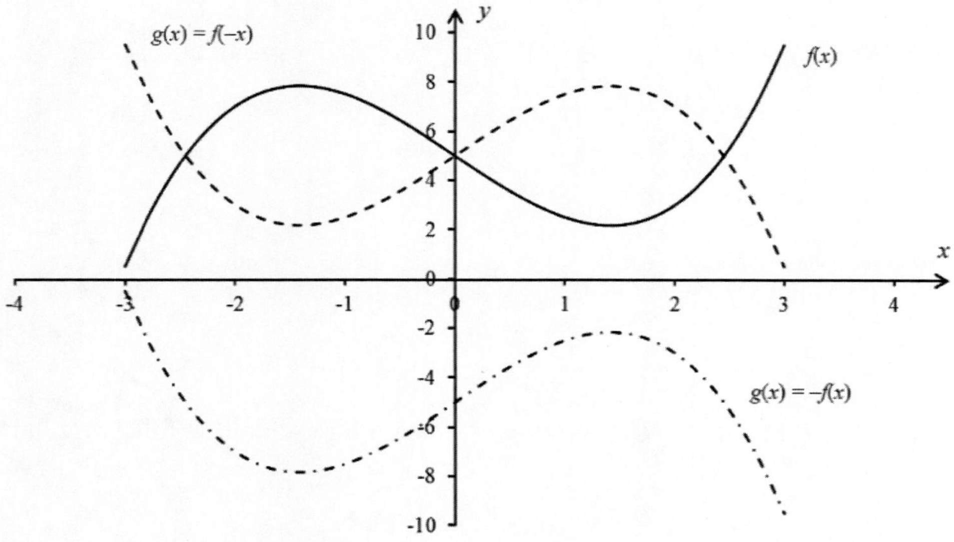

Figure 4.8: Graphs of the original and reflected functions in Example 4.13

Exercises 4.1

1. Plot $y = x^2 + x + 2$ as a graph in the range [–4, 4] for x.

2. Determine the domain for each of the following functions respectively.

   a) $y = \sqrt{x+2}$
   b) $y = -x^2 + 4$
   c) $y = \dfrac{x^2 + x - 2}{x^2 - 3x + 2}$
   d) $y = \sqrt{\sqrt[3]{27x^3}}$

3. Find the combined function and determine its domain for each of the following functions.

   a) Given $f(x) = -2x^2$ and $g(x) = \sqrt{x^3 - 2x^2}$, determine $3f(x) + 2g(x)$.
   b) Given $f(x) = -2x^2$ and $g(x) = \sqrt{x^3 - 2x^2}$, determine $f(x) \times g(x)$.
   c) Given $f(x) = \sqrt{x-2}$ and $g(x) = \sqrt{x+2}$, determine $\dfrac{f(x)}{g(x)}$.
   d) Given $f(x) = \sqrt{x-2}$ and $g(x) = \sqrt{x+2}$, determine $h(x) = g(f(x))$.
   e) Given $f(x) = \sqrt{x-2}$ and $g(x) = \sqrt{x+2}$, determine $h(x) = g(f(x)^2)$.

4. Given function $y = f(x) = x^3$, find the following translated functions and plot their graphs.

   a) $g(x) = f(x) - 9$;
   b) $g(x) = f(x+1)$;
   c) $g(x) = \dfrac{1}{3} f(x)$.

5. Given function $f(x) = \dfrac{1}{5}x^3 - 3$, find the following functions by reflecting $f(x)$ and plot their graphs.

   a) $g(x) = f(-x)$,
   b) $g(x) = -f(x)$.

4.2 Special Functions and Inverse Functions

4.2.1 Symmetric functions

Even functions

Some functions may have special characteristics that other functions do not have. For example, assigning 5 and –5 to x in function $g(x) = x^2$ produces the same value of 25. Such feature of $g(x) = g(-x)$ will not happen for function $f(x) = \sqrt{24-4x}$ since $f(5) = \sqrt{24-4\times 5} = 2$ and $f(-5) = \sqrt{24-4\times(-5)} = \sqrt{44} = 2\sqrt{11}$.

Any function that satisfies $f(x) = f(-x)$ is called an ***even function***. An even function $y = f(x)$ is symmetric about the y-axis in the graph, or mirrored about the y-axis. Figure 4.9 is the graph of even function $y = -x^2 + 10$. The curve is symmetric about the y-axis, i.e., the left side of the curve (or $x < 0$) is the reflection of the right side of the curve (or $x > 0$) about the y-axis, and vice versa. For example, $y(-1) = 9$ and $y(1) = 9$, $y(2) = 6$ and $y(-2) = 6$.

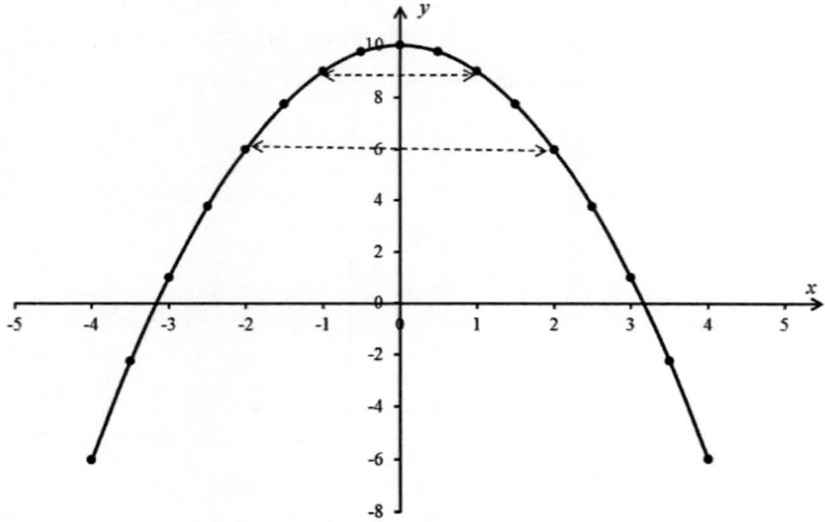

Figure 4.9: The graph of even function $y = -x^2 + 10$ symmetric about the y-axis

Odd functions

Assigning 3 and –3 to x in function $f(x) = x^3 - 15x$, it produces $f(-3) = 18$ and $f(3) = -18$. Such feature of $f(-x) = -f(x)$ will not happen for even function $g(x) = x^2$.

Any function that satisfies $f(-x) = -f(x)$ is called an ***odd function***. An odd function $y = f(x)$ is symmetric about the origin in the graph, or mirrored about the origin. Letting $x = 0$, any odd function must come with $f(0) = 0$. Figure 4.10 is the graph for odd function $f(x) = x^3 - 15x$. The curve is symmetric about the origin, i.e., any point on the left side of the curve (or $x < 0$) is projected to the point on the right side of the curve (or $x > 0$) through the origin, and vice versa. For example, $y(-3) = 18$ and $y(3) = -18$.

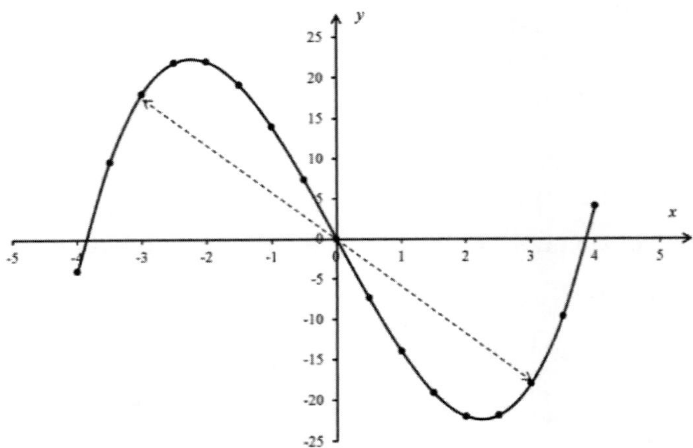

Figure 4.10: The graph of odd function $f(x) = x^3 - 15x$ symmetric about the origin

Combinations of even and odd functions

Combining two even and/or odd functions may create another even or odd function. The common combinations of even and/or odd functions are summarised in Table 4.4. Note that division of two even and/or odd functions is dependable on different circumstances and thus not included in this table.

Table 4.4 Symmetry of some combined functions

Original functions	*Combined function*	*Example*
If both $f(x)$ and $g(x)$ are even,	$f(x) \pm g(x)$: even	$f(x) = 2x^2,\ g(x) = x^4 \to 2x^2 - x^4$
If both $f(x)$ and $g(x)$ are odd,	$f(x) \pm g(x)$: odd	$f(x) = 3x,\ g(x) = x^3 \to x^3 - 3x$
If both $f(x)$ and $g(x)$ are even,	$f(x) \times g(x)$: even	$f(x) = 2x^2,\ g(x) = x^4 \to 2x^2 \bullet x^4 = 2x^6$
If both $f(x)$ and $g(x)$ are odd,	$f(x) \times g(x)$: even	$f(x) = 3x,\ g(x) = x^3 \to 3x \bullet x^3 = 3x^4$
If $f(x)$ is even and $g(x)$ is odd,	$f(x) \times g(x)$: odd	$f(x) = 2x^2,\ g(x) = x \to 2x^2 \bullet x = 2x^3$
If $f(x)$ is odd and $g(x)$ is even,	$f(x) \times g(x)$: odd	$f(x) = -3x,\ g(x) = x^2 \to -3x \bullet x^2 = -3x^3$

4.2.2 Special functions

Constant functions

If a function takes a constant value c in its domain, such function $f(x) = c$ is called a ***constant function***. For example, given $f(x) = 2$ defined in domain $x \geq -10$, it represents a horizontal line cross $y = 2$ in its graph (Figure 4.11). This constant function means at any point within $[-10, \infty)$, y keeps unchanged as 2, i.e., $f(30) = 2$, $f(500) = 2$, and $f(-8) = 2$. Given $g(x) = -3$ defined within $(-\infty, \infty)$, y is fixed as -3 over the whole x-domain.

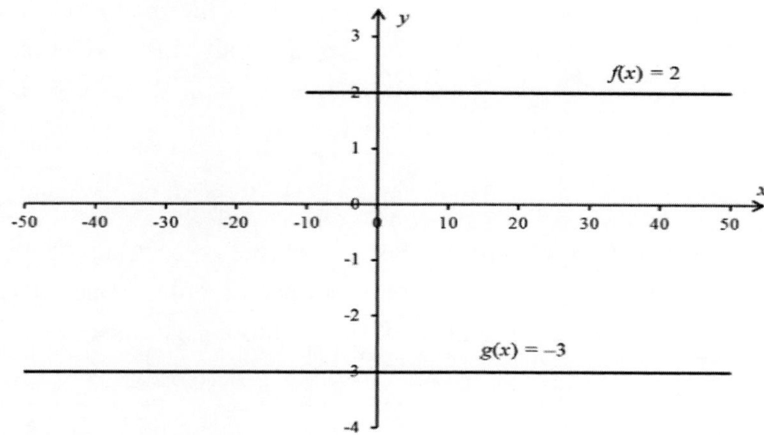

Figure 4.11: Graphs of constant functions $f(x) = 2$ and $g(x) = -3$ in their domains respectively

Absolute-value functions

A function defined by $y = |f(x)|$ is called an ***absolute function***. Similar to the absolute operation for $|x|$, absolute function $y = |f(x)|$ can be expressed by the combination of two sections as

$$y = |f(x)| = \begin{cases} f(x) & \text{if } f(x) \geq 0 \\ -f(x) & \text{if } f(x) < 0 \end{cases}. \quad (4.3)$$

For example, given $f(x) = |x + 2|$ defined in domain $-10 \leq x \leq 10$, it can be expressed by

$$y = |x + 2| = \begin{cases} x + 2 & -2 \leq x \leq 10 \\ -(x + 2) & -10 \leq x < -2 \end{cases}.$$

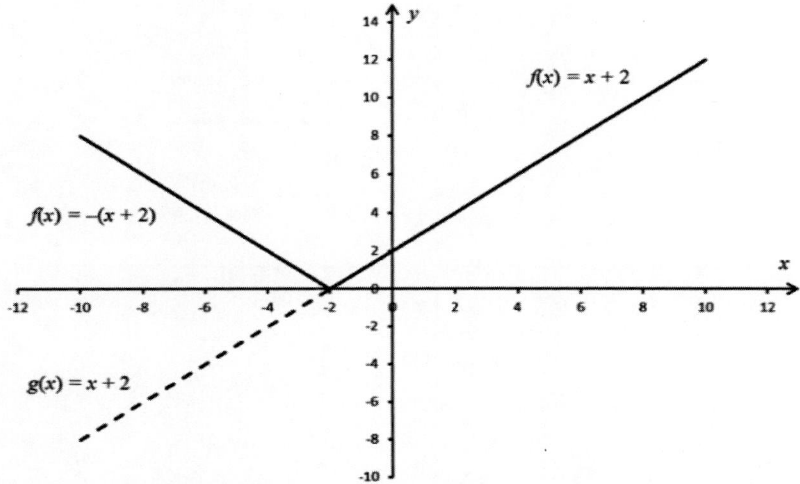

Figure 4.12: Graphs of absolute functions $f(x) = |x + 2|$ and function $g(x) = x + 2$

This is plotted in Figure 4.12 along with $g(x) = x + 2$ that is the function without applying the absolute operation. The absolute function has non-negative values for y. Compared with $g(x) = x + 2$ without applying the absolute operation, $f(x)$ shares the non-negative section with $g(x)$, but flips up $g(x)$ vertically about the x-axis for positive values.

Segmented functions

If a function takes different forms in different sub-domains or sections of the x-domain, such function is called a ***segmented function*** or case-defined, or piecewise function. Segmented functions can be expressed by the combination of two or more sections as

$$\boxed{y = f(x) = \begin{cases} f_1(x) & x_0 < x \le x_1 \\ f_2(x) & x_1 < x \le x_2 \\ \cdots & \\ f_n(x) & x_{n-1} < x \le x_n \end{cases}}. \tag{4.4}$$

Example 4.14: Given the following segmented function

$$y = \begin{cases} -1 & -3 \le x < -1 \\ x & -1 \le x < 1 \\ 1 & 1 \le x \le 3 \end{cases},$$

plot the graph of this function.

Solution

This segmented function is defined in domain $-3 \le x \le 3$. The first section is a horizontal line one unit below the x-axis within $[-3, 0)$. The second section is an oblique line from -1 to 1. The third section is a horizontal line one unit above the x-axis within $(1, 3]$ (Figure 4.13).

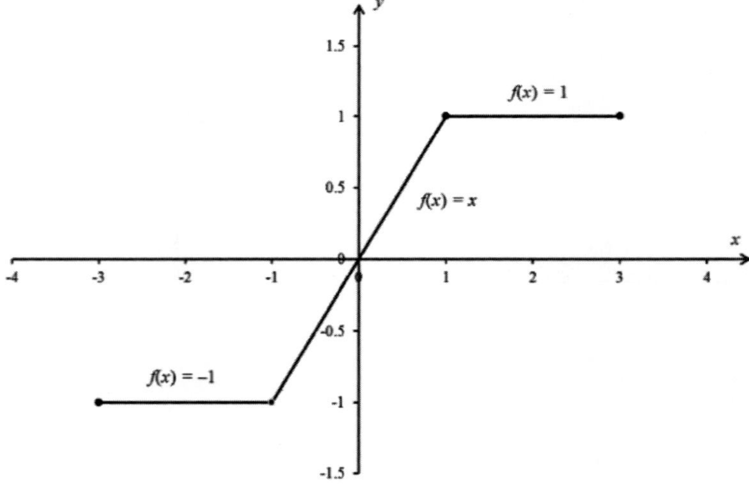

Figure 4.13: Graph of segmented function in Example 4.14

Example 4.15: Assume the tax rates for personal income are summarised as follows:

$$r = \begin{cases} 0\% & income \le \$10,000 \\ 15\% & \$10,001 \le income \le \$50,000 \\ 30\% & \$50,001 \le income \le \$100,000 \\ 40\% & \$100,001 \le income \le \$150,000 \\ 48\% & income > \$150,000 \end{cases}.$$

In the last financial year, Andrew's taxable income was $85,000 and Alison's was $180,000. How much tax would Andrew and Alison need to pay in the last financial year?

Solution

Let the annual personal income and tax be x and y respectively. The relationship between x and y can be established through the tax rates as

$$y = \begin{cases} 0 & x \le \$10,000 \\ 0.15(x-10000) & \$10,001 \le x \le \$50,000 \\ 6000+0.3(x-50000) & \$50,001 \le x \le \$100,000 \\ 21000+0.4(x-100000) & \$100,001 \le x \le \$150,000 \\ 41000+0.48(x-150000) & x > \$150,000 \end{cases}.$$

Since Andrew's taxable income was $85,000 last year, he would need to pay tax of

$$y = 6000+0.3(x-50000) = 6000+0.3(85000-50000) = 6000+0.3\times 35000 = \$16500.$$

As Alison's taxable income was $180,000 last year, much more than Andrew's, she would need to pay more tax of

$$y = 41000+0.48(x-150000) = 41000+0.48(180000-150000) = 41000+0.48\times 30000 = \$55400.$$

Periodic functions

If a function repeats itself in a confined interval of the independent variable, such function is called a ***periodic function***. Periodic functions can be expressed by specifying the functional expression AND the range (or confined interval) for repeating such as

$$\boxed{y = f(x), \qquad x_1 < x \le x_2}. \tag{4.5}$$

The difference of the specified range, $p = x_2 - x_1$, is called the *period* of the periodic function $f(x)$.

Example 4.16: Given periodic function $y = x^2$, $-1 \le x \le 1$, plot its graph and find its period.

Solution

This periodic function is defined in domain $-1 \le x \le 1$ and repeats periodically over the whole x-domain, as shown in Figure 4.14. The curve from –1 to 1 in the x-axis repeats itself from –3 to –1, and from 1 to 3 too. Such can be extended to infinities at the both ends. Its period is $p = 1 - (-1) = 2$, i.e., the curve repeats over every 2 units.

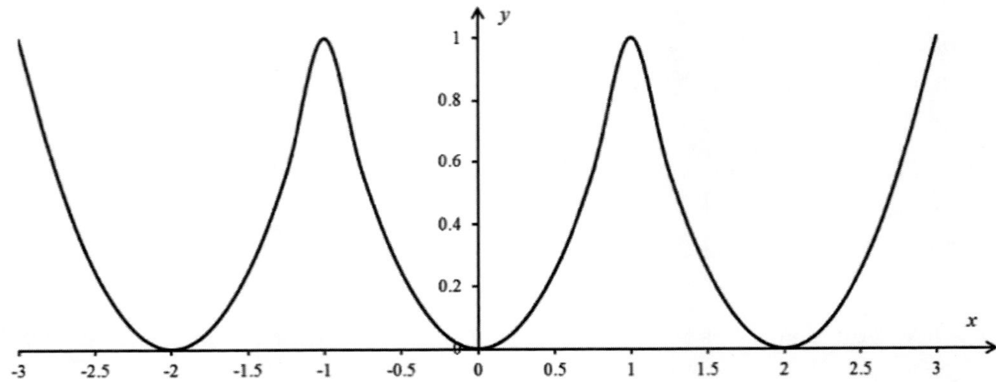

Figure 4.14: Graph of the periodic function in Example 4.16

4.2.3 Implicit and parametric functions

Implicit functions

By far, we have assumed that generic functions take the form of $y = f(x)$, such as $y = f(x) = 2x^2$. In many situations, a function between x and y may be given as a general equation, for example, $x^2 + y^2 = 1$ and $x\sqrt[3]{y} + y^2 = 2$. Compared with $y = f(x)$ that is an ***explicit function*** between x and y, a function defined by a general equation $f(x, y) = c$ (c is constant) is called an ***implicit function*** between x and y.

Sometimes by reorganising a general equation, an implicit function can be converted to an explicit function. For instance, $x^2 + y^2 = 1$ can be reorganised as an explicit function $y = \pm\sqrt{1-x^2}$. However, in most cases, such conversion is not possible. Graphs for implicit functions can still be created if the *x*-*y* data pairs are available.

Example 4.17: Given $x^2 + y^2 = r^2$, plot its graph for r = 1, 2, and 3 respectively.

Solution

The given function is an implicit function that can be reorganised as an explicit function as follows

$$x^2 + y^2 = r^2 \rightarrow y^2 = r^2 - x^2 \rightarrow y = \pm\sqrt{r^2 - x^2}.$$

Note both the positive and negative values of $\sqrt{r^2 - x^2}$ satisfy $x^2 + y^2 = r^2$.

By substituting r = 1, 2, and 3 into $y = \pm\sqrt{r^2 - x^2}$ respectively, three sets of *x*-*y* data pairs are obtained by using an interval of 0.1 for *x* (the smaller the better). Plotting these *x*-*y* data pairs on the *xy*-plane creates the graphs for the three incidents respectively shown in Figure 4.15. It can be seen that the implicit function $x^2 + y^2 = r^2$ represents circles with different radiuses determined by constant *r*, just like *r* = 1, 2, and 3 representing three circles with radius of 1, 2, and 3 units respectively.

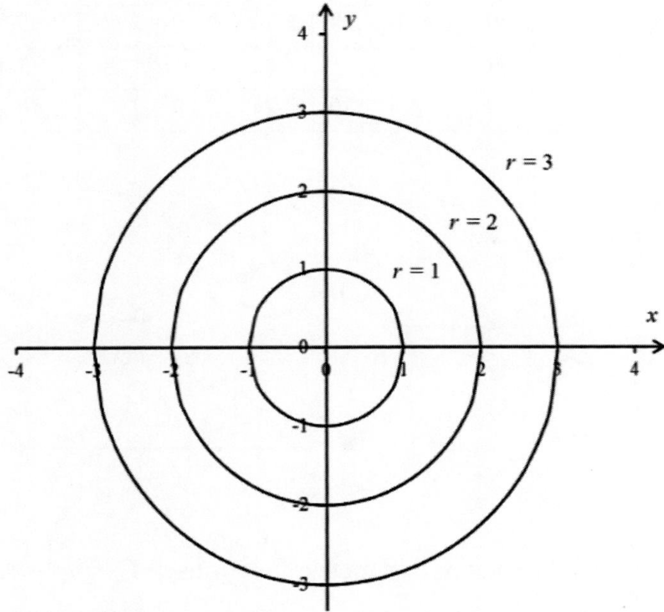

Figure 4.15: Graph of implicit function $x^2 + y^2 = r^2$ in Example 4.17

Parametric functions

If a relationship exists between x and y but such relationship is implicitly defined through another variable t commonly shared by both x and y, i.e.,

$$\boxed{\begin{cases} x = f(t) \\ y = g(t) \end{cases}}, \qquad (4.6)$$

the pair together defines a function called ***parametric function*** through parameter t. For example, the pair of $x = t + 2$ and $y = 2t^2 - 1$ together defines a parametric function.

Example 4.18: A parametric function is defined by $x = t - 2$ and $y = \sqrt{t^2 + 5}$. Draw its graph within the range [–3, 3] for t.

Solution

We can calculate the values for both x and y with respect to parameter t within [–3, 3] with a constant interval of 0.5. These associated data sets are tabulated in the table below (top 3 rows). We first plot both x and y against t to show their graphs individually (Figure 4.16a), and then use the x-y data pairs to create a graph of this parametric function on the xy-plane shown in Figure 4.16b.

t	-3	-2.5	-2	-1.5	-1	-0.5	0	0.5	1	1.5	2	2.5	3
x	-5	-4.5	-4	-4.5	-3	-2.5	-2	-1.5	-1	-0.5	0	0.5	1
y	4.74	4.35	3	2.69	2.45	2.29	2.24	2.29	2.45	2.69	3	4.35	4.74
x	-5	-4.5	-4	-4.5	-3	-2.5	-2	-1.5	-1	-0.5	0	0.5	1
$y = F(x)$	4.74	4.35	3	2.69	2.45	2.29	2.24	2.29	2.45	2.69	3	4.35	4.74

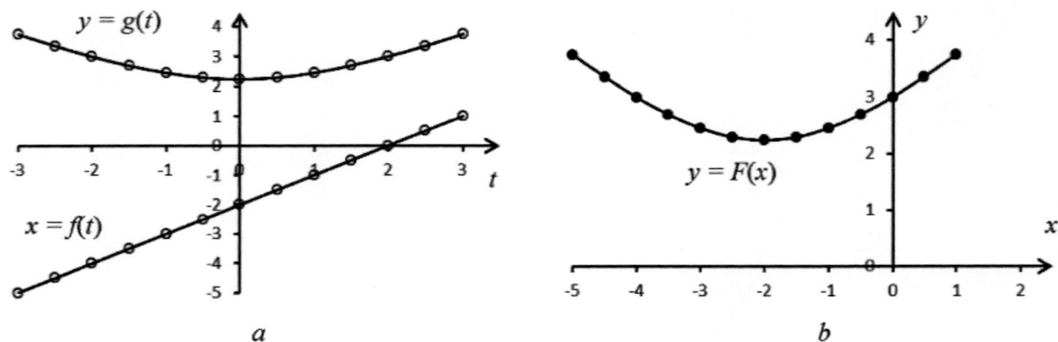

Figure 4.16: Graphs of parametric functions in Example 4.18

As the given parametric function is simple enough, we can do some simple substitutions as follows.

$$x = t - 2 \;\rightarrow\; t = x + 2;$$

$$y = \sqrt{t^2 + 5} \longrightarrow y = \sqrt{(x+2)^2 + 5} \longrightarrow y = \sqrt{x^2 + 4x + 9} = F(x).$$

This converts the given parametric function to an explicit function $y = F(x)$. We can use $y = F(x)$ to calculate x-y data pairs within [–5, 1] for x. The results are the same as their corresponding parametric values shown in the bottom two rows in the table above.

Note that not all parametric functions can be converted to their corresponding explicit functions.

4.2.4 Inverse functions

The concept of inverse functions

In generic function $y = f(x)$, x is the independent variable and y is the dependent variable by default. Usually we use values of x to determine values for y. For example, if we know the average speed of an airplane, we can use the travelling time to work out the distance the airplane has flown. In this case, time t is the independent variable and the distance is the dependent variable. On the other hand, given the average speed of a type of aircraft, if we want to estimate how long a one-way fly takes from Sydney to each of the other major cities in Australia, time t becomes the dependent variable to the independent variable distance d. This is because the distances from Sydney to other cities are known and vary city by city.

In mathematical terms, this becomes determining x from y given $y = f(x)$, which is generally expressed as $x = f^{-1}(y)$. As we are familiar with x as independent variable and y as the dependent variable, the notation is rewritten as

$$y = f^{-1}(x), \tag{4.7}$$

and called the ***inverse function*** to the original function $f(x)$.

The inverse function is derived from its original function. However, both are different functions and may have different characteristics. They are not necessarily matched one-to-one, and may be defined in different domains too.

Example 4.19: Given $y = f(x) = \frac{1}{5}x^3 + 1$, determine its inverse function and domain.

Solution

We first work out x from $y = f(x)$, and then exchange x and y to determine the inverse function.

$$\because y = f(x) = \frac{1}{5}x^3 + 1 \longrightarrow 5y = x^3 + 5 \longrightarrow x^3 = 5y - 5 \longrightarrow x = \sqrt[3]{5y - 5};$$

$$\therefore y = f^{-1}(x) = \sqrt[3]{5x - 5}.$$

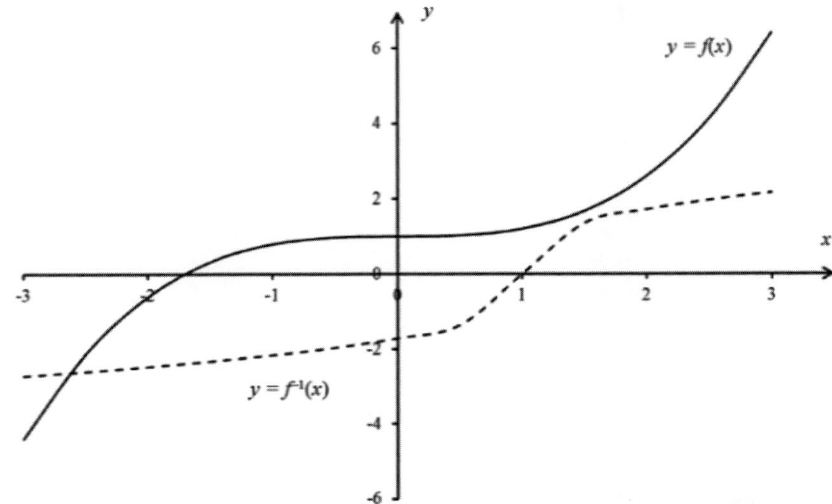

Figure 4.17: Graphs of the original and inverse functions in Example 4.19

This inverse function is defined in the whole x-domain, same as for the original function $f(x) = x^3 - 1$. Both functions are shown in Figure 4.17.

Example 4.20: Given $y = f(x) = x^2 - 2x - 3$, determine its inverse function and domain.

Solution

$$y = f(x) = x^2 - 2x - 3 = x^2 - 2x + 1 - 1 - 3 = (x^2 - 2x + 1) - 4 = (x-1)^2 - 4$$

$$y = (x-1)^2 - 4 \longrightarrow (x-1)^2 = y + 4 \longrightarrow x - 1 = \pm\sqrt{y+4} \longrightarrow x = 1 \pm \sqrt{y+4};$$

$$\therefore y_1 = f^{-1}(x) = 1 + \sqrt{x+4} \text{ and } y_2 = f^{-1}(x) = 1 - \sqrt{x+4};$$

$$\text{Domain}: x + 4 \ge 0 \longrightarrow x \ge -4.$$

The inverse function has two solutions $y_1 = f^{-1}(x) = 1 + \sqrt{x+4}$ and $y_2 = f^{-1}(x) = 1 - \sqrt{x+4}$ and both are defined in the x-domain $x \ge -4$. The original function $f(x) = x^2 - 2x - 3$ is defined in the whole x-domain. This is not a case of one-to-one match. These functions are shown in Figure 4.18.

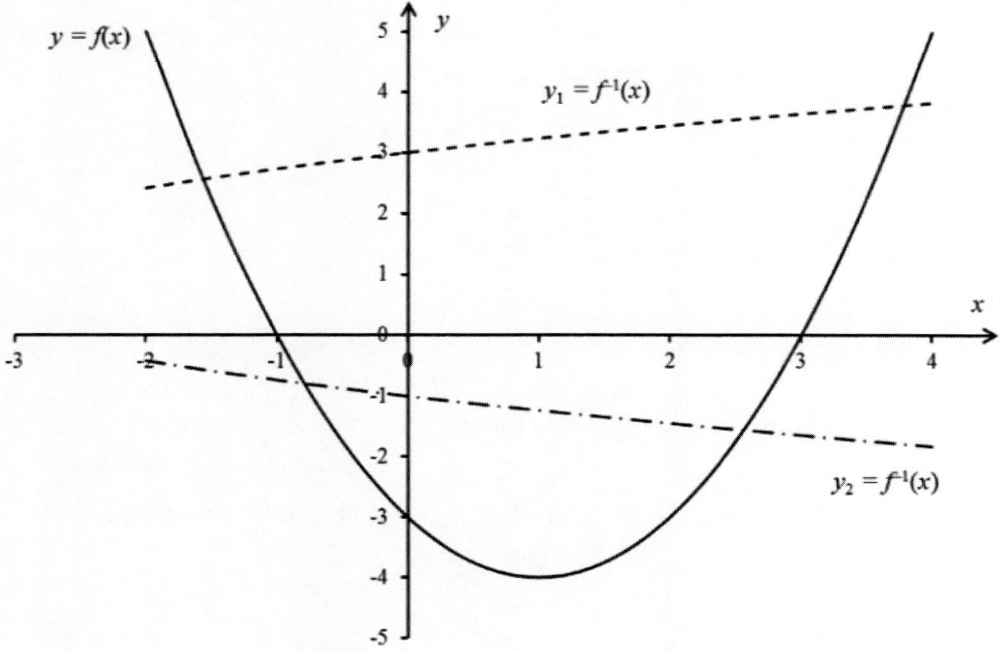

Figure 4.18: Graphs of the original and inverse functions in Example 4.20

Exercises 4.2

1. Given the following segmented function

$$y=\begin{cases}-x & -3\le x<-1\\ x^2 & -1\le x<1\\ x & 1\le x\le 3\end{cases},$$

plot the graph of this function.

2. Given periodic function $y=x^3$, $-1\le x\le 1$, plot its graph and find its period.

3. Given periodic function $y=1-x^2$, $-1\le x\le 1$, plot its graph and find its period.

4. Given $\frac{x^2}{16}+\frac{y^2}{9}=1$, plot its graph.

5. Given $\sqrt{x}-\sqrt{y}=1$, plot its graph.

6. A parametric function is defined by $x=2t-1$ and $y=t^2+1$. Draw its graph in [–3, 3] for t.

7. A parametric function is defined by $x=\sqrt{3t^2+4}$ and $y=3-2t$. Draw its graph in [–2, 2] for t.

8. A parametric function is defined by $x=2\sin t$ and $y=2\cos t$. Draw its graph in [–π, π] for t.

9. A parametric function is defined by $x=4\cos t$ and $y=3\sin t$. Draw its graph in [–π, π] for t.

10. Given $y=f(x)=2\sqrt{x}-1$, determine its inverse function and domain.

11. Given $y=f(x)=x^3-10$, determine its inverse function and domain.

4.3 Multivariable Functions and the Cartesian System

4.3.1 Basics of multivariable functions and the Cartesian system

The concept of multivariable functions

In an elementary function $y = f(x)$ or $f(x,y) = c$, the dependent variable y changes with respect to only one independent variable x. If a dependent variable u changes with respect to two or more independent variables, such as x_1, x_2, ... x_n, the relationship between the dependent variable and its independent variables is called a ***multivariable function***, written as $u = f(x_1, x_2, \ldots x_n)$ for explicit functions, or $f(x_1, x_2, \ldots x_n, u) = c$ for implicit functions. It is easy to understand that elementary functions are the special case of multivariable functions where there is only one independent variable.

The Grade Point Average (GPA) of a student is an example of multivariable functions:

$$GPA = f(n_{HD}, n_D, n_C, n_P, n_F)$$
$$= (w_{HD}n_{HD} + w_D n_D + w_C n_C + w_P n_P + w_F n_F) / (n_{HD} + n_D + n_C + n_P + n_F)$$

in which w_{HD}, w_D, w_C, w_P and w_F are constants of credit weights for grades High Distinction (HD), Distinction (D), Credit (C), Pass (P), and Fail (F) respectively whereas n_{HD}, n_D, n_C, n_P and n_F are the number of each of the grades HD, D, C, P and F received by a student during the course. Different students have different numbers for n_{HD}, n_D, n_C, n_P and n_F; thus n_{HD}, n_D, n_C, n_P and n_F are variables.

Locations of points in a 3-dementional (3D) space and the Cartesian system

Given a data pair (x, y), or the 2D coordinates (x, y), we can locate this point on the xy-plane using values of x and y. In a 3-dementional (3D) space, given a three-value set (x, y, z), or the 3D coordinate (x, y, z), we can locate this point in a 3D space using values of x, y and z. A 3D space can be defined by different coordinating systems. The ***Cartesian system*** is a commonly used system that uses three mutually perpendicular linear axes (x, y, z) to represent two horizontal directions and one vertical direction of the 3D space. As long as these three directions follow the *right-hand rule* illustrated in Figure 4.3, any combination of orientations constitutes a Cartesian system.

Figure 4.19 is the mostly used Cartesian system representing a 3D space. Regarding the three axes having the same scale, we can locate points A – J in this 3D space using values for x, y, and z respectively. To locate point $E(6,8,8)$, first find the intersection point of $x = 6$ units and $y = 8$ units on the xy-plane, i.e., $D(6,8,0)$, then move up vertically from point D for 8 units where point E is located. Another way to locate a point in this 3D space, for example $J(1,3,5)$, is to move from the origin $O(0,0,0)$ to the projected point $I(1,3,0)$ of point J on the xy-plane, then move up vertically 5 units from point I to find point J. The simplest way is to move straightway from the origin $O(0,0,0)$ to $J(1,3,5)$.

The 3D Cartesian system can also be seen as the intersection of three mutually perpendicular planes, the *xy*-plane, the *yz*-plane, and the *zx*-plane following the right-hand rule. Elementary function $y = f(x)$ is the projection of a 3D function onto the *xy*-plane.

The distance between two points in the Cartesian system

The distance between any two points $P_1(x_1, y_1, z_1)$ and $P_2(x_2, y_2, z_2)$ in the Cartesian system is determined by

$$\boxed{d = \sqrt{(x_2 - x_1)^2 + (y_2 - y_1)^2 + (z_2 - z_1)^2}}. \qquad (4.8)$$

This formula also applies to any two points on a 2D plane, in which the third coordinate (or the perpendicular dimension) is regarded as zero.

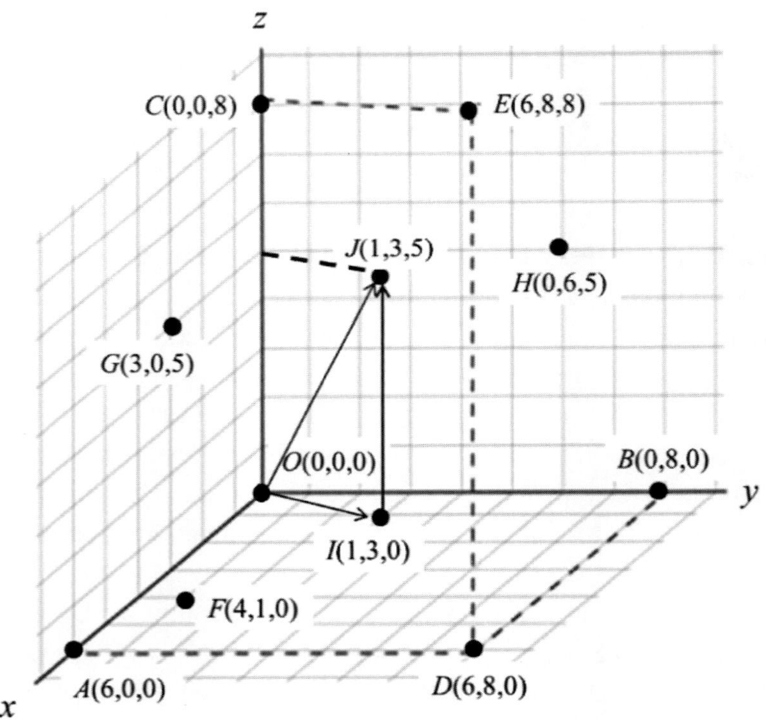

Figure 4.19: Locations of points in the 3D Cartesian space

Example 4.21: Referring to Figure 4.19, determine the distance between the following two points:

a) $O(0,0,0)$, $J(1,3,5)$; b) $A(6,0,0)$, $B(0,8,0)$; c) $A(6,0,0)$, $G(3,0,5)$;

d) $C(0,0,8)$, $H(0,6,5)$; e) $E(6,8,8)$, $J(1,3,5)$.

Solution

a) $d[O,J]=\sqrt{(x_2-x_1)^2+(y_2-y_1)^2+(z_2-z_1)^2}=\sqrt{(1-0)^2+(3-0)^2+(5-0)^2}=\sqrt{1+9+25}=\sqrt{35}$

b) $d[A,B]=\sqrt{(x_2-x_1)^2+(y_2-y_1)^2+(z_2-z_1)^2}=\sqrt{(0-6)^2+(8-0)^2+(0-0)^2}$

$=\sqrt{36+64+0}=\sqrt{100}=10$ (A and B are on the xy - plane.)

c) $d[A,G]=\sqrt{(x_2-x_1)^2+(y_2-y_1)^2+(z_2-z_1)^2}=\sqrt{(3-6)^2+(0-0)^2+(5-0)^2}$

$=\sqrt{9+0+25}=\sqrt{34}$ (A and G are on the zx - plane.)

d) $d[C,H]=\sqrt{(x_2-x_1)^2+(y_2-y_1)^2+(z_2-z_1)^2}=\sqrt{(0-0)^2+(6-0)^2+(5-8)^2}$

$=\sqrt{0+36+9}=\sqrt{45}=3\sqrt{5}$ (C and H are on the yz - plane.)

e) $d[E,J]=\sqrt{(x_2-x_1)^2+(y_2-y_1)^2+(z_2-z_1)^2}=\sqrt{(1-6)^2+(3-8)^2+(5-8)^2}=\sqrt{25+25+9}=\sqrt{59}$

4.3.2 3D functions and graphs in the Cartesian system

A relationship between a dependent variable and TWO independent variables defines a 3D function, i.e., $z = f(x,y)$ as an explicit function, or $f(x,y,z) = c$ as an implicit function. A 3D function can be displayed as a 3D graph in the Cartesian system. Similar to plotting function $y = f(x)$ on the xy-plane using the x-y data pairs, we can use $z = f(x,y)$ to calculate an array of three-value datasets (x, y, z), and then plot each of these datasets to its location in the 3D space similar to what shown in Figure 4.19. All these datasets together should form a surface (rather than a curve) in the 3D space.

Example 4.22: Plot the 3D graph of $z=2x+y-3$ in ranges $0 \le x \le 10$ and $0 \le y \le 10$ in the Cartesian system.

Solution

This is a 3D linear function and simple enough to calculate an array of (x, y, z) datasets as listed in the table below. A 3D surface (a plane) is created using this array of (x, y, z) datasets shown in Figure 4.20. Note this graph follows the 'right-hand rule'.

		y					
	z	0	2	4	6	8	10
	0	-3	-1	1	3	5	7
	2	1	3	5	7	9	11
x	4	5	7	9	11	13	15
	6	9	11	13	15	17	19
	8	13	15	17	19	21	23
	10	17	19	21	23	25	27

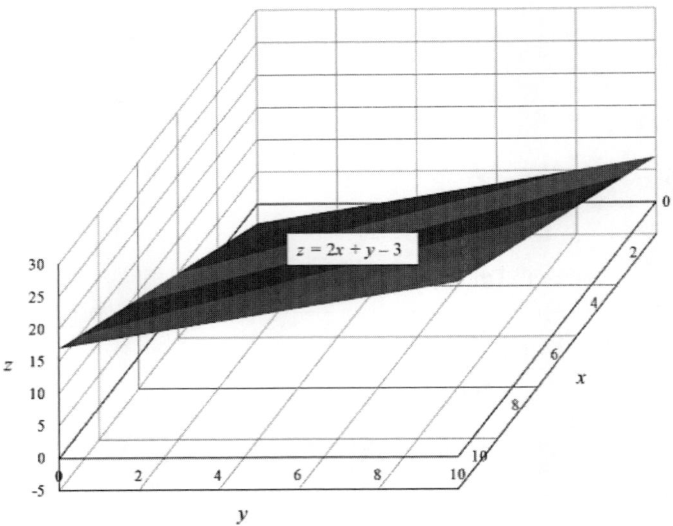

Figure 4.20: The 3D graph for $z = 2x + y - 3$

Example 4.23: Plot the 3D graph of $z = x^2 + y^2$ in ranges $-3 \le x \le 3$ and $-3 \le y \le 3$ in the Cartesian system.

Solution

For complicated 3D functions, it is more appropriate to use software to create the 3D graphs. Figure 4.21 is the graph of $z = x^2 + y^2$ created using Excel following the 'right-hand rule'.

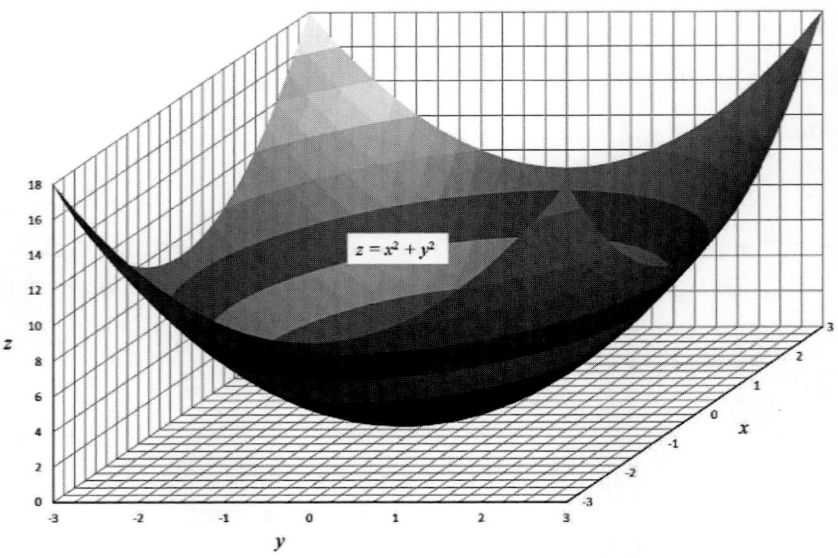

Figure 4.21: The 3D graph for $z = x^2 + y^2$

Exercises 4.3

1. In the 3D Cartesian space, determine the distance between the following two points.

   a) $O(0,0,0)$, $J(-1,3,-5)$;
   b) $A(2,0,1)$, $B(0,3,-1)$;
   c) $A(-3,1,2)$, $G(0,0,-3)$;
   d) $C(0,2,0)$, $H(2,0,-5)$;
   e) $E(4,3,-1)$, $J(1,5,2)$.

2. Plot the 3D graph of $z^2 = x^2 + y^2$ in the ranges $-2 \le x \le 2$ and $-2 \le y \le 2$ in the Cartesian system.

Chapter 4: Exercises Answers

Exercises 4.1

2. a) $[-2,\infty)$; b) $(-\infty,\infty)$; c) $\begin{cases} x \neq 2 \\ x \neq 1 \end{cases}$; d) $[0,\infty)$.

3. a) $[2,\infty)$; b) $[2,\infty)$; c) $[2,\infty)$; d) $[2,\infty)$; e) $[2,\infty)$ or $[0,\infty)$.

4. a) $g(x)=x^3-9$; b) $g(x)=x^3+3x^2+3x+1$; c) $g(x)=\frac{1}{3}x^3$.

5. a) $g(x)=-\frac{1}{5}x^3-3$, b) $g(x)=-\frac{1}{5}x^3+3$.

Exercises 4.2

10. $y=f^{-1}(x)=\frac{1}{4}(x+1)^2$. The domain is the whole real umber line as x can take any number.

11. $y=f^{-1}(x)=\sqrt[3]{x+10}$. The domain is the whole real umber line as x can take any number.

Exercises 4.3

1. a) $d_{OJ}=\sqrt{35}$, b) $d_{AB}=\sqrt{17}$, c) $d_{AG}=\sqrt{35}$, d) $d_{CH}=\sqrt{33}$, e) $d_{EJ}=\sqrt{22}$

CHAPTER 5

5 Polynomial Functions

CHAPTER OBJECTIVES

- Introduce the concept of polynomial functions
- Study linear functions and properties
- Study quadratic functions and properties
- Use quadratic functions to solve problems
- Introduce higher order polynomials and properties

Essential statements on polynomial functions:

- Polynomial functions are the mostly used functions in applied mathematics.
- Two known points can uniquely define a linear function or a straight line.
- A quadratic function is symmetric about the axis of symmetry and has either a maximum or a minimum value (point).
- Quadratic functions have a wide range of real-world applications.

Key topics:

- Slope or gradient of linear function of
- Different forms of linear functions
- Perpendicular and parallel lines
- Vertex of a quadratic function
- Solutions of quadratic equations

Flowchart of mathematical knowledge building

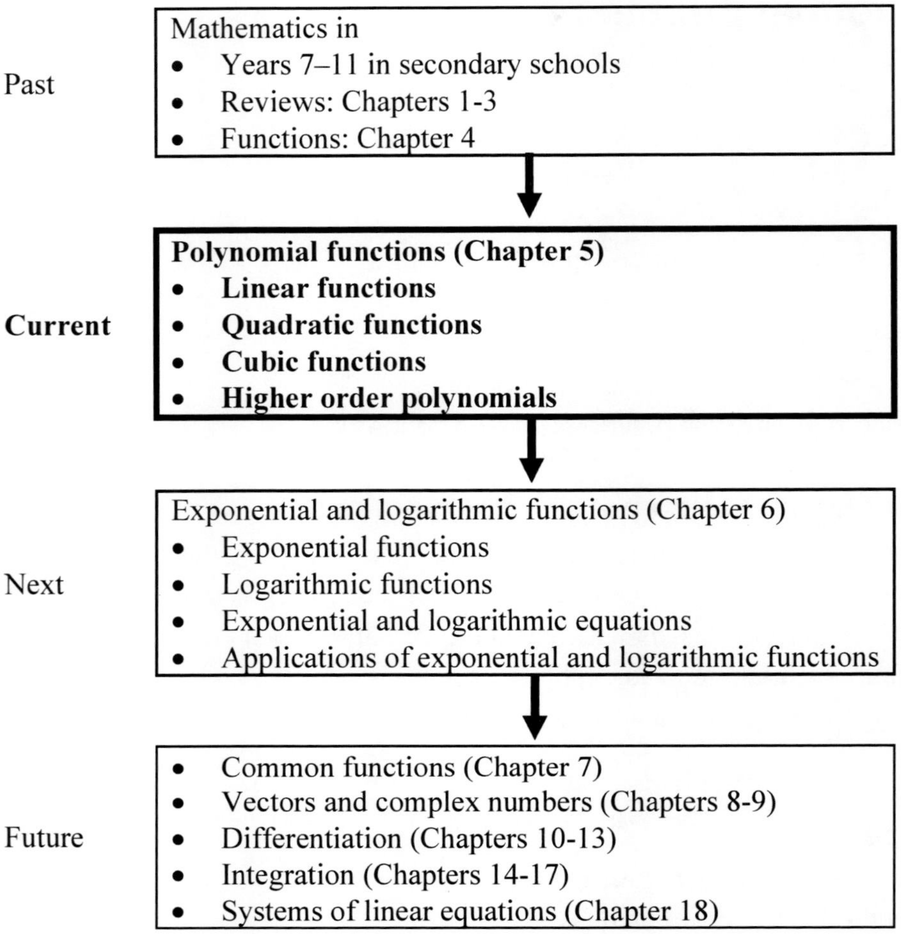

Past	Mathematics in • Years 7–11 in secondary schools • Reviews: Chapters 1-3 • Functions: Chapter 4
	↓
Current	**Polynomial functions (Chapter 5)** • **Linear functions** • **Quadratic functions** • **Cubic functions** • **Higher order polynomials**
	↓
Next	Exponential and logarithmic functions (Chapter 6) • Exponential functions • Logarithmic functions • Exponential and logarithmic equations • Applications of exponential and logarithmic functions
	↓
Future	• Common functions (Chapter 7) • Vectors and complex numbers (Chapters 8-9) • Differentiation (Chapters 10-13) • Integration (Chapters 14-17) • Systems of linear equations (Chapter 18)

Chapter 5: Polynomial Functions

5.1 Linear Functions

The concept of polynomial functions

Polynomial functions, or simply polynomials, are defined as

$$\boxed{y = a_0 + a_1x + a_2x^2 + a_3x^3 + \cdots + a_nx^n \quad (n : non-negative\ integer)}, \tag{5.1}$$

where $a_0 \sim a_n$ are called the coefficients of the polynomial and can be negative, zero or positive real numbers. The highest exponent in a polynomial determines the *order* of the polynomial. For example, in $y = 2x^5 - 3x^4 + x^2 - 12$, the highest exponent is 5; hence it is called a 5th-order polynomial. If the highest exponent is zero, the polynomial becomes a constant function that was discussed as a special function in Chapter 4.

The *first-order polynomial* (commonly known as the *linear function*) and the second-order polynomial (commonly known as the *quadratic function*) are widely used in real-world applications, followed by the third-order polynomial (commonly known as the *cubic function*). Therefore, we focus on the linear and quadratic functions with a brief introduction to the cubic functions in this chapter.

5.1.1 Expressions of linear functions

5.1.1.1 The standard form of linear functions

Interceptions of linear functions

In formula (5.1), the polynomial becomes linear functions if n = 1, i.e., $y = a_0 + a_1x$. More commonly linear functions are expressed as

$$\boxed{y = ax + b,} \tag{5.2}$$

where a and b are constants. A linear function represents a straight line on the xy-plane. Logically a straight line can be uniquely drawn by connecting any two known points in the line. Thus in formula (5.2), let $x = 0$, $y = b$; let $y = 0$, $x = -b/a$ ($a \neq 0$). This means that points $A(0, b)$ and $B(-b/a, 0)$ are in the linear function $y = ax + b$, and hence we can use points A and B to create its graph on the xy-plane. For any linear function expressed by formulas (5.2), y must equal b if $x = 0$. Therefore, b is called the y-intercept of $y = ax + b$, and located at point $(0, b)$ in the line. Similarly, x must equal $-b/a$ if y = 0. Therefore, $-b/a$ is called the x-intercept of $y = ax + b$, and located at point $(-b/a, 0)$ in the line.

For example, given two lines $y = 2x + 3$ and $y = -x - 2$, their graphs, x-intercepts and y-intercepts can be determined as follows.

For $y = 2x + 3$, we find two points in this line, and then draw a straight line across these two points to create its graph.

$$x = 0,\ y = 2 \times 0 + 3 = 3 \longrightarrow B(0,3);$$
$$x = 1,\ y = 2 \times 1 + 3 = 5 \longrightarrow C(1,5).$$

A straight line crossing points *B* and *C* (with solid dots) is drawn in Figure 5.1.

Similarly we can find two points in $y = -x - 2$,

$$x = 0,\ y = 0 - 2 = -2 \quad \rightarrow \quad D(0,-2);$$
$$x = 2,\ y = -2 - 2 = -4 \quad \rightarrow \quad F(2,-4).$$

Another straight line across points *D* and *F* (with circles) is drawn in Figure 5.1 as well.

Point *B*(0, 3) shows the *y*-intercept for $y = 2x + 3$, i.e., $y = 2x + 3$ crosses the *y*-axis 3 units above the origin. The *x*-intercept for $y = 2x + 3$ is $x = -b/a = -3/2 = -1.5$, i.e., $y = 2x + 3$ crosses the *x*-axis 1.5 units on the left of the origin. Point *D*(0, –2) shows the *y*-intercept for $y = -x - 2$, i.e., $y = -x - 2$ crosses the *y*-axis 2 units below the origin. The *x*-intercept for $y = -x - 2$ is $x = -(-2)/(-1) = -2$ so $y = -x - 2$ crosses the *x*-axis 2 units on the left of the origin.

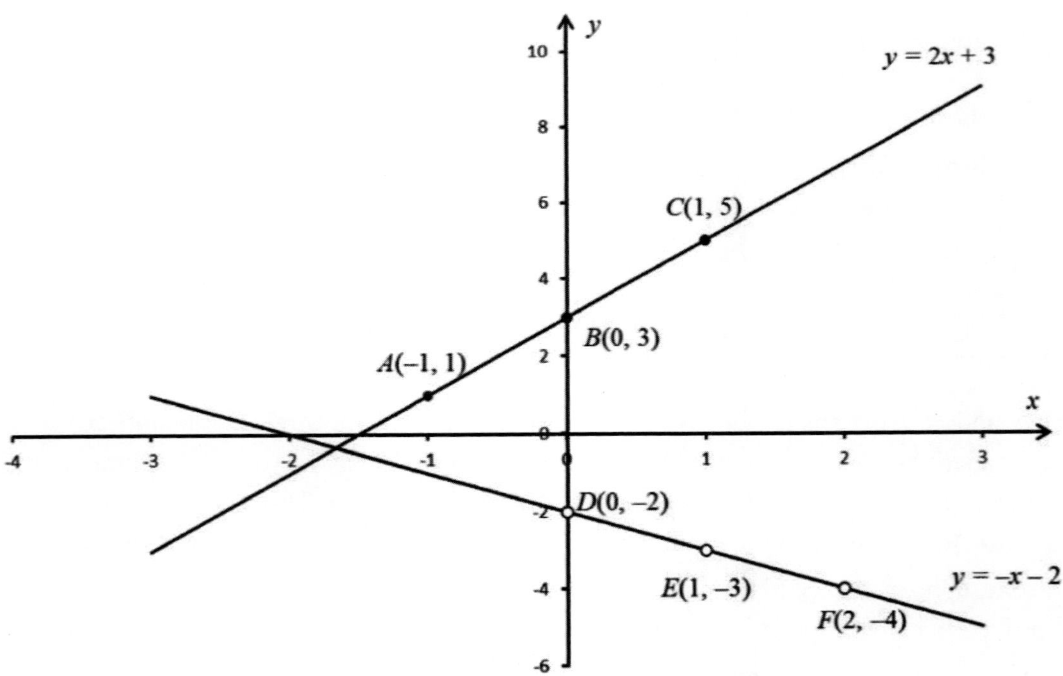

Figure 5.1: Graphs for linear functions $y = 2x + 3$ and $y = -x - 2$

The slope or gradient of linear functions

In Figure 5.1, y increases with an increase in x by $y = 2x + 3$ whereas y decreases with an increase in x by $y = -x - 2$. Such difference is determined by the coefficient a in $y = ax + b$, and *hence a is called the slope or gradient of linear function* $y = ax + b$. The slope or gradient of linear function $y = ax + b$ determines the rate of change in y with respect to any change in x, generally described by

If $a > 0 \rightarrow$ y increases positively with an increase in x or moving upwards along the x-axis;
If $a < 0 \rightarrow$ y decreases with an increase in x or moving downwards along the x-axis.

In $y = 2x + 3$, $a = 2 > 0$ or with a positive gradient, hence y increases with any increase in x or moves upwards along the x-axis. In $y = -x - 2$, $a = -1 < 0$ or with a negative gradient, hence y decreases with any increase in x or moves downwards along the x-axis.

The magnitude of slope $|a|$ determines how fast the line moves upwards (for positive a) or downwards (for negative a). The larger the magnitude; the faster the line moves. In $y = 2x + 3$, from $B(0, 3)$ to $C(1, 5)$, x moved 1 unit but y increased $5 - 3 = 2$ units. In $y = -x - 2$, if x moved 1 unit from $D(0, -2)$ to $E(1, -3)$, y decreased by 1 unit only. This is because the magnitude of slope for $y = 2x + 3$ is double that for $y = -x - 2$.

The slope (or gradient) of a linear function is constant, i.e., for any section in a straight line, its slope keeps unchanged. In other words, given two known points $A(x_1, y_1)$ and $B(x_2, y_2)$ in the same line, we can uniquely determine its slope a by

$$\boxed{a = \frac{change\ in\ y}{change\ in\ x} = \frac{y_2 - y_1}{x_2 - x_1}}. \qquad (5.3)$$

Example 5.1: Verify formula (5.3) by determining the slopes for the two lines in Figure 5.1.

Solution

For $y = 2x + 3$, we can use any two points from $A(-1, 1)$, $B(0, 3)$ and $C(1, 5)$ to obtain its slope using formula (5.3)

$$A(-1,1) \text{ and } B(0,3) \longrightarrow a = \frac{y_2 - y_1}{x_2 - x_1} = \frac{3-1}{0-(-1)} = \frac{2}{1} = 2$$

$$A(-1,1) \text{ and } C(1,5) \longrightarrow = \frac{y_2 - y_1}{x_2 - x_1} = \frac{5-1}{1-(-1)} = \frac{4}{2} = 2$$

$$B(0,3) \text{ and } C(1,5) \longrightarrow a = \frac{y_2 - y_1}{x_2 - x_1} = \frac{5-3}{1-0} = \frac{2}{1} = 2.$$

The slope is fixed as $a = 2$ for $y = 2x + 3$.

For $y=-x-2$, we can use any two points from $D(0, -2)$, $E(1, -3)$ and $F(2, -4)$ to obtain its slope using formula (5.3)

$$D(0,-2) \text{ and } E(1,-3) \longrightarrow a=\frac{y_2-y_1}{x_2-x_1}=\frac{-3-(-2)}{1-0}=\frac{-3+2}{1}=-1$$

$$D(0,-2) \text{ and } F(2,-4) \longrightarrow a=\frac{y_2-y_1}{x_2-x_1}=\frac{-4-(-2)}{2-0}=\frac{-4+2}{2}=\frac{-2}{2}=-1$$

$$E(1,-3) \text{ and } F(2,-4) \longrightarrow a=\frac{y_2-y_1}{x_2-x_1}=\frac{-4-(-3)}{2-1}=\frac{-4+3}{1}=-1.$$

The slope is fixed as $a=-1$ for $y=-x-2$.

Since $y=ax+b$ is made up by its slope a and y-intercept b, $y=ax+b$ is also commonly known as the *slope-intercept form* of linear functions subject to $a \neq 0$.

Example 5.2: A straight line crosses points $A(2, -3)$ and $B(-1, 3)$. Determine this linear function by the slope-intercept form.

Solution

Assuming the function being $y=ax+b$, it goes through points A and B, hence both should satisfy $y=ax+b$, i.e.,

$$A(2,-3) \longrightarrow -3=a\times 2+b \longrightarrow 2a+b=-3$$

$$B(-1,3) \longrightarrow 3=a\times(-1)+b \longrightarrow -a+b=3.$$

These form a system of linear equations, which can be solved using elimination as follows:

$$\begin{cases} 2a+b=-3 \\ -a+b=3 \end{cases} \xrightarrow[\times 2]{\rightarrow} \begin{cases} 2a+b=-3 \\ -2a+2b=6 \end{cases} \xrightarrow{(+)} 3b=3 \rightarrow b=1.$$

$$-a+b=3 \longrightarrow a=b-3=1-3=-2.$$

Therefore, the linear function is $y=-2x+1$. It is going downwards along the x-axis as its slope is negative. You can verify the correctness of this function using points A and B.

5.1.1.2 Other forms of linear functions

The point-slope form of linear functions

Now that the slope of a line keeps unchanged in any section of the line, we can use this property to derive another expression for linear functions if we are given the slope a and one known point $A(x_1, y_1)$ in the line. Suppose another point $P(x, y)$ is also in the line (Figure 5.2). By formula (5.3), we should have

$$a = \frac{y - y_1}{x - x_1} \text{ or}$$

$$\boxed{y - y_1 = a(x - x_1)}. \tag{5.4}$$

Formula (5.4) is called the *point-slope form* of linear functions.

Example 5.3: If the slope of a linear function is 3 and the line passes through point $A(-2, 1)$, find the expression of this linear function.

Solution

As $a = 3$, $x_1 = -2$ and $y_1 = 1$, by formula (5.4),

$$y - 1 = 3[x - (-2)] \longrightarrow y - 1 = 3(x + 2) \longrightarrow y - 1 = 3x + 6 \longrightarrow y = 3x + 7.$$

By substituting $x = -2$ into $y = 3x + 7$, $y = 3 \times (-2) + 7 = -6 + 7 = 1$, which verifies the correctness of this linear function.

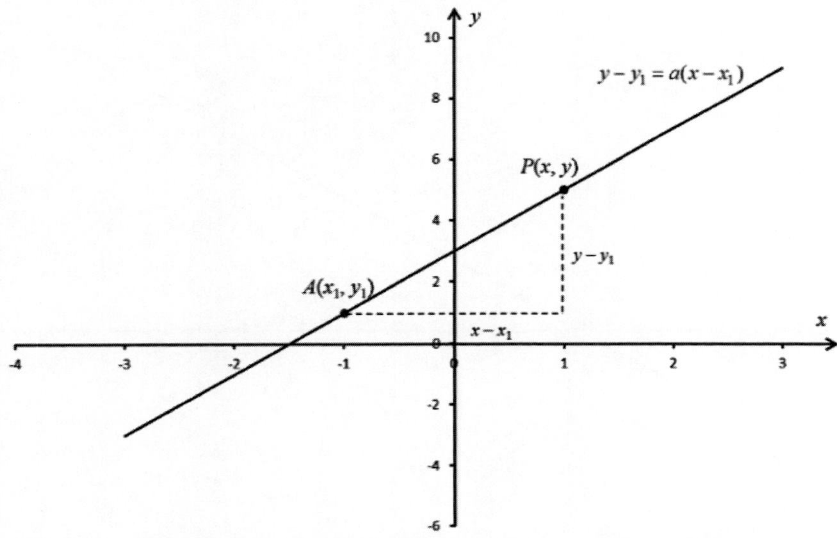

Figure 5.2: Graph for the point-slope form of linear functions

The two-point form of linear functions

Since the slope of a line keeps unchanged in any section over the line, we can also derive another form for linear functions if given two known points $A(x_1, y_1)$ and $B(x_2, y_2)$ in the line. Suppose another point $P(x, y)$ is also in the line (Figure 5.3). By formula (5.3), we should have

$$a = \frac{y_2 - y_1}{x_2 - x_1} = \frac{y - y_1}{x - x_1} = \frac{y - y_2}{x - x_2}.$$

This equals either of the two forms below:

$$\frac{y_2 - y_1}{x_2 - x_1} = \frac{y - y_1}{x - x_1} \text{ or } \frac{y_2 - y_1}{x_2 - x_1} = \frac{y - y_2}{x - x_2}$$

or

$$\boxed{y - y_1 = \frac{y_2 - y_1}{x_2 - x_1}(x - x_1) \text{ or } y - y_2 = \frac{y_2 - y_1}{x_2 - x_1}(x - x_2).} \tag{5.5}$$

Both expressions in formula (5.5) are called the *two-point form* of linear functions.

$$\begin{array}{|l|} y - y_1 = \dfrac{y_2 - y_1}{x_2 - x_1}(x - x_1) \leftarrow \text{parallel to the } x\text{-axis} \\ x - x_1 = \dfrac{x_2 - x_1}{y_2 - y_1}(y - y_1) \leftarrow \text{parallel to the } y\text{-axis.} \end{array} \tag{5.6}$$

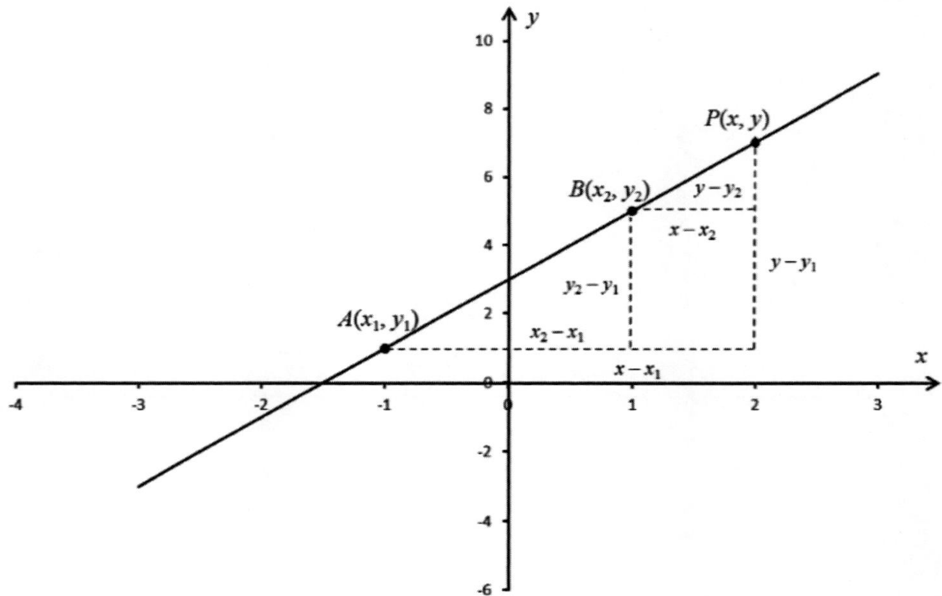

Figure 5.3: Graph for the two-point form of linear functions

The two-point form of linear functions has a unique advantage in determining special linear functions parallel to either the x-axis or the y-axis. Any linear function parallel to the x-axis implies a zero slope ($a = 0$) in the slope-intercept form, i.e., $y = ax + b \rightarrow y = b$. It is a line parallel to the x-axis and crosses the y-intercept (Figure 5.4).

Any linear function parallel to the y-axis implies an infinite value for the slope ($a = \pm\infty$) in the slope-intercept form, but this cannot be expressed in a normal formula. The two-point form (5.5) can resolve these special cases in a slightly twisted form as follows.

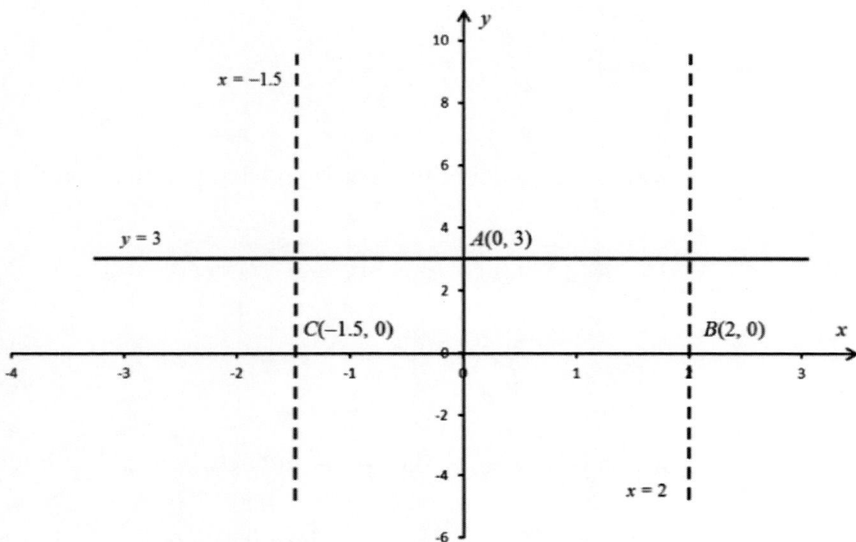

Figure 5.4: Linear functions parallel to the x-axis ($y = b = 3$) and the y-axis ($x = 2$ and $x = -1.5$)

Example 5.4: A linear function passes points $A(0, 3)$ and $B(2, 1)$. Find the expression of this linear function.

Solution

As $x_1 = 0$ and $y_1 = 3$, and $x_2 = 2$ and $y_2 = 1$, by formula (5.5),

$$y - y_1 = \frac{y_2 - y_1}{x_2 - x_1}(x - x_1) \longrightarrow y - 3 = \frac{1-3}{2-0}(x - 0) \longrightarrow y - 3 = -x \longrightarrow y = -x + 3$$

or

$$y - y_2 = \frac{y_2 - y_1}{x_2 - x_1}(x - x_2) \longrightarrow y - 1 = \frac{1-3}{2-0}(x - 2) \longrightarrow y - 1 = -(x - 2) \longrightarrow y = -x + 3.$$

Both come with the same function $y = -x + 3$.

Example 5.5: Find the expression of each of the following linear functions:

a) The linear function crosses points $A(0, 3)$ and $D(3, 3)$;
b) The linear function crosses points $B(2, 0)$ and $E(2, 6)$;
c) The linear function crosses points $C(-1.5, 0)$ and $F(-1.5, -4)$.

Solution

a) As $x_1 = 0$ and $y_1 = 3$, and $x_2 = 3$ and $y_2 = 3$, by formula (5.6),

$$y - y_1 = \frac{y_2 - y_1}{x_2 - x_1}(x - x_1) \longrightarrow y - 3 = \frac{3-3}{3-0}(x-0) \longrightarrow y - 3 = 0 \longrightarrow y = 3.$$

This linear function is $y = 3$, which is parallel to the x-axis and across $A(0, 3)$ in Figure 5.4.

b) As $x_1 = 2$ and $y_1 = 0$, and $x_2 = 2$ and $y_2 = 6$, by formula (5.6),

$$x - x_1 = \frac{x_2 - x_1}{y_2 - y_1}(y - y_1) \longrightarrow x - 2 = \frac{2-2}{6-0}(y-0) \longrightarrow x - 2 = 0 \longrightarrow x = 2.$$

This linear function is $x = 2$, which is parallel to the y-axis and across $B(2, 0)$ in Figure 5.4.

c) As $x_1 = -1.5$ and $y_1 = 0$, and $x_2 = -1.5$ and $y_2 = -4$, by formula (5.6),

$$x - x_1 = \frac{x_2 - x_1}{y_2 - y_1}(y - y_1) \longrightarrow x - (-1.5) = \frac{-1.5-(-1.5)}{-4-0}(y-0) \longrightarrow x + 1.5 = 0 \longrightarrow x = -1.5.$$

This linear function is $x = -1.5$, parallel to the y-axis and crossing $C(-1.5, 0)$ in Figure 5.4.

The general form of linear functions

If we express a linear function using an implicit function, it becomes the *general form of linear functions* as

$$\boxed{\alpha x + \beta y = \gamma \quad \text{or} \quad \frac{\alpha}{\gamma}x + \frac{\beta}{\gamma}y = 1 \quad \text{or} \quad \underline{Ax + By = 1}}, \qquad (5.7)$$

where α, β, γ ($\neq 0$), A and B are constants. Formula (5.7) can be rewritten as

$$\boxed{Ax + By = 1 \;\rightarrow\; \underline{\frac{x}{c} + \frac{y}{b} = 1}}, \qquad (5.8)$$

where c and b are constants. Formula (5.8) is also called the *two-intercept form of linear functions* as c is the x-intercept and b is the y-intercept of the linear function.

Example 5.6: Use formulas (5.7) and (5.8) to find the linear function that passes through points $C(0, 3)$ and $D(2, 1)$ respectively.

Solution

As both points C and D are in the line, substitute both points into $Ax + By = 1$,

$$A \times 0 + B \times 3 = 1 \longrightarrow 3B = 1 \longrightarrow B = \frac{1}{3},$$

$$A \times 2 + B \times 1 = 1 \longrightarrow 2A + B = 1 \longrightarrow 2A + \frac{1}{3} = 1 \longrightarrow 2A = 1 - \frac{1}{3} = \frac{2}{3} \longrightarrow A = \frac{1}{3},$$

$$Ax + By = 1 \longrightarrow \frac{x}{3} + \frac{y}{3} = 1 \longrightarrow x + y = 3 \longrightarrow y = -x + 3.$$

Substitute both points C and D into formula (5.8),

$$\frac{0}{c} + \frac{3}{b} = 1 \longrightarrow \frac{3}{b} = 1 \longrightarrow b = 3,$$

$$\frac{2}{c} + \frac{1}{b} = 1 \longrightarrow \frac{2}{c} + \frac{1}{3} = 1 \longrightarrow \frac{2}{c} = 1 - \frac{1}{3} \longrightarrow \frac{2}{c} = \frac{2}{3} \longrightarrow c = 3,$$

$$\frac{x}{3} + \frac{y}{3} = 1 \longrightarrow x + y = 3 \longrightarrow y = -x + 3.$$

These two are the same as that obtained in Example 5.4 using the two-point form.

As linear functions may appear in different forms, Table 5.1 summaries these forms along with the properties of each of these forms. In real applications, we are likely to use a mixed approach that may not be clearly identified as any single method listed in this table.

Table 5.1 Forms and features of linear functions

Form	Expression and feature
Standard or slope-intercept	$y = ax + b$, slope $a \neq 0$, y-intercept: b
Point-slope	$y - y_1 = a(x - x_1)$. If $a = 0$, a line $y = y_1$ parallel to the x-axis.
Two-point	$y - y_1 = \frac{y_2 - y_1}{x_2 - x_1}(x - x_1)$ or $y - y_2 = \frac{y_2 - y_1}{x_2 - x_1}(x - x_2)$
General	$Ax + By = 1$
Two-intercept	$\frac{x}{c} + \frac{y}{b} = 1$, x-intercept: c, y-intercept: b
A line parallel to the x-axis	$y = b \rightarrow b$: y-intercept
A line parallel to the y-axis	$x = c \rightarrow c$: x-intercept

5.1.2 Properties of linear functions

5.1.2.1 Parallel and perpendicular lines

Parallel lines

All linear functions that share the same slope but with different y-intercepts are parallel to each other. If the slope is zero, the linear function represents a series of horizontal lines with respect to different values of y-intercept b. If the slope is infinite, the linear function represents a series of vertical lines with respect to different values of x-intercept c.

Example 5.7: Plot the following linear functions on a graph to observe the parallel property.

a) Linear functions $y = 2x - 5$, $y = 2x$, $y = 2x + 7$;
b) Linear functions $y = -3x + 5$, $y = -3x$, $y = -3x - 4$;

Solution

a) We need two points in each line to draw a line. Let us choose $x_1 = 0$ and $x_2 = 1$ to locate the corresponding values of y by $y = 2x - 5$, $y = 2x$, $y = 2x + 7$ respectively.

$$x_1 = 0 \longrightarrow y = 2x - 5 = -5, \quad y = 2x = 0, \quad y = 2x + 7 = 7;$$
$$x_2 = 1 \longrightarrow y = 2x - 5 = -3, \quad y = 2x = 2, \quad y = 2x + 7 = 9.$$

We have (0, –5) and (1, –3) for $y = 2x - 5$, (0, 0) and (1, 2) for $y = 2x$, and (0, 7) and (1, 9) for $y = 2x + 7$. These lines are drawn in Figure 5.5 (solid lines). Since they share the same slope 2, they are upward and parallel to each other and crossing the y-axis at –5, 0, and 7 respectively.

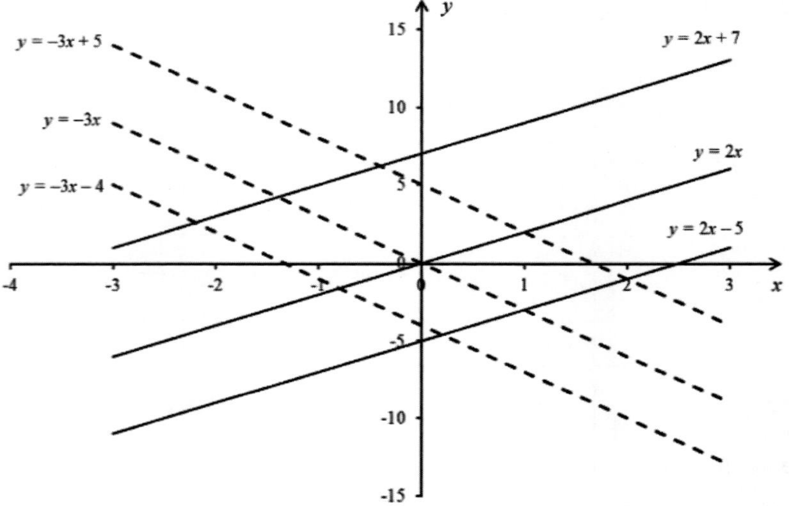

Figure 5.5: Graphs of parallel linear functions in Example 5.7.

b) Similarly we choose $x_1 = 0$ and $x_2 = 1$ to locate the corresponding values of y by $y=-3x+5,\ y=-3x,\ y=-3x-4$ respectively,

$$x_1 = 0 \longrightarrow y=-3x+5=5,\ \ y=-3x=0,\ \ y=-3x-4=-4;$$
$$x_2 = 1 \longrightarrow y=-3x+5=2,\ \ y=-3x=-3,\ \ y=-3x-4=-7.$$

We have (0, 5) and (1, 2) for $y=-3x+5$, (0, 0) and (1, –3) for $y=-3x$, and (0, –4) and (1, –7) for $y=-3x-4$. These lines are drawn in Figure 5.5 (dashed). All three share the same slope –3, so they are downward, parallel to each other, and crossing the y-axis at 5, 0, and –4 respectively.

Perpendicular lines

If two lines are perpendicular to each other, the product of the two slopes of these linear functions must be –1, i.e., $y = ax + b$ is perpendicular to $y = cx + d$ if

$$\boxed{ac=-1,\ \text{or}\ c=-\frac{1}{a},\ \text{or}\ a=-\frac{1}{c}.} \tag{5.9}$$

Example 5.8: We know the following two pairs of linear functions are perpendicular to each other as shown in Figure 5.6. Use these two pairs to check the relationship (5.9).

a) Linear functions $y=x$ and $y=-x$;
b) Linear functions $y=-2x+1$ and $y=0.5x+3$;

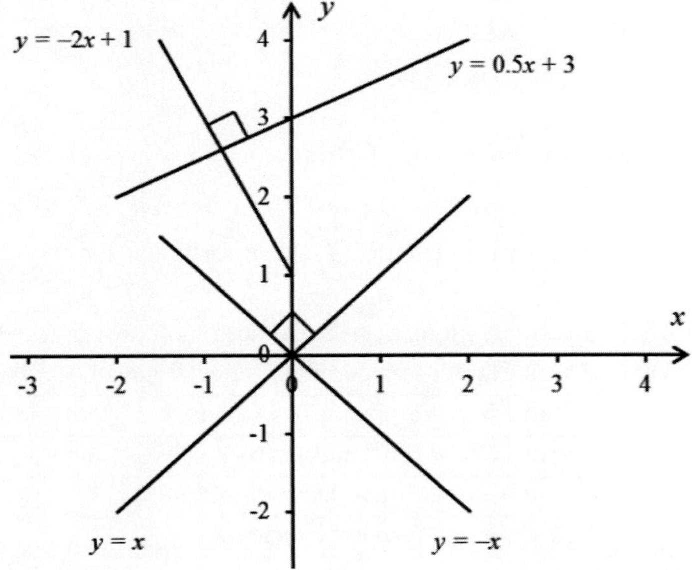

Figure 5.6: Graphs of perpendicular linear functions in Example 5.8.

Solution

a) The slopes of $y = x$ and $y = -x$ are $a = 1$ and $c = -1$ respectively. Thus $ac = 1\times(-1) = -1$, which verifies the relationship (5.9).

b) The slopes of $y = -2x + 1$ and $y = 0.5x + 3$ are $a = -2$ and $c = 0.5$ respectively. Therefore, we have $ac = (-2)\times 0.5 = -1$, which proves the relationship (5.9) too.

Example 5.9: One straight line crosses (1, 2) and (2, 5). Find

a) the expression of this linear function;
b) the expression of the linear function that is normal to this line and passes through (0, 2).

Solution

a) Let L1 be the line crossing (1, 2) and (2, 5). We can use the two-point form to determine this linear function. As $x_1 = 1$ and $y_1 = 2$, and $x_2 = 2$ and $y_2 = 5$, by formula (5.5),

$$y - y_1 = \frac{y_2 - y_1}{x_2 - x_1}(x - x_1) \longrightarrow y - 2 = \frac{5-2}{2-1}(x-1) \longrightarrow y - 2 = 3x - 3 \longrightarrow y = 3x - 1.$$

b) Let L2 be the line normal to L1. By relationship (5.9), the slope of L2 must be

$$c = -\frac{1}{a} = -\frac{1}{3}.$$

As the slope of L2 is $-\frac{1}{3}$, $x_1 = 0$ and $y_1 = 2$, by the point-slope form

$$y - y_1 = c(x - x_1) \longrightarrow y - 2 = -\frac{1}{3}(x - 0) \longrightarrow y - 2 = -\frac{1}{3}x \longrightarrow y = -\frac{1}{3}x + 2.$$

5.1.2.2 Translations and reflections of lines

Translations of lines

By translations of functions discussed in Chapter 4, those for linear functions are in Table 5.2.

Table 5.2 Transformations of linear functions (constant $c > 0$)

New linear function	Graphic transform of original function $y = f(x) = ax + b$
$g(x) = f(x) + c$	Shift $y = ax + b$ upwards by c units
$g(x) = f(x) - c$	Shift $y = ax + b$ downwards by c units
$g(x) = f(x - c)$	Shift $y = ax + b$ to right by c units
$g(x) = f(x + c)$	Shift $y = ax + b$ to left by c units
$g(x) = cf(x)$ $(c > 1)$	Stretch $y = ax + b$ vertically away from the x-axis by a factor of c
$g(x) = cf(x)$ $(c < 1)$	Shrink $y = ax + b$ vertically toward the x-axis by a factor of c

Example 5.10: Given $y = f(x) = 2x - 1$, find the following functions by translating $f(x)$ and plot their graphs.

a) $y = f(x) + 10$; b) $y = f(x-3)$; c) $y = 2f(x)$; d) $y = \frac{1}{2}f(x)$.

Solution

a) $y = f(x) + 10 = (2x - 1) + 10 = 2x + 9.$

Its graph can be obtained by shifting the graph of $f(x)$ upwards vertically by 10 units.

b) $y = f(x-3) = 2(x-3) - 1 = 2x - 6 - 1 = 2x - 7.$

Its graph can be obtained by shifting the graph of $f(x)$ 3 units to the right horizontally.

c) $y = 2f(x) = 2(2x - 1) = 4x - 2.$

Its graph can be obtained by stretching the graph of $f(x)$ vertically away from the x-axis by a factor of 2.

d) $y = \frac{1}{2}f(x) = \frac{1}{2}(2x-1) = x - \frac{1}{2} = x - 0.5.$

Its graph can be obtained by shrinking the graph of $f(x)$ vertically toward the x-axis by a factor of 0.5.

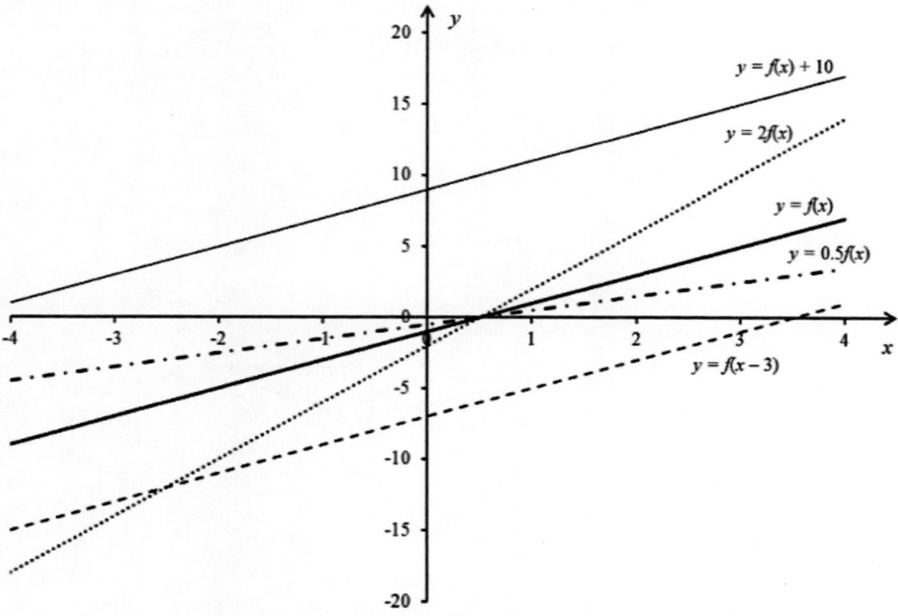

Figure 5.7: Graphs for linear functions in Example 5.10

Symmetry and reflections of lines

The properties of symmetry and reflections for general functions discussed in Chapter 4 can also be applied to linear functions. These are summarised in Table 5.3.

Table 5.3 Symmetry and reflections of linear functions

New linear function	Graphic transform of original function $f(x) = ax + b$
$g(x) = -f(x) = -ax - b$	Reflect $y = ax + b$ about the x-axis, or symmetric about the x-axis
$g(x) = f(-x) = -ax + b$	Reflect $y = ax + b$ about the y-axis, or symmetric about the y-axis
$y = ax \quad (b = 0)$	$y = ax$ and $y = -ax$ are odd functions and symmetric about the origin.

Example 5.11: Given $f(x) = -x + 3,$ find the following translations of $f(x)$ and plot their graphs.

a) $y = -f(x);$ b) $y = f(-x).$

Solution

a) $y = -f(x) = -(-x+3) = x - 3.$

Its graph can be obtained by reflecting the graph of $f(x)$ about the x-axis.

b) $y = f(-x) = -(-x) + 3 = x + 3.$

Its graph can be obtained by reflecting the graph of $f(x)$ about the y-axis.

The original function $f(x)$ is symmetric with $-f(x)$ about the x-axis; function $f(-x)$ is symmetric with the original $f(x)$ about the y-axis (Figure 5.8).

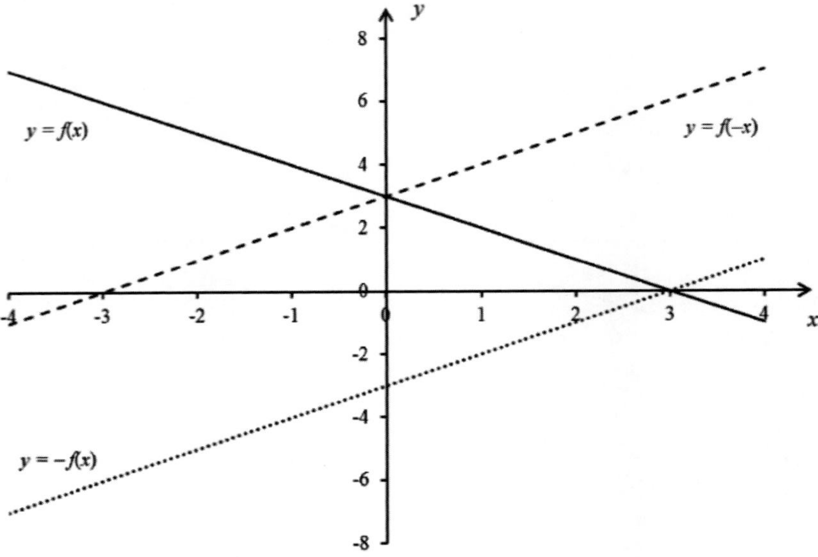

Figure 5.8: Graphs for linear functions in Example 5.11

5.1.3 Applications of linear functions

Rescaling exam scores

Example 5.12: In a course study, assignments are counted for 60% and the final exam is counted for 40% in the final score achieved by a student. In the exam, all allocated marks to questions are added up to 100. Therefore the exam mark a student obtained must be rescaled back to the score by a factor of 40% or 0.4.

a) Derive a formula to convert the exam mark in the 100% scale to the 40% scale;
b) Rescale the following exam marks to their scores in the 40% scale: 30, 50, 75, 85, and 95 respectively.

Solution

a) This conversion is a linear transform that must rescale 0 and 100 marks in the exam (x) to 0 and 40 scores in the 40% scale (y) correspondingly. This means that the linear function crosses (0, 0) and (100, 40). We can use the two-point form to determine this linear function. As $x_1 = 0$ and $y_1 = 0$, and $x_2 = 100$ and $y_2 = 40$, by formula (5.5),

$$y - y_1 = \frac{y_2 - y_1}{x_2 - x_1}(x - x_1) \longrightarrow y - 0 = \frac{40-0}{100-0}(x-0) \longrightarrow y = \frac{2}{5}x = 0.4x. \quad (0 \le x \le 100)$$

b) The 30, 50, 75, 85, and 95 marks obtained in the exam are rescaled to the following scores in the 40% scale:

$$30 \longrightarrow 0.4 \times 30 = 12, \quad 50 \longrightarrow 0.4 \times 50 = 20, \quad 75 \longrightarrow 0.4 \times 75 = 30,$$

$$85 \longrightarrow 0.4 \times 85 = 34, \quad 95 \longrightarrow 0.4 \times 95 = 38.$$

The charge rates of electricity

Example 5.13: The charge rates of electricity for residential customers are based on the 32.5 cents per kilowatt-hour plan plus a base charge each month. Suppose a monthly bill for a customer is \$240.5 for 432 kilowatt-hours. Find a linear function to describe this bill between the electricity consumption (x) and the monthly charge (y).

Solution

For this case, the slope of the linear function is the 32.5 cents per kilowatt-hour plan, i.e., $a =$ 0.325 (\$/KWH). This function must produce \$240.5 for 432 kilowatt-hours, i.e., going through (432, 240.5). We can use the point-slope form $y = ax + b$ to determine this linear function.

$$y = ax + b \longrightarrow 240.5 = 0.325 \times 432 + b \longrightarrow b = 240.5 - 0.325 \times 432 = 100.1.$$

$$\therefore\ y(\$) = 0.325x(KWH) + 100.1$$

Appreciation

Example 5.14: A house is purchased for \$255,000 now. It is expected the value of this house will be doubled in 10 years. Find a linear function to describe the value (v) of this house with time t (in year).

Solution

Suppose the current year is $t = 0$. The value of the house is \$255,000. In 10 years or $t = 10$, the value of this house is doubled to \$510,000. Thus this linear function passes (0, 255000) and (10, 510000). We can use the two-point form to determine this linear function. As $t_1 = 0$ and $v_1 = 255000$, and $t_2 = 10$ and $v_2 = 510000$, by formula (5.5),

$$v - v_1 = \frac{v_2 - v_1}{t_2 - t_1}(t - t_1) \longrightarrow v - 255000 = \frac{510000 - 255000}{10 - 0}(t - 0) \longrightarrow v = 25500t + 255000.$$

$$\therefore v(\$) = 25500t(year) + 255000.$$

Travel planning

Example 5.15: Two trucks deliver goods from the same depot to the same destination of 100 km away. Truck A has an average speed of 80 km/h whereas truck B has an average speed of 100 km/h. If two trucks must arrive at the destination at the same time, how earlier should truck A depart the depot than truck B?

Solution

Since truck A goes slower than truck B, truck A must depart earlier than truck B so as to arrive the destination at the same time. Suppose truck B departs at $t = 0$. The distance truck B travelled should be $s = 100t$. To travel the same distance, truck A needs extra time t_0 in addition to time t travelled by truck B, i.e., $s = 80(t + t_0)$. Thus we have a system of two linear equations as follows:

$$\begin{cases} s = 100t \\ s = 80(t + t_0) \end{cases}.$$

Solve this system to find t_0.

$$\begin{cases} s = 100t \\ s = 80(t + t_0) \end{cases} \longrightarrow \begin{cases} 100 = 100t \\ 100 = 80(t + t_0) \end{cases} \longrightarrow \begin{cases} t = 1\ (h) \\ t + t_0 = 1.25\ (h) \end{cases} \longrightarrow t_0 = 1.25 - 1 = 0.25\ (h) = 15\ (min).$$

Therefore, truck A should depart the depot 15 minutes earlier than truck B.

Grid of transport network

Example 5.16: One straight road R1 intersects two other straight roads R2 and R3 at different locations. On the local grid of transport map (network), R1 crosses (0, 2) and (2, 5); R2 goes through (0, 1) and (1, 2); R3 passes (0, 3) and (3, 1).

a) Find the expressions of these three roads in terms of linear functions;
b) Examine if there are parallel or normal roads among the three roads;
c) Find the grid locations of intersections among these three roads respectively;
d) Plot these three roads on a map.

Solution

a) Since R1 crosses (0, 2) and (2, 5), we can use the two-point form to determine this linear function. As $x_1 = 0$ and $y_1 = 2$, and $x_2 = 2$ and $y_2 = 5$, by formula (5.5),

$$y - y_1 = \frac{y_2 - y_1}{x_2 - x_1}(x - x_1) \longrightarrow y - 2 = \frac{5-2}{2-0}(x-0) \longrightarrow y - 2 = \frac{3}{2}x \longrightarrow y = \frac{3}{2}x + 2.$$

As R2 goes through (0, 1) and (1, 2), $x_1 = 0$ and $y_1 = 1$, and $x_2 = 1$ and $y_2 = 2$,

$$y - y_1 = \frac{y_2 - y_1}{x_2 - x_1}(x - x_1) \longrightarrow y - 1 = \frac{2-1}{1-0}(x-0) \longrightarrow y - 1 = x \longrightarrow y = x + 1.$$

R3 crosses (0, 3) and (3, 1) so $x_1 = 0$ and $y_1 = 3$, and $x_2 = 3$ and $y_2 = 1$.

$$y - y_1 = \frac{y_2 - y_1}{x_2 - x_1}(x - x_1) \longrightarrow y - 3 = \frac{1-3}{3-0}(x-0) \longrightarrow y - 3 = -\frac{2}{3}x \longrightarrow y = -\frac{2}{3}x + 3.$$

b) Since the slopes of these three lines are different from each other, these three roads are not parallel to each other.

- The product of slopes of R1 and R2 is $\frac{3}{2} \times 1 \neq -1$, R1 and R2 not normal to each other.
- The product of slopes of R1 and R3 is $\frac{3}{2} \times (-\frac{2}{3}) = -1$, R1 and R3 normal to each other.
- The product of slopes of R2 and R3 is $1 \times (-\frac{2}{3}) \neq -1$, R2 and R3 not normal to each other.

c) Solve the following systems of linear equations by substitution to find the grid of intersection between any two of the three roads.

$$\begin{cases} y = \frac{3}{2}x + 2 \\ y = x + 1 \end{cases} \longrightarrow \frac{3}{2}x + 2 = x + 1 \longrightarrow \frac{1}{2}x = -1 \longrightarrow \begin{cases} x = -2 \\ y = x + 1 = -2 + 1 = -1 \end{cases}$$

Therefore, R1 and R2 intersect at grid (–2, –1).

$$\begin{cases} y = \frac{3}{2}x + 2 \\ y = -\frac{2}{3}x + 3 \end{cases} \longrightarrow \frac{3}{2}x + 2 = -\frac{2}{3}x + 3 \longrightarrow \frac{9+4}{6}x = 1 \longrightarrow \begin{cases} x = \frac{6}{13} \\ y = \frac{3}{2}x + 2 = \frac{3}{2} \times \frac{6}{13} + 2 = \frac{9}{13} + \frac{26}{13} = \frac{35}{13} \end{cases}$$

Therefore, R1 and R3 intersect at grid $(\frac{6}{13}, \frac{35}{13})$.

$$\begin{cases} y = x + 1 \\ y = -\frac{2}{3}x + 3 \end{cases} \longrightarrow x + 1 = -\frac{2}{3}x + 3 \longrightarrow \frac{3+2}{3}x = 2 \longrightarrow \begin{cases} x = \frac{6}{5} \\ y = x + 1 = \frac{6}{5} + 1 = \frac{6}{5} + \frac{5}{5} = \frac{11}{5} \end{cases}$$

Therefore, R2 and R3 intersect at grid $(\frac{6}{5}, \frac{11}{5})$.

d) These three roads are plotted in the map shown in Figure 5.9. Remember any two points in a line can uniquely define the line.

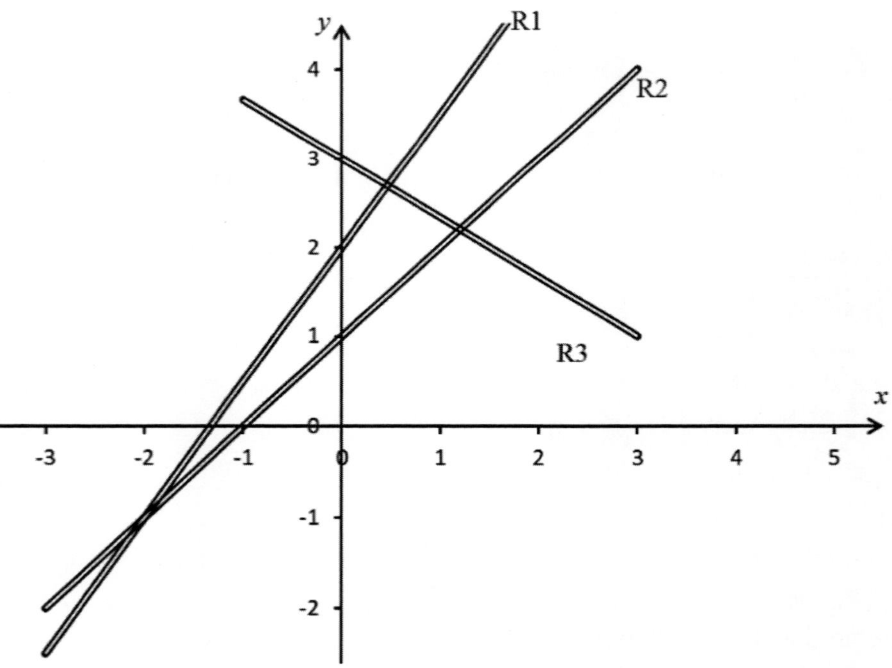

Figure 5.9: The map of the three roads determined in Example 5.16

Exercises 5.1

1. A straight line crosses points $A(-3, 2)$ and $B(3, -1)$. Determine this linear function by the slope-intercept form, two-point form, and general form respectively.

2. A straight line has a slope of 0.5 and crosses point $A(4, 3)$. Determine this linear function by the slope-intercept form, and slope-point form respectively.

3. One straight line is parallel to $f(x) = -x + 3$. Find the expression of the linear function that is normal to this line and passes through (1, 5).

4. One straight line crosses (2, 4) and (4, 1).

   a) Find the expression of this linear function;
   b) Find the expression of the linear function that is normal to this line and passes (0, 7).

5. Given $y = f(x) = -x + 5$, find the following functions by translating $f(x)$ and plot their graphs.

   a) $y = f(x) + 10$;
   b) $y = f(x - 3)$;
   c) $y = 2f(x)$;
   d) $y = \frac{1}{2}f(x)$;
   e) $y = -f(x)$;
   f) $y = f(-x)$.

6. The charge rates of water for residential customers are based on the 17.5 cents per kilolitre plan plus a base charge each month. Suppose a monthly bill for a customer is $55 for 60 kilolitres, find a linear function to describe this bill between the water usage (x) and the monthly charge (y).

7. In a quiz competition, each participant is assigned a base of 80 points in the beginning. There are ten quizzes and each quiz attracts 10 points for the correct answer and zero for a wrong answer. Find a linear equation to describe the final points that any participant can achieve.

   a) Alan had a final score of 140 points. How many quizzes did Alan get correct?
   b) Annie had a final score of 170 points. How many quizzes did Annie get correct?

8. A car is purchased for $25,000. It is expected the value of this car will be decreasing by 10% annually without compound effect. Find a linear function to describe the value (v) of this car with time t (in year).

5.2 Quadratic Functions

5.2.1 Expressions and features of quadratic functions

The standard expression of quadratic functions

In formula (5.1), the polynomial becomes quadratic functions if $n = 2$, i.e., $y = a_0 + a_1x + a_2x^2$. More commonly quadratic functions are expressed as

$$\boxed{y = ax^2 + bx + c,} \tag{5.10}$$

where a ($\neq 0$), b and c are constants. This is the *standard expression* of quadratic functions. A quadratic function is also called a *parabola* because of its U-shaped graph as shown in Figure 5.10 for $y = 3x^2 - 12x + 9$. We can see that the parabola extends upwards infinitely if $a > 0$ whereas it goes downwards infinitely if $a < 0$.

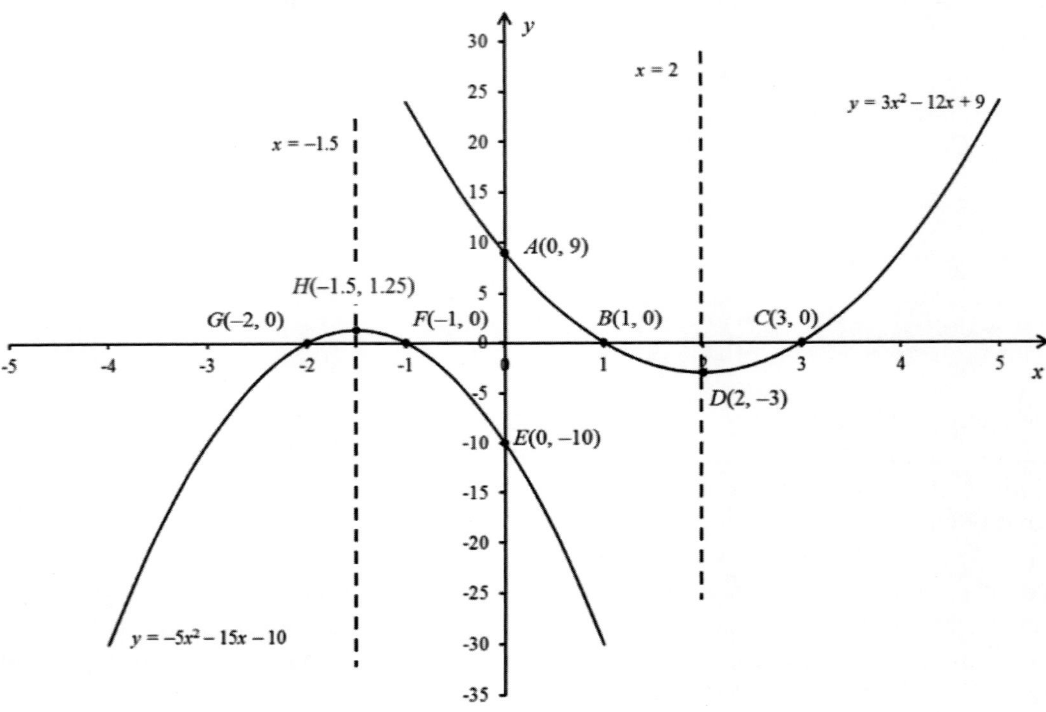

Figure 5.10: Graphs of quadratic functions $y = 3x^2 - 12x + 9$ and $y = -5x^2 - 15x - 10$

If $x = 0$, $y = c$. Thus, c is the y-intercept of $y = ax^2 + bx + c$, which is located at point $(0, c)$ in the parabola. In Figure 5.10 such is $A(0, 9)$ for $y = 3x^2 - 12x + 9$ and $E(0, -10)$ for $y = -5x^2 - 15x - 10$.

If $y = 0$, the quadratic function becomes the quadratic equation $ax^2 + bx + c = 0$. Known from Chapter 1, it has two solutions:

$$x_1 = \frac{-b+\sqrt{b^2-4ac}}{2a} \text{ and } x_2 = \frac{-b-\sqrt{b^2-4ac}}{2a} \qquad (a \neq 0).$$

Depending on the values of *a*, *b* and *c*, a quadratic function may have none (no real solution), one $(b^2-4ac=0)$, or two $(b^2-4ac>0)$ *x*-intercepts in the parabola. In Figure 5.10, $y=3x^2-12x+9$ intersects the *x*-axis at *B*(1, 0) and *C*(3, 0), or has two *x*-intercepts of 1 and 3 respectively. Meanwhile, $y=-5x^2-15x-10$ intersects the *x*-axis at *F*(–1, 0) and *G*(–2, 0), or has two *x*-intercepts of –1 and –2 respectively. You can verify these by substitute the real values for *a*, *b* and *c* into x_1 and x_2 respectively.

The vertex and axis of symmetry of quadratic functions

A parabola has either a maximum point if $a < 0$ or a minimum point if $a > 0$ in the graph. This maximum or minimum point is called the *vertex* of the parabola or quadratic function. In Figure 5.10, point *D*(2, –3) is the vertex for $y=3x^2-12x+9$ whereas point *H*(–1.5, 1.25) is the vertex for $y=-5x^2-15x-10$.

Since the vertex is either a maximum point or a minimum point, we can find the location of the vertex by manipulating the standard quadratic function to the form of a complete square, i.e.,

$$y = ax^2+bx+c = a(x^2+\frac{b}{a}x)+c = a[x^2+2\bullet\frac{b}{2a}x+(\frac{b}{2a})^2-(\frac{b}{2a})^2]+c$$

$$= a[(x+\frac{b}{2a})^2-\frac{b^2}{4a^2}]+c = a(x+\frac{b}{2a})^2-\frac{b^2}{4a}+c,$$

or

$$\boxed{y = a(x+\frac{b}{2a})^2+c-\frac{b^2}{4a}.} \qquad (5.11)$$

Since $(x+\frac{b}{2a})^2 \geq 0$ and $c-\frac{b^2}{4a}$ is a constant, $y=c-\frac{b^2}{4a}$ if $x=-\frac{b}{2a}$. If $a > 0$, any $x \neq -\frac{b}{2a}$ makes $a(x+\frac{b}{2a})^2>0$, hence $y>c-\frac{b^2}{4a}$. This means that point $(-\frac{b}{2a},c-\frac{b^2}{4a})$ is a minimum point. If $a <$ 0, any $x \neq -\frac{b}{2a}$ makes $a(x+\frac{b}{2a})^2<0$, hence $y<c-\frac{b^2}{4a}$, or point $(-\frac{b}{2a},c-\frac{b^2}{4a})$ is a maximum point. In either case, we know the vertex of a quadratic function is located at

$$\boxed{vertex: \quad x=-\frac{b}{2a} \text{ and } y=c-\frac{b^2}{4a}.} \qquad (5.12)$$

We can use formula (5.12) to verify the vertexes for $y=3x^2-12x+9$ and $y=-5x^2-15x-10$ as follows.

$$y=3x^2-12x+9:\quad a=3,\ b=-12 \text{ and } c=9$$

$$x=-\frac{b}{2a}=-\frac{-12}{2\times 3}=2 \text{ and } y=c-\frac{b^2}{4a}=9-\frac{(-12)^2}{4\times 3}=9-12=-3;$$

$$y=-5x^2-15x-10:\quad a=-5,\ b=-15 \text{ and } c=-10$$

$$x=-\frac{b}{2a}=-\frac{-15}{2\times(-5)}=-\frac{3}{2}=-1.5 \text{ and } y=c-\frac{b^2}{4a}=-10-\frac{(-15)^2}{4\times(-5)}=-10+\frac{45}{4}=1.25.$$

These are the same as points $D(2, -3)$ and $H(-1.5, 1.25)$ in Figure 5.10.

The quadratic function or parabola $y=ax^2+bx+c$ is symmetric about a vertical line that crosses the vertex, which is called the *axis of symmetry* of the parabola. In Figure 5.10, $x = 2$ is the axis of symmetry for $y=3x^2-12x+9$ whereas $x = -1.5$ is that for $y=-5x^2-15x-10$.

Example 5.17: Given $y=f(x)=x^2-6x+5$, find the vertex and intercepts of this quadratic function.

Solution

For $y=x^2-6x+5$, $a = 1$, $b = -6$ and $c = 5$. By formula (5.12), the vertex is located at

$$x=-\frac{b}{2a}=-\frac{-6}{2\times 1}=3 \text{ and } y=c-\frac{b^2}{4a}=5-\frac{(-6)^2}{4\times 1}=5-9=-4 \text{ or } V(3,-4).$$

The y-intercept is $c=5$. The x-intercepts can be found by solving equation $x^2-6x+5=0$:

$$x^2-6x+5=0 \longrightarrow (x-1)(x-5)=0 \longrightarrow x=1 \text{ and } x=5.$$

Example 5.18: The vertex of a quadratic function is located at $V(2, 1)$ and its y-intercept is 3. Find the expression of this quadratic function.

Solution

Suppose the function is $y=ax^2+bx+c$. As its y-intercept is 3, thus $c = 3$. Since its vertex is located at $V(2, 1)$, by formula (5.12) and substitution,

$$x=-\frac{b}{2a} \longrightarrow 2=-\frac{b}{2a} \longrightarrow b=-4a \longrightarrow a=-\frac{b}{4};$$

$$y=c-\frac{b^2}{4a} \longrightarrow 1=3-\frac{b^2}{4a} \longrightarrow b^2=8a \longrightarrow b^2=8(-\frac{b}{4})=-2b.$$

$$b^2 = -2b \longrightarrow b^2 + 2b = 0 \longrightarrow b(b+2) = 0 \longrightarrow b = -2 \quad (b = 0 \text{ is not meaningful.})$$
$$a = \frac{-b}{4} = \frac{2}{4} = 0.5.$$

Thus, the quadratic function is $y = 0.5x^2 - 2x + 3$. This can be verified by checking the vertex and y-intercept as follows:

$$x = -\frac{b}{2a} = -\frac{-2}{2\times 0.5} = 2, \text{ and } y = c - \frac{b^2}{4a} = 3 - \frac{(-2)^2}{4\times 0.5} = 3 - 2 = 1.$$
$$c = f(0) = 0.5(0)^2 - 3(0) + 3 = 3.$$

Example 5.19: A quadratic function has a minimum value –1 and crosses the x-axis at $A(0, 0)$ and $B(-2, 0)$. Find the expression of this quadratic function.

Solution

Suppose the function is $y = ax^2 + bx + c$. Since the function passes $A(0, 0)$ and $B(-2, 0)$,

$$0 = a(0)^2 + b(0) + c \longrightarrow c = 0$$
$$0 = a(-2)^2 + b(-2) + 0 \longrightarrow 4a - 2b = 0 \longrightarrow 2a = b \longrightarrow a = \frac{b}{2}.$$

By formula (5.12), the minimum value should be

$$y = c - \frac{b^2}{4a} \longrightarrow -1 = 0 - \frac{b^2}{4a} \longrightarrow 4a = b^2 \longrightarrow a = \frac{b^2}{4}. \text{ Thus}$$
$$\frac{b^2}{4} = \frac{b}{2} \longrightarrow b^2 - 2b = 0 \longrightarrow b(b-2) = 0 \longrightarrow b = 2 \quad (b = 0 \text{ is not meaningful.})$$
$$a = \frac{b}{2} = 1.$$

Hence the quadratic function is $y = x^2 + 2x$. This can be verified by checking the minimum value and x-intercepts as follows:

$$y = c - \frac{b^2}{4a} = 0 - \frac{2^2}{4\times 1} = -1;$$
$$0 = x^2 + 2x \longrightarrow x(x+2) = 0 \longrightarrow x = 0 \text{ and } x = -2.$$

Translations and reflections of quadratic functions

In a special case where $b = c = 0$, $y = ax^2$ is symmetric about the y-axis, or $y = ax^2$ is an even function about the y-axis. Such example is shown in Figure 5.11 as $y = -4x^2$. If $b = 0$ but $c \neq 0$, $y = ax^2 + c$ is still symmetric about the y-axis. In the graph, $y = ax^2 + c$ can be obtained through moving $y = ax^2$ vertically by c units or by the y-intercept, examples being the parabolas of $y = -4x^2 + 20$ and $y = -4x^2 - 10$ shown in Figure 5.11.

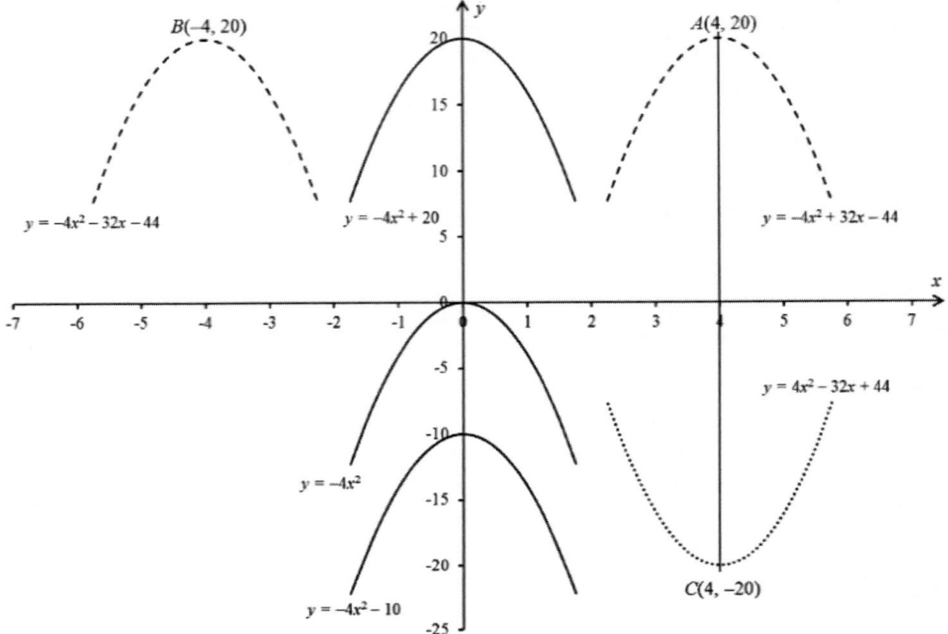

Figure 5.11: Graphic presentations of translations and reflections of quadratic functions

The general quadratic function $y = ax^2 + bx + c$ can be rewritten as formula (5.11)

$$y = a(x + \frac{b}{2a})^2 + c - \frac{b^2}{4a} = a(x - d)^2 + e,$$

where $d = -\frac{b}{2a}$ and $e = c - \frac{b^2}{4a}$ are two constants. This implies that we can obtain the graph of $y = ax^2 + bx + c$ by firstly moving the graph of $y = ax^2$ vertically by $e = c - \frac{b^2}{4a}$ units, and then moving horizontally by further $d = -\frac{b}{2a}$ units. By formula (5.12), Point (d, e) or $(-\frac{b}{2a}, c - \frac{b^2}{4a})$ is the vertex of $y = ax^2 + bx + c$. Thus, $y = ax^2 + bx + c$ is a result of the combined translations of $y = ax^2$ by a horizontal shift of $-\frac{b}{2a}$ units PLUS a vertical shift of $c - \frac{b^2}{4a}$ units. In Figure 5.11,

$y=-4x^2+32x-44$ equals the combined transition of shifting left by 4 units ($-\frac{b}{2a}=-\frac{32}{-2\times4}=4$) and vertical shifting by 20 units ($c-\frac{b^2}{4a}=-44-\frac{32^2}{-4\times4}=-44+64=20$) applied to $y=-4x^2$. Similarly, $y=-4x^2-32x-44$ is equivalent to the combined transition of a rightward shift of 4 units and a vertical shift of 20 units applied to $y=-4x^2$.

The general rules of function translations are applicable to all quadratic functions, i.e., $y=f(x-d)$ moves $y=f(x)$ horizontally by d units. For example, for $f(x)=-4x^2+20$, $f(x-4)$ is to shift $f(x)$ to the right by 4 units. This is the case of moving $y=-4x^2+20$ to the right by 4 units in Figure 5.11, which is coincidently the same as $y=-4x^2+32x-44$. This is because

$$f(x-4)=-4(x-4)^2+20=-4(x^2-8x+16)+20=-4x^2+32x-64+20=-4x^2+32x-44.$$

The general rules of function reflections are also applicable to quadratic functions. The rule of $g(x)=-f(x)$ reflects $f(x)$ about the x-axis. For example, if $f(x)=-4x^2+32x-44$, $g(x)=-f(x)=-[-4x^2+32x-44]=4x^2-32x+44$ is to mirror $f(x)$ about the x-axis, which is shown in Figure 5.11 as the dotted parabola with respect to the dashed parabola both sharing the same axis of symmetry $x = 4$.

The rule of $g(x)=f(-x)$ reflects $f(x)$ about the y-axis. For $f(x)=-4x^2+32x-44$, $g(x)=f(-x)=-4(-x)^2+32(-x)-44=-4x^2-32x-44$ is to mirror $f(x)$ about the y-axis, which is shown in Figure 5.11 as the two dashed parabolas on the opposite sides of the y-axis.

These translational and reflective features of quadratic functions are summarised in Table 5.4. Note that the stretching and shrinking translations of quadratic functions are not included.

Table 5.4 The translational and reflective features of quadratic functions

Function	Feature
$y=f(x)=ax^2+bx+c$	1. U-shaped parabolas 2. Upward parabolas if $a>0$ and downward parabolas if $a<0$ 3. The y-intercept: c 4. The vertex (maximum or minimum point): $(-\frac{b}{2a}, c-\frac{b^2}{4a})$ 5. The axis of symmetry: $x=-\frac{b}{2a}$ 6. $b=c=0$, $y=ax^2$ symmetric about the y-axis
$g(x)=f(x-d)$	Moving $y=f(x)$ horizontally by d units
$g(x)=-f(x)$	Reflecting $y=f(x)$ about the x-axis
$g(x)=f(-x)$	Reflecting $y=f(x)$ about the y-axis

Example 5.20: A quadratic function was derived by shifting $f(x) = -x^2 + 3$ to the left by 2 units. Find the expression of this quadratic function.

Solution

Since the original function is $f(x) = -x^2 + 3$, by shifting it 2 units to the left, it means $y = f(x+2)$. Thus the current function is

$$y = f(x+2) = -(x+2)^2 + 3 = -(x^2 + 4x + 4) + 3 = -x^2 - 4x - 4 + 3 = -x^2 - 4x - 1.$$

This can be verified by checking if the vertex is located at (–2, 3), moved from (0, 3) originally as follows:

$$x = -\frac{b}{2a} = -\frac{-4}{-2\times 1} = -2, \text{ and } y = c - \frac{b^2}{4a} = -1 - \frac{(-4)^2}{-4\times 1} = -1 + 4 = 3.$$

Example 5.21: The vertex of a quadratic function is located at *V*(2, 3). This parabola was translated from $f(x) = -2x^2$ originally. Find the expression of this quadratic function.

Solution

Suppose the function is $y = ax^2 + bx + c$. Since the translation does not change coefficient *a*, the function sought becomes $y = -2x^2 + bx + c$. As its vertex is located at *V*(2, 3), by formula (5.12),

$$x = -\frac{b}{2a} \longrightarrow 2 = -\frac{b}{-2\times 2} \longrightarrow b = 8;$$

$$y = c - \frac{b^2}{4a} \longrightarrow 3 = c - \frac{8^2}{-4\times 2} \longrightarrow 3 = c + 8 \longrightarrow c = -5.$$

Thus the quadratic function is $y = -2x^2 + 8x - 5$. This can be verified by checking the vertex:

$$x = -\frac{b}{2a} = -\frac{8}{-2\times 2} = 2, \text{ and } y = c - \frac{b^2}{4a} = -5 - \frac{8^2}{-4\times 2} = -5 + 8 = 3.$$

Example 5.22: The quadratic function $g(x) = 3x^2 - 2x - 5$ was originally translated from *f*(*x*) by the following three consecutive moves (or steps):

1) shifting *f*(*x*) leftwards by 3 units;
2) then shifting upwards by 10 units;
3) finally reflecting about the *x*-axis.

Find the expression of the original quadratic function *f*(*x*).

Solution

The recovering process for $f(x)$ from $g(x)$ goes in the inverse order, i.e., from Step 3 to Step 1.

- Step 3: reflecting $g(x)$ about the x-axis for $f_2(x)$

  $$f_2(x) = -g(x) = -(3x^2 - 2x - 5y) = -3x^2 + 2x + 5$$

- Step 2: shifting $f_2(x)$ downwards by 10 units

  $$f_1(x) = f_2(x) - 10 = -3x^2 + 2x + 5 - 10 = -3x^2 + 2x - 5$$

- Step 1: shifting $f_1(x)$ rightwards by 3 units

  $$f(x) = f_1(x-3) = -3(x-3)^2 + 2(x-3) - 5 = -3(x^2 - 6x + 9) + 2x - 6 - 5$$
  $$= -3x^2 + 18x - 27 + 2x - 11 = -3x^2 + 20x - 38.$$

Therefore, $g(x)$ was translated from $f(x) = -3x^2 + 20x - 38$ originally.

Example 5.23: Find the points of intersection between $y = 2x^2 - x - 5$ and $y = x - 1$. Plot the graphs of these functions.

Solution

These two functions intersect where both have the same value for y, i.e.,

$$2x^2 - x - 5 = x - 1 \longrightarrow 2x^2 - 2x - 4 = 0 \longrightarrow x^2 - x - 2 = 0 \longrightarrow (x+1)(x-2) = 0 \longrightarrow x = -1, x = 2.$$

The corresponding y-values are:

$$x = -1 \longrightarrow y = x - 1 = -1 - 1 = -2, \text{ and } x = 2 \longrightarrow y = x - 1 = 2 - 1 = 1.$$

Thus the intersection points are $A(-1, -2)$ and B(2, 1) as shown in Figure 5.12.

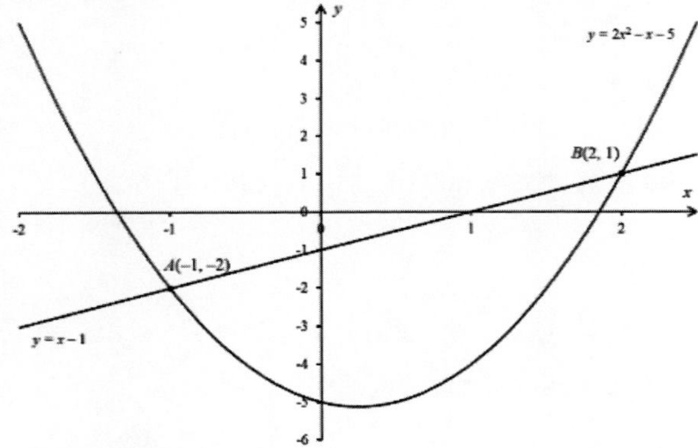

Figure 5.12: Graphs of $y = 2x^2 - x - 5$ and $y = x - 1$ in Example 5.23

5.2.2 Applications of quadratic functions

Area of rectangle with a fixed perimeter

Example 5.24: Given a rope of p meters long to make a closed rectangular shape on the ground,

a) find the area (A) of the rectangle with respect to its length (l);
b) find the length (l_{max}) and width (w_{max}) that produce the maximum area (A_{max});
c) determine the l_{max}, w_{max} and A_{max} given $p = 80$ meters.

Solution

a) Relationship between the perimeter and length and width of any rectangle is $p = 2(l + w)$. Thus the width is $w = \frac{1}{2}p - l$. The area is: $A = lw = l(\frac{1}{2}p - l) = -l^2 + \frac{1}{2}pl$.

b) $A = -l^2 + \frac{1}{2}pl$ is a quadratic function with $a = -1$ and $b = \frac{1}{2}p$. Thus the vertex of this function gives the maximum area where the length (and hence width) takes the following size:

$$l_{max} = -\frac{b}{2a} = -\frac{\frac{1}{2}p}{-2\times 1} = \frac{1}{4}p, \longrightarrow w_{max} = \frac{p}{2} - l = \frac{p}{2} - \frac{1}{4}p = \frac{1}{4}p;$$

$$A_{max} = -l_{max}^2 + \frac{1}{2}pl_{max} = -(\frac{1}{4}p)^2 + \frac{1}{2}p(\frac{1}{4}p) = -\frac{1}{16}p^2 + \frac{1}{8}p^2 = \frac{1}{16}p^2.$$

Since $l_{max} = w_{max}$, it means that a square produces the largest area if the perimeter is fixed.

c) Given $p = 80$ meters,

$$l_{max} = w_{max} = \frac{1}{4}p = \frac{80}{4} = 20\ (m), \qquad A_{max} = l_{max} \times w_{max} = 20 \times 20 = 400\ (m^2).$$

Height of a free-fall object

Example 5.25: If an object is thrown up vertically, the height (h) the object travels with time (t) follows the quadratic rule below:

$$h = -4.9t^2 + 49t + 12.5\ (m).$$

Find the time (t_{max}) when the object reaches the highest point (h_{max}).

Solution

For $h = -4.9t^2 + 49t + 12.5$, $a = -4.9$, $b = 49$ and $c = 12.5$. Thus the vertex of this function gives the maximum height when it comes to time t_{max}.

$$t_{max} = -\frac{b}{2a} = -\frac{49}{-2\times 4.9} = 5\ (s), \longrightarrow h_{max} = c - \frac{b^2}{4a} = 12.5 - \frac{49^2}{-4\times 4.9} = 12.5 + \frac{2401}{19.6} = 135\ (m).$$

Design of a parabolic bridge arch

Example 5.26: A parabolic bridge arch is shown in Figure 5.13. The vertex of this parabola is located at $V(0, -2)$ and the height of the arch is 8 meters. The span of this arch is 5 times of the height. Find the formula of this parabolic arch.

Solution

Since the parabolic arch is symmetric about the y-axis, its formula must be $y = ax^2 + c$. As the vertex $V(0, -2)$ is located in the y-axis, it means $c = -2$. Thus $y = ax^2 - 2$.

Since the span of this arch is 5 times of the height that is 8 meters, the distance from the left end to the right end of this arch is 40 meters. This means the right end of this parabola is located at point (20, –10), i.e.,

$$y = ax^2 - 2 \longrightarrow -10 = a(20)^2 - 2 \longrightarrow -8 = 400a \longrightarrow a = -\frac{8}{400} = -0.02.$$

$$\therefore y = -0.02x^2 - 2.$$

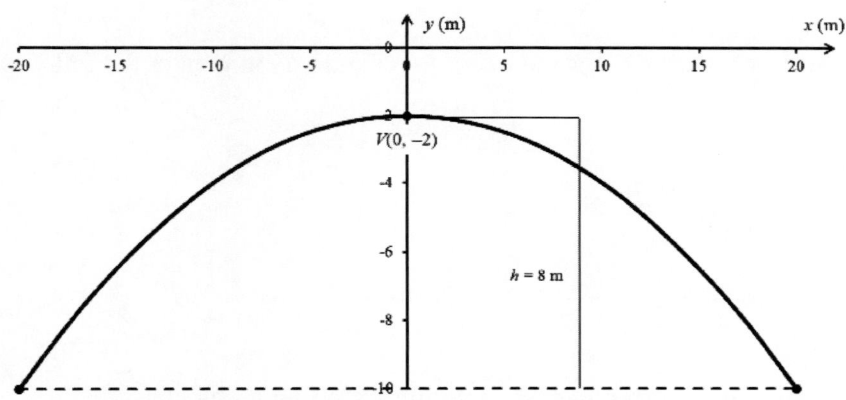

Figure 5.13: Graph of the parabolic bridge arch for Example 5.25

Bending moment of a simply supported beam

Example 5.27: A simply supported beam of length l with a uniformly distributed load w (kg/m) is shown in Figure 5.14.

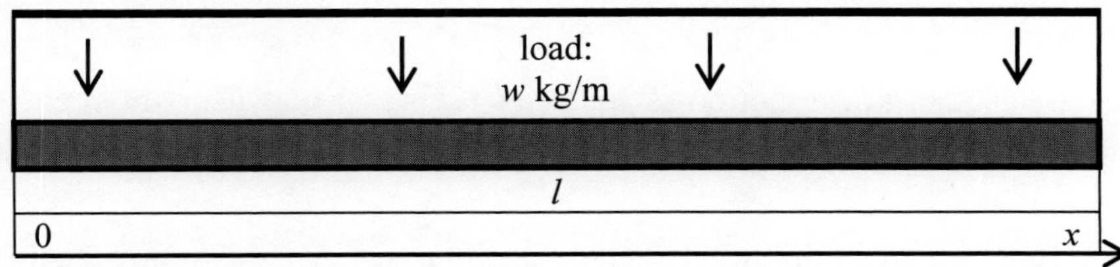

Figure 5.14: The simply supported beam with a uniformly distributed load for Example 5.26

The bending moment (M) at any point x is given by

$$M = -\frac{1}{2}wx^2 + \frac{1}{2}wlx.$$

a) find the locations on the beam where the bending moment is zero;
b) find the location where the bending moment is the maximum;

Solution

a) The locations on the beam where the bending moment is zero can be obtained by solving the following quadratic equation:

$$0 = -\frac{1}{2}wx^2 + \frac{1}{2}wlx \longrightarrow x(l-x) = 0 \longrightarrow \begin{cases} x = 0 \\ l - x = 0 \end{cases} \longrightarrow x = 0 \text{ and } x = l.$$

The two ends of this beam have no bending moment.

b) The vertex of this function gives the maximum bending moment at the following location:

$$x_{\max} = -\frac{b}{2a} = -\frac{\frac{1}{2}wl}{-2 \times \frac{1}{2}w} = \frac{1}{2}l;$$

$$M_{\max} = -\frac{1}{2}wx_{\max}^2 + \frac{1}{2}wlx_{\max} = -\frac{1}{2}w(\frac{1}{2}l)^2 + \frac{1}{2}wl(\frac{1}{2}l) = -\frac{1}{8}wl^2 + \frac{1}{4}wl^2 = \frac{1}{8}wl^2.$$

The middle point of the beam has the maximum bending moment.

Exercises 5.2

1. Given $y = f(x) = x^2 - 5x + 6$, find the vertex and intercepts of this quadratic function. Plot this function in a Cartesian system.

2. The vertex of a quadratic function is located at $V(1, 2)$ and its y-intercept is 3. Find the expression of this quadratic function.

3. A quadratic function has a minimum value of –3 and crosses the x-axis at $A(-1, 0)$ and $B(3, 0)$. Determine the expression of this quadratic function.

4. Given quadratic function $f(x) = x^2 + 1$, find its translated functions after single transformation as follows (not in sequence, i.e., each single transformation comes with one new function):

   a) shifting $f(x)$ rightwards by 4 units.
   b) shifting $f(x)$ downwards by 6 units.
   c) reflecting $f(x)$ about the x-axis.

5. The quadratic function $g(x) = -2x^2 + 4x + 3$ was originally translated from $f(x)$ by the following three consecutive moves (or steps):

   1) shifting $f(x)$ rightwards by 5 units;
   2) then reflecting about the x-axis;
   3) finally shifting downwards by 3 units.

   Find the expression of the original quadratic function $f(x)$.

6. Find the points of intersection between $y = x^2 - 3x + 5$ and $y = 2x - 1$.

7. Find the points of intersection between $y = -3x^2 + x - 1$ and $y = -x - 2$.

8. Find the points of intersection between $y = x^2 - 3x - 5$ and $y = -x^2$.

9. The sum of two positive integers is 100. To make the product of these two numbers maximum, what should the two numbers be?

10. If 40 mango trees are planted in a fixed area of land, each tree yields 200 mangos on average. The yield will decrease about 4 mangos per tree for each additional tree planted in the land. How many trees should be planted for the maximum total yield?

11. A simply supported beam of 4 metres with a uniformly distributed load w kg/m (refer to Figure 5.14). The bending moment (M) at any point x is given by

$$M = -\frac{1}{2}wx^2 + \frac{1}{2}wlx.$$

   a) find the location where the bending moment is the maximum;
   b) find the location(s) on the beam where the bending moment is half of the maximum.

5.3 Higher Order Polynomials

Cubic functions

In formula (5.1), the polynomial becomes cubic functions if $n = 3$, i.e., $y = a_0 + a_1x + a_2x^2 + a_3x^3$. More commonly cubic functions are expressed as

$$\boxed{y = ax^3 + bx^2 + cx + d,} \tag{5.13}$$

where a ($\neq 0$), b, c and d are constants.

$y = ax^3$ is called the *characteristic cubic function*, which is an S-shaped graph ($a < 0$) or an inversed S-shaped graph ($a > 0$) as shown in Figure 5.15 for quadratic functions $y = -x^3$ and $y = x^3$ respectively. This characteristic function passes through the origin and is symmetric about the origin, and extends to positive and negative infinities. It crosses the x-axis at the origin only ONCE and hence the equation $ax^3 = 0$ has ONE solution $x = 0$. This characteristic function has no vertex, i.e., no local minimum/maximum point.

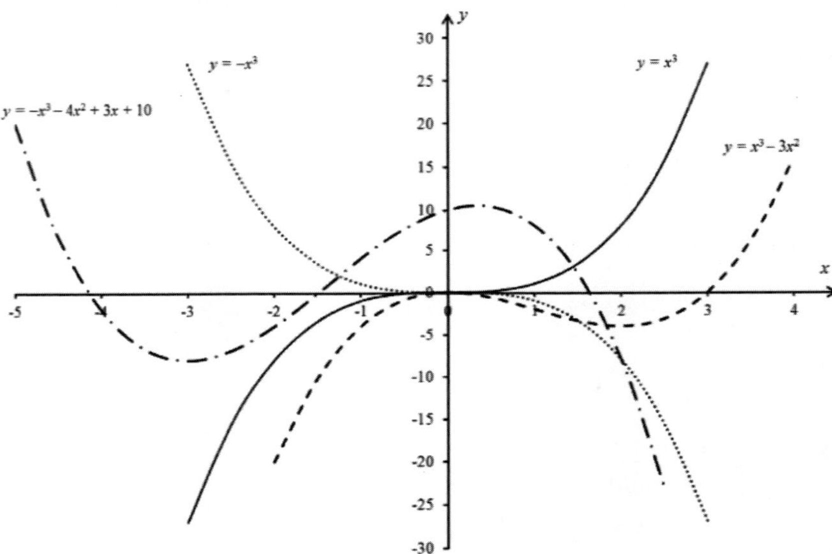

Figure 5.15: Graphs of some cubic functions

In general, a cubic function may cross the x-axis at least once and at most three times. This means cubic equation $ax^3 + bx^2 + cx + d = 0$ has at least one real solution and at most three real solutions. For example in Figure 5.15, $x^3 = 0$ has only one solution $x = 0$; $x^3 - 3x^2 = 0$ has two real solutions of $x = 0$ and $x = 3$; $-x^3 - 4x^2 + 3x + 10 = 0$ has three real solutions around –4, –1.5 and 1.7 (dash-dotted curve in Figure 5.15). Some simple cubic equations can be solved by factoring the equation, for example,

$$x^3 - 3x^2 = 0 \longrightarrow x^2(x-3) = 0 \longrightarrow x = 0,\ x = 3.$$

However, the *accurate* solutions for complicated cubic equations may not always be obtained by factoring the equations. Other means, such as calculators and computers will be required to solve complicated equations.

A cubic function may have no vertex or two vertexes at most. Generally if $y = ax^3 + bx^2 + cx + d$ crosses any horizontal line (the x-axis is a special case) only once, it has no vertex, like $y = -x^3$ and $y = x^3$ in Figure 5.15. If $y = ax^3 + bx^2 + cx + d$ crosses any horizontal line three times, it has two vertexes. For instance, although $x^3 - 3x^2 = 0$ passes the x-axis twice at $x = 0$ and $x = 3$, it crosses the horizontal line $y = -1$ three times. Therefore, $y = x^3 - 3x^2$ has two vertexes at $x = 0$ and $x = 2$ respectively. Similarly $y = -x^3 - 4x^2 + 3x + 10$ has two vertexes as it crosses the x-axis three times. Determining the locations of these vertexes requires other more advanced techniques such as calculus.

Example 5.28: Solve cubic equation $2x^3 + 3x^2 - 8x + 3 = 0$.

Solution

Since

$$\begin{aligned}2x^3 + 3x^2 - 8x + 3 &= x^3 + 3x^2 + x^3 - 9x + x + 3 = x^2(x+3) + x(x^2 - 9) + (x+3) \\ &= x^2(x+3) + x(x-3)(x+3) + (x+3) = (x+3)[x^2 + x(x-3) + 1] = (x+3)(2x^2 - 3x + 1) \\ &= (x+3)(2x-1)(x-1),\end{aligned}$$

$$(x+3)(2x-1)(x-1) = 0 \longrightarrow \begin{cases} x+3=0 \\ 2x-1=0 \\ x-1=0 \end{cases} \longrightarrow \begin{cases} x=-3 \\ x=\dfrac{1}{2} \\ x=1 \end{cases}.$$

Example 5.29: A spherical container has a volume of 100 cubic metres. What is the surface area of this sphere?

Solution

Since the volume of a sphere is $V = \frac{4}{3}\pi r^3$, a volume of 100 cubic metres gives its radius as

$$100 = \frac{4}{3}\pi r^3 \longrightarrow r^3 = \frac{100 \times 3}{4\pi} \longrightarrow r \approx \sqrt[3]{\frac{300}{4 \times 3.14}} \approx 2.88\ (m).$$

The surface area of the sphere is

$$S = 4\pi r^2 \approx 4 \times 3.14 \times 2.88^2 \approx 104.18\ (m^2).$$

The general properties of function transformation and reflection are also applicable to cubic functions, which are demonstrated by the following examples.

Example 5.30: Find the new function by shifting $f(x)=x^3-x^2-2x+5$ rightwards by 2 units.

Solution

$$g(x)=f(x-2)=(x-2)^3-(x-2)^2-2(x-2)+5$$
$$=x^3-6x^2+12x-8-x^2+4x-4-2x+4+5=x^3-7x^2+14x-3.$$

Example 5.31: Find the translated function after shifting $f(x)=2x^3+5x-3$ upwards by 5 units and then reflecting about the y-axis.

Solution

Shifting upwards by 5 units:

$$g(x)=f(x)+5=2x^3+5x-3+5=2x^3+5x+2$$

Then reflecting $g(x)$ about the y-axis:

$$h(x)=g(-x)=2(-x)^3+5(-x)+2=-2x^3-5x+2$$

Example 5.32: Find the points of intersection between $y=x^3+2x^2+3x-3$ and $y=3x^2+4x-4$.

Solution

$$x^3+2x^2+3x-3=3x^2+4x-4 \longrightarrow x^3-x^2-x+1=0 \longrightarrow x^2(x-1)-(x-1)=0$$

$$(x-1)(x^2-1)=0 \longrightarrow \begin{cases}(x-1)^2=0\\ x+1=0\end{cases} \longrightarrow \begin{cases}x=1\\ x=-1\end{cases}$$

Other higher-order polynomials

The fourth-order polynomial is

$$y=a_0+a_1x+a_2x^2+a_3x^3+a_4x^4 \quad (a_4\neq 0). \tag{5.14}$$

Some examples of the fourth-order polynomials are shown in Figure 5.16.

Higher-order polynomials are not widely used in real applications compared with the linear, quadratic and cubic functions. The higher the order of a polynomial, the more complicated their features. Thus, finding the vertexes of higher-order polynomials and solving higher-order polynomial equations will require more advanced mathematical techniques and assistance with computers.

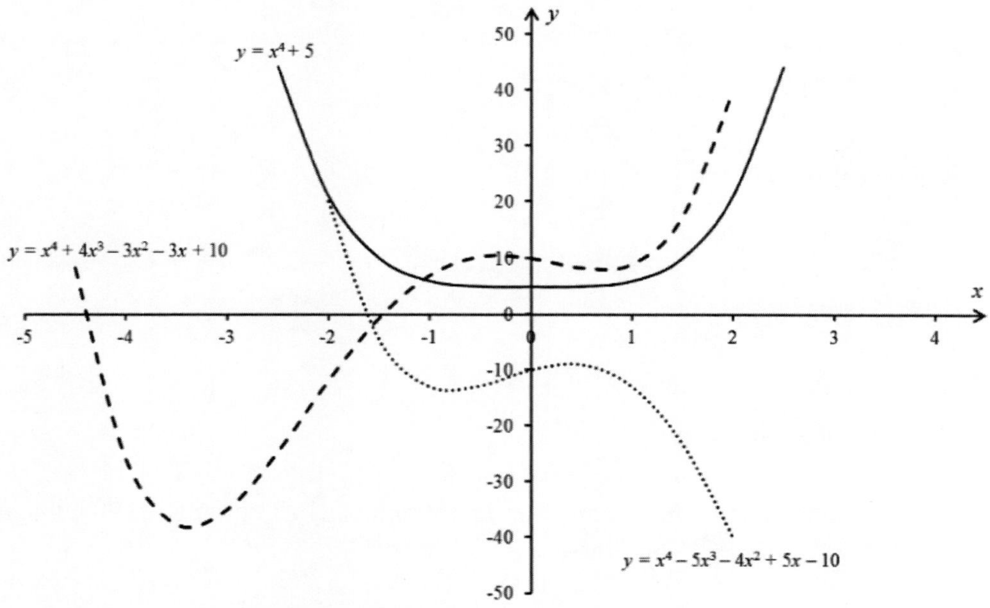

Figure 5.16: Graphs of some fourth-order polynomials

Example 5.33: Solve $x^4 - 13x^2 + 36 = 0$. Infer how many vertexes $f(x) = x^4 - 13x^2 + 36$ can have.

Solution

Let $z = x^2$. $x^4 - 13x^2 + 36 = z^2 - 13z + 36 = (z-4)(z-9)$. Thus

$$x^4 - 13x^2 + 36 = 0 \longrightarrow (z-4)(z-9) = 0 \longrightarrow z = 4 \text{ and } z = 9.$$

$$z = 4 \Rightarrow x^2 = 4 \longrightarrow x = 2 \text{ and } x = -2;$$

$$z = 9 \Rightarrow x^2 = 9 \longrightarrow x = 3 \text{ and } x = -3.$$

Three internal subsections are confined by the four x-intercepts: [–3, –2], [–2, 2] and [2, 3]. This means that $f(x)$ enters a section from the left x-intercept and then goes out the section from the right x-intercept. Hence, $f(x)$ must reach a high point or a low point within the section, i.e., there is one vertex within each internal section. Thus, $f(x) = x^4 - 13x^2 + 36$ should have three vertexes.

Exercises 5.3

1. Solve cubic equation $x^3-3x^2-6x+8=0$.

2. Solve cubic equation $x^3+6x^2+3x-10=0$.

3. Find the translated function after reflecting $f(x)=x^3+2x^2-7x+8$ about the y-axis and then shifting downwards by 3 units.

4. Find the translated function after shifting $f(x)=-2x^3-8x^2-32x-26$ rightwards by 3 units and then reflecting about the x-axis.

5. A right circular cone is equal in its height and base radius. If its volume is 100 cubic metres. What is the height of this right circular cone?

6. Find the points of intersection between $y=x^3-3x^2+x-4$ and $y=2x-7$.

7. Find the points of intersection between $y=3x^3-x^2-3x-2$ and $y=x^2-4$.

8. Solve $x^4-15x^2-10x+24=0$.

Chapter 5: Exercises Answers

Exercises 5.1

1. $y = -\frac{1}{2}(x-1)$
2. $y = \frac{1}{2}(x+7)$
3. $y = -x+2$
4. a) $y = -\frac{3}{2}x+7$, b) $y = \frac{2}{3}x+7$
5. a) $y = -x+15$, b) $y = -x+8$, c) $y = -2x+10$, d) $y = -\frac{1}{2}x+\frac{5}{2}$, e) $y = x-5$, f) $y = x+5$
6. $y = 0.175x + 44.5$ *(dollar)*
7. a) $x = 6$, b) $x = 9$
8. $y = -2500x + 25000$ *(dollar)*

Exercises 5.2

1. $V(\frac{5}{2}, -\frac{1}{4})$, $y-intercept: (0,6)$, $x-intercept: (2,0)$ *and* $(3,0)$.
2. $y = x^2 - 2x + 3$
3. $y = \frac{3}{4}(x^2 - 2x - 3)$
4. a) $g(x) = x^2 - 8x + 17$. b) $g(x) = x^2 - 5$. c) $g(x) = -x^2 - 1$
5. $f(x) = 2x^2 + 16x + 24$
6. $(2,3)$ and $(3,5)$
7. $(1,-3)$ and $(-\frac{1}{3}, -\frac{5}{3})$
8. $(-1,-1)$ and $(\frac{5}{2}, -\frac{25}{4})$
9. 50 and 50 $\longrightarrow p_{max} = 50 \times 50 = 2500$
10. 45 trees $\longrightarrow y_{max} = 8100$
11. a) $(2, 2w)$, b) $(2-\sqrt{2}, w)$ and $(2-\sqrt{2}, w)$.

Exercises 5.3

1. $x = 1, x = 4, x = -2$
2. $x = 1, x = -2, x = -5$
3. $g(x) = -x^3 + 2x^2 + 7x + 5$
4. $g(x) = 2x^3 - 10x^2 + 38x - 52$
5. $r = h \approx 4.57$ m
6. $x = 1, x = 3, x = -1$
7. $x = 1, x = 2/3, x = -1$
8. $x = 1, x = 4, x = -2, x = -3$

CHAPTER 6

6 Exponential and Logarithmic Functions

CHAPTER OBJECTIVES

- Introduce exponential functions
- Explain the properties of special exponential functions
- Introduce logarithmic functions
- Explain the properties of special logarithmic functions
- Introduce logarithmic scales
- Use exponential and logarithmic functions to solve real-world problems

Essential statements on exponential and logarithmic functions:

- Exponential and logarithmic functions are commonly encountered in applied mathematics.
- The common, binary and natural exponential and logarithmic functions play an important role in scientific and engineering applications.

Key topics:

- The common, binary and natural exponential functions
- The common, binary and natural logarithmic functions

Flowchart of mathematical knowledge building

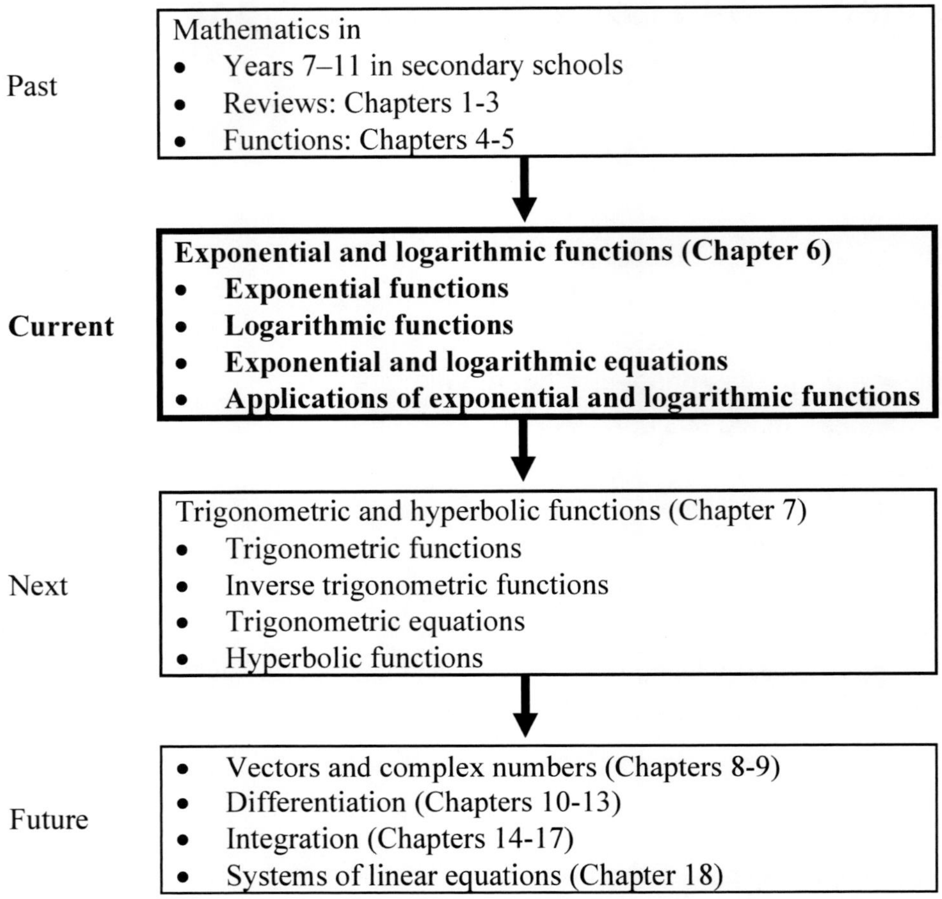

Past	Mathematics in • Years 7–11 in secondary schools • Reviews: Chapters 1-3 • Functions: Chapters 4-5
	↓
Current	**Exponential and logarithmic functions (Chapter 6)** • **Exponential functions** • **Logarithmic functions** • **Exponential and logarithmic equations** • **Applications of exponential and logarithmic functions**
	↓
Next	Trigonometric and hyperbolic functions (Chapter 7) • Trigonometric functions • Inverse trigonometric functions • Trigonometric equations • Hyperbolic functions
	↓
Future	• Vectors and complex numbers (Chapters 8-9) • Differentiation (Chapters 10-13) • Integration (Chapters 14-17) • Systems of linear equations (Chapter 18)

Chapter 6: Exponential and Logarithmic Functions

6.1 Exponential Functions

6.1.1 Exponential functions and properties

Definition

The exponential function is defined as

$$\boxed{y = f(x) = b^x \qquad (b > 0 \ \ \& \ \ b \neq 1)} \tag{6.1}$$

where b is called the base of exponent x and x can be any real number. For example, both $y = 2^x$ and $y = 0.2^{-x}$ are exponential functions as the base for both functions are positive and not equal to 1.

An exponential function can be regarded as the expansion of both exponents of integers and radicals of fractions to *continuous exponents* with any *positive* base except 1. Therefore, the rules for exponents and radicals are also applicable to exponential functions. Some commonly used rules are recaptured in Table 6.1.

Table 6.1 Common rules for continuous exponents

$b > 0$ and $c > 0$; $b \neq 1$ and $c \neq 1$			
1	$b^0 = 1$ and $b^1 = b$	6	$b^x b^y = b^{x+y}$ or $b^{x+y} = b^{y+x} = b^x b^y$
2	$b^{-1} = \frac{1}{b^1} = \frac{1}{b}, b^{-x} = \frac{1}{b^x}$ or $b^x = \frac{1}{b^{-x}}$	7	$b^{x-y} = b^x b^{-y} = \frac{b^x}{b^y}$ and $\frac{b^x}{b^x} = b^{x-x} = b^0 = 1$
3	$(bc)^x = b^x c^x$	8	$b^{xy} = (b^x)^y = (b^y)^x$
4	$(\frac{b}{c})^x = \frac{b^x}{c^x} = b^x c^{-x}$	9	$b^{-xy} = \frac{1}{b^{xy}}$
5	$(\frac{b}{c})^{-x} = b^{-x} c^x = (\frac{c}{b})^x$	10	$(b+c)^{-x} = \frac{1}{(b+c)^x}$ or $(b+c)^x = \frac{1}{(b+c)^{-x}}$

Properties of exponential functions

Exponential functions have the following properties:

1. The domain of any exponential function is the entire x-axis, or $(-\infty, \infty)$; the range of the dependent variable of any exponential function is $(0, \infty)$.

   This is because any x-value always produces a positive real number from b^x as base $b > 0$. The four exponential functions, $y = 2^x$, $y = 2^{-x} = (\frac{1}{2})^x$, $y = 4^x$ and $y = 4^{-x} = (\frac{1}{4})^x$, all appear above the x-axis in Figure 6.1.

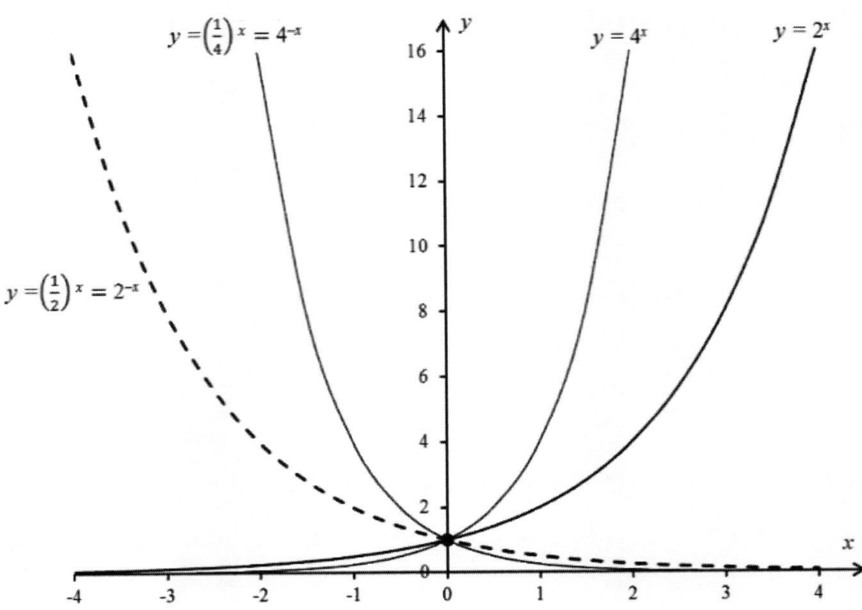

Figure 6.1: Graphs of some exponential functions

2. Every exponential function goes through point (0, 1), i.e., having a y-intercept of 1, but has no x-intercept.

   Because $b^0 = 1$, and $y = b^x > 0$ for any x-value, in Figure 6.1 all curves cross point (0, 1).

3. $y = b^x$ increases from zero towards infinity with increase in x from negative infinity towards positive infinity if $b > 1$; $y = b^x$ decreases from infinity towards zero with increase in x from negative infinity towards positive infinity if $b < 1$.

   In Figure 6.1, both $y = 2^x$ and $y = 4^x$ rise from left to right as the base is greater than 1. Both $y = (\frac{1}{2})^x$ and $y = (\frac{1}{4})^x$ fall from left to right as the base is smaller than 1.

4. $y = b^x$ approaches the x-axis when x moves to negative infinity if $b > 1$; $y = b^x$ approaches the x-axis when x moves to positive infinity if $b < 1$.

   In Figure 6.1, both $y = 2^x$ and $y = 4^x$ approach the x-axis when x moves to negative infinity as the base is greater than 1. Both $y = (\frac{1}{2})^x$ and $y = (\frac{1}{4})^x$ approach the x-axis when x moves to positive infinity as the base is smaller than 1.

Reflections and translations of exponential functions

The reflective rules for general functions are also applicable to exponential functions. For instance,

$f(x) = b^x \longrightarrow g(x) = f(-x) = b^{-x}$: reflecting $f(x)$ about the y-axis.

In Figure 6.1, $y = 2^x$ and $y = (\frac{1}{2})^x = 2^{-x}$ are mutually reflective to each other about the y-axis. For instance, for $x = 2$, two units to the right of the origin, $2^2 = 4$; for $x = -2$, two units to the left of the origin, $(\frac{1}{2})^{-2} = \frac{1}{2^{-2}} = 2^2 = 4$, the same number for y but on the opposite sides of the origin.

$f(x) = b^x \longrightarrow g(x) = -f(x) = -b^x$: reflecting $f(x)$ about the x-axis.

In Figure 6.2, $y = -2^x$ can be obtained by reflecting $y = 2^x$ about the x-axis.

$f(x) = b^x \longrightarrow g(x) = f(x) + c = b^x + c$: shifting $f(x)$ vertically by c units;

$f(x) = b^x \longrightarrow g(x) = f(x - c) = b^{x-c}$: shifting $f(x)$ horizontally by c units.

In Figure 6.2, $y = 2^x - 4$ is obtained by shifting $y = 2^x$ vertically downwards by 4 units ($c = -4$). $y = 2^{x-2}$ can be obtained by rightly shifting $y = 2^x$ by 2 units ($c = 2$).

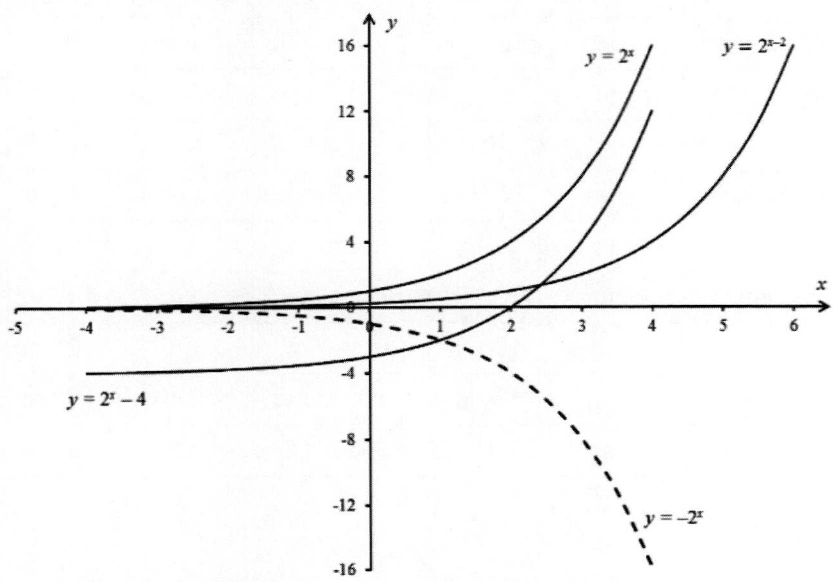

Figure 6.2: Reflection and translation of exponential functions

6.1.2 Special exponential functions

The common exponential function

The common exponential function refers to the exponential function with base 10, i.e.,

$$\boxed{y = f(x) = 10^x.} \tag{6.2}$$

It is called the *common* exponential function because the base of 10 is the basic factor of the *decimal system* that has been widely used in the world.

Example 6.1: Use the common exponential function to describe the scale of length in International System (IS) of units.

Solution

In the International System of units, the scale of length is defined by multiplication or division of the base unit by factor 10, i.e., it follows the pattern of the common exponential function 10^n. The base unit for length is *metre* (m), being the value of 10^n where $n = 0$, i.e., $10^0 = 1$. Other units can be derived using 10^n by choosing different integers for n. Some commonly encountered units for length are summarised in the table below.

Dividing by factor 10			*Multiplying by factor 10*		
		Base unit:	**1 metre = 10^0 m**		
Name	**SI symbol**	**Value**	**Name**	**SI symbol**	**Value**
yoctometre	ym	10^{-24} m	decametre	dam	10^1 m
zeptometre	zm	10^{-21} m	hectometre	hm	10^2 m
attometre	am	10^{-18} m	kilometre	km	10^3 m
femtometre	fm	10^{-15} m	megametre	Mm	10^6 m
picometre	pm	10^{-12} m	gigametre	Gm	10^9 m
nanometre	nm	10^{-9} m	terametre	Tm	10^{12} m
micrometre	μm	10^{-6} m	petametre	Pm	10^{15} m
millimetre	mm	10^{-3} m	exametre	Em	10^{18} m
centimetre	cm	10^{-2} m	zettametre	Zm	10^{21} m
decimetre	dm	10^{-1} m	yottametre	Ym	10^{24} m

Example 6.2: Given decimal numbers 21598, 501 and 23.764, use common exponential function to

a) decompose these numbers by individual digits;
b) find the order of these numbers.

Solution

a) A decimal number can be decomposed as the sum of individual digits multiplied by their corresponding *order of factor* 10.

$$\underset{4\;3\;2\;1\;0}{\underline{24598}} = 20000 + 4000 + 500 + 90 + 8 = 2\times10^4 + 4\times10^3 + 5\times10^2 + 9\times10^1 + 8\times10^0$$

$$\underset{2\;1\;0}{\underline{501}} = 500 + 1 = 5\times10^2 + 0\times10^1 + 1\times10^0$$

$$\underset{1\;0\;-1\;-2\;-3}{\underline{23.764}} = 20 + 3 + 0.7 + 0.06 + 0.004 = \underset{tens}{2\times10^1} + \underset{ones}{3\times10^0} + \underset{tenths}{7\times10^{-1}} + \underset{hundredths}{6\times10^{-2}} + \underset{thousandths}{4\times10^{-3}}$$

If we use the decimal point (.) as a reference, the digit immediately to the left of the decimal point is in the order of zero to factor 10, or commonly known as in the order of *ones*; the order increases by 1 if moving one more position leftwards one after another, namely *tens*, *hundreds*, *thousands*, and so on. The digit immediately to the right of the decimal point is in the order of –1 to factor 10, or commonly known as in the order of *tenths*; the order decreases by 1 if moving one more position rightwards one after another, namely *hundredths*, *thousandths*, and so on.

b) For a decimal number, we use its highest order of factor 10 as the order of this number. Thus, the order of 24598 is 10^4 or 24598 is in the order of *tens of thousands*; the order of 501 is 10^2 or 501 is in the *order of hundreds*; the order of 23.764 is 10^1 or 23.763 is in the *order of tens*. This leads to the scientific form of decimal numbers as

$$24598 = 2.4598\times10^4$$
$$501 = 5.01\times10^2$$
$$23.764 = 2.3764\times10$$

Example 6.3: Use the common exponential function to describe the commonly used coins and notes of Australian currency.

Solution

The base unit in Australian currency is \$1. All other commonly used coins and notes of Australian currency can be derived by multiplication or division of this base unit by factor 10, i.e., it follows the pattern of the common exponential function 10^n. The base unit for Australian currency is the value of $10^0 = 1$. Other common coins and notes in Australian currency are summarised in the table below.

	Cent (c)				Dollar (\$)						
Coins/Notes	5	10	20	50	1	2	5	10	20	50	100
Exponent (n)	*-1.3*	*-1*	*-0.7*	*-0.301*	*0*	*0.301*	*0.699*	*1*	*1.30104*	*1.69897*	*2*
10^n	0.05	0.1	0.2	0.5	1	2	5	10	20	50	100

This example shows that the exponent can take any real value (integers or decimals) in exponential functions.

The binary exponential function

The binary exponential function refers to the exponential function with base 2, i.e.,

$$\boxed{y = f(x) = 2^x.} \tag{6.3}$$

The binary exponential function is commonly used in computers and control engineering where only two statuses, "0" or "1", are possible in the whole domain. For example, the memory of most digital devices consists of a number of the *basic electronic unit* that has indicators (called bits in computer science) either 0 (meaning free) or 1 (occupied). In other words, each basic electronic unit has a capacity of 2 bits. Instead of using bits directly for storing data (too small!), normally a *Byte* formed by 8 bits is used as the *base storage unit* to store real numbers, letters, characters *etc* in many digital devices. Similarly a Byte has only two statuses, "0" or "1" to signal either 'free' or 'occupied'. Hence a device with 10 such base storage units should have a total capacity of 2^{10} = 1,024 Bytes, roughly 1 KB; a device with 20 such units will have a total capacity of 2^{20} = 1,048,576 Bytes, roughly 1 MB.

Example 6.4: Convert 1110 and 10110001 to the corresponding decimal numbers respectively.

Solution

A binary number can be decomposed as the sum of individual digits multiplied by their corresponding *order of factor* 2, from which a binary number can be converted to the corresponding decimal number.

$$\underset{3\ 2\ 1\ 0}{\underline{1\ 1\ 1\ 0}} = 1\times 2^3 + 1\times 2^2 + 1\times 2^1 + 0\times 2^0 = 8+4+2+0 = 14$$

$$\underset{7\ 6\ 5\ 4\ 3\ 2\ 1\ 0}{\underline{1\ 0\ 1\ 1\ 0\ 0\ 0\ 1}} = 1\times 2^7 + 0\times 2^6 + 1\times 2^5 + 1\times 2^4 + 0\times 2^3 + 0\times 2^2 + 0\times 2^1 + 1\times 2^0 = 128+32+16+1 = 177$$

Example 6.5: Convert decimal numbers 14 and 501 to the corresponding binary numbers respectively.

Solution

Converting a decimal number to the corresponding binary number is done by dividing the decimal number by factor 2 repeatedly until reducing the decimal number to 0. At each step of division, the value at that step is 0 if there is no remainder; otherwise the value is 1. The fraction part in the quotient (or remainder of 1) is not carried to the next division. The first division correlates to the lowest digit in the binary number and the last division correlates to the highest order (digit) in the binary number.

For number 14, the dividing process is shown in the table below, which converts decimal number 14 to 1110 in the binary scale.

Division by 2	Remainder	Binary value	Binary order
14/2 = 7	0	0	2^0
7/2 = 3	1	1	2^1
3/2 = 1	1	1	2^2
1/2 = 0	1	1	2^3

For number 501, the dividing process is shown in the table below, which converts decimal number 501 to 111110101 in the binary scale.

Division by 2	Remainder	Binary value	Binary order
501/2 = 250	1	1	2^0
250/2 = 125	0	0	2^1
125/2 = 62	1	1	2^2
62/2 = 31	0	0	2^3
31/2 = 15	1	1	2^4
15/2 = 7	1	1	2^5
7/2 = 3	1	1	2^6
3/2 = 1	1	1	2^7
1/2 = 0	1	1	2^8

Example 6.6: Use the binary exponential function to describe the scale of digital storages.

Solution

Digital storages are quantified by Byte. The scale of digital storages is defined by multiplication of Byte by factor 2, i.e., it follows the pattern of 2^n. Some commonly encountered capacities of digital storages are summarised in the table below.

Name	Symbol	Binary value	Decimal value
kilobyte	KB/kB	2^{10}	10^3
megabyte	MB	2^{20}	10^6
gigabyte	GB	2^{30}	10^9
terabyte	TB	2^{40}	10^{12}
petabyte	PB	2^{50}	10^{15}
exabyte	EB	2^{60}	10^{18}
zettabyte	ZB	2^{70}	10^{21}
yottabyte	YB	2^{80}	10^{24}

The natural exponential function

Constant e that approximately equals 2.71828 is commonly encountered in scientific and engineering applications. Strictly in mathematical terms, e is defined by the limit

$$e = \lim_{n\to\infty}(1+\frac{1}{n})^n. \tag{6.4}$$

This constant, first discovered by the Swiss mathematician Jacob Bernoulli, is known as *Euler's number*, or *Napier's constant*. If Euler's number is chosen as the base, the exponential function is known as the natural exponential function, i.e.,

$$\boxed{y = f(x) = e^x.} \tag{6.5}$$

Example 6.7: Compute the approximate values of e by formula (6.4) using n = 10, 100, 1000, 10000, 100000 and 1000000 respectively. Keep 5 decimal (5D) places.

Solution

The approximate values are summarised in the table below. The value of $e \approx 2.71828$ is the result of using n = 1,000,000.

n	Formula	Decimal value	n	Formula	Decimal value
10	$(1+\frac{1}{10})^{10}$	2.59374	10000	$(1+\frac{1}{10000})^{10000}$	2.71815
100	$(1+\frac{1}{100})^{100}$	2.70481	100000	$(1+\frac{1}{100000})^{100000}$	2.71827
1000	$(1+\frac{1}{1000})^{1000}$	2.71692	1000000	$(1+\frac{1}{1000000})^{1000000}$	2.71828

Example 6.8: Plot the natural exponential function $y = e^x$ in range [–4, 4], from which then plot the graph for $y = e^{-x}$.

Solution

We use formula (6.5) to calculate the *x*-*y* pairs for $y = e^x$ in [–4, 4] by an interval of 0.5. These data pairs are tabulated below and plotted in Figure 6.3 as the curve with solid dots.

By the property of exponential functions, $y = e^{-x}$ is the reflection of $y = e^x$ about the *y*-axis. Therefore by mirroring the curve of $y = e^x$ about the *y*-axis, the graph of $y = e^{-x}$ is shown in Figure 6.3 as the curve with circles.

x	-4	-3.5	-3	-2.5	-2	-1.5	-1	-0.5	0
$y = e^x$	0.0183	0.0302	0.0498	0.0821	0.1353	0.2231	0.3679	0.6065	1
x	0	0.5	1	1.5	2	2.5	3	3.5	4
$y = e^x$	1	1.6487	2.7183	4.4817	7.3891	12.1825	20.0855	33.1155	54.5982

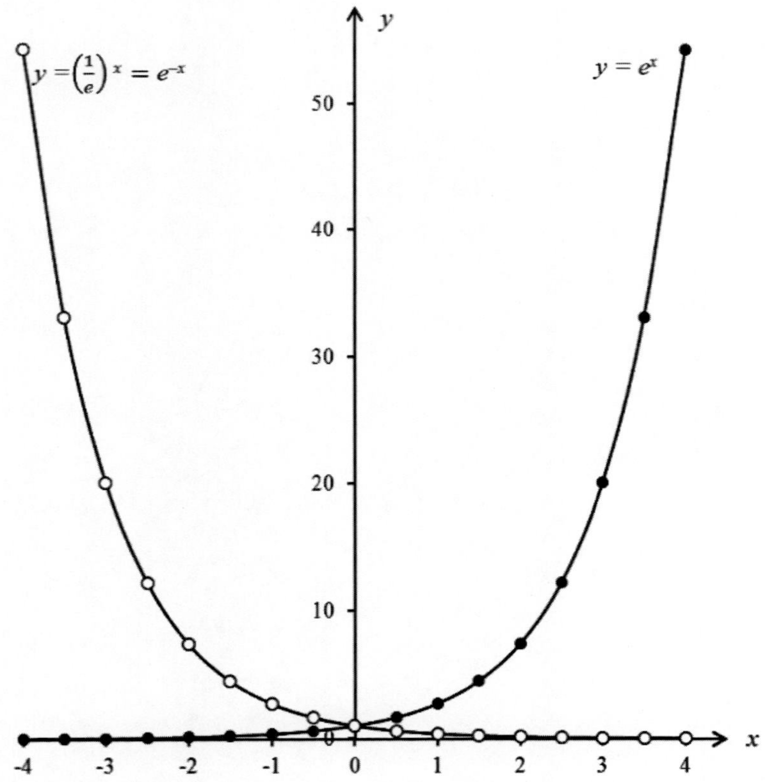

Figure 6.3: Graphs of natural exponential functions $y = e^x$ and $y = e^{-x}$

Exercises 6.1

1. Given decimal numbers 21507, 711 and 412.542, use the common exponential function to

   a) decompose these numbers by individual digits;
   b) find the order of these numbers.

2. Given binary numbers 11101 and 10110101, use the binary exponential function to convert them to the corresponding decimal numbers respectively.

3. Convert decimal numbers 19 and 499 to the corresponding binary numbers respectively.

4. Plot the natural exponential function $y = \frac{1}{100}e^{2x}$ in range [–2, 2], from which then plot the graph for $y = \frac{1}{100}e^{-2x}$.

6.2 Logarithmic Functions

6.2.1 Logarithmic functions and properties

Definition

Logarithm is the inverse operation to exponentiation. In mathematical terms, logarithm of a positive number x is the exponent y to another fixed number (base b) so that the exponent of the base must produce that number x. This is expressed as

$$y = \log_b x \qquad (b > 0 \ \& \ b \neq 1 \ \& \ x > 0) \tag{6.6}$$

where b is called the base of the logarithm and can be any real positive number except 1.

Special logarithmic functions

The ***common logarithmic function*** is the inverse of the common exponential function $y = 10^x$,

$$y = \log_{10} x = \log x. \qquad (x > 0) \tag{6.7}$$

The ***binary logarithmic function*** is the inverse of the binary exponential function $y = 2^x$:

$$y = \log_2 x. \qquad (x > 0) \tag{6.8}$$

The ***natural logarithmic function*** is the inverse of the natural exponential function $y = e^x$:

$$y = \log_e x = \ln x. \qquad (x > 0) \tag{6.9}$$

Figure 6.4 shows the graphs of these three logarithmic functions along with $y = \log_{0.5} x$ using the xy-data pairs listed in the table below.

x	0.00001	0.01	0.1	1	2	4	8	10	20	40	60
$\log x$	-6.00	-2.00	-1.00	0.00	0.30	0.60	0.90	1.00	1.30	1.60	1.78
$\log_2 x$	-16.61	-6.64	-3.32	0.00	1.00	2.00	3.00	3.32	4.32	6.32	6.91
$\ln x$	-11.51	-4.61	-2.30	0.00	0.69	1.39	2.08	2.30	3.00	3.69	4.09
$\log_{0.5} x$	16.61	6.64	3.32	0.00	-1.00	-2.00	-3.00	-3.32	-4.32	-6.32	-6.91

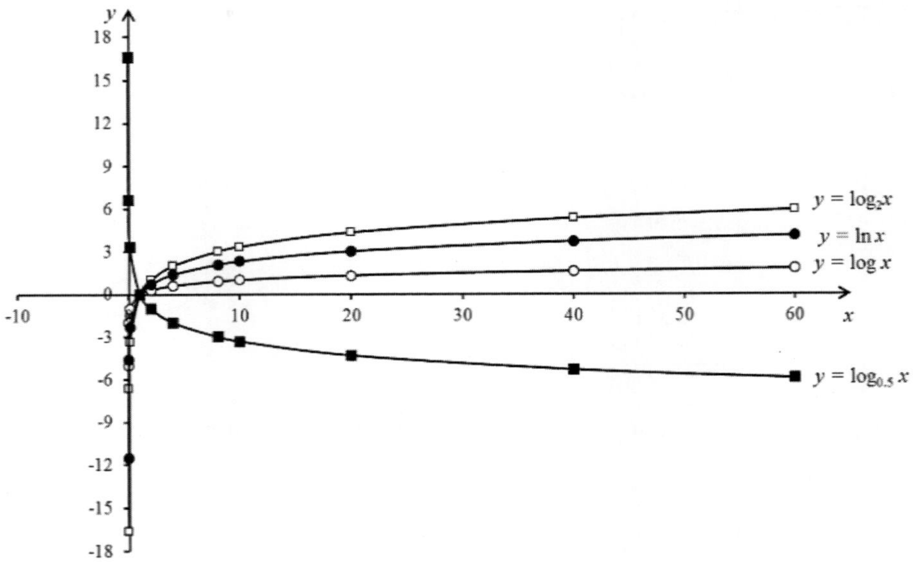

Figure 6.4: Graphs of the common, binary, natural logarithmic functions and $y = \log_{0.5} x$

Properties of logarithmic functions

Logarithmic functions have the following properties:

1. The domain of any logarithmic function is the positive half of the *x*-axis, or (0, ∞); the range of dependent variable *y* of any exponential function is (–∞, ∞).

   In Figure 16.4, the four logarithmic functions all appear to the right half of the *x*-axis.

2. Every logarithmic function goes through point (1, 0), i.e., having an *x*-intercept of 1, but has no *y*-intercept.

   This is because $b^0 = 1$ or $\log_b 1 = 0$ for all logarithmic functions with any base value, which is shown in Figure 6.4 as all curves cross point (1, 0).

3. If $b > 1$, all logarithmic functions increase from negative infinity to positive infinity whilst *x* increases from an infinitely small positive value to positive infinity. If $0 < b < 1$, the logarithmic function decreases from positive infinity to negative infinity whilst *x* increases from an infinitely small positive value to positive infinity.

   In Figure 6.4, the *y*-values of the common, binary and natural logarithmic functions rise from the smallest negative value to zero at point (1, 0), then become positive and increase steadily whilst moving rightwards. For $y = \log_{0.5} x$, it decreases from the largest positive value to zero at point (1, 0), then becomes negative and further decreases steadily whilst moving rightwards.

4. $y = \log_b x$ approaches the y-axis when x approaches the origin.

   In Figure 6.4, all four curves move closer to the y-axis when x approaches the origin.

5. The base value controls the rate of change of a logarithmic function. If $b > 1$, the smaller the base value, the higher the rate of change.

   In Figure 6.4, when x moves from 1 to 2, $y = \log_2 x$ increases from 0 to 1, the largest change with the smallest base value of 2 among the three logarithmic functions; $y = \ln x$ increases from 0 to 0.69; $y = \log x$ increases from 0 to 0.3, the smallest change with largest base value of 10 among the three logarithmic functions.

Operational rules of logarithm

Logarithmic operations with positive numbers follow the rules summarised in Table 1.6, which is replicated here as Table 6.2. Some examples of using these rules are presented as follows.

Example 6.9: Given log2 ≈ 0.3010, log3 ≈ 0.4771, and log7 ≈ 0.8451, use the rules in Table 6.2 to calculate the common logarithmic values for 4, 5, 6, 8, 9, 40, 81, 98, 3/28 and $\sqrt[3]{3\sqrt{5}}$.

Solution

$$\log 4 = \log 2^2 = 2\log 2 \approx 2\times 0.3010 = 0.6020$$

$$\log 5 = \log(\frac{10}{2}) = \log 10 - \log 2 \approx 1 - 0.3010 = 0.6990$$

$$\log 6 = \log(2\times 3) = \log 2 + \log 3 \approx 0.3010 + 0.4771 = 0.7781$$

$$\log 8 = \log 2^3 = 3\log 2 \approx 3\times 0.3010 = 0.9030$$

$$\log 9 = \log 3^2 = 2\log 3 \approx 2\times 0.4771 = 0.9542$$

$$\log 40 = \log(4\times 10) = \log 4 + \log 10 \approx 0.6020 + 1 = 1.6020$$

$$\log 81 = \log 9^2 = \log 3^4 = 4\log 3 \approx 4\times 0.4771 = 1.9084$$

$$\log 98 = \log(2\times 49) = \log 2 + \log 7^2 = \log 2 + 2\log 7 \approx 0.3010 + 2\times 0.8451 = 1.9912$$

$$\log(\frac{3}{28}) = \log 3 - \log 28 = \log 3 - \log(4\times 7) = \log 3 - \log 4 - \log 7 \approx 0.4771 - 0.6020 - 0.8451 = -0.97$$

$$\log\sqrt[3]{3\sqrt{5}} = \log(3\sqrt{5})^{\frac{1}{3}} = \frac{1}{3}(\log 3 + \log\sqrt{5}) = \frac{1}{3}(\log 3 + \frac{1}{2}\log 5) \approx \frac{1}{3}(0.4771 + \frac{1}{2}\times 0.6990) = 0.2755.$$

Table 6.2 Common rules for logarithmic operations

$a > 0$ and $a \neq 1$; $b > 0$ and $b \neq 1$; $x > 0$ and $y > 0$		
	Rules	***Examples***
1	$\log_b 1 = 0$	$\log 1 = 0,\ \log_2 1 = 0,\ \ln 1 = 0,\ \log_{15.2} 1 = 0,\ \log_{0.5} 1 = 0$
2	$\log_b b = 1$	$\log 10 = 1,\ \ln e = 1,\ \log_{15.2} 15.2 = 1,\ \log_{0.5} 0.5 = 1$
3	$\log_b b^x = x$ (b^x can be exactly divided by b for x times.)	$\log 10^2 = 2,\ \log_2 2^{8.1} = 8.1,\ \ln e^{\frac{2}{3}} = \frac{2}{3},\ \log_{0.5} 0.5^{\sqrt{2}} = \sqrt{2}$
4	$\log_b x^n = n\log_b x$ or $n\log_b x = \log_b x^n$	$\log x^2 = 2\log x,\ \log_2 4^x = x\log_2 4 = x\log_2 2^2 = 2x$ $\frac{1}{2}\log_{0.5} 0.25 = \log_{0.5}(0.25)^{\frac{1}{2}} = \log_{0.5}[(0.5)^2]^{\frac{1}{2}} = \log_{0.5}(0.5)^{\frac{2}{2}} = 1$
5	$\log_b(xy) = \log_b x + \log_b y$ or $\log_b x + \log_b y = \log_b(xy)$	$\log(10\sqrt{10}) = \log 10 + \log\sqrt{10} = \log 10 + \log 10^{\frac{1}{2}} = 1 + \frac{1}{2} = \frac{3}{2}$ $\log_{0.5} 0.25 + \log_{0.5} 2 = \log_{0.5}(0.25 \times 2) = \log_{0.5} 0.5 = 1$
6	$\log_b(xy)^n = n(\log_b x + \log_b y)$ or $n(\log_b x + \log_b y) = \log_b(xy)^n$	$\log_2(4 \times 2)^3 = 3(\log_2 4 + \log_2 2) = 3[(\log_2 2^2) + 1] = 3(2+1) = 9$ $\frac{1}{3}(\log 3 + \log 9) = \log(3 \times 9)^{\frac{1}{3}} = \log[(3^3)^{\frac{1}{3}}] = \log(3)^{\frac{3}{3}} = \log 3$
7	$\log_b(\frac{x}{y}) = \log_b x - \log_b y$ or $\log_b x - \log_b y = \log_b(\frac{x}{y})$	$\log_2(\frac{1}{2}) = \log_2 1 - \log_2 2 = 0 - 1 = -1$ $\log 300 - \log 3 = \log\frac{300}{3} = \log 100 = \log 10^2 = 2$
8	$\log_b(\frac{x}{y})^n = n(\log_b x - \log_b y)$ or $n(\log_b x - \log_b y) = \log_b(\frac{x}{y})^n$	$\log_2(\frac{1}{4})^3 = 3(\log_2 1 - \log_2 4) = 3[0 - (\log_2 2^2)] = 3(0-2) = -6$ $\frac{1}{2}(\log 50 - \log 2) = \frac{1}{2}\log(\frac{50}{2}) = \frac{1}{2}\log 25 = \frac{1}{2}\log 5^2 = \log 5$
9	$b^{\log_b x} = x$	$10^{\log 3} = 3,\ 2^{\log_2 10} = 10,\ e^{\ln(3y)} = 3y,\ 0.5^{\log_{0.5} 8.2} = 8.2$
10	$\log_b x = \frac{\log_a x}{\log_a b}$ (base exchange)	$\log_{0.1} 10 = \frac{\log 10}{\log 0.1} = \frac{1}{\log(\frac{1}{10})} = \frac{1}{\log 1 - \log 10} = \frac{1}{0-1} = -1$

Example 6.10: Determine the domain of function $y = 10^{\log(x+2)}$.

Solution

According to Rule 9 in Table 6.2,

$$y = 10^{\log(x+2)} = x + 2.$$

Although any value of x can satisfy $y = x + 2$, however, $\log(x+2)$ requires $x + 2 > 0$. Thus the domain of $y = 10^{\log(x+2)}$ is $x > -2$.

Example 6.11: Use the rules in Table 6.2 to evaluate:

a) $2^{\log_4 8}$,

b) $\dfrac{(\sqrt[3]{4\sqrt{8}})(\sqrt[6]{2})}{\sqrt[5]{32}}$.

Solution

a) By Rule 10 in Table 6.2, apply binary logarithm to the exponent $\log_4 8$

$$\log_4 8 = \frac{\log_2 8}{\log_2 4} = \frac{\log_2 2^3}{\log_2 2^2} = \frac{3}{2}.$$

Thus,

$$2^{\log_4 8} = 2^{\frac{3}{2}} = \sqrt{8} = 2\sqrt{2}.$$

b) Let $A = \dfrac{(\sqrt[3]{4\sqrt{8}})(\sqrt[6]{2})}{\sqrt[5]{32}}$. Apply binary logarithm to both sides,

$$\log_2 A = \log_2 \frac{(\sqrt[3]{4\sqrt{8}})(\sqrt[6]{2})}{\sqrt[5]{32}} = \log_2[(\sqrt[3]{4\sqrt{8}})(\sqrt[6]{2})] - \log_2 \sqrt[5]{32} = \log_2(\sqrt[3]{4\sqrt{8}}) + \log_2 \sqrt[6]{2} - \log_2 \sqrt[5]{32}$$

$$= \frac{1}{3}(\log_2 4 + \log_2 \sqrt{8}) + \frac{1}{6}\log_2 2 - \frac{1}{5}\log_2 32 = \frac{1}{3}(\log_2 2^2 + \frac{1}{2}\log_2 8) + \frac{1}{6}\times 1 - \frac{1}{5}\log_2 2^5$$

$$= \frac{1}{3}(2 + \frac{1}{2}\log_2 2^3) + \frac{1}{6} - \frac{1}{5}\times 5 = \frac{1}{3}(2 + \frac{1}{2}\times 3) + \frac{1}{6} - 1 = \frac{2}{3} + \frac{1}{2} + \frac{1}{6} - 1 = \frac{4}{6} + \frac{3}{6} + \frac{1}{6} - \frac{6}{6} = \frac{2}{6} = \frac{1}{3}.$$

Thus,

$$A = 2^{\frac{1}{3}} = \sqrt[3]{2}, \text{ or } \frac{(\sqrt[3]{4\sqrt{8}})(\sqrt[6]{2})}{\sqrt[5]{32}} = \sqrt[3]{2}.$$

6.2.2 Logarithmic scales

The single logarithmic scale

The graphs of functions introduced by far have been plotted onto the xy-plane using '*linear scales*' for both the x-axis and the y-axis, i.e., the interval between any two adjacent major units keeps the same value. For example, the interval between the two adjacent units x_1 (or y_1) = 1 and x_2 (or y_2) = 2 is 1, so be that between x_{10} (or y_{10}) = 10 and x_{11} (or y_{11}) = 11. The linear scale presents graphs well for many functions like linear and quadratic functions. However, for exponential and logarithmic functions, the linear scale is incapable of presenting their graphs in a wider range without losing some details. For instance, if x increases from 0 to 4, the y-value of the common exponential function $y = 10^x$ will increase from 1 to 10,000. If x further increases to 6, y increases to 1 million, which will compress the details of smaller y-values in the graph in linear scales (Figure 6.5a).

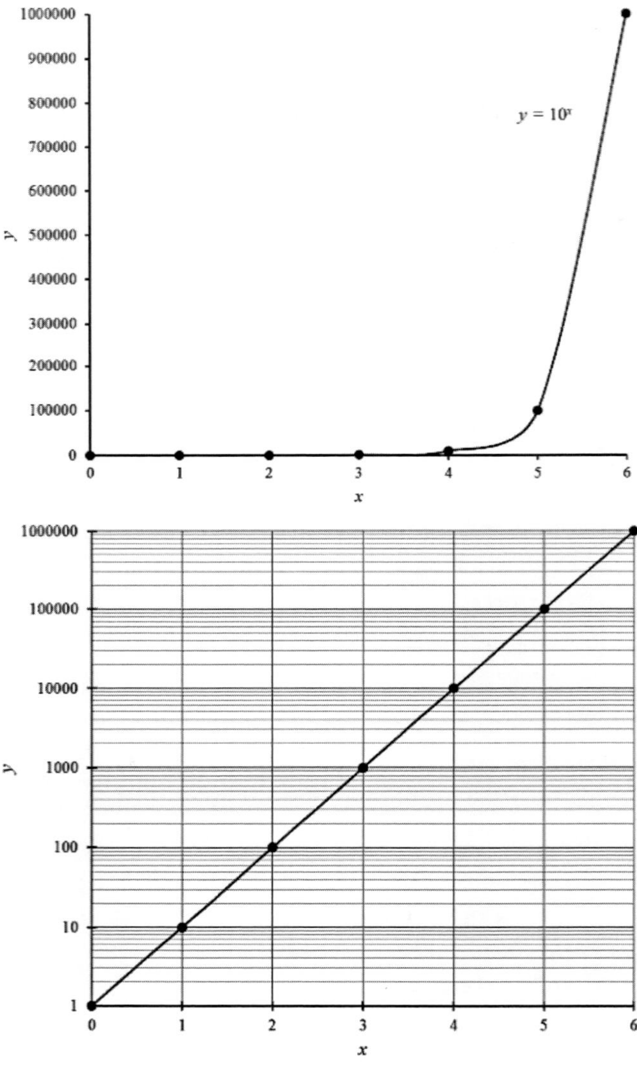

Figure 6.5: Graphs of $y = 10^x$ in linear scales (a) and the single logarithmic scale (b)

To give the smaller y-values the same '*weight of exposure*' on the xy-plane, we can use a logarithmic scale to represent the y-values of an exponential function, such as using the common logarithmic scale to represent $y = 10^x$ whereas still keeping x in linear scale. Such action will create a graph for the same function (such as $y = 10^x$) in a (single) logarithmic scale shown in Figure 6.5b. This time all values of y are given the same '*weight of exposure*' regardless of 10 or 1 million. This is achieved by increasing the interval between any two consecutive major units by the base factor of 10 for $y = 10^x$ for example. The interval between the two adjacent units in the x-axis is still 1. Since logarithm is the inverse of exponent, the same principle should be applicable to logarithmic functions by adopting a logarithmic scale in the x-axis whereas keeping a linear scale in the y-axis.

Example 6.12: Plot the graph for function $y = \log\sqrt{x} = \frac{1}{2}\log x$ using both the linear and single logarithmic scales respectively.

Solution

The common logarithm $\log x$ compresses x-values by factors of 10. For example, 100 has 2 factors of 10, i.e., $100 = 10\times10 = 10^2$ so $y = \log 100 = 2$. Letting $X = \log x$, the given function becomes $y = 0.5X$, effectively a liner function between y and X. This offers a way to create the graph for $y = \frac{1}{2}\log x$ using two sets of data pairs: the x-y data pairs for $y = \frac{1}{2}\log x$ and X-y data pairs for $y = \frac{1}{2}X$, tabulated in the table below. The graph in linear scales is shown in Figure 6.6a, a curve that clusters the smaller x-values and stretches the larger x-values. The graph in (single) logarithmic scale is a line, evenly distributed along the x-axis (Figure 6.6b).

x	0.001	0.01	0.1	1	20	40	80	100	300	600	1000
$X = \log x$	-3.00	-2.00	-1.00	0	1.301	1.602	1.903	2.00	2.477	2.778	3.00
$y = 0.5X$	-1.50	-1.00	-0.50	0	0.651	0.801	0.952	1.00	1.239	1.389	1.50

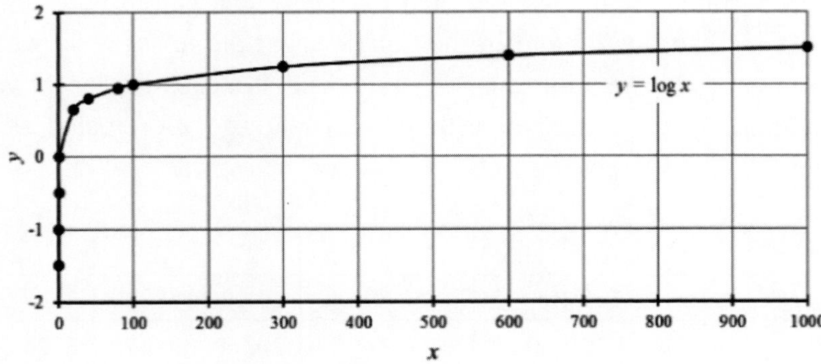

Figure 6.6a: Graph of $y = 0.5\log x$ in linear scales

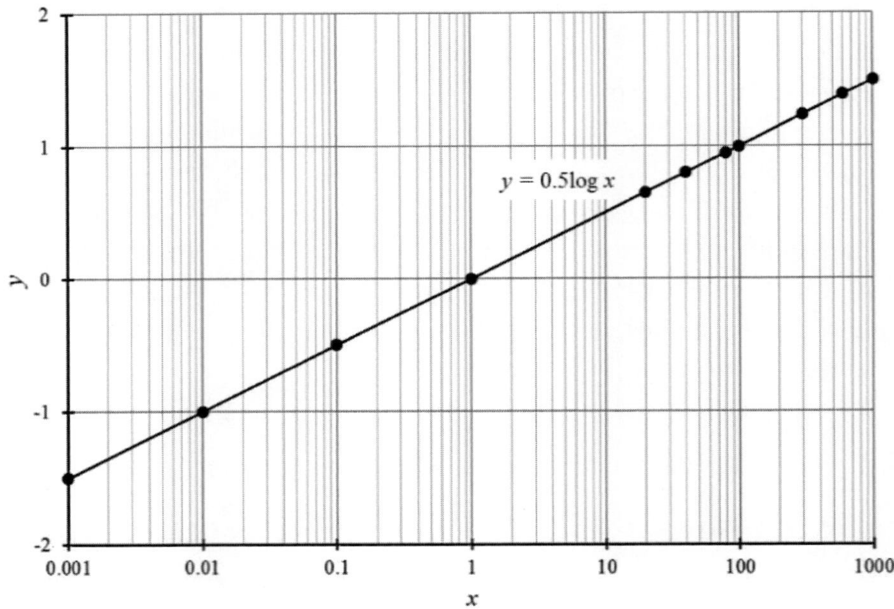

Figure 6.6b: Graph of $y = 0.5\log x$ in the single logarithmic scale

The double logarithmic scale

In some circumstances, logarithmic scales must be adopted for both the x-axis and the y-axis, which leads to the double logarithmic scale. The following example shows an application of the double logarithmic scale.

Example 6.13: Plot the graph of the function defined by $x = 0.5 \times 10^{2t}$ and $y = 0.2 \times 10^{0.5t^2}$ in range $t[0, 3]$ using both the linear and double logarithmic scales respectively.

Solution

The relationship between x and y is defined by another common parameter t. Similar to the process adopted in Example 6.12, we first calculate x, y, $\log x$ and $\log y$ within range $t[0, 3]$, which are tabulated in the table below. We then use these data pairs to create the x-y graph in linear scales shown in Figure 6.7a, and in the double logarithmic scale shown in Figure 6.7b. The graph in linear scales clusters the smaller values and stretches the larger values for both x and y whereas the graph in the double logarithmic scale shows a more balanced distribution for both x and y.

t	0	0.5	1	1.5	2	2.5	3
$x = 0.5 \times 10^{2t}$	0.5	5	50	500	5000	50000	500000
$y = 0.2 \times 10^{0.5t^2}$	0.20	0.27	0.63	2.67	20.00	266.70	6324.56
$\log x$	-0.30	0.70	1.70	2.70	3.70	4.70	6.70
$\log y$	-0.70	-0.57	-0.20	0.43	1.30	2.43	3.80

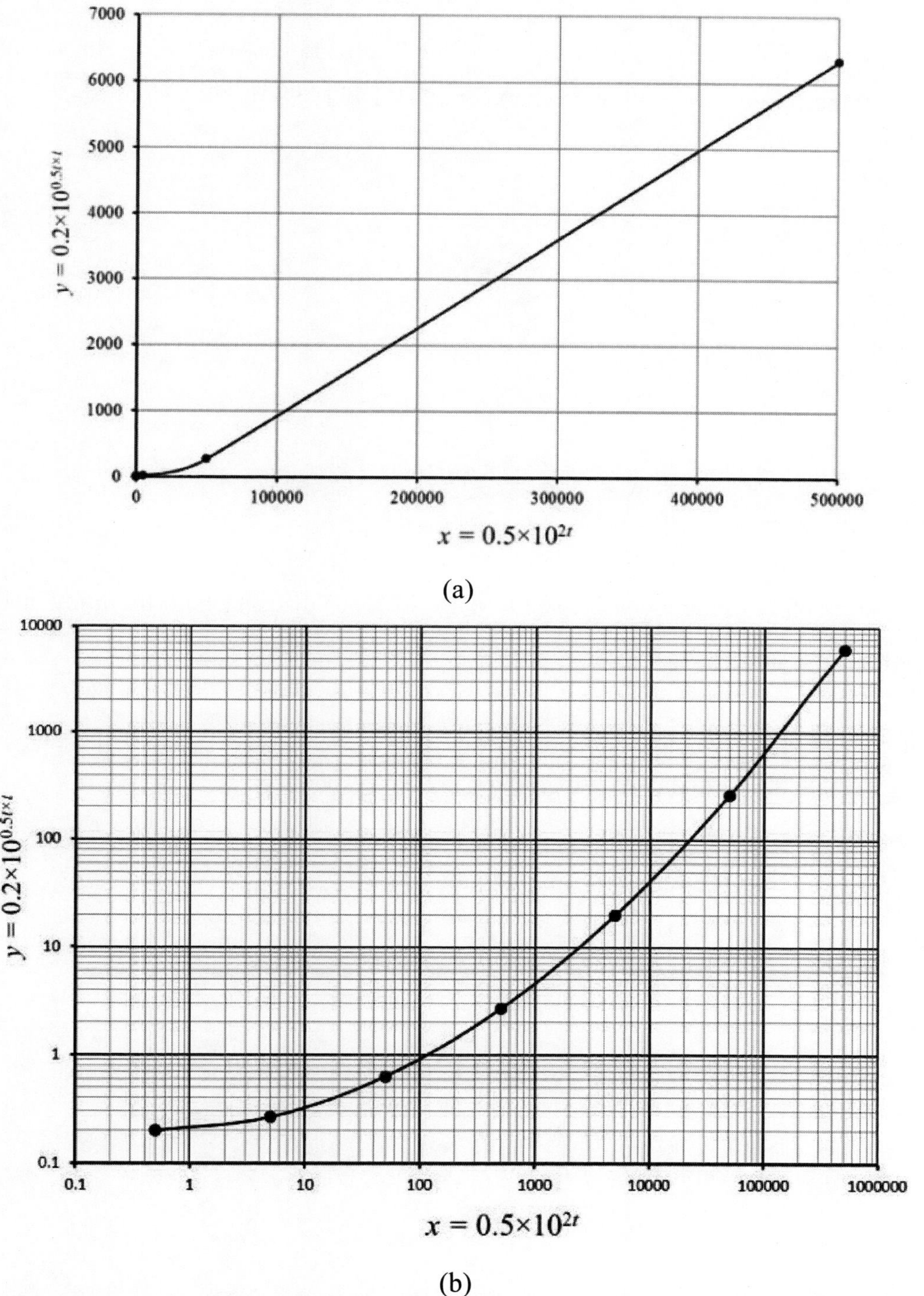

Figure 6.7: Graphs of functions in Example 6.13 in linear scales (a) and double logarithmic scales (b)

Exercises 6.2

1. Given log2 $\approx$ 0.3010, log3 $\approx$ 0.4771, and log7 $\approx$ 0.8451, use the rules in Table 6.2 to calculate the common logarithmic values for 12, 15, 49, 1/98, 1/280 and $\sqrt{14\sqrt[3]{420}}$.

2. Evaluate

   a) $2^{-\log_8 \sqrt{8}}$,

   b) $y = 10^{\log(x^2-5)}$,

   c) $\dfrac{(\sqrt[3]{4\sqrt{8}})(\sqrt[5]{4})}{\sqrt[6]{128}}$.

3. Plot the graph for function $y = \log_2 x^{3/2}$ in [0.2, 128] using both the linear and single logarithmic scales respectively.

4. Plot the graph for function $y = \log\sqrt[3]{x} = \dfrac{1}{3}\log x$ in [1, 100] using both the linear and single logarithmic scales respectively.

5. Plot the graph of function defined by $x = 2\times 10^{0.2t}$ and $y = 0.5\times 10^{0.1t^2}$ in the range t[0, 5] using both the linear and double logarithmic scales respectively.

6.3 Applications of Exponential and Logarithmic Functions

6.3.1 Exponential and logarithmic equations

An equation including at least an unknown variable involving exponential and/or logarithmic forms is called an exponential and/or logarithmic equation. For instance, $50 = 0.5 \times 10^{2x}$ is an exponential equation whereas $2(\log x - 1) = \log(2x + 1)$ is a logarithmic equation. Not all exponential and logarithmic equations can be solved exactly by hand. Many sophisticated exponential and logarithmic equations may be solvable only by numeric computation.

Solving an exponential or a logarithmic equation requires firstly converting both sides of the equation to the similar forms with the same base using the general properties of exponents or logarithms. With the same base, it may be possible to equate the corresponding parts on both sides of the equation to form an equivalent algebraic equation. Potential solutions can be obtained by solving such algebraic equation. However, these potential solutions must be examined so as to confirm if meeting the fundamental requirements for the exponential and/or logarithmic functions. The following examples demonstrate this general procedure for solving some simple exponential and logarithmic equations.

Example 6.14: Solve

a) $50 = 0.5 \times 10^{2x}$ and
b) $2(\log x - 1) = \log(2x)$.

Solution

a) Solve exponential equation $50 = 0.5 \times 10^{2x}$.

Step 1: Converting both sides of the equation to exponents with the same base of 10

$$50 = 0.5 \times 10^{2x} \longrightarrow 10^{2x} = \frac{50}{0.5} = 100 \xrightarrow{or} 10^{2x} = 10^2$$

Step 2: Equating the exponents with the same base of 10 on both sides of the converted terms

Let $2x = 2 \longrightarrow x = 1$.

Step 3: Examining if $x = 1$ is in the domain of the given exponential equation

Since 10^x accepts any value for x, $x = 1$ is the solution to $50 = 0.5 \times 10^{2x}$.

b) Solve logarithmic equation $2(\log x - 1) = \log(2x)$.

Step 1: Converting both sides of the equation to logarithms with the same base of 10

Left side: $2(\log x - 1) = 2(\log x - \log 10) = 2\log(\frac{x}{10}) = \log(\frac{x}{10})^2$

No change to the right side as it is already in the common logarithm: $\log(2x)$

Thus: $2(\log x - 1) = \log(2x) \longrightarrow \log(\frac{x}{10})^2 = \log(2x)$

Step 2: Equating the terms upon which the common logarithm was applied on the both sides

Let $(\frac{x}{10})^2 = 2x \longrightarrow \frac{x^2}{100} = 2x \longrightarrow x^2 = 200x \longrightarrow x^2 - 200x = 0 \longrightarrow x(x-200) = 0$

Thus $x = 0$ and $x = 200$.

Step 3: Examining if $x = 0$ and $x = 200$ are in the domain of the given logarithmic equation

Since logarithm is only for positive values, $x = 0$ is not a solution. The solution is $x = 200$.

If both sides of an exponential or a logarithmic equation cannot be converted to the terms with the same base, other approaches, such as substitution and approximation, should be considered. The following examples show how these can be applied to solving exponential and logarithmic equations.

Example 6.15: Solve $2^x + 12 = 4^x$.

Solution

Since $4^x = (2^2)^x = 2^{2x}$, thus $2^x + 12 = 2^{2x}$.

Let $2^x = t$, then $2^{2x} = t^2$.

Hence $2^x + 12 = 2^{2x} \longrightarrow t + 12 = t^2 \longrightarrow t^2 - t - 12 = 0 \longrightarrow (t-4)(t+3) = 0$.

Thus $t = 4$ and $t = -3 \longrightarrow 2^x = 4$ and $2^x = -3$ (*improper*).

As $2^x = 4 = 2^2$, thus $x = 2$.

Example 6.16: Solve $e^{x^2+1} = 5e^{-x}$.

Solution

Apply natural logarithm to both sides of the equations

$\ln(e^{x^2+1}) = \ln(5e^{-x}) \longrightarrow x^2 + 1 = \ln 5 + \ln e^{-x} \longrightarrow x^2 + 1 = \ln 5 - x \longrightarrow x^2 + x + 1 - \ln 5 = 0$

$$x = \frac{-1 \pm \sqrt{1^2 - 4(1 - \ln 5)}}{2} \approx \frac{-1 \pm \sqrt{1 + 4 \times 0.6094}}{2} = \frac{-1 \pm \sqrt{3.4376}}{2} \approx \frac{-1 \pm 1.854}{2}.$$

Thus $x \approx 0.427$ and $x \approx -1.427$, both being the solutions to $e^{x^2+1} = 5e^{-x}$.

Example 6.17: Solve $\log_2(3x) - 1 = \log_8 x$.

Solution

Since $\log_2(3x) - 1 = \log_2 3 + \log_2 x - 1$, $\log_8 x = \frac{\log_2 x}{\log_2 8} = \frac{\log_2 x}{\log_2 2^3} = \frac{\log_2 x}{3}$,

thus $\log_2(3x) - 1 = \log_8 x \longrightarrow \log_2 3 + \log_2 x - 1 = \frac{\log_2 x}{3} \longrightarrow \log_2 x - \frac{1}{3}\log_2 x = 1 - \log_2 3$

$\frac{2}{3}\log_2 x = \log_2 2 - \log_2 3 \longrightarrow \log_2 x^{\frac{2}{3}} = \log_2 \frac{2}{3}$, thus,

$x^{\frac{2}{3}} = \frac{2}{3} \longrightarrow x = (\frac{2}{3})^{\frac{3}{2}} = \sqrt{\frac{8}{27}}$.

6.3.2 Compound interest

If you deposit a certain amount of money (called principal P) into a term account in a bank for n years with an annual interest rate of r, the annual interest earned in the previous year will be added to the principal to earn extra interest for the following year. This compounding process repeats itself for the duration of the term. This is commonly known as "*interest on interest*".

In the beginning, suppose the principal is P and the annual interest is r. By the end of the first year, there is an amount of $P + rP = P(1 + r)$ in the account. This is the new principal for the second year, which attracts an interest at rate r in the second year. This repeating process can be described mathematically as follows.

Year	Principal	Interest	Compounded principal
1	P	rP	$P + rP = P(1+r)$
2	$P(1+r)$	$rP(1+r)$	$P(1+r) + rP(1+r) = P(1+r)(1+r) = P(1+r)^2$
3	$P(1+r)^2$	$rP(1+r)^2$	$P(1+r)^2 + rP(1+r)^2 = P(1+r)^2(1+r) = P(1+r)^3$
…	…	…	…
n	$P(1+r)^{n-1}$	$rP(1+r)^{n-1}$	$P(1+r)^{n-1} + rP(1+r)^{n-1} = P(1+r)^{n-1}(1+r) = P(1+r)^n$

At the end of n years, the compound amount (S) of the initial principal P at the rate of r compounded annually is

$$S = P(1+r)^n. \tag{6.10}$$

The *compounded interest* (*CI*) earned is

$$CI = S - P = P(1+r)^n - P = P[(1+r)^n - 1]. \tag{6.11}$$

Example 6.18: Alison deposited \$1000 into a 5-year term account at the rate 5.5% compounded annually.

a) Find the compound amount;
b) Find the compounded interest;
c) How long would it be required to reach a compound amount of \$1200?

Solution

a) Use formula (6.10) to find the compound amount:

$$S = P(1+r)^n = 1000(1+0.055)^5 = \$1306.96$$

b) Use formula (6.11) to find the compound interest:

$$CI = S - P = 1306.96 - 1000 = \$306.96\ .$$

c) Use formula (6.10) to find the time n when the compound amount reaching \$1200:

$$S = P(1+r)^n \longrightarrow 1200 = 1000(1+0.055)^n \longrightarrow 1.055^n = \frac{1200}{1000} = 1.2$$

$$\text{Apply common logarithm to both sides: } \log 1.055^n = \log 1.2 \longrightarrow n \log 1.055 = \log 1.2$$

$$\text{Thus: } n = \frac{\log 1.2}{\log 1.055} \approx \frac{0.0792}{0.0233} \approx 3.4 \text{ (years).}$$

Example 6.19: Bill also deposited \$1000 into a 5-year term account in another bank at a higher rate compounded annually. By the end of the term, Bill received a compounded interest of \$402.56.

a) Find the compounded annual rate for this term deposit;
b) How long would it be required to reach a compound amount of \$1200?

Solution

a) Use formula (6.11) to work out the compounded annual rate of interest:

$$CI = P[(1+r)^n - 1] \longrightarrow 402.55 = 1000[(1+r)^5 - 1] \longrightarrow (1+r)^5 - 1 = 0.40255$$

$$(1+r)^5 = 1.40255 \longrightarrow 1+r = \sqrt[5]{1.40255} \longrightarrow r = \sqrt[5]{1.40255} - 1 \approx 1.07 - 1 = 0.07 = 7\%.$$

b) Use formula (6.10) to find the time n when the compound amount reaching \$1200:

$$S = P(1+r)^n \longrightarrow 1200 = 1000(1+0.07)^n \longrightarrow 1.07^n = \frac{1200}{1000} = 1.2$$

Apply common logarithm to both sides: $\log 1.07^n = \log 1.2 \longrightarrow n \log 1.07 = \log 1.2$

Thus: $n = \dfrac{\log 1.2}{\log 1.07} \approx \dfrac{0.0792}{0.0294} \approx 2.7$ (years).

6.3.3 Population growth

Formula (6.10) can also be used to predict population growth for human, cells, bacteria and others in a certain period. In such applications, P is the initial population in the beginning of the period; n is the final count of the period (maybe in years, months, weeks, days, hours, and so on for different circumstances); r is the growth rate with respect to the same unit of time.

Example 6.20: A city currently has a population of 1 million people. Due to stagnation of economy, more people leave the city for jobs elsewhere. It is estimated the annual growth rate of the city for the next 10 years would be –0.1% per year or losing one per one thousand people annually.

a) Forecast the population of this city in 5 years and 10 years respectively;
b) What would be the required annual growth rate if the City Council wants the population to reach 1.1 million in 10 years?

Solution

a) Use formula (6.10) to predict the population of this city with $P = 1$ m, $r = -0.1\% = -0.001$:

$$n = 5: \quad S = P(1+r)^n = 1000000(1-0.001)^5 \approx 995{,}010;$$

$$n = 10: \quad S = P(1+r)^n = 1000000(1-0.001)^{10} \approx 990{,}045.$$

b) Use formula (6.10) to find the required growth rate r in 10 years:

$$S = P(1+r)^n \longrightarrow 1100000 = 1000000(1+r)^{10} \longrightarrow (1+r)^{10} = \frac{1100000}{1000000} = 1.1$$

$$1+r = (1.1)^{1/10} \approx 1.00958 \longrightarrow r \approx 1.00958 - 1 \approx 0.0096 = 0.96\%.$$

Example 6.21: Suppose a certain type of cell can double itself in every 4 hours within the first 48 hours. If in the beginning there are 5 such cells, find the total number of cells after 48 hours.

Solution

We can use formula (6.10) to calculate the total number of cells ever lived after 48 hours. The cell reproduces itself in every 4 hours, i.e., one more from one cell, or r = 100% = 1 *per 4 hours*. Within 48 hours, there are 48 /4 = 12 units of 4 hours so $n = 12$. Given $P_0 = 5$, $r = 1$ and $n = 12$,

$$S_{12} = P_0(1+r)^n = 5(1+1)^{12} = 5\times 2^{12} = 20,480.$$

Once a cell is split, the original cell disappears. The number of cells alive after 48 hours will be

$$P_{12} = S_{12} - S_{11} = 5\times 2^{12} - 5\times 2^{11} = 5\times 2^{11}(2-1) = 10,240.$$

Note that converting r = 1 *per 4 hours* into r = 1/4 = 0.25 *per hour* is not correct because cells double themselves ONLY in every 4 hours.

6.3.4 Earthquake energy

Earthquake releases energy that is related to the magnitude of the earthquake. One of many ways to estimate the energy released by an earthquake is by the following formula

$$E = 10^{1.5M+4.8}\text{ (J)}, \tag{6.12}$$

where E is the energy released by the earthquake in joules and M represents the magnitude determined in the Richter scale. Joule (= Nm) is a unit of measure for energy.

Example 6.22: Christchurch in New Zealand was shaken by an earthquake of magnitude 7.1 on 4 September 2010. Another earthquake of magnitude 6.3 occurred in that region again on 22 February 2011. Estimate the energy released by the two earthquakes respectively.

Solution

Use formula (6.12) to estimate the energy released by the two earthquakes respectively:

$$Ms = 7.1:\quad E_{7.1} = 10^{1.5M+4.8} = 10^{1.5\times 7.1+4.8} = 10^{15.45} \approx 2.82\times 10^{15}\text{(J)} = 2.82\text{ (PJ)};$$
$$Ms = 6.3:\quad E_{6.3} = 10^{1.5M+4.8} = 10^{1.5\times 6.3+4.8} = 10^{14.25} \approx 1.78\times 10^{14}\text{(J)} = 0.178\text{ (PJ)};$$
$$\frac{E_{7.1}}{E_{6.3}} = \frac{2.82}{0.178} \approx 15.8.$$

The ratio $E_{7.1}/E_{6.3}$ = 15.8 means the energy released by the quake of magnitude 7.1 is about 16 times more than that released by the quake of magnitude 6.3.

6.3.5 Radioactive decay

The amount of a radioactive element is self-decreasing with time. This is known as *radioactive decay* that follows a law of natural exponential function

$$\boxed{N = N_0 e^{-\lambda t}}, \tag{6.13}$$

here N is the amount of a radioactive material at time t; N_0 is the initial amount of such material in the beginning (t=0); λ is a positive constant called the *decay constant* for the material.

By the time the amount of a radioactive element is reduced to half of the initial amount, i.e., $N = \frac{1}{2}N_0$, that moment is called the *half-life* (T) of the element, which can be derived by

$$\frac{1}{2}N_0 = N_0 e^{-\lambda T} \longrightarrow \frac{1}{2} = e^{-\lambda T} \longrightarrow \ln e^{-\lambda T} = \ln\frac{1}{2} \longrightarrow -\lambda T = \ln 1 - \ln 2 = -\ln 2, \text{ hence}$$

$$\boxed{T = \frac{\ln 2}{\lambda}.} \tag{6.14}$$

Example 6.23: A radioactive element has a decay constant 0.007/day. Initially the amount of the element is 10 grams.

a) Find the amount remains after 50 days;
b) Find the half-life of this radioactive element.

Solution

a) Use formula (6.13) to determine the remaining amount after 50 days with N_0 = 10 g, λ = 0.007 and t = 50:

$$N = N_0 e^{-\lambda t} = 10e^{-0.007\times 50} = 10e^{-0.35} \approx 7.05\ (g).$$

b) By formula (6.14) the half-life of this element is:

$$T = \frac{\ln 2}{\lambda} \approx \frac{0.6931}{0.007} \approx 99\ (days).$$

Example 6.24: A radioactive element has 9 grams left after 100 days and a half-life of 117.5 days. Find the initial amount of this element.

Solution

As T = 117.5 days, the decay constant λ can be determined by formula (6.14) as follows:

$$T = \frac{\ln 2}{\lambda} \longrightarrow \lambda = \frac{\ln 2}{T} \approx \frac{0.6931}{117.5} \approx 0.0059\ (per\ day).$$

As $\lambda = 0.0059$, $t = 100$ and $N(100) = 9$, by formula (6.13),

$$N = N_0 e^{-\lambda t} \longrightarrow 9 = N_0 e^{-0.0059\times100} \longrightarrow 9 = N_0 e^{-0.59} \longrightarrow 9 \approx 0.5543 N_0$$

$$N_0 \approx \frac{9}{0.5543} \approx 16.237\ (g).$$

The initial amount of this element is about 16.237 grams.

6.3.6 Charging capacitors

The voltage across the capacitor V_C at time t when charging a capacitor with capacitance C and resistance R is determined by

$$V_C = V_0(1 - e^{-\frac{t}{RC}}), \tag{6.15}$$

where V_0 is the supply voltage. This is a natural exponential function.

Example 6.25: Assume the supply voltage is 36 voltages. The capacitance is 5×10^{-4} farads, and the resistance is 2×10^5 ohms. Plot the graph of V_C from 0 to 600 seconds.

Solution

Using formula (6.15), we can calculate the voltage across the capacitor V_C at any time from 0 to 600 seconds. Since $V_0 = 36$ voltages, $R = 2\times10^5$ ohms, and $C = 5\times10^{-4}$ farads,

$$RC = 2\times10^5 \times 5\times10^{-4} = 10\times10^{5-4} = 100,$$

$$V_C = V_0(1 - e^{-\frac{t}{RC}}) = 36(1 - e^{-\frac{t}{100}}) = 36(1 - e^{-0.01t}).$$

The t-V_C data pairs with an interval of 60 seconds are calculated and listed in the table below. The graph of this charging process is plotted in Figure 6.8.

t (s)	0	60	120	180	240	300	360	420	480	540	600
V_C (v)	0	16.24	26.16	30.05	32.73	34.21	36.02	36.46	36.70	36.84	36.91

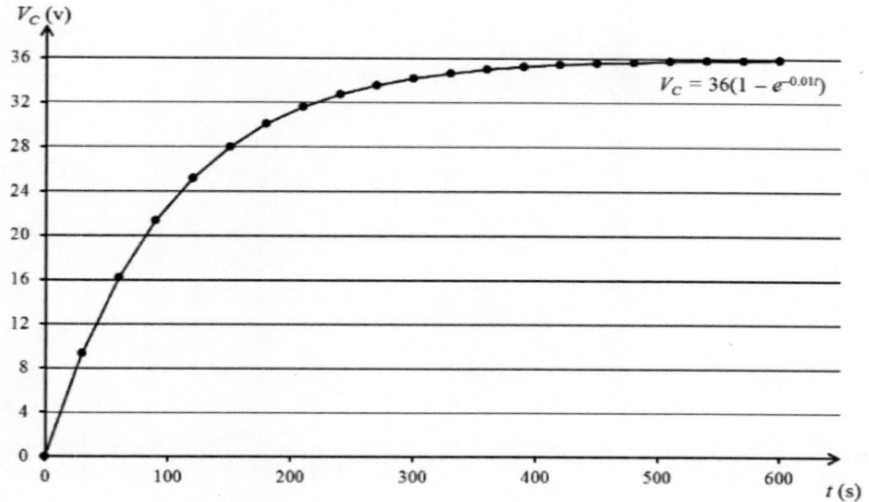

Figure 6.8: The graph of charging a capacitor from 0 to 600 seconds in Example 6.25

Example 6.26: In Example 6.25, when would the capacitor be charged to half of the supply voltage?

Solution

In Example 6.25, the voltage across the capacitor V_C at any time is $V_C = 36(1 - e^{-0.01t})$. When the voltage across the capacitor is half of the supply voltage, i.e., 18 voltages, it means

$$18 = 36(1 - e^{-0.01t}) \longrightarrow \frac{18}{36} = 1 - e^{-0.01t} \longrightarrow 0.5 = 1 - e^{-0.01t} \longrightarrow e^{-0.01t} = 0.5.$$

Apply natural logarithm to both sides of $e^{-0.01t} = 0.5$,

$$-0.01t = \ln 0.5 \longrightarrow t = \frac{\ln 0.5}{-0.01} = \frac{-0.6931}{-0.01} \approx 69.31\ (s).$$

The capacitor would be charged to half of the supply voltage in 69.31 seconds; a little over 1 minute after charging was begun.

Exercises 6.3

1. Solve the following equations:

   a. $0.5 = 50 \times 10^{-3x}$
   b. $\log(x^2 + 100) - 1 = \log(2x)$
   c. $2^{x+3} = 4^x$
   d. $\log_2(4x) + 1 = \log_4 x^3$
   e. $e^{-x^2+1} = 3e^{-2x}$

2. Lisa deposited \$3500 into a 3-year term account at the rate 5.5% compounded annually.

   a. Find the compound amount;
   b. Find the compounded interest;
   c. How long would it be required to reach a compound amount of \$5000?

3. Bill deposited \$4000 into a 3-year term account in another bank at a higher rate compounded annually. By the end of the term, Bill received a compounded interest of \$911.55.

   a. Find the compounded annual rate for this termed deposit;
   b. How long would it be required to reach a compound amount of \$6000?

4. A city currently has a population of 300,000 people. It is estimated the annual growth rate of the city for the next 10 years would be 0.75% per year.

   a. Forecast the population of this city in 5 years and 10 years respectively;
   b. What would be the required annual growth rate if the City Council wants the population to reach 400,000 in 10 years?

5. A radioactive element has 21 grams left after 30 days and a half-life of 91 days. Find the initial amount of this element.

6. A radioactive element has 20 grams left after 10 days. After another 10 days, it remains with 16 grams. Find the decay constant, half-life, and initial amount of this element.

7. Estimate the energy released by two earthquakes of magnitudes 8.5 and 7.3 respectively.

8. It takes 20 seconds to charge a capacitor to two-thirds of the supply voltage. What is the product of the capacitance and resistance of this capacitor?

Chapter 6: Exercises Answers

Exercises 6.1

1. a)

$21507 = 20000 + 1000 + 500 + 2 = 2\times10^4 + 1\times10^3 + 5\times10^2 + 0\times10 + 2\times10^0$

$711 = 700 + 10 + 1 = 7\times10^2 + 1\times10 + 1\times10^0$

$412.542 = 400 + 10 + 2 + 0.5 + 0.04 + 0.002 = 4\times10^2 + 1\times10 + 2\times10^0 + 5\times10^{-1} + 4\times10^{-2} + 2\times10^{-3}$

b) The order of 21507 is 4; that of 711 is 2; that of 421.542 is 2 as well.

2. $11101 = 29$, $10110101 = 181$
3. 19 → 10011 and 499 → 111110011

Exercises 6.2

1. $\log 12 \approx 1.0791$, $\log 15 \approx 1.1761$, $\log 49 \approx 1.6902$, $\log\frac{1}{98} \approx -1.9912$

   $\log\frac{1}{280} \approx -2.4471$, $\log\sqrt{14\sqrt[3]{420}} \approx 1.0103$

2. a) $2^{-\log_8\sqrt{8}} = \frac{1}{\sqrt{2}}$, b) $y = x^2 - 5$, c) $\frac{(\sqrt[3]{4\sqrt{8}})(\sqrt[5]{4})}{\sqrt[6]{128}} = 2^{2/5} = \sqrt[5]{4}$

Exercises 6.3

1. a) $x = \frac{2}{3}$, b) $x = 10$, c) $x = 3$, d) $x = 64$, e) $x \approx 1.9494$ and $x \approx 0.0506$
2. a) $S = \$4109.84$, b) $CI = \$609.84$, c) $n \approx 6.65 \approx 7$ (years)
3. a) $r \approx 7.1\%$, b) $n \approx 5.9 \approx 6$ (years)
4. a) $n = 5$: $P \approx 311420$; $n = 10$: $P \approx 323275$; b) $r \approx 2.919\%$
5. $N_0 \approx 26.379$ (g)
6. $\lambda \approx 0.0223$ days, T ≈ 31.08 days, $N_0 \approx 25$ g
7. $E_{8.5} \approx 3.55\times10^{17}$ (J) = 355 (PJ), $E_{7.3} \approx 5.62\times10^{15}$ (J) = 5.62 (PJ), $\frac{E_{8.5}}{E_{7.3}} \approx 63.17$
8. $RC \approx 18.2$ seconds

CHAPTER 7

7 Trigonometric and Hyperbolic Functions

CHAPTER OBJECTIVES

- Study characteristics of trigonometric functions
- Analyse characteristics of combinative functions by sine and cosine functions
- Introduce inverse trigonometric functions and principal-value ranges
- Solve simple trigonometric equations
- Introduce hyperbolic functions and properties

Essential statements on trigonometric functions:

- Periodic characteristics are the key to understand behaviours of trigonometric functions.
- Combinative functions by sine and cosine functions are commonly encountered in scientific and engineering applications.

Key topics:

- Characteristics of trigonometric functions
- Generic sine and cosine functions and combinations
- Inverse trigonometric functions and principal-value ranges
- Trigonometric equations
- Hyperbolic functions

Flowchart of mathematical knowledge building

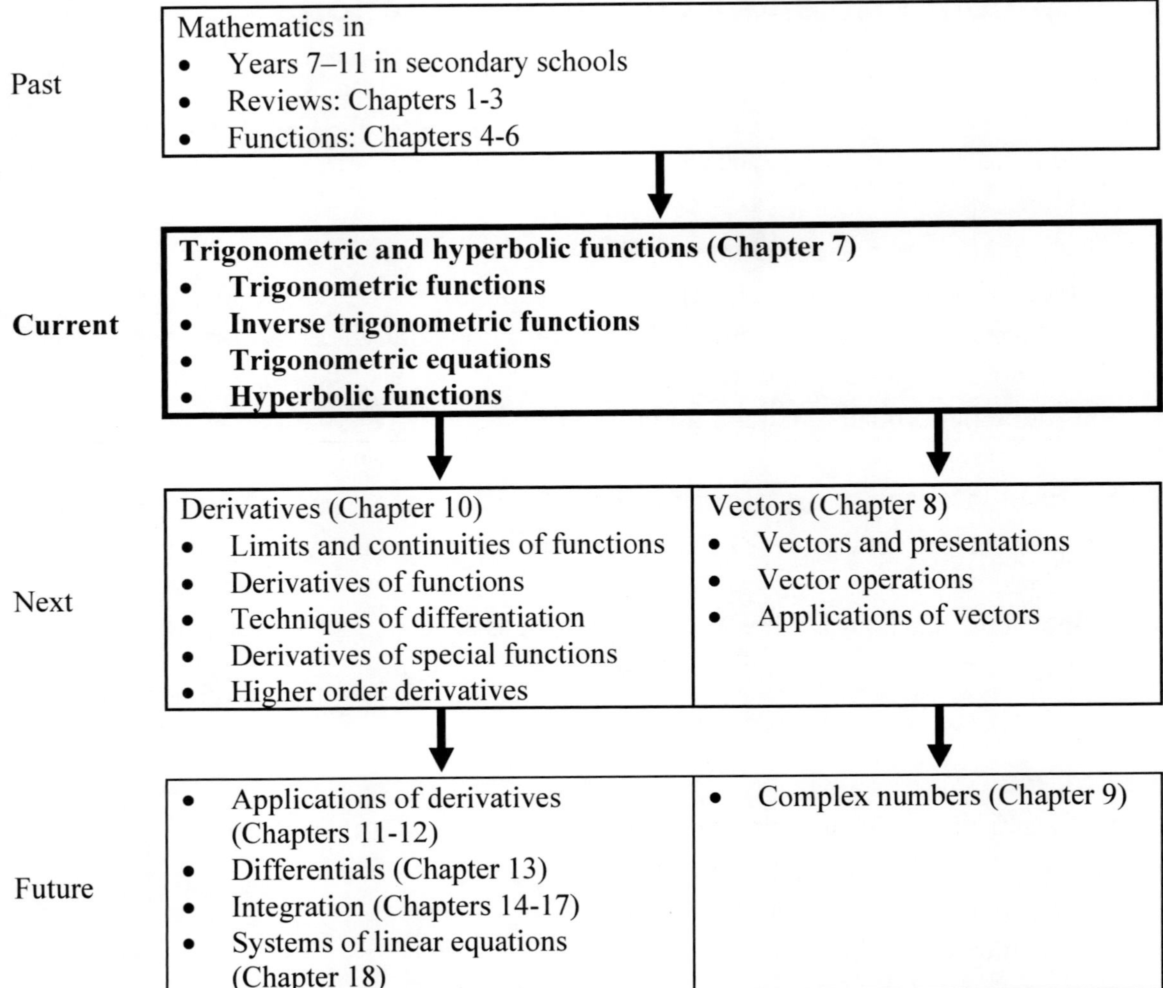

Chapter 7: Trigonometric and Hyperbolic Functions

7.1 General Trigonometric Functions

7.1.1 Characteristics of general trigonometric functions

Unit circle and periodicity of trigonometric functions

Although trigonometric functions were introduced using right triangles in Chapter 2, trigonometric functions are independent of triangles. Therefore, the trigonometric identities, relationships, formulae, and values of special angles are universally valid. Functions like linear, quadratic, exponential and logarithmic functions have an independent variable taking '*normal*' real numbers as its values. Trigonometric functions, however, only take '*angles*' either in radians or degrees as the values to independent variable θ (or α, β etc.). Values for any angular variable are naturally repeated in every full turn of 360° or 2π. Accordingly the values of trigonometric functions must repeat in every circle of 360° or 2π. Since the values of trigonometric functions only rely on the values of an angle, regardless of the length of sides that form the angle, a *unit circle* is commonly used to introduce the concept of general sine and cosine functions, the primary trigonometric functions.

A unit circle has a radius of 1. The independent variable 'angle' is represented by the usual symbol x, which begins from the easterly horizontal direction as its *initial angular position* (Figure 7.1). Rotating counter-clockwise sees x take positive values whereas rotating clockwise results in negative values for x. The negative values of x can be regarded as being generated by either rotating clockwise or keeping rotating counter-clockwise from the '*past*' to '*present*' and going further. With these assumptions in mind, the graphs of general functions $y = \sin x$ and $y = \cos x$ are shown in Figure 7.1, each as a result of rotating the unit circle.

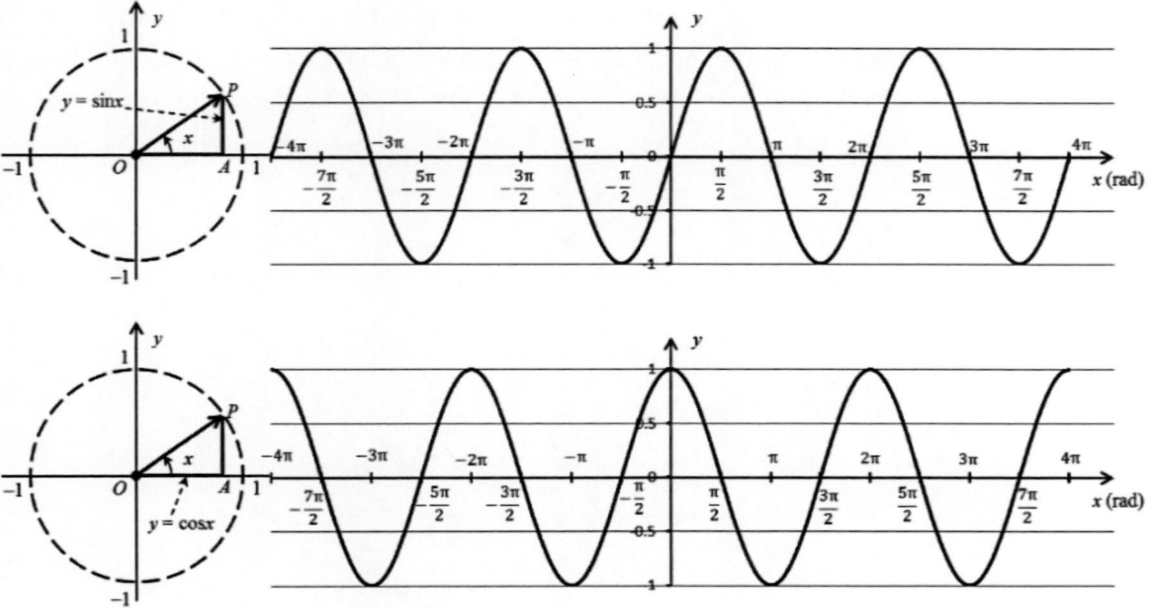

Figure 7.1: The general sine (a) and cosine (b) waves generated by rotating a unit circle

Both graphs of sin$x$ and cos$x$ look like '*waves*' and the waveforms repeat themselves over any interval of 2π. For example, the waveform of sin$x$ is the same within $[-3\pi/2, \pi/2]$ and $[\pi/2, 5\pi/2]$; the waveform of cos$x$ is the same within $[-2\pi, 0]$ and $[2\pi, 4\pi]$. This repetition of waveform is called *periodicity of sine and cosine functions*. In other words, sine and cosine functions are periodic over a period of 2π.

In Figure 7.1, $y = \sin x$ is the vertical projection of the unit radius at angle x whereas $y = \cos x$ is the horizontal projection of the unit radius at the same angle. Thus sin$x$ and cos$x$ are 90° or $\pi/2$ apart. This means shifting sin$x$ backwards or shifting cos$x$ forwards 90° or $\pi/2$ will make both sin$x$ and cos$x$ completely overlapped with each other.

Secondary functions tan$x$ and cot$x$, derived from sin$x$ and cos$x$, are periodic too but with a different period of π. Both tan$x$ and cot$x$ are mutually reciprocal and not defined in the whole angular domain (Figure 7.2). Function tan$x$ is undecided where x approaching $n\pi/2$ (n is any odd integer) because $\cos(n\pi/2) = 0$; cot$x$ is undecided where x approaching $n\pi$ (n is any integer) as $\sin(n\pi) = 0$.

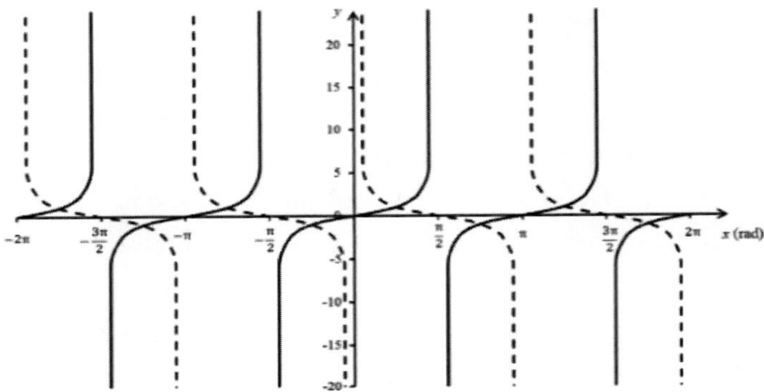

Figure 7.2: Graphs of the general tangent (solid curve) and cotangent (dashed curve) functions

Similarly sec$x$ and csc$x$ are reciprocal functions of cos$x$ and sin$x$ respectively with a period of 2π (Figure 7.3). Function sec$x$ is undecided where x approaching $n\pi/2$ (n is any odd integer) whereas csc$x$ is undecided where x approaching $n\pi$ (n is any integer).

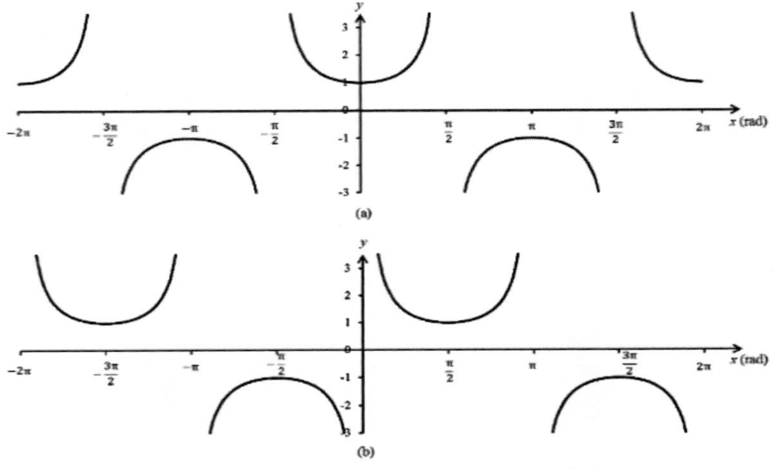

Figure 7.3: Graphs of the general secant (top) and cosecant (bottom) functions

Although the six trigonometric functions have different variations within one period, each has a specific pattern of variation with increases in angular values. These features are summarised in Table 7.1.

Table 7.1: Variations of trigonometric functions from 0 to 2π

Increase in x	0 to $\frac{\pi}{2}$	$\frac{\pi}{2}$ to π	π to $\frac{3\pi}{2}$	$\frac{3\pi}{2}$ to 2π	Period
sin$x$	↑ from 0 to 1	↓ from 1 to 0	↓ from 0 to –1	↑ from –1 to 0	2π
cos$x$	↓ from 1 to 0	↓ from 0 to –1	↑ from –1 to 0	↑ from 0 to 1	2π
tan$x$	↑ from 0 to ∞	↑ from –∞ to 0	↑ from 0 to ∞	↑ from –∞ to 0	π
cot$x$	↓ from ∞ to 0	↓ from 0 to –∞	↓ from ∞ to 0	↓ from 0 to –∞	π
sec$x$	↑ from 1 to ∞	↑ from –∞ to –1	↓ from –1 to –∞	↓ from ∞ to 1	2π
csc$x$	↓ from ∞ to 1	↑ from 1 to ∞	↑ from –∞ to –1	↓ from –1 to –∞	2π

Reflections of trigonometric functions

By using the angular variable x, formulae (2.22) in Chapter 2 can be expressed as

$$\sin(-x) = -\sin x, \quad \csc(-x) = -\csc x \xleftarrow{f(-x)=-f(x)} \text{odd functions}$$

$$\cos(-x) = \cos x, \quad \sec(-x) = \sec x \xleftarrow{f(-x)=f(x)} \text{even functions}$$

$$\tan(-x) = -\tan x, \quad \cot(-x) = -\cot x \xleftarrow{f(-x)=-f(x)} \text{odd functions}$$

These indicate that sine, cosecant, tangent and cotangent functions follow pattern $f(x) = -f(x)$. Thus sin$x$, csc$x$, tan$x$ and cot$x$ are odd functions and symmetric about the origin. However, both cosine and secant functions follow pattern $f(-x) = f(x)$, which means cos$x$ and sec$x$ are even function and symmetric about the y-axis.

Translations of trigonometric functions

Adding a constant number to or subtracting a constant number from any trigonometric function shifts the original function vertically upwards or downwards by that number. For example, y = sin$x$ + 2 shifts sin$x$ upwards by 2 units whereas y = sin$x$ – 2 moves sin$x$ downwards by 2 units (Figure 7.4).

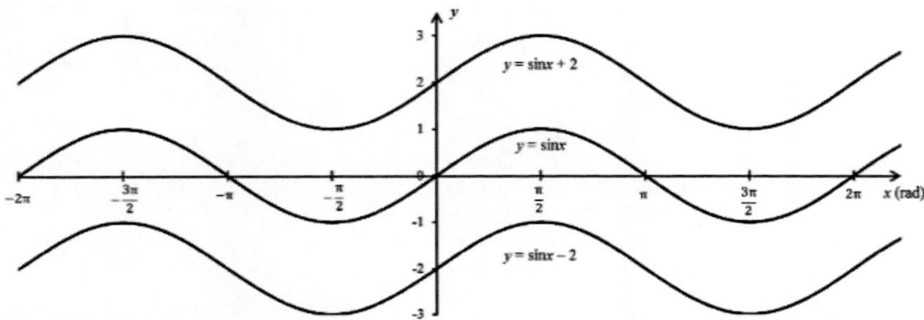

Figure 7.4: Vertical translations of sin$x$ by adding ±2 units to sin$x$ respectively

Adding a constant angular value (called initial angle) to the independent variable x pushes backwards (or delays) the original trigonometric function by a unit of the initial angle. Subtracting a constant angular value (or initial angle) from the independent variable x pushes forwards (or advances) the original trigonometric function by a unit of the initial angle. For instance, $y = \cos(x + \pi/4)$ pushes $\cos x$ backwards by $\pi/4$ or 45°; similarly $y = \cos(x - \pi/2)$ advances $\cos x$ forwards by $\pi/2$ or 90°, which becomes the curve of $\sin x$ (Figure 7.5). This is because $\cos(x - 90°) = \cos[-(90° - x)] = \cos(90° - x) = \sin x$. Therefore, sine is 90° ahead of cosine as shown in Figure 7.1.

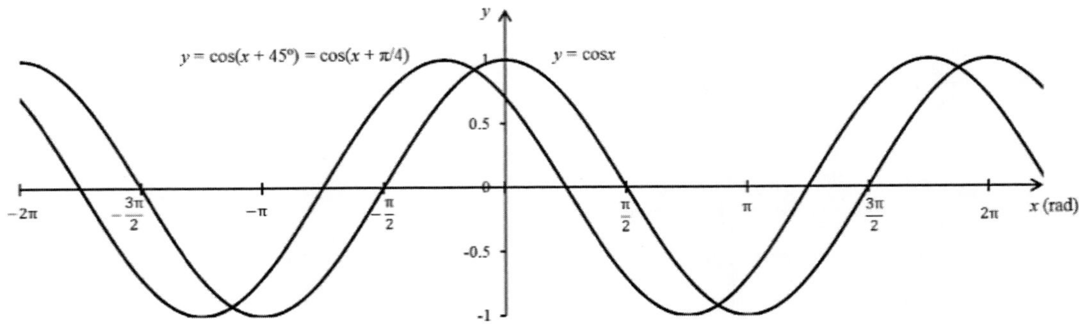

Figure 7.5: Translation of $\cos x$ backwards by $\pi/4$ through $\cos(x + \pi/4)$

Multiplying any trigonometric function by a positive factor great than 1 stretches the original function away from the x-axis whereas multiplying any trigonometric function by a positive factor smaller than 1 compresses the original function toward the x-axis. This is shown in Figure 7.6 using sine functions.

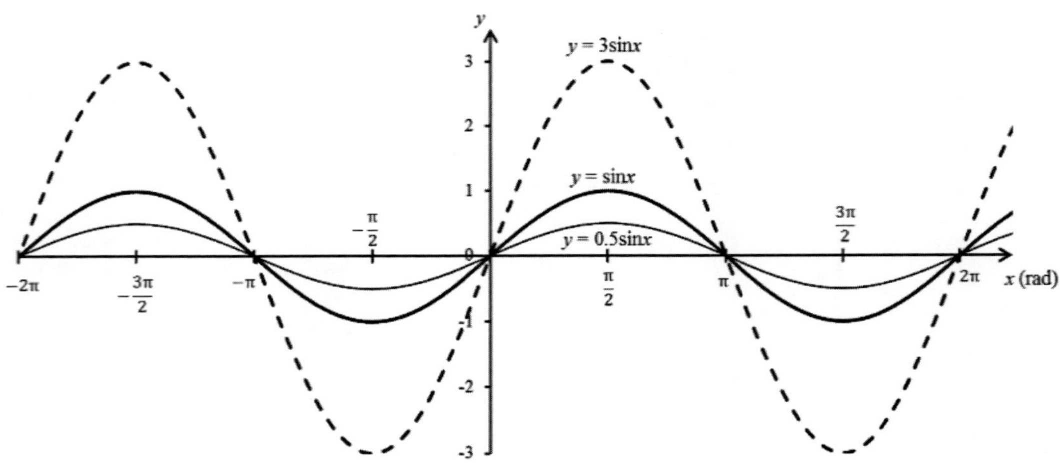

Figure 7.6: Stretching and compressing $\sin x$ by factors of 3 and 0.5 respectively

Example 7.1: Given the values of sin$x$ in [–π, π] in the table below, use the translational properties of sine function to derive values for sin(x – π/3) and sin(x + π/3) in the same range. Plot their graphs to show these translations.

x	–π	–5π/6	–2π/3	–π/2	–π/3	–π/6	0	π/6	π/3	π/2	2π/3	5π/6	π
sin$x$	0	–1/2	–√3/2	–1	–√3/2	–1/2	0	1/2	√3/2	1	√3/2	1/2	0

Solution

As shown in the table below, values for sin(x – π/3) can be obtained by shifting the entire values of sin$x$ (the middle row) forwards by π/3; values for sin(x + π/3) can be obtained by shifting the entire values of sin$x$ backwards by π/3. These movements are indicated by the positions of corresponding sin0 in the table.

x	Property	$-\pi$	$-\frac{5\pi}{6}$	$-\frac{2\pi}{3}$	$-\frac{\pi}{2}$	$-\frac{\pi}{3}$	$-\frac{\pi}{6}$	0	$\frac{\pi}{6}$	$\frac{\pi}{3}$	$\frac{\pi}{2}$	$\frac{2\pi}{3}$	$\frac{5\pi}{6}$	π
$\sin(x-\frac{\pi}{3})$	Advance by $\frac{\pi}{3}$	$\frac{\sqrt{3}}{2}$	$\frac{1}{2}$	0	$-\frac{1}{2}$	$-\frac{\sqrt{3}}{2}$	-1	$-\frac{\sqrt{3}}{2}$	$-\frac{1}{2}$	**0**	$\frac{1}{2}$	$\frac{\sqrt{3}}{2}$	1	$\frac{\sqrt{3}}{2}$
sin$x$	Original	0	$-\frac{1}{2}$	$-\frac{\sqrt{3}}{2}$	-1	$-\frac{\sqrt{3}}{2}$	$-\frac{1}{2}$	**0**	$\frac{1}{2}$	$\frac{\sqrt{3}}{2}$	1	$\frac{\sqrt{3}}{2}$	$\frac{1}{2}$	0
$\sin(x+\frac{\pi}{3})$	Delay by $\frac{\pi}{3}$	$-\frac{\sqrt{3}}{2}$	-1	$-\frac{\sqrt{3}}{2}$	$-\frac{1}{2}$	**0**	$\frac{1}{2}$	$\frac{\sqrt{3}}{2}$	1	$\frac{\sqrt{3}}{2}$	$\frac{1}{2}$	0	$-\frac{1}{2}$	$-\frac{\sqrt{3}}{2}$

The graphs of these three sine curves are plotted in Figure 7.7. Using sin$x$ as the reference, it is obvious that sin(x – π/3) is ahead of sin$x$ by π/3 radians and sin(x + π/3) is behind sin$x$ by π/3 radians.

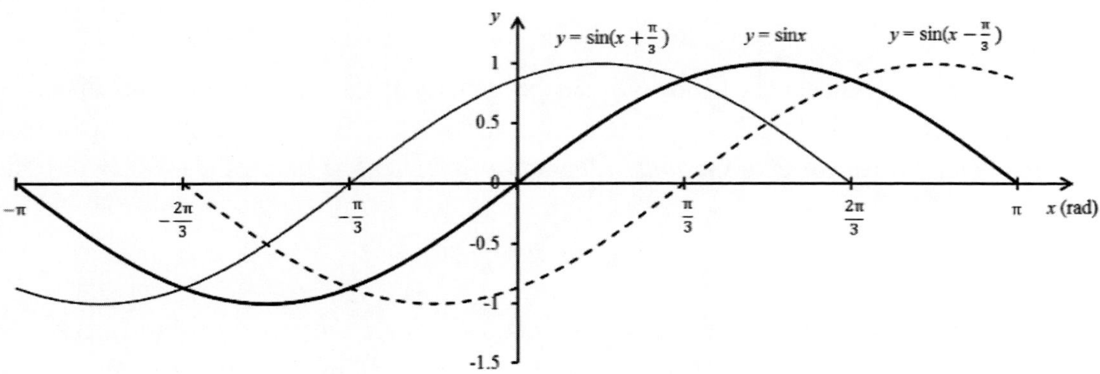

Figure 7.7: Curves of sin$x$, sin(x – π/3) and sin(x + π/3)

7.1.2 Generic sine and cosine functions and combinations

Generic sine and cosine functions (sinusoid)

Since a cosine curve is simply a backward translation of the corresponding sine curve by 90°, both sine and cosine curves or waves are commonly called *sinusoid*. Both sine and cosine functions share the same generic expression in applications of science, engineering and technology:

$$\boxed{\begin{array}{l} y(x) = A\sin(ax+b) \quad \text{or} \\ y(t) = A\sin(\omega t+\varphi) = A\sin(2\pi ft+\varphi). \end{array}} \tag{7.1}$$

- A is the amplitude or the peak deviation of the function from zero. $|A| > 1$ means the peak of the oscillation is greater than that of $\sin x$ whereas $|A| < 1$ means the peak of the oscillation is smaller than that of $\sin x$.
- a or ω is the angular frequency and defined as the rate of change of the independent variable in units of radian per full turn or 2π. The *unit value* (or 1) for angular frequency represents exactly one cycle per full turn. A value of great than 1 for angular frequency means more than one cycle per full turn whereas a value of smaller than 1 for angular frequency indicates a full cycle of the waveform cannot be completed within 2π radians.
- f is the ordinary frequency and defined as the number of cycles or full turns per second of time. The higher the ordinary frequency, the more the number of cycles per second of time.
- φ is the phase that is defined as the initial angle when the rotation or oscillation starts. A positive phase delays the standard curve whereas a negative phase advances the standard wave.

These formulae indicate that sinusoid can be generated in either spatial domain (x is an angular variable) or time domain (t is a temporal variable). It is obvious that $\sin x$ is a special case where A and a (or ω) take the *unit value* and φ is *zero*. The case of $\varphi = 90°$ will transfer $\sin x$ to $\cos x$.

Relationships between frequency and period in generic sine and cosine functions

Formulae (7.1) also bring the formal definition of *period* (T) for any sine or cosine function as

$$\boxed{\begin{array}{l} T = \dfrac{2\pi}{a} = \dfrac{2\pi}{\omega} = \dfrac{1}{f} \quad \text{or} \\ f = \dfrac{1}{T} = \dfrac{\omega}{2\pi}. \end{array}} \tag{7.2}$$

For $\sin x$ or $\cos x$, since $a = 1$, its period must be $T = 2\pi/1 = 2\pi$.

Example 7.2: Given $y(x) = 0.5\sin(2x - \pi/3)$ and $y(t) = -3\sin(0.5t + \pi/4)$, determine the amplitude, phase, angular and ordinary frequencies, and period of each of these functions, and plot their graphs respectively.

Solution

For $y(x) = 0.5\sin(2x - \pi/3)$, according to formulae (7.1) and (7.2),

- Amplitude: $A = 0.5$. This means the range of the oscillation is half of that for $\sin 2x$.
- Phase: $\varphi = -\pi/3$. This advances $\sin 2x$ by $\pi/3$.
- Angular frequency: $\omega = 2$. This means the waveform repeats two times in 2π radians.
- Ordinary frequency: $f = \omega/2\pi = 2/2\pi = 1/\pi$. This means about 0.3 cycles per second.
- Period: $T = 1/f = \pi$. This means one full cycle in every π radians or 180°.

For $y(t) = -3\sin(0.5t + \pi/4)$,

- Amplitude: $A = |-3| = 3$. This means the range of the oscillation is 3 times of that for $\sin 0.5t$.
- Phase: $\varphi = \pi/4$. This delays $\sin 0.5x$ by $\pi/4$.
- Angular frequency: $\omega = 0.5$. The waveform is only half of the full form in 2π radians.
- Ordinary frequency: $f = \omega/2\pi = 0.5/2\pi = 1/4\pi$. This means about 0.08 cycles per second.
- Period: $T = 1/f = 4\pi$. This means one full cycle in every 4π or 720°.

The curves of these two functions and $\sin x$ as the reference are shown in Figure 7.8 with features represented by their parameters discussed above.

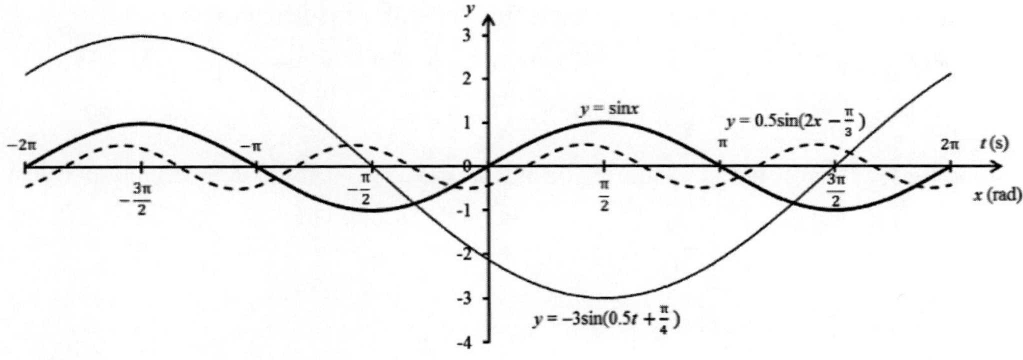

Figure 7.8: Curves of $\sin x$, $0.5\sin(2x - \pi/3)$ and $y(t) = -3\sin(0.5t + \pi/4)$

Combinations of generic sine and cosine functions with a common frequency

By applying some of the trigonometric relationships reviewed in Chapter 2 to generic sine and cosine functions with the same frequency, the following combinative function can be derived:

$$\boxed{\begin{array}{l} A\sin ax + B\cos ax = C\sin(ax+\varphi_0) \\ \left.\begin{array}{l} A = C\cos\varphi_0 \\ B = C\sin\varphi_0 \end{array}\right\} \longrightarrow \varphi_0 = \tan^{-1}\dfrac{B}{A} \;\;\&\;\; C = \begin{cases} \sqrt{A^2+B^2} & \text{if } A>0 \\ -\sqrt{A^2+B^2} & \text{if } A<0 \end{cases} \end{array}} \qquad (7.3)$$

where C and φ_0 are the amplitude and the initial phase of the combined sinusoid. This implies that addition (and subtraction) of sine and cosine functions with the same frequency results in another sine function. This resultant sine function has the original angular frequency but differs in amplitude and phase from those of the original functions.

Example 7.3: Derive formulae (7.3) using trigonometric relationships.

Solution

As A and B in $A\sin ax + B\cos ax$ are constants, assuming $A = C\cos\varphi_0$ and $B = C\sin\varphi_0$,

$$\underline{A\sin ax + B\cos ax} = C\sin ax\cos\varphi_0 + C\cos ax\sin\varphi_0 = C(\sin ax\cos\varphi_0 + \cos ax\sin\varphi_0) = \underline{C\sin(ax+\varphi_0)};$$

$$A^2+B^2 = (C\cos\varphi_0)^2 + (C\sin\varphi_0)^2 = C^2(\cos^2\varphi_0 + \sin^2\varphi_0) = C^2.\ \text{Thus,}$$

$$\underline{C = \pm\sqrt{A^2+B^2}};\quad \frac{B}{A} = \frac{C\sin\varphi_0}{C\cos\varphi_0} = \tan\varphi_0 \longrightarrow \underline{\varphi_0 = \tan^{-1}\frac{B}{A}}.$$

Example 7.4: Given 3sin2x and 4cos2x, determine the combined functions 3sin2x + 4cos2x and 3sin2x – 4cos2x respectively. Plot these functions on the same diagram.

Solution

By formulae (7.3), the combined function for 3sin2x + 4cos2x can be determined as follows.

$$C = \sqrt{A^2+B^2} = \sqrt{3^2+4^2} = \sqrt{9+16} = \sqrt{25} = 5$$

$$\varphi_0 = \tan^{-1}\frac{B}{A} = \tan^{-1}\frac{4}{3} = 0.9273$$

$$3\sin 2x + 4\cos 2x = C\sin(ax+\varphi_0) = 5\sin(2x+0.9273).$$

Since 3sin2x – 4cos2x = 3sin2x + (–4cos2x), A = 3 and B = –4.

$$C = \sqrt{A^2+B^2} = \sqrt{3^2+(-4)^2} = \sqrt{9+16} = \sqrt{25} = 5,\quad \varphi_0 = \tan^{-1}\frac{B}{A} = \tan^{-1}\frac{-4}{3} = -0.9273$$

$$\therefore 3\sin 2x - 4\cos 2x = C\sin(ax+\varphi_0) = 5\sin(2x-0.9273).$$

The two combined functions share the same waveform but 3sin2x + 4cos2x is behind 3sin2x – 4cos2x by 0.9273×2 = 1.8546 radians. The curves of these functions are shown in Figure 7.9.

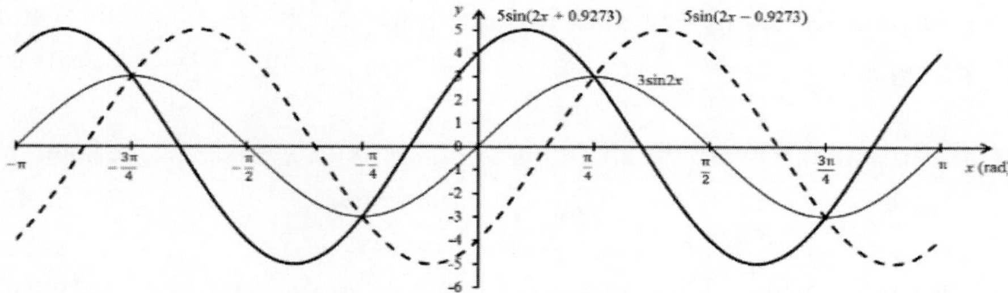

Figure 7.9: Curves of 3sin2x, 3sin2x + 4cos2x, and 3sin2x – 4cos2x

Example 7.5: Given $f(x)$ = 5cos4x and $g(x)$ = 3sin4x, find the function $h(x) = 2f(x) - 3g(x)$.

Solution

$$h(x) = 2f(x) - 3g(x) = 2[5\cos 4x] - 3[3\sin 4x] = 10\cos 4x - 9\sin 4x = -9\sin 4x + 10\cos 4x$$
$$\therefore A = -9,\ B = 10$$

By formulae (7.3), the combined function for $h(x)$ can be determined as follows.

$$C = -\sqrt{A^2 + B^2} = -\sqrt{(-9)^2 + 10^2} = -\sqrt{81 + 100} = -\sqrt{181} \longrightarrow \varphi_0 = \tan^{-1}\frac{B}{A} = \tan^{-1}\frac{10}{-9} \approx -0.8380$$
$$\therefore h(x) = 2f(x) - 3g(x) = C\sin(ax + \varphi_0) = -\sqrt{181}\sin(4x - 0.8380).$$

Example 7.6: Function $y = 10\sin(3x + 0.75)$ was combined from $y_1 + y_2 = A\sin 3x + B\cos 3x$. Find A and B that formed this combined function.

Solution

Since $y_1 + y_2 = A\sin 3x + B\cos 3x$, or $y_1 = A\sin 3x$ and $y_2 = B\cos 3x$, by formulae (7.3),

$$\varphi_0 = \tan^{-1}\frac{B}{A} \longrightarrow 0.75 = \tan^{-1}\frac{B}{A} \longrightarrow \tan 0.75 = \frac{B}{A}$$
$$C = \sqrt{A^2 + B^2} \longrightarrow 10 = \sqrt{A^2 + B^2} \longrightarrow 100 = A^2 + B^2$$
$$\begin{cases} A^2 + B^2 = 100 \\ B = A\tan 0.75 \end{cases} \longrightarrow A^2 + (A\tan 0.75)^2 = 100 \longrightarrow A^2(1 + \tan^2 0.75) = 100$$
$$A^2 = \frac{100}{1 + \tan^2 0.75} = \frac{100}{\sec^2 0.75} = 100\cos^2 0.75 = (10\cos 0.75)^2 \longleftarrow (\sec^2\theta = 1 + \tan^2\theta)$$
$$A = \sqrt{(10\cos 0.75)^2} = 10\cos 0.75 \approx 7.3169 \longrightarrow B = A\tan 0.75 = 10\cos 0.75 \times \frac{\sin 0.75}{\cos 0.75} \approx 6.8164$$
$$\therefore y_1 = 7.3169\sin 3x,\ y_2 = 6.8164\cos 3x.$$

Combinations of generic sine and cosine functions with different frequencies

Combining two generic sine and/or cosine functions with different frequencies together is much more difficult and has no universal analytical way to deal with all possible variations. Most such cases can be resolved by using pairwise numeric operations. This is demonstrated by the following two examples.

Example 7.7: If $\sin x$ and $\cos 2x$ are defined in $[-\pi, \pi]$, plot $\sin x - \cos 2x$ in the same range.

Solution

We can calculate values for both $\sin x$ and $\cos 2x$ using a fixed interval of x within $[-\pi, \pi]$; then use these paired values to work out the corresponding values for $\sin x - \cos 2x$. These data are listed in the table below and the graphs of these three functions are plotted in Figure 7.10.

x	$-\pi$	$-\frac{5\pi}{6}$	$-\frac{2\pi}{3}$	$-\frac{\pi}{2}$	$-\frac{\pi}{3}$	$-\frac{\pi}{6}$	0	$\frac{\pi}{6}$	$\frac{\pi}{3}$	$\frac{\pi}{2}$	$\frac{2\pi}{3}$	$\frac{5\pi}{6}$	π
$\sin x$	0	−0.5	−0.866	−1	−0.866	−0.5	0	0.5	0.866	1	0.866	0.5	0
$\cos 2x$	1	0.5	−0.5	−1	−0.5	0.5	1	0.5	−0.5	−1	−0.5	0.5	1
$\sin x - \cos 2x$	−1	−1	−0.366	0	−0.366	−1	−1	0	1.366	2	1.366	0	−1

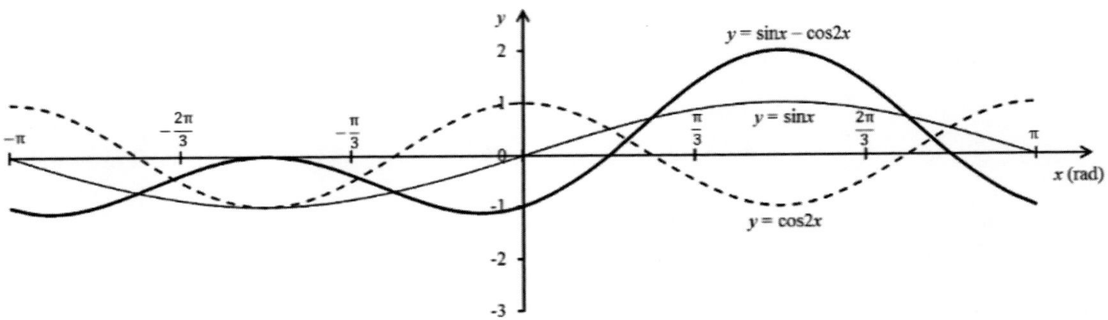

Figure 7.10: Curves of $\sin x$, $\cos 2x$ and $\sin x - \cos 2x$ in $[-\pi, \pi]$

Example 7.8: If $y_1 = 2\sin 3x$ and $y_2 = 3\cos 2x$ are defined in $[0, 2\pi]$, plot $y_1 + y_2$ in the same range.

Solution

Following the similar process shown in Example 7.7, the paired values of y_1 and y_2 and the corresponding values for $y_1 + y_2$ in $[0, 2\pi]$ are obtained and listed in the table below and the graphs of these three functions are plotted in Figure 7.11.

x	0	$\frac{\pi}{6}$	$\frac{\pi}{3}$	$\frac{\pi}{2}$	$\frac{2\pi}{3}$	$\frac{5\pi}{6}$	π	$\frac{7\pi}{6}$	$\frac{4\pi}{3}$	$\frac{3\pi}{2}$	$\frac{5\pi}{3}$	$\frac{11\pi}{6}$	2π
$2\sin 3x$	0	2	0	−2	0	2	0	−2	0	2	0	−2	0
$3\cos 2x$	3	1.5	−1.5	−3	−1.5	1.5	3	1.5	−1.5	−3	−1.5	1.5	3
$y_1 + y_2$	3	3.5	−1.5	−5	−1.5	3.5	3	−0.5	−1.5	−1	−1.5	−0.5	3

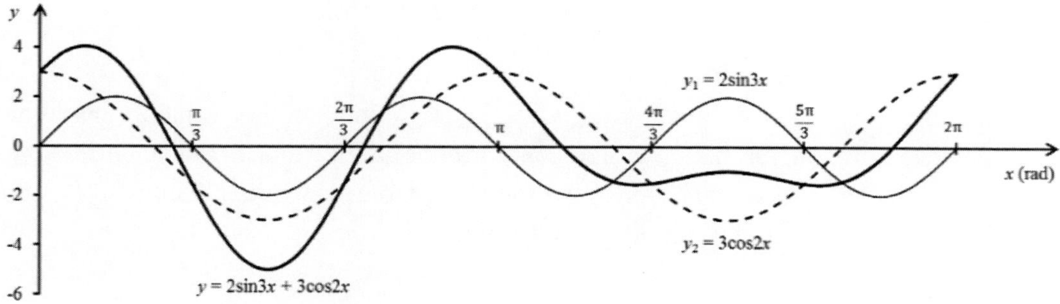

Figure 7.11: Curves of 2sin3x, 3cos2x and 2sin3x + 3cos2x in [0, 2π]

Exercises 7.1

1. Use the values of cos$x$ in [–π, π] and the translational properties of cosine function to derive values for cos(x – π/4) and cos(x + π/4) in the same range. Plot their graphs to show these translations.

2. Given 2sin3x and 5cos3x, determine the combined functions 2sin3x + 5cos3x and 2sin3x – 5cos3x respectively. Plot these functions on the same diagram.

3. Given $f(x)$ = –2sin2x and $g(x)$ = 4cos2x, determine the combined function 3$f(x)$ + 2$g(x)$. Plot this function.

4. Given $y(x)$ = –2sin(3x + π/4) and $y(t)$ = 3sin(0.5t – π/6), determine the amplitude, phase, angular and ordinary frequencies, and period of each of these functions, and plot their graphs respectively.

5. Given $y(x)$ = 4sin(0.2x – 2π/3) and $y(t)$ = –5sin(4t + 3π/4), determine the amplitude, phase, angular and ordinary frequencies, and period of each of these functions, and plot their graphs respectively.

6. Given $f(x)$ = –2sin3x and $g(x)$ = 3cos$x$, determine the combined function 3$f(x)$ + 4$g(x)$. Plot this function.

7. Function $y = 8\sin(2x - 1.25)$ was combined from $y_1 + y_2 = A\sin 2x + B\cos 2x$. Find A and B that formed this combined function.

8. If sin2x and cos0.5x are defined in [–π, π], plot 2sin2x – 3cos0.5x in the same range.

9. If sin$x$ and cos3x are defined in [–π, π], plot 4sin$x$ + 2cos3x in the same range.

10. If y_1 = –2sin3x, y_2 = 3cos$x$, and y_3 = 3cos0.5x are defined in [0, 2π], plot $y_1 + y_2 + y_3$ in the same range.

7.2 Inverses of Trigonometric Functions

Inverse trigonometric functions

Given an angle, a trigonometric function transfers the angle into another value in the domain of that trigonometric function. On the other hand, any trigonometric value (x) must correlate to one or more angular values (y) that can produce the same trigonometric value. Such relationships are called the inverse trigonometric functions, and denoted as

$$\boxed{\begin{cases} y = \arcsin x = \sin^{-1} x \\ y = \arccos x = \cos^{-1} x \\ y = \arctan x = \tan^{-1} x \\ y = \operatorname{arccot} x = \cot^{-1} x \\ y = \operatorname{arcsec} x = \sec^{-1} x \\ y = \operatorname{arccsc} x = \csc^{-1} x \end{cases}} \tag{7.4}$$

Note the common confusion between $\sin^{-1} x = \arcsin x$ and $(\sin x)^{-1} = \dfrac{1}{\sin x} = \csc x$. In trigonometric functions, $\sin^{-1} x$ (similar to other trigonometric functions) is to specially denote the inverse of sin$x$, not the reciprocal of sin$x$. To avoid any potential confusion, it is always a good practice to use the 'arc-' form for inverse trigonometric functions.

The graphs of inverse sine and cosine functions are shown in Figure 7.12. Note the angular values (y) extend periodically to both infinities if x swings between –1 and 1.

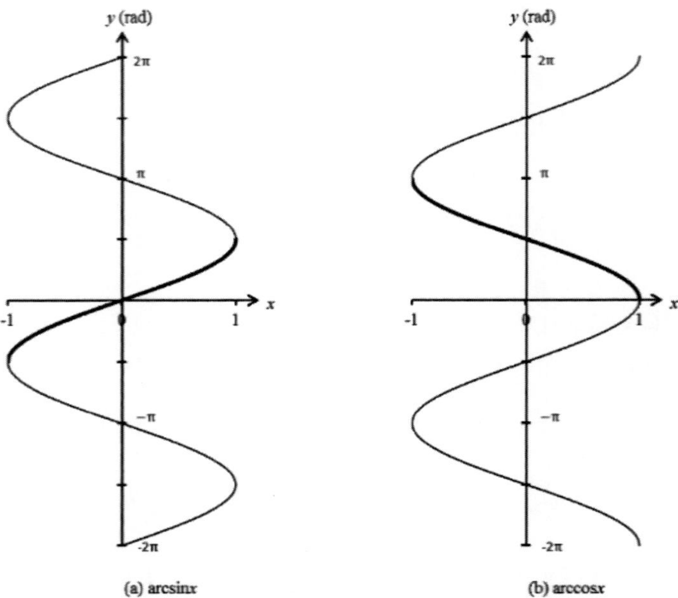

Figure 7.12: The graph of inverse sine function (a) and inverse cosine function (b)

Similarly, the angular values (y) of inverse tangent, cotangent, secant and cosecant functions also extend periodically to both infinities if x swings between $-\infty$ and $+\infty$ (Figure 7.13). Therefore, without specifying the range for an angle, any x will result in infinitely many angles that all produce the same x value for any inverse trigonometric function. This is called the *one-to-many scenario*.

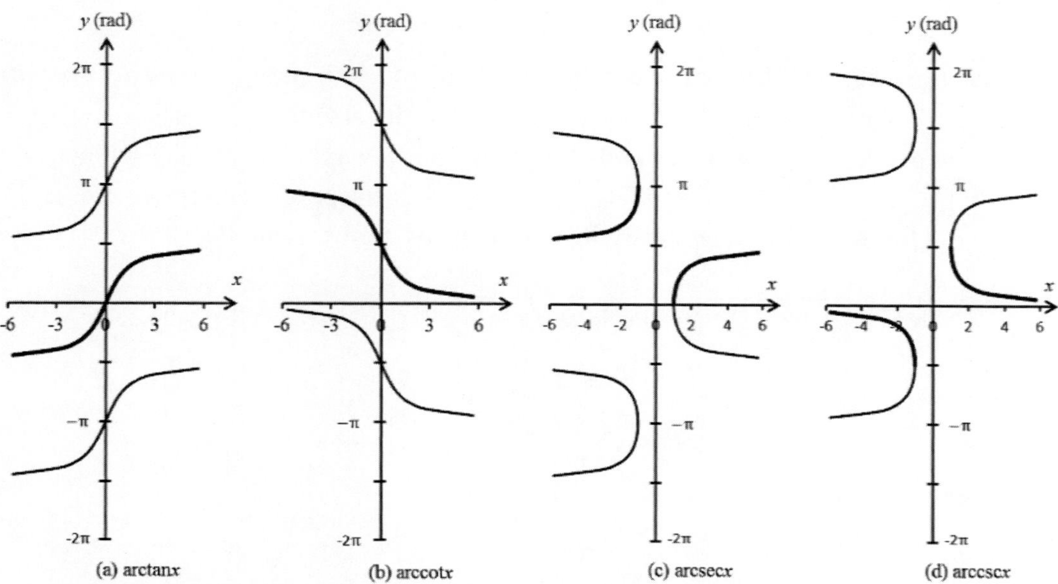

Figure 7.13: The graph of arctan$x$ (a), arccot$x$ (b), arcsec$x$ (c) and arccsc$x$ (d)

Example 7.9: Evaluate the angles in $[0, \pi]$ returned by the following inverse trigonometric functions respectively.

$$\text{(a)}\ \arcsin\frac{\sqrt{3}}{2};\quad \text{(b)}\ \arccos(-\frac{\sqrt{2}}{2});\quad \text{(c)}\ \arctan\sqrt{3};$$

$$\text{(d)}\ \operatorname{arccot}(-\frac{\sqrt{3}}{3});\quad \text{(e)}\ \operatorname{arcsec}2;\quad \text{(f)}\ \operatorname{arccsc}(-\sqrt{2}).$$

<u>**Solution**</u>

Since the angle must be in $[0, \pi]$, it narrows our assessment within quadrants I and II.

$$\text{(a)}\ \text{In}\ [0, \pi],\ \text{both}\ \sin\frac{\pi}{3} = \frac{\sqrt{3}}{2}\ \text{and}\ \sin\frac{2\pi}{3} = \frac{\sqrt{3}}{2},\quad \therefore \arcsin\frac{\sqrt{3}}{2} = \frac{\pi}{3}\ \text{or}\ \frac{2\pi}{3};$$

$$\text{(b)}\ \text{In}\ [0, \pi],\ \cos\frac{3\pi}{4} = -\frac{\sqrt{2}}{2},\quad \therefore \arccos(-\frac{\sqrt{2}}{2}) = \frac{3\pi}{4};$$

$$\text{(c)}\ \text{In}\ [0, \pi],\ \tan\frac{\pi}{3} = \sqrt{3},\quad \therefore \arctan\sqrt{3} = \frac{\pi}{3};$$

(d) In $[0, \pi]$, $\cot\frac{2\pi}{3} = -\frac{\sqrt{3}}{3}$, $\therefore \operatorname{arccot}(-\frac{\sqrt{3}}{3}) = \frac{2\pi}{3}$;

(e) In $[0, \pi]$, $\sec\frac{\pi}{3} = 2$, $\therefore \operatorname{arcsec} 2 = \frac{\pi}{3}$;

(f) In $[0, \pi]$, $\csc x \geq 0$, $\therefore \operatorname{arccsc}(-\sqrt{2})$ is not in $[0, \pi]$.

Operations between trigonometric functions and inverse trigonometric functions

Applying the trigonometric function to its corresponding inverse trigonometric function results in the independent variable itself. Equally, applying the inverse trigonometric function to its corresponding trigonometric function also produces the independent variable itself. This is captured by the following formulae involving sine and cosine functions.

$$\begin{cases} \sin(\arcsin x) = x \\ \sin(\arccos x) = \pm\sqrt{1-x^2} \\ \cos(\arccos x) = x \\ \cos(\arcsin x) = \pm\sqrt{1-x^2} \end{cases} \Longleftrightarrow \begin{cases} \arcsin(\sin x) = x \\ \arcsin(\cos x) = \frac{\pi}{2} - x \\ \arccos(\cos x) = x \\ \arccos(\sin x) = \frac{\pi}{2} - x \end{cases} \tag{7.5}$$

Example 7.10: Evaluate

(a) $\sin(\arcsin\frac{1}{3})$; (b) $\cos[\arccos(-0,6654)]$; (c) $\sin(\arccos\frac{\sqrt{3}}{2})$; (d) $\arccos(\cos\frac{\pi}{5})$;

(e) $\arcsin[\sin(-\frac{\pi}{6})]$; (f) $\arccos[\sin(-\frac{\pi}{4})]$.

Solution

(a) $\sin(\arcsin\frac{1}{3}) = \frac{1}{3}$;

(b) $\cos[\arccos(-0,6654)] = -0,6654$;

(c) $\sin(\arccos\frac{\sqrt{3}}{2}) = \pm\sqrt{1-(\frac{\sqrt{3}}{2})^2} = \pm\sqrt{1-\frac{3}{4}} = \pm\sqrt{\frac{1}{4}} = \pm\frac{1}{2}$;

(d) $\arccos(\cos\frac{\pi}{5}) = \frac{\pi}{5}$;

(e) $\arcsin[\sin(-\frac{\pi}{6})] = -\frac{\pi}{6}$;

(f) $\arccos[\sin(-\frac{\pi}{4})] = \frac{\pi}{2} - (-\frac{\pi}{4}) = \frac{\pi}{2} + \frac{\pi}{4} = \frac{3\pi}{4}$.

Principal-value ranges of inverse trigonometric functions

In order to make any inverse trigonometric function produce one-to-one solution between x and y, a principal-value range for each of the six inverse trigonometric functions is set to narrow down the options for the angular value y. There are different selections for these principal-value ranges; here in Table 7.2 we choose the following ranges for the convenience in studying calculus in the future.

Note: the angle of any inverse trigonometric function obtained from your calculator is the angle within the principal-value range of the corresponding inverse trigonometric function.

Table 7.2: Principal-value ranges of inverse trigonometric functions

Inverse function	Principal-value range
$y = \arcsin x$	$-\frac{\pi}{2} \le y \le \frac{\pi}{2}$
$y = \arccos x$	$0 \le y \le \pi$
$y = \arctan x$	$-\frac{\pi}{2} < y < \frac{\pi}{2}$
$y = \operatorname{arc}\cot x$	$0 < y < \pi$
$y = \operatorname{arc}\sec x$	$0 \le y \le \pi$, $y \ne \frac{\pi}{2}$
$y = \operatorname{arc}\csc x$	$-\frac{\pi}{2} \le y \le \frac{\pi}{2}$, $y \ne 0$

Example 7.11: Find the angles within the principal-value ranges for the following inverse trigonometric functions respectively.

(a) $\arcsin\frac{\sqrt{2}}{2}$; (b) $\arcsin(-\frac{\sqrt{3}}{2})$; (c) $\arccos(-\frac{1}{2})$; (d) $\arctan(-1)$.

Solution

(a) In $[-\frac{\pi}{2}, \frac{\pi}{2}]$, $\sin\frac{\pi}{4} = \frac{\sqrt{2}}{2}$, $\therefore \arcsin\frac{\sqrt{2}}{2} = \frac{\pi}{4}$;

(b) In $[-\frac{\pi}{2}, \frac{\pi}{2}]$, $\sin(-\frac{\pi}{3}) = -\frac{\sqrt{3}}{2}$, $\therefore \arcsin(-\frac{\sqrt{3}}{2}) = -\frac{\pi}{3}$;

(c) In $[0, \pi]$, $\cos\frac{2\pi}{3} = -\frac{1}{2}$, $\therefore \arccos(-\frac{1}{2}) = \frac{2\pi}{3}$;

(d) In $(-\frac{\pi}{2}, \frac{\pi}{2})$, $\tan(-\frac{\pi}{4}) = -1$, $\therefore \arctan(-1) = -\frac{\pi}{4}$.

General values of inverse trigonometric functions

The solution in the principal-value range for inverse trigonometric function can be used to define all other cyclic solutions in the whole range of y. For example, the general values for arcsin$x$, arccos$x$ and arctan$x$ can be derived using the rules presented in Table 2.7 in Chapter 2 and in Table 7.2, which are listed in Table 7.3.

Table 7.3: General values of the inverse sine, cosine and tangent functions

Principal solution*	Range of θ	Two solutions in $[0, 2\pi]$	All general solutions
$\theta = \arcsin x$	$0 < \theta < \pi/2$	$y_1 = \theta$ & $y_2 = \pi - \theta$	$y_1 + 2n\pi$ & $y_2 + 2n\pi$ $n = 0, \pm 1, \pm 2, \ldots$
	$-\pi/2 < \theta < 0$	$y_1 = \pi + \lvert\theta\rvert$ & $y_2 = 2\pi - \lvert\theta\rvert$	
$\theta = \arccos x$	$0 < \theta < \pi$	$y_1 = \theta$ & $y_2 = 2\pi - \theta$	
$\theta = \arctan x$	$0 < \theta < \pi/2$	$y_1 = \theta$ & $y_2 = \pi + \theta$	
	$-\pi/2 < \theta < 0$	$y_1 = \pi - \lvert\theta\rvert$ & $y_2 = 2\pi - \lvert\theta\rvert$	

* θ is the angle from applying the inverse trigonometric function to x using your calculator.

Example 7.12: Find all general solutions for the inverse functions in Example 7.11.

Solution

(a) Since $\theta = \arcsin\dfrac{\sqrt{2}}{2} = \dfrac{\pi}{4}$, $y_1 = \theta = \dfrac{\pi}{4}$, $y_2 = \pi - \dfrac{\pi}{4} = \dfrac{3\pi}{4}$.

Thus, $y = y_1 + 2n\pi = \dfrac{\pi}{4} + 2n\pi$, and $y = y_2 + 2n\pi = \dfrac{3\pi}{4} + 2n\pi$. $(n = 0, \pm1, \pm2, \ldots)$

(b) Since $\theta = \arcsin(-\dfrac{\sqrt{3}}{2}) = -\dfrac{\pi}{3}$, $y_1 = \pi + |\theta| = \pi + \dfrac{\pi}{3} = \dfrac{4\pi}{3}$, $y_2 = 2\pi - |\theta| = 2\pi - \dfrac{\pi}{3} = \dfrac{5\pi}{3}$.

Thus, $y = y_1 + 2n\pi = \dfrac{4\pi}{3} + 2n\pi$, and $y = y_2 + 2n\pi = \dfrac{5\pi}{3} + 2n\pi$. $(n = 0, \pm1, \pm2, \ldots)$

(c) Since $\theta = \arccos(-\dfrac{1}{2}) = \dfrac{2\pi}{3}$, $y_1 = \theta = \dfrac{2\pi}{3}$, $y_2 = 2\pi - \dfrac{2\pi}{3} = \dfrac{4\pi}{3}$.

Thus, $y = y_1 + 2n\pi = \dfrac{2\pi}{3} + 2n\pi$, and $y = y_2 + 2n\pi = \dfrac{4\pi}{3} + 2n\pi$. $(n = 0, \pm1, \pm2, \ldots)$

(d) Since $\theta = \arctan(-1) = -\dfrac{\pi}{4}$, $y_1 = \pi - |\theta| = \pi - \dfrac{\pi}{4} = \dfrac{3\pi}{4}$, $y_2 = 2\pi - |\theta| = 2\pi - \dfrac{\pi}{4} = \dfrac{7\pi}{4}$.

Thus, $y = y_1 + 2n\pi = \dfrac{3\pi}{4} + 2n\pi$, and $y = y_2 + 2n\pi = \dfrac{7\pi}{4} + 2n\pi$. $(n = 0, \pm1, \pm2, \ldots)$

Example 7.13: Given $\alpha = \arcsin\dfrac{1}{2}$ and $\beta = \arccos(-\dfrac{3}{5})$ in their principal-value ranges, evaluate

(a) $\sin(\alpha + \beta)$, (b) $\cos(\alpha - \beta)$, (c) $\tan 2\beta$, (d) $\cos\dfrac{\beta}{2}$.

Solution

First, we need to find which quadrant α and β are located. From there we can evaluate the signs for the trigonometric functions involved.

The principal value for $\alpha = \arcsin\frac{1}{2}$ should be in quadrant I (Figure 7.14a). Therefore, $\sin\alpha = \sin(\arcsin\frac{1}{2}) = \frac{1}{2}$, and $\cos\alpha$ should be positive.

The principal value for $\beta = \arccos(-\frac{3}{5})$ should be in quadrant II (Figure 7.14b). Therefore, $\cos\beta = \cos[\arccos(-\frac{3}{5})] = -\frac{3}{5}$ and $\sin\beta$ and $\cos\frac{\beta}{2}$ are positive.

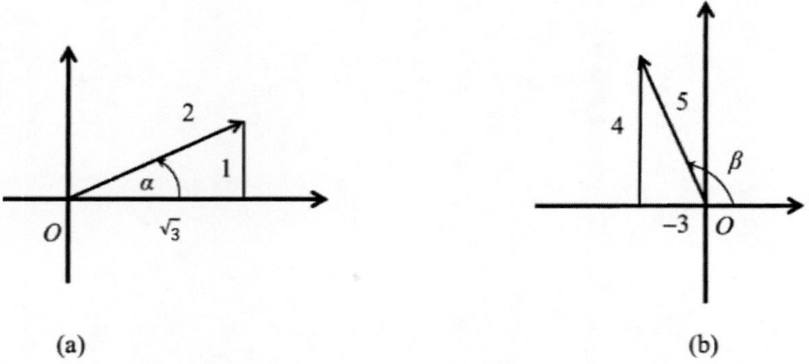

Figure 7.14: Locations of α (a) and β (b) in their principal-value range for Example 7.10

$$\sin\alpha = \sin(\arcsin\frac{1}{2}) = \frac{1}{2} \longrightarrow \cos\alpha = \sqrt{1-\sin^2\alpha} = \sqrt{1-(\frac{1}{2})^2} = \sqrt{\frac{4}{4}-\frac{1}{4}} = \sqrt{\frac{3}{4}} = \frac{\sqrt{3}}{2}$$

$$\cos\beta = \cos[\cos(-\frac{3}{5})] = -\frac{3}{5} \longrightarrow \sin\beta = \sqrt{1-\cos^2\beta} = \sqrt{1-(-\frac{3}{5})^2} = \sqrt{\frac{25}{25}-\frac{9}{25}} = \sqrt{\frac{16}{25}} = \frac{4}{5}$$

$$\tan\beta = \frac{\sin\beta}{\cos\beta} = \frac{\frac{4}{5}}{-\frac{3}{5}} = -\frac{4}{3}.$$

(a) $\sin(\alpha+\beta) = \sin\alpha\cos\beta + \cos\alpha\sin\beta = \frac{1}{2}\bullet(-\frac{3}{5}) + \frac{\sqrt{3}}{2}\bullet\frac{4}{5} = \frac{-3}{10} + \frac{4\sqrt{3}}{10} = \frac{4\sqrt{3}-3}{10}$;

(b) $\cos(\alpha-\beta) = \cos\alpha\cos\beta + \sin\alpha\sin\beta = \frac{\sqrt{3}}{2}\bullet(-\frac{3}{5}) + \frac{1}{2}\bullet\frac{4}{5} = \frac{-3\sqrt{3}}{10} + \frac{4}{10} = \frac{4-3\sqrt{3}}{10}$;

(c) $\tan 2\beta = \frac{2\tan\beta}{1-\tan^2\beta} = \frac{2\times(-\frac{4}{3})}{1-(-\frac{4}{3})^2} = \frac{-\frac{8}{3}}{\frac{9}{9}-\frac{16}{9}} = \frac{-\frac{8}{3}}{-\frac{7}{9}} = \frac{24}{7}$;

(d) $\cos\frac{\beta}{2} = \sqrt{\frac{1+\cos\beta}{2}} = \sqrt{\frac{1-\frac{3}{5}}{2}} = \sqrt{\frac{\frac{2}{5}}{2}} = \sqrt{\frac{1}{5}} = \frac{\sqrt{5}}{5}$.

Exercises 7.2

1. Evaluate the angles in $[0, \pi]$ returned by the following inverse trigonometric functions respectively.

   (a) $\arcsin\frac{1}{2}$; (b) $\arccos(-\frac{\sqrt{3}}{2})$; (c) $\arctan\frac{\sqrt{3}}{3}$;

   (d) $\operatorname{arc}\cot(-1)$; (e) $\operatorname{arc}\sec\sqrt{2}$; (f) $\operatorname{arc}\csc(-2)$.

2. Prove

   (a) $\sin(\arccos x) = \mp\sqrt{1-x^2}$; (b) $\cos(\arcsin x) = \pm\sqrt{1-x^2}$; (c) $\arcsin(\cos x) = \frac{\pi}{2} - x$;

   (d) $\cos(\arcsin x) = \arcsin(\cos x) = \frac{\pi}{2} - x$

3. Evaluate

   (a) $\cos(\arccos 0.4351)$; (b) $\cos[\arcsin(-\frac{1}{3})]$; (c) $\sin(\arccos\frac{4}{5})$; (d) $\arcsin(\cos\frac{\pi}{3})$;

   (e) $\arcsin[\cos(-\frac{\pi}{8})]$; (f) $\arccos(\sin\frac{4\pi}{3})]$.

4. Find the angles within the principal-value ranges for the following inverse trigonometric functions respectively.

   (a) $\arcsin\frac{\sqrt{3}}{2}$; (b) $\arccos 0.1736$; (c) $\arccos(-\frac{\sqrt{3}}{2})$; (d) $\arctan(-\sqrt{3})$.

5. Find all general solutions for the inverse functions below.

   (a) $\arcsin\frac{1}{2}$; (b) $\arccos(-\frac{\sqrt{3}}{2})$; (c) $\arctan\frac{\sqrt{3}}{3}$;

   (d) $\operatorname{arc}\cot(-1)$; (e) $\operatorname{arc}\sec\sqrt{2}$; (f) $\operatorname{arc}\csc(-2)$.

6. Given $\alpha = \arcsin\frac{\sqrt{2}}{2}$ and $\beta = \arccos(-\frac{1}{2})$ in their principal-value ranges, evaluate the following expressions by their exact values.

   (a) $\sin(\alpha+\beta)$, (b) $\cos(\alpha-\beta)$, (c) $\tan\frac{\beta}{2}$, (d) $\cos 2\beta$.

7.3 Trigonometric Equations

Trigonometric equations are those equations involving trigonometric functions of unknown angles. There is no general method for solving trigonometric equations. All trigonometric relationships could be used to translate any given trigonometric equation to an equivalent expression that may be easier to solve. We use the following examples to demonstrate some useful techniques in solving trigonometric equations.

Example 7.14: Solve $\cos x + 2\sin x\cos x = 0$ for x in $[-\pi/2, \pi/2]$.

Solution

$$\cos x + 2\sin x\cos x = 0 \longrightarrow \cos x(1+2\sin x) = 0 \longrightarrow \begin{cases} \cos x = 0 \\ 1+2\sin x = 0 \end{cases}$$

$$\cos x = 0 \longrightarrow x = -\frac{\pi}{2} \text{ and } \frac{\pi}{2}; \quad 1+2\sin x = 0 \longrightarrow \sin x = -\frac{1}{2} \longrightarrow x = -\frac{\pi}{6}.$$

Thus, the solutions in $[-\pi/2, \pi/2]$ are $x = -\pi/6$, $-\pi/2$ and $\pi/2$.

Check:

For $x = -\frac{\pi}{6}$,

$$\cos x + 2\sin x\cos x = \cos x + \sin 2x = \cos(-\frac{\pi}{6}) - \sin(2\times\frac{\pi}{6}) = \cos\frac{\pi}{6} - \sin\frac{\pi}{3} = \frac{\sqrt{3}}{2} - \frac{\sqrt{3}}{2} = 0$$

For $x = \pm\frac{\pi}{2}$,

$$\cos x + 2\sin x\cos x = \cos x + \sin 2x = \cos(\pm\frac{\pi}{2}) + \sin(\pm 2\times\frac{\pi}{2}) = 0 + \sin(\pm\pi) = 0 + 0 = 0.$$

Example 7.15: Solve $\tan x + \sec x = 1$ for x in $[0, \pi]$.

Solution

$$\tan x + \sec x = 1 \longrightarrow \frac{\sin x}{\cos x} + \frac{1}{\cos x} = 1 \xrightarrow{\times\cos x} \sin x + 1 = \cos x \longrightarrow \sin x = \cos x - 1$$

$$\text{Let } \cos x = z, \longrightarrow \sin x = \sqrt{1-\cos^2 x} = \sqrt{1-z^2}$$

$$\sin x = \cos x - 1 \longrightarrow \sqrt{1-z^2} = z-1 \longrightarrow 1-z^2 = (z-1)^2 \longrightarrow 1-z^2 = z^2-2z+1$$

$$2z^2 - 2z = 0 \longrightarrow 2z(z-1) = 0 \longrightarrow \begin{cases} 2z = 0 \\ z-1 = 0 \end{cases}$$

$$2z = 0 \longrightarrow z = 0 \longrightarrow \cos x = 0 \longrightarrow x = \frac{\pi}{2}; \qquad z-1 = 0 \longrightarrow z = 1 \longrightarrow \cos x = 1 \longrightarrow x = 0.$$

Since $\tan(\pi/2)$ is undecided, thus, the solution in $[0, \pi]$ is $x = 0$. Check by yourself.

Example 7.16: Solve $\sin x + \cos^2 x = 1$ for x in $[0, \pi]$.

Solution

$$\sin x + \cos^2 x = 1 \longrightarrow \sin x = 1 - \cos^2 x \longrightarrow \sin x = \sin^2 x \longrightarrow \sin^2 x - \sin x = 0$$

$$\sin x(\sin x - 1) = 0 \longrightarrow \begin{cases} \sin x = 0 \\ \sin x - 1 = 0 \end{cases}$$

$$\sin x = 0 \longrightarrow x = 0,\ \pi; \qquad \sin x - 1 = 0 \longrightarrow \sin x = 1 \longrightarrow x = \frac{\pi}{2}.$$

These are solutions in $[0, \pi]$ for $\sin x + \cos^2 x = 1$.

Example 7.17: Solve $3\sin^2 x + \sin x - 2 = 0$ for x in $[-\pi/2, \pi/2]$.

Solution

$$3\sin^2 x + \sin x - 2 = 0 \longrightarrow (3\sin x - 2)(\sin x + 1) = 0 \longrightarrow \begin{cases} 3\sin x - 2 = 0 \\ \sin x + 1 = 0 \end{cases}$$

$$3\sin x - 2 = 0 \longrightarrow x = \arcsin\frac{2}{3} \approx 0.73\,(\text{rad}); \qquad \sin x + 1 = 0 \longrightarrow \sin x = -1 \longrightarrow x = -\frac{\pi}{2}.$$

These are solutions in $[-\pi/2, \pi/2]$ for $3\sin^2 x + \sin x - 2 = 0$.

Example 7.18: Solve $\cos 2x - 3\cos x + 2 = 0$ for x in $[-\pi/2, \pi/2]$.

Solution

$$\cos 2x - 3\cos x + 2 = 0 \longrightarrow 2\cos^2 x - 1 - 3\cos x + 2 = 0 \longrightarrow 2\cos^2 x - 3\cos x + 1 = 0$$

$$(2\cos x - 1)(\cos x - 1) = 0 \longrightarrow \begin{cases} 2\cos x - 1 = 0 \\ \cos x - 1 = 0 \end{cases}$$

$$2\cos x - 1 = 0 \longrightarrow \cos x = \frac{1}{2} \longrightarrow x = \pm\frac{\pi}{3}; \qquad \cos x - 1 = 0 \longrightarrow \cos x = 1 \longrightarrow x = 0.$$

These are solutions in $[-\pi/2, \pi/2]$ for $\cos 2x - 3\cos x + 2 = 0$.

Exercises 7.3

1. Solve $\cos 2x - 2\sin x\cos x = 0$ for x in $[-\pi/2, \pi/2]$.

2. Solve $3\tan x - 4\sin x = 0$ for x in $[0, \pi]$.

3. Solve $\cos^2 x - 2\sin x = -1$ for x in $[0, \pi]$.

4. Solve $3\tan x - \cot x = 0$ for x in $[-\pi/2, \pi/2]$.

5. Solve $\sin x\cos x - 1 = \sin x - \cos x$ for x in $[-\pi/2, \pi/2]$.

6. Solve $2\sin^2 \frac{x}{2} = \cos x$ for x in $[-\pi/2, \pi/2]$.

7. Solve $\cos^2 x + 4\cos x = 1$ for x in $[0, \pi]$.

8. Solve $2\csc x = \tan x + \cot x$ for x in $[-\pi/2, \pi/2]$.

7.4 Hyperbolic Functions

7.4.1 Hyperbolic functions

Hyperbolic functions are defined by combinations of exponential functions through similar forms of trigonometric functions as follows:

$$\begin{aligned}
&\text{hyperbolic sine function: } \sinh x = \frac{e^x - e^{-x}}{2} \\
&\text{hyperbolic cosine function: } \cosh x = \frac{e^x + e^{-x}}{2} \\
&\text{hyperbolic tangent function: } \tanh x = \frac{\sinh x}{\cosh x} = \frac{e^x - e^{-x}}{e^x + e^{-x}} \\
&\text{hyperbolic cotangent function: } \coth x = \frac{\cosh x}{\sinh x} = \frac{e^x + e^{-x}}{e^x - e^{-x}} \\
&\text{hyperbolic secant function: } \operatorname{sech} x = \frac{1}{\cosh x} = \frac{2}{e^x + e^{-x}} \\
&\text{hyperbolic cosecant function: } \operatorname{csch} x = \frac{1}{\sinh x} = \frac{2}{e^x - e^{-x}}
\end{aligned} \tag{7.6}$$

All these functions are defined in the whole range of x and *not periodic*.

Hyperbolic sine and cosecant functions are mutually reciprocal and symmetric about the original (Figure 7.15). In other words, both hyperbolic sine and cosecant functions are odd functions.

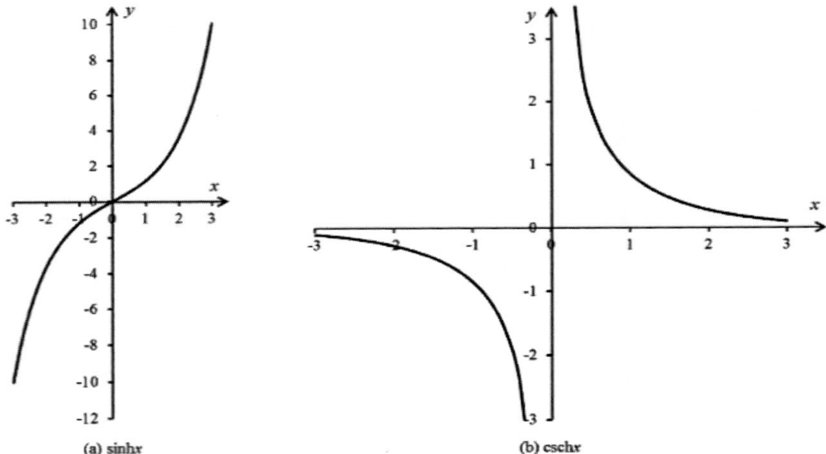

Figure 7.15: Graphs of hyperbolic sine (a) and cosecant (b) functions

Hyperbolic cosine and secant functions are mutually reciprocal and symmetric about the y-axis (Figure 7.16). In other words, both hyperbolic cosine and secant functions are even functions.

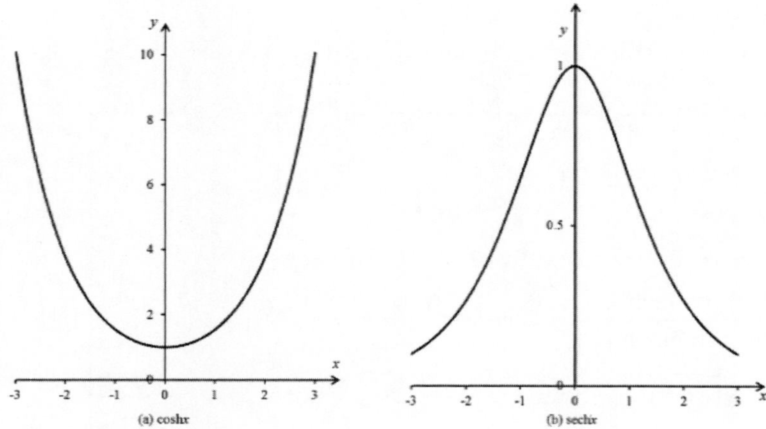

Figure 7.16: Graphs of hyperbolic cosine (a) and secant (b) functions

Hyperbolic tangent and cotangent functions are mutually reciprocal and symmetric about the original (Figure 7.17). Hence both hyperbolic tangent and cotangent functions are odd functions.

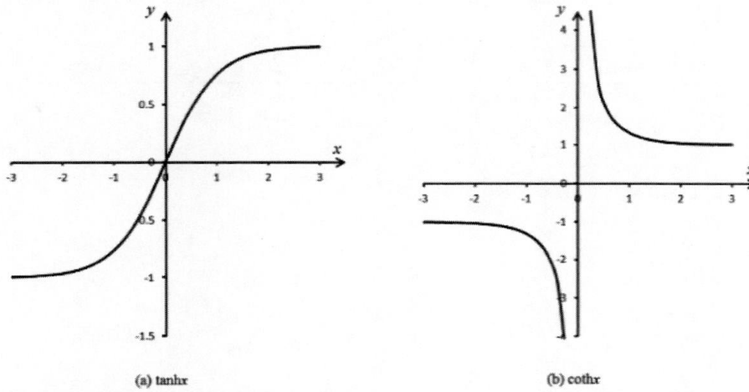

Figure 7.17: Graphs of hyperbolic tangent (a) and cotangent (b) functions

7.4.2 Hyperbolic Identities and Relationships

Hyperbolic identities

Since hyperbolic functions take similar forms as trigonometric functions, similar hyperbolic identities can be drawn as follows.

$$\boxed{\begin{aligned} &\cosh^2 x - \sinh^2 x = 1 \\ &1 - \tanh^2 x = \operatorname{sech}^2 x \\ &\coth^2 x - 1 = \operatorname{csch}^2 x \end{aligned}} \qquad (7.7)$$

Hyperbolic relationships

Similar hyperbolic relationships exist among different hyperbolic functions as follows.

$$\begin{array}{|l|}\hline \sinh(x \pm y) = \sinh x \cosh y \pm \cosh x \sinh y \\ \cosh(x \pm y) = \cosh x \cosh y \pm \sinh x \sinh y \\ \sinh 2x = 2\sinh x \cosh x \\ \cosh 2x = \cosh^2 x + \sinh^2 x = 2\cosh^2 x - 1 = 1 + 2\sinh^2 x \\ \tanh(x \pm y) = \dfrac{\tanh x \pm \tanh y}{1 \pm \tanh x \tanh y}, \qquad \tanh 2x = \dfrac{2\tanh x}{1 + \tanh^2 x} \\ \sinh^{-1} x = \ln(x + \sqrt{1 + x^2}), \ \ \cosh^{-1} x = \ln(x + \sqrt{x^2 - 1}), \ \ \tanh^{-1} x = \dfrac{1}{2}\ln\dfrac{1+x}{1-x} \\ \hline \end{array} \tag{7.8}$$

Example 7.19: Prove $\cosh^2 x - \sinh^2 x = 1$.

Solution

$$\cosh^2 x = (\frac{e^x + e^{-x}}{2})^2 = \frac{e^{2x} + 2e^x e^{-x} + e^{-2x}}{4} = \frac{e^{2x} + 2 + e^{-2x}}{4}, \quad \sinh^2 x = (\frac{e^x - e^{-x}}{2})^2 = \frac{e^{2x} - 2 + e^{-2x}}{4}$$

$$\cosh^2 x - \sinh^2 x = \frac{e^{2x} + 2 + e^{-2x}}{4} - \frac{e^{2x} - 2 + e^{-2x}}{4} = \frac{e^{2x} + 2 + e^{-2x} - e^{2x} + 2 - e^{-2x}}{4} = \frac{4}{4} = 1.$$

Example 7.20: Solve $\cosh x + \sinh x = e^{x^2 - 2x}$.

Solution

$$\cosh x + \sinh x = \frac{e^x + e^{-x}}{2} + \frac{e^x - e^{-x}}{2} = \frac{e^x + e^{-x} + e^x - e^{-x}}{2} = \frac{2e^x}{2} = e^x. \text{ Thus}$$

$$e^x = e^{x^2 - 2x} \longrightarrow x = x^2 - 2x \longrightarrow x^2 - 3x = 0 \longrightarrow x(x-3) = 0 \longrightarrow \begin{cases} x = 0 \\ x = 3 \end{cases}$$

The solutions are $x = 0$ and $x = 3$.

Exercises 7.4

1. Prove $1 - \tanh^2 x = \operatorname{sech}^2 x$.

2. Prove $\coth^2 x - 1 = \operatorname{csch}^2 x$.

3. Prove $\sinh 2x = 2\sinh x \cosh x$.

4. Solve $\cosh x - \sinh x = e^{-x^2 + 3x}$.

Chapter 7: Exercises Answers

Exercises 7.1

1. Values for cos(x – π/4) can be obtained by shifting the entire values of cos$x$ forwards by π/4; values for sin(x + π/4) can be obtained by shifting the entire values of cos$x$ backwards by π/4.
2. $2\sin 3x + 5\cos 3x = \sqrt{29}\sin(3x + 1.1903)$ and $2\sin 3x - 5\cos 3x = \sqrt{29}\sin(3x - 1.1903)$.
3. $3f(x) + 2g(x) = -10\sin(2x - 0.9273)$
4. $y(x) = -2\sin(3x + \pi/4)$: $A = 2$, $\varphi = \pi/4$, $\omega = 3$, $f = 3/(2\pi)$, $T = 1/f = 2\pi/3$.
   $y(t) = 3\sin(0.5t - \pi/6)$: $A = 3$, $\varphi = -\pi/6$, $\omega = 0.5$, $f = \omega/2\pi = 1/(4\pi)$, $T = 1/f = 4\pi$.
5. $y(x) = 4\sin(0.2x - 2\pi/3)$: $A = 4$, $\varphi = -2\pi/3$, $\omega = 0.2$, $f = 0.1/\pi$, $T = 10\pi$.
   $y(t) = -5\sin(4t + 3\pi/4)$; $A = 5$, $\varphi = 3\pi/4$, $\omega = 4$, $f = 2/\pi$, $T = 1/f = \pi/2$.
6. $3f(x) + 4g(x) = -6\sin 3x + 12\cos x$
7. $A \approx 2.5226$, $B \approx -7.5919$
8. $h(x) = 2f(x) - 3g(x) = 2\sin 2x - 3\cos 0.5x$
9. $h(x) = 4f(x) + 2g(x) = 4\sin x + 2\cos 3x$
10. $f(x) = y_1 + y_2 + y_3 = -2\sin 3x + 3\cos x + 3\cos 0.5x$

Exercises 7.2

1. a) $\frac{\pi}{6}$ or $\frac{5\pi}{6}$; b) $\frac{5\pi}{6}$; c) $\frac{\pi}{6}$; d) $\frac{3\pi}{4}$; e) $\frac{\pi}{4}$; f) arc csc(−2) is not in [0, π].

3. a) 0.4351; b) $\pm\frac{2\sqrt{2}}{3}$; c) $\pm\frac{3}{5}$; d) $\frac{\pi}{6}$; e) $\frac{5\pi}{8}$; f) $-\frac{5\pi}{6}$

4. a) $\frac{\pi}{3}$; b) $\approx 80°$; c) $\frac{5\pi}{6}$; d) $-\frac{\pi}{3}$.

5. a) $y = \frac{\pi}{6} + 2n\pi$, and $y = \frac{5\pi}{6} + 2n\pi$. $(n = 0, \pm1, \pm2, ...)$

   b) $y = \frac{5\pi}{6} + 2n\pi$, and $y = \frac{7\pi}{6} + 2n\pi$. $(n = 0, \pm1, \pm2, ...)$

   (c) $y = \frac{\pi}{6} + 2n\pi$, and $y = \frac{7\pi}{6} + 2n\pi$. $(n = 0, \pm1, \pm2, ...)$

   d) $y = \frac{3\pi}{4} + 2n\pi$, and $y = \frac{7\pi}{4} + 2n\pi$. $(n = 0, \pm1, \pm2, ...)$

   e) $y = \frac{\pi}{4} + 2n\pi$, and $y = \frac{7\pi}{4} + 2n\pi$. $(n = 0, \pm1, \pm2, ...)$

   f) $y = \frac{7\pi}{6} + 2n\pi$, and $y = \frac{11\pi}{6} + 2n\pi$. $(n = 0, \pm1, \pm2, ...)$

6. a) $\frac{\sqrt{2}(\sqrt{3}-1)}{4}$, b) $\frac{\sqrt{2}(\sqrt{3}-1)}{4}$; c) $\sqrt{3}$, d) $-\frac{1}{2}$

Exercises 7.3

1. $x = 22.5°$ and $-67.5°$
2. $x \approx 41.41°$
3. $x \approx 47.06°$
4. $x = 30°$ and $-30°$
5. $x = -90°$
6. $x = 60°$ and $-60°$
7. $x \approx 76.35°$
8. $x = 60°$

Exercises 7.4

4. $x = 0$ and $x = 4$

CHAPTER 8

8 Essentials of Vectors

CHAPTER OBJECTIVES

- Introduce plane vectors and representations
- Study different methods in vector operations
- Use vectors to solve scientific and engineering problems

Essential statements on vectors:

- A vector is a value that has both magnitude and direction.
- A plane vector can have different forms: two-point form, polar form, and rectangular form.
- Vectors are widely used in engineering and science.

Key topics:

- Plane vectors, representations, and operations
- Additions, subtractions and multiplications of plane vectors in the Cartesian system
- Solving real problems by vectors

Flowchart of mathematical knowledge building

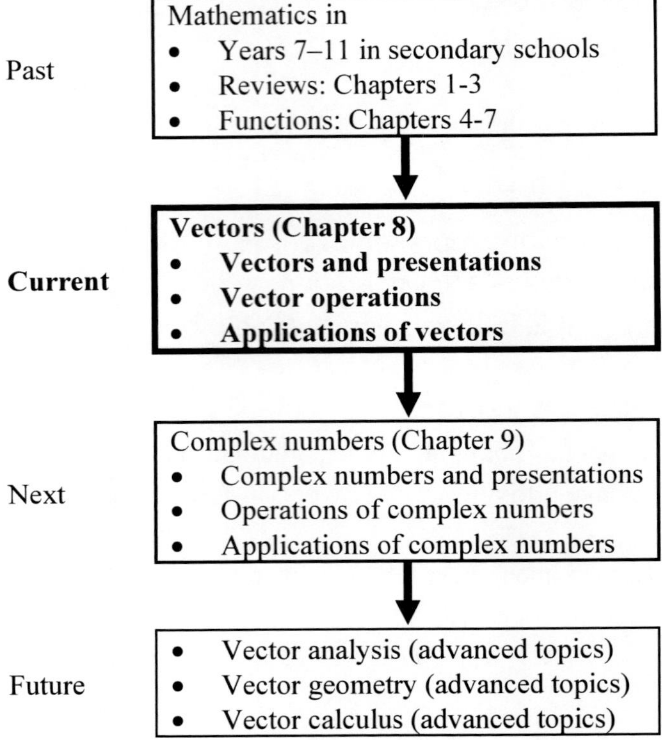

Chapter 8: Essentials of Vectors

8.1 Vectors and Operations

8.1.1 The concept and properties of vectors

Vectors and representations

The common values dealt with in the previous chapters only have quantities without any concern about how these values are oriented. Such values are called *scalars* which can be used to represent quantities like volume, surface area, length, time, mass, work, etc. *A vector is a value that has both magnitude and direction*, for example, velocity with direction (where it is moving towards) and speed or magnitude (how fast it is travelling), force with direction (where it is applied) and magnitude (how strong it is). *If a vector has a magnitude of 1, such vector is called a unit vector.* Unit vector is commonly used to represent the direction of other vectors that share the same direction but differ in magnitude.

A vector is commonly expressed using a directed line segment from A to B, denoted as boldface **AB**, or a boldface letter **V** (Figure 8.1a). In practice, the magnitude of vector **V** is represented by *italic* letter V or $|\mathbf{V}|$ and its direction is represented by a unit vector $\hat{\mathbf{v}}$ in the same direction of **V**. Thus, a generic vector **V** can be represented as $\mathbf{V} = V\hat{\mathbf{v}}$ where the unit vector $\hat{\mathbf{v}}$ is determined by

$$\boxed{\hat{\mathbf{v}} = \frac{\mathbf{V}}{|\mathbf{V}|} = \frac{\mathbf{V}}{V}.} \tag{8.1}$$

A plane vector can be represented in either the polar form or the rectangular form. The polar form combines the magnitude (V) of vector **V** with the angle θ of **V** measured normally from the easterly horizontal direction counter-clockwise (Figure 8.1b), denoted as $V\angle\theta$. For instance, $5\angle 50^\circ$ is a vector of magnitude of 5 units and 50° counter-clockwise from the easterly horizontal direction.

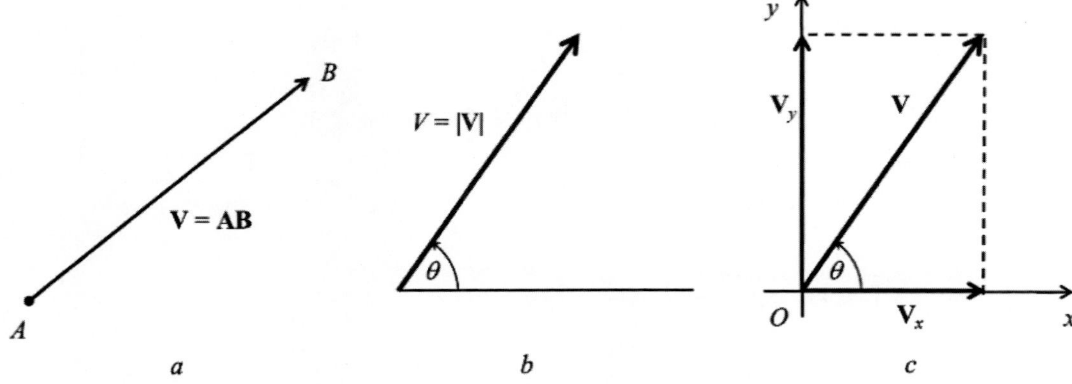

Figure 8.1: Vector in two-point form (a), polar form (b) and rectangular form (c)

Another commonly used representation for plane vectors is the rectangular form, in which vector **V** is decomposed to the horizontal (the x-axis) and the vertical (the y-axis) components $\mathbf{V}_x$ and $\mathbf{V}_y$ respectively (Figure 8.1c). Traditionally, the unit vectors in the x-axis and the y-axis are denoted by **i** and **j** respectively (**k** for the z-axis). Thus $\mathbf{V}_x = V_x\mathbf{i}$ and $\mathbf{V}_y = V_y\mathbf{j}$ and **V**, $\mathbf{V}_x$ and $\mathbf{V}_y$ form a right triangle (Figure 8.1c).

Any vector in either the polar form or the rectangular form can be mutually converted by

$$\boxed{\text{rectangular} \rightarrow \mathbf{V} = V_x\mathbf{i} + V_y\mathbf{j} \leftrightarrow \begin{cases} V_x = V\cos\theta \\ V_y = V\sin\theta \end{cases} \rightleftharpoons \begin{rcases} V = \sqrt{V_x^2 + V_y^2} \\ \theta = \tan^{-1}\dfrac{V_y}{V_x} \end{rcases} \leftrightarrow V\angle\theta \leftarrow \text{polar}}. \quad (8.2)$$

In handwriting, vector **V** = **AB** is often expressed as an arrow over the letter(s), i.e., $\vec{V}$ or $\overrightarrow{AB}$. Note **AB** and **BA** represent two different vectors which have the same magnitude but opposite to each other. Thus if **V** = **AB**, **BA** = –**AB** = –**V**.

Example 8.1: A plane vector is drawn from the origin of the Cartesian system to point (6, 8). Express this vector by means of its rectangular form and its polar form respectively. Determine the unit vector and find the vector that has the same magnitude but opposite to this vector.

Solution

Refer to Figure 8.2. The two components of vector **V** are $V_x = 6$, and $V_y = 8$.

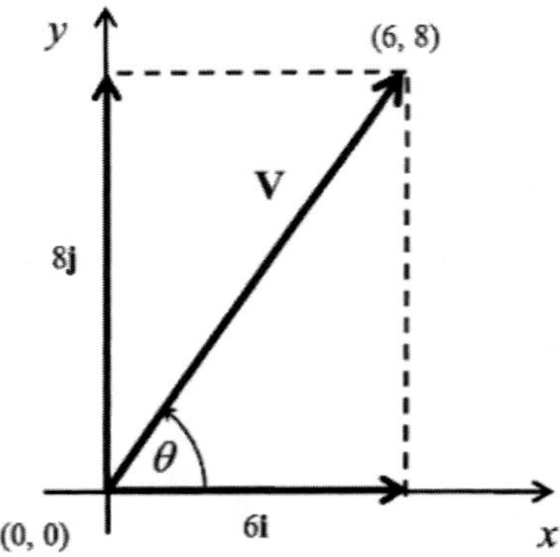

Figure 8.2: Vector in rectangular form

By formula (8.2), the magnitude and angle of **V** are

$$V = \sqrt{6^2 + 8^2} = \sqrt{36 + 64} = \sqrt{100} = 10, \text{ and } \theta = \tan^{-1}\frac{8}{6} = \tan^{-1}\frac{4}{3} \approx 53.13°.$$

Thus, this vector can be expressed as

$$\mathbf{V} = 6\mathbf{i} + 8\mathbf{j} = 2(3\mathbf{i} + 4\mathbf{j}) \text{ or } 10\angle 53.13°$$

$$\hat{\mathbf{v}} = \frac{\mathbf{V}}{V} = \frac{2(3\mathbf{i}+4\mathbf{j})}{10} = \frac{3\mathbf{i}+4\mathbf{j}}{5} = \frac{3\mathbf{i}}{5} + \frac{4\mathbf{j}}{5} \text{ or } \hat{\mathbf{v}} = \frac{\mathbf{V}}{V} = \frac{10\angle 53.13°}{10} = 1\angle 53.13°$$

The vector that has the same magnitude but opposite to vector **V** is –**V**, i.e.,

$$-\mathbf{V} = -6\mathbf{i} - 8\mathbf{j} = -2(3\mathbf{i} + 4\mathbf{j}) \text{ or } 10\angle(180° + 53.13°) \longrightarrow 10\angle 233.13°.$$

Example 8.2: A plane vector has a magnitude of 4 units and 30º clockwise from the horizontal direction. Express this vector by means of the polar form and the rectangular form respectively. Draw a diagram to show this vector in the Cartesian system.

Solution

As the vector has $V = 4$ and $\theta = -30°$ (clockwise), its polar form is $4\angle -30°$.

By formula (8.2),

$$V_x = 4\cos(-30°) = 4 \times \frac{\sqrt{3}}{2} = 2\sqrt{3}, \text{ and } V_y = 4\sin(-30°) = 4 \times (-\frac{1}{2}) = -2.$$

Thus its rectangular form is

$$\mathbf{V} = 2\sqrt{3}\mathbf{i} - 2\mathbf{j} = 2(\sqrt{3}\mathbf{i} - \mathbf{j}).$$

This vector is shown in Figure 8.3.

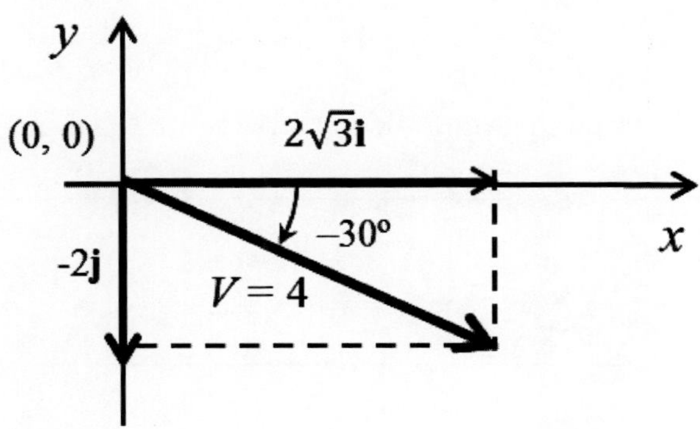

Figure 8.3: Vector in the Cartesian system

Properties of vectors

If **a**, **b** and **c** are vectors and m and n are scalars, the following rules are applicable to vectors

$$\begin{array}{|l|}\hline \mathbf{a}+\mathbf{b}=\mathbf{b}+\mathbf{a} \\ (\mathbf{a}+\mathbf{b})+\mathbf{c}=\mathbf{a}+(\mathbf{b}+\mathbf{c}) \\ m(\mathbf{a}+\mathbf{b})=m\mathbf{a}+m\mathbf{b} \\ m\mathbf{a}+n\mathbf{b}=n\mathbf{b}+m\mathbf{a} \\ \mathbf{a}-\mathbf{b}=\mathbf{a}+(-\mathbf{b})\neq\mathbf{b}-\mathbf{a} \\ \hline \end{array}. \tag{8.3}$$

8.1.2 Addition and subtraction of plane vectors

Given n vectors (Figure 8.4), each of these vectors is decomposed to the x-axis as x_k and the y-axis as y_k respectively. The resultant x-component X is the arithmetic summation of all x_k and the resultant y-component Y is the arithmetic summation of all y_k. The resultant vector becomes

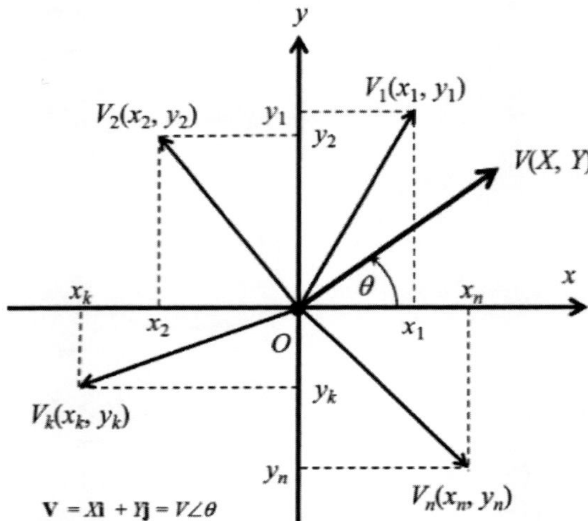

Figure 8.4: Adding and subtracting multiple vectors on the same plane in Cartesian system

$$\begin{array}{|l|}\hline \mathbf{V}=X\mathbf{i}+Y\mathbf{j} \\ X=x_1+x_2+...+x_k+...+x_n=\sum_{k=1}^{n}x_k, \text{ and } Y=y_1+y_2+...+y_k+...+y_n=\sum_{k=1}^{n}y_k. \\ \hline \end{array} \tag{8.4}$$

Or in polar form,

$$\begin{array}{|l|}\hline \mathbf{V}=V\angle\theta \\ V=\sqrt{X^2+Y^2}=\sqrt{(\sum_{k=1}^{n}x_k)^2+(\sum_{k=1}^{n}y_k)^2}, \text{ and } \theta=\arctan\frac{Y}{X}. \\ \hline \end{array} \tag{8.5}$$

For only two vectors $\mathbf{a} = x_1\mathbf{i} + y_1\mathbf{j}$, and $\mathbf{b} = x_2\mathbf{i} + y_2\mathbf{j}$, formulae (8.4-8.5) become

$$\mathbf{c} = \mathbf{a}+\mathbf{b} = (x_1+x_2)\mathbf{i} + (y_1+y_2)\mathbf{j} \rightleftarrows \mathbf{c} = c\angle\theta \leftarrow \begin{cases} c = |\mathbf{a}+\mathbf{b}| = \sqrt{(x_1+x_2)^2+(y_1+y_2)^2} \\ \theta = \arctan\dfrac{y_1+y_2}{x_1+x_2} \end{cases}$$

$$\mathbf{d} = \mathbf{a}-\mathbf{b} = (x_1-x_2)\mathbf{i} + (y_1-y_2)\mathbf{j} \rightleftarrows \mathbf{d} = d\angle\theta \leftarrow \begin{cases} c = |\mathbf{a}-\mathbf{b}| = \sqrt{(x_1-x_2)^2+(y_1-y_2)^2} \\ \theta = \arctan\dfrac{y_1-y_2}{x_1-x_2} \end{cases} \quad (8.6)$$

Example 8.3: Given two vectors **a** = 4∠30° and **b** = 3∠60°, use the rectangular method to determine **c** = **a** + **b**, and **d** = **a** – **b** respectively.

Solution

We first decompose **a** = 4∠30° and **b** = 3∠60° to the *x*-axis and the *y*-axis respectively as follows:

$$x_a = 4\cos 30° = 4\times\frac{\sqrt{3}}{2} = 2\sqrt{3}, \text{ and } y_a = 4\sin 30° = 4\times\frac{1}{2} = 2;$$

$$x_b = 3\cos 60° = 3\times\frac{1}{2} = \frac{3}{2}, \text{ and } y_b = 3\sin 60° = 3\times\frac{\sqrt{3}}{2} = \frac{3\sqrt{3}}{2}.$$

This also means both vectors are $\mathbf{a} = x_a\mathbf{i} + y_a\mathbf{j} = 2\sqrt{3}\mathbf{i} + 2\mathbf{j}$ and $\mathbf{b} = x_b\mathbf{i} + y_b\mathbf{j} = \frac{3}{2}\mathbf{i} + \frac{3\sqrt{3}}{2}\mathbf{j}$ in rectangular form. The resultant vector **c** is

$$X = x_a + x_b = 2\sqrt{3} + \frac{3}{2} = \frac{4\sqrt{3}+3}{2}, \text{ and } Y = y_a + y_b = 2 + \frac{3\sqrt{3}}{2} = \frac{4+3\sqrt{3}}{2};$$

$$\mathbf{c} = X\mathbf{i} + Y\mathbf{j} = \frac{4\sqrt{3}+3}{2}\mathbf{i} + \frac{4+3\sqrt{3}}{2}\mathbf{j}.$$

In polar form, the resultant vector **c** is

$$c = \sqrt{X^2+Y^2} = \sqrt{(\frac{4\sqrt{3}+3}{2})^2 + (\frac{4+3\sqrt{3}}{2})^2} = \sqrt{\frac{1}{4}[(4\sqrt{3})^2 + 2\times 4\sqrt{3}\times 3 + 3^2 + 4^2 + 2\times 4\times 3\sqrt{3} + (3\sqrt{3})^2]}$$

$$= \sqrt{\frac{1}{4}(48 + 24\sqrt{3} + 9 + 16 + 24\sqrt{3} + 27)} = \sqrt{\frac{1}{4}(100 + 48\sqrt{3})} = \sqrt{25 + 12\sqrt{3}},$$

$$\theta = \arctan\frac{Y}{X} = \arctan\frac{\frac{4+3\sqrt{3}}{2}}{\frac{4\sqrt{3}+3}{2}} = \arctan\frac{4+3\sqrt{3}}{4\sqrt{3}+3} \approx 42.81°.$$

For **d** = **a** – **b**, it can be regarded as and **d** = **a** + (–**b**). As $\mathbf{a} = 2\sqrt{3}\mathbf{i} + 2\mathbf{j}$, and $\mathbf{b} = \frac{3}{2}\mathbf{i} + \frac{3\sqrt{3}}{2}\mathbf{j}$,

$$\mathbf{d} = \mathbf{a} + (-\mathbf{b}) = (2\sqrt{3}\mathbf{i} + 2\mathbf{j}) - (\frac{3}{2}\mathbf{i} + \frac{3\sqrt{3}}{2}\mathbf{j}) = (2\sqrt{3} - \frac{3}{2})\mathbf{i} + (2 - \frac{3\sqrt{3}}{2})\mathbf{j} = \frac{4\sqrt{3}-3}{2}\mathbf{i} + \frac{4-3\sqrt{3}}{2}\mathbf{j}$$

$$d = \sqrt{X^2 + Y^2} = \sqrt{(\frac{4\sqrt{3}-3}{2})^2 + (\frac{4-3\sqrt{3}}{2})^2}$$

$$= \sqrt{\frac{1}{4}[(4\sqrt{3})^2 - 2\times 4\sqrt{3}\times 3 + 3^2 + 4^2 - 2\times 4\times 3\sqrt{3} + (3\sqrt{3})^2]}$$

$$= \sqrt{\frac{1}{4}(48 - 24\sqrt{3} + 9 + 16 - 24\sqrt{3} + 27)} = \sqrt{\frac{1}{4}(100 - 48\sqrt{3})} = \sqrt{25 - 12\sqrt{3}} \approx 2.0531$$

$$\theta = \arctan\frac{Y}{X} = \arctan\frac{\frac{4-3\sqrt{3}}{2}}{\frac{4\sqrt{3}-3}{2}} = \arctan\frac{4-3\sqrt{3}}{4\sqrt{3}-3} \approx -16.94^\circ \longrightarrow V = 2.0531\angle -16.94^\circ$$

The angle is negative, which means the resultant vector is in quadrant IV.

Example 8.4: Given three vectors **a** = 4∠30°, **b** = 3∠150°, and **c** = 2∠300°, use the rectangular method to determine **d** = **a** + **b** + **c**.

Solution

We first decompose **a** = 4∠30°, **b** = 3∠150°, and **c** = 2∠300° to the *x*-axis and the *y*-axis respectively as follows:

$$x_a = 4\cos 30^\circ = 4\times\frac{\sqrt{3}}{2} = 2\sqrt{3}, \text{ and } y_a = 4\sin 30^\circ = 4\times\frac{1}{2} = 2;$$

$$x_b = 3\cos 150^\circ = -3\cos 30^\circ = -3\times\frac{\sqrt{3}}{2} = -\frac{3\sqrt{3}}{2}, \text{ and } y_b = 3\sin 150^\circ = 3\sin 30^\circ = 3\times\frac{1}{2} = \frac{3}{2};$$

$$x_c = 2\cos 300^\circ = 2\cos 60^\circ = 2\times\frac{1}{2} = 1, \text{ and } y_c = 2\sin 300^\circ = -2\sin 60^\circ = -2\times\frac{\sqrt{3}}{2} = -\sqrt{3}.$$

$$X = x_a + x_b + x_c = 2\sqrt{3} - \frac{3\sqrt{3}}{2} + 1 = \frac{4\sqrt{3} - 3\sqrt{3} + 2}{2} = \frac{\sqrt{3}+2}{2},$$

$$Y = y_a + y_b + y_c = 2 + \frac{3}{2} - \sqrt{3} = \frac{7-2\sqrt{3}}{2} \longrightarrow \mathbf{d} = X\mathbf{i} + Y\mathbf{j} = \frac{\sqrt{3}+2}{2}\mathbf{i} + \frac{7-2\sqrt{3}}{2}\mathbf{j};$$

$$V = \sqrt{(\frac{\sqrt{3}+2}{2})^2 + (\frac{7-2\sqrt{3}}{2})^2} = \sqrt{\frac{1}{4}[(\sqrt{3})^2 + 2\times\sqrt{3}\times 2 + 2^2 + 7^2 - 2\times 7\times 2\sqrt{3} + (2\sqrt{3})^2]}$$

$$= \sqrt{\frac{1}{4}(3 + 4\sqrt{3} + 4 + 49 - 28\sqrt{3} + 12)} = \sqrt{\frac{1}{4}(68 - 24\sqrt{3})} = \sqrt{17 - 6\sqrt{3}} \approx 2.5705$$

$$\theta = \arctan\frac{Y}{X} = \arctan\frac{\frac{7-2\sqrt{3}}{2}}{\frac{\sqrt{3}+2}{2}} = \arctan\frac{7-2\sqrt{3}}{\sqrt{3}+2} \approx 43.45^\circ \longrightarrow V = 2.5705\angle 43.45^\circ$$

8.1.3 Multiplications of two plane vectors

Dot multiplication of two plane vectors

The dot multiplication (or product) of two plane vectors $\mathbf{a} = a\angle\alpha$ and $\mathbf{b} = b\angle\beta$ with an included angle θ is a scalar defined as

$$c = \mathbf{a}\bullet\mathbf{b} = ab\cos(\alpha-\beta) = ab\cos(\beta-\alpha) = ab\cos\theta. \tag{8.7}$$

In rectangular form, $\mathbf{a} = a_x\mathbf{i} + a_y\mathbf{j}$ and $\mathbf{b} = b_x\mathbf{i} + b_y\mathbf{j}$. Thus

$$c = \mathbf{a}\bullet\mathbf{b} = (a_x\mathbf{i} + a_y\mathbf{j})\bullet(b_x\mathbf{i} + b_y\mathbf{j}) = a_x b_x + a_y b_y. \tag{8.8}$$

If $\mathbf{b} = \mathbf{a}$, formula (8.7) provides another way of obtaining the magnitude of vector $\mathbf{a}$ below,

$$\mathbf{a}\bullet\mathbf{a} = a \times a\cos 0° = a^2, \text{ or } a = \sqrt{\mathbf{a}\bullet\mathbf{a}} = \sqrt{a_x^2 + a_y^2} \tag{8.9}$$

The dot multiplication of two plane vectors is the product of the magnitude of the first vector with the projected magnitude of the second vector to the first vector (Figure 8.5a). If $\mathbf{a}\bullet\mathbf{b} = 0$, the two vectors $\mathbf{a}$ and $\mathbf{b}$ are perpendicular to each other.

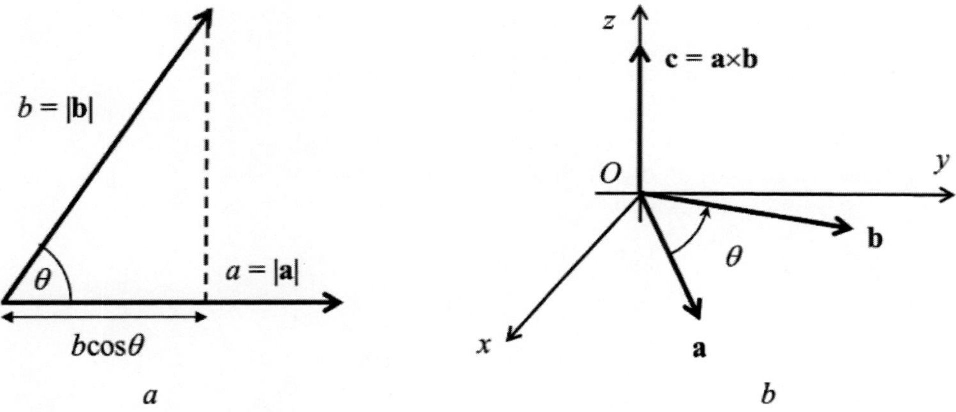

Figure 8.5: Dot multiplication (a) and cross multiplication (b) of two plane vectors

Cross multiplication of two plane vectors

The cross multiplication (or product) of two plane vectors $\mathbf{a} = a\angle\alpha$ and $\mathbf{b} = b\angle\beta$ is a new vector $\mathbf{c}$ that is perpendicular to the plane formed by $\mathbf{a}$ and $\mathbf{b}$ defined as (Figure 8.5b)

$$\mathbf{c} = \mathbf{a}\times\mathbf{b} = ab(\hat{\mathbf{a}}\times\hat{\mathbf{b}}) = ab\sin(\beta-\alpha)\hat{\mathbf{c}} \tag{8.10}$$

Note $\mathbf{a}\times\mathbf{b}\neq\mathbf{b}\times\mathbf{a}$. The direction of the resultant vector follows the 'right-hand rule'. Hence $\mathbf{a}\times\mathbf{b}$ and $\mathbf{b}\times\mathbf{a}$ have the same magnitude but are opposite to each other.

In rectangular form, $\mathbf{a}=a_x\mathbf{i}+a_y\mathbf{j}$ and $\mathbf{b}=b_x\mathbf{i}+b_y\mathbf{j}$. Thus

$$\boxed{\mathbf{c}=\mathbf{a}\times\mathbf{b}=(a_x\mathbf{i}+a_y\mathbf{j})\times(b_x\mathbf{i}+b_y\mathbf{j})=(a_xb_y-a_yb_x)(\mathbf{i}\times\mathbf{j})=(a_xb_y-a_yb_x)\mathbf{k}.} \tag{8.11}$$

Here **k** is the unit vector in the z-axis. <u>If $\mathbf{a}\times\mathbf{b}=0$, the two vectors are parallel to each other.</u>

Example 8.5: Given two vectors $\mathbf{a}=4\angle 30^\circ$ and $\mathbf{b}=3\angle 60^\circ$, determine $c=\mathbf{a}\bullet\mathbf{b}$ and $\mathbf{d}=\mathbf{a}\times\mathbf{b}$ respectively.

<u>Solution</u>

As $c=\mathbf{a}\bullet\mathbf{b}$ is a scalar, by formula (8.7),

$$c=\mathbf{a}\bullet\mathbf{b}=ab\cos(\alpha-\beta)=4\times3\cos(30^\circ-60^\circ)=12\cos(-30^\circ)=12\cos30^\circ=12\times\frac{\sqrt{3}}{2}=6\sqrt{3}.$$

Decomposing $\mathbf{a}=4\angle 30^\circ$ and $\mathbf{b}=3\angle 60^\circ$ to the x-axis and the y-axis was done in Example 8.3, which gave $\mathbf{a}=2\sqrt{3}\mathbf{i}+2\mathbf{j}$ and $\mathbf{b}=\frac{3}{2}\mathbf{i}+\frac{3\sqrt{3}}{2}\mathbf{j}$. Thus by formula (8.8),

$$c=\mathbf{a}\bullet\mathbf{b}=a_xb_x+a_yb_y=2\sqrt{3}\times\frac{3}{2}+2\times\frac{3\sqrt{3}}{2}=3\sqrt{3}+3\sqrt{3}=6\sqrt{3}.$$

This is the same as the result obtained using formula (8.7).

For $\mathbf{d}=\mathbf{a}\times\mathbf{b}$, we use formula (8.11) as follows:

$$\mathbf{d}=\mathbf{a}\times\mathbf{b}=(a_xb_y-a_yb_x)\mathbf{k}=(2\sqrt{3}\times\frac{3\sqrt{3}}{2}-2\times\frac{3}{2})\mathbf{k}=(9-3)\mathbf{k}=6\mathbf{k}.$$

Using formula (8.10),

$$\mathbf{d}=\mathbf{a}\times\mathbf{b}=ab\sin(\beta-\alpha)\hat{\mathbf{c}}=4\times3\sin30^\circ\mathbf{k}=12\times\frac{1}{2}\mathbf{k}=6\mathbf{k}$$

This is the same as the result obtained using formula (8.11).

Example 8.6: Assume $\mathbf{a} = 4\mathbf{i} - 5\mathbf{j}$, $\mathbf{b} = A\mathbf{i} + B\mathbf{j}$, and $\mathbf{c} = B\mathbf{i} + A\mathbf{j}$. If $\mathbf{a} \bullet \mathbf{b} = 8$ and $\mathbf{a} \times \mathbf{b} = 0$, determine **b** and **c**.

Solution

$$\because \mathbf{a} \bullet \mathbf{b} = 8 \longrightarrow (4\mathbf{i} - 5\mathbf{j}) \bullet (A\mathbf{i} + B\mathbf{j}) = 4A - 5B = 8; \quad \mathbf{a} \times \mathbf{c} = 0 \longrightarrow (4\mathbf{i} - 5\mathbf{j}) \times (B\mathbf{i} + A\mathbf{j}) = 4A + 5B = 0$$

$$\therefore \begin{cases} 4A - 5B = 8 \\ 4A + 5B = 0 \end{cases} \xrightarrow{(+)} 8A = 8 \longrightarrow \begin{cases} A = 1 \\ B = -\dfrac{4}{5}A = -\dfrac{4}{5} \end{cases}$$

$$\therefore \mathbf{b} = A\mathbf{i} + B\mathbf{j} = \mathbf{i} - \frac{4}{5}\mathbf{j}, \quad \mathbf{c} = B\mathbf{i} + A\mathbf{j} = -\frac{4}{5}\mathbf{i} + \mathbf{j}$$

Example 8.7: Given three vectors $\mathbf{a} = 3\mathbf{i} - \mathbf{j}$, $\mathbf{b} = -2\mathbf{i} + 3\mathbf{j}$, and $\mathbf{c} = -5\mathbf{i} + 2\mathbf{j}$, determine $(\mathbf{a} \times \mathbf{b}) \bullet \mathbf{c}$, $\mathbf{a} \times \mathbf{b} \times \mathbf{c}$, and $\mathbf{c} \times \mathbf{b} \times \mathbf{a}$.

Solution

$$(\mathbf{a} \times \mathbf{b}) \bullet \mathbf{c} = [(3\mathbf{i} - \mathbf{j}) \times (-2\mathbf{i} + 3\mathbf{j})] \bullet (-5\mathbf{i} + 2\mathbf{j}) = [(9-2)\mathbf{k}] \bullet (-5\mathbf{i} + 2\mathbf{j}) = 7\mathbf{k} \bullet (-5\mathbf{i} + 2\mathbf{j})$$

$$= -35\mathbf{k} \bullet \mathbf{i} + 14\mathbf{k} \bullet \mathbf{j} = 0 \longleftarrow \mathbf{k} \bullet \mathbf{i} = \mathbf{k} \bullet \mathbf{j} = \cos 90° = 0$$

$$\mathbf{a} \times \mathbf{b} \times \mathbf{c} = (3\mathbf{i} - \mathbf{j}) \times (-2\mathbf{i} + 3\mathbf{j}) \times (-5\mathbf{i} + 2\mathbf{j}) = (7\mathbf{k}) \times (-5\mathbf{i} + 2\mathbf{j}) = -35\mathbf{k} \times \mathbf{i} + 14\mathbf{k} \times \mathbf{j}$$

$$= -35\mathbf{j} - 14\mathbf{i} = -7(2\mathbf{i} + 5\mathbf{j}) \longleftarrow \mathbf{k} \times \mathbf{i} = \mathbf{j},\ \mathbf{k} \times \mathbf{j} = -\mathbf{i}$$

$$\mathbf{c} \times \mathbf{b} \times \mathbf{a} = (-5\mathbf{i} + 2\mathbf{j}) \times (-2\mathbf{i} + 3\mathbf{j}) \times (3\mathbf{i} - \mathbf{j}) = [(-15+4)\mathbf{k}] \times (3\mathbf{i} - \mathbf{j}) = -11\mathbf{k} \times (3\mathbf{i} - \mathbf{j})$$

$$= -11(3\mathbf{k} \times \mathbf{i} - \mathbf{k} \times \mathbf{j}) = -11(3\mathbf{j} + \mathbf{i}) = -11(\mathbf{i} + 3\mathbf{j}) \longleftarrow \mathbf{k} \times \mathbf{i} = \mathbf{j},\ \mathbf{k} \times \mathbf{j} = -\mathbf{i}$$

Note that $\mathbf{a} \times \mathbf{b} \times \mathbf{c}$ and $\mathbf{c} \times \mathbf{b} \times \mathbf{a}$ are two different vectors.

Exercises 8.1

1. Convert the following vectors in polar form to their corresponding rectangular form respectively.

   a) $\mathbf{a} = 5\angle 60^\circ$
   b) $\mathbf{b} = 4\angle 315^\circ$
   c) $\mathbf{c} = 6\angle 150^\circ$
   d) $\mathbf{d} = 7\angle -30^\circ$

2. Convert the following vectors in rectangular form to their corresponding polar form respectively.

   a) $\mathbf{a} = \sqrt{3}\mathbf{i} + 2\mathbf{j}$.
   b) $\mathbf{b} = -3\mathbf{i} + 2\mathbf{j}$
   c) $\mathbf{c} = 6\mathbf{i} - 8\mathbf{j}$
   d) $\mathbf{d} = -4\mathbf{i} - 3\mathbf{j}$

3. Use the rectangular method to solve the following problems of vector addition and/or subtractions.

   a) Given $\mathbf{a} = 5\angle 60^\circ$ and $\mathbf{b} = 3\angle 45^\circ$, determine $\mathbf{c} = \mathbf{a} + \mathbf{b}$.
   b) Given $\mathbf{a} = 8\angle 30^\circ$ and $\mathbf{b} = 4\angle 315^\circ$, determine $\mathbf{c} = \mathbf{a} + \mathbf{b}$ and $\mathbf{d} = \mathbf{b} - \mathbf{a}$.
   c) Given $\mathbf{a} = 6\angle 150^\circ$ and $\mathbf{b} = 3\angle 210^\circ$, determine $\mathbf{c} = \mathbf{a} + \mathbf{b}$.
   d) Given $\mathbf{a} = 4\angle -90^\circ$ and $\mathbf{b} = 7\angle -30^\circ$, determine $\mathbf{c} = \mathbf{a} + \mathbf{b}$ and $\mathbf{d} = \mathbf{a} - \mathbf{b}$.

4. Use the rectangular method to solve the following problems of vector addition and/or subtractions.

   a) $\mathbf{a} = 4\angle 30^\circ$, $\mathbf{b} = 3\angle 150^\circ$, and $\mathbf{c} = 2\angle 300^\circ$, what is $\mathbf{d} = \mathbf{a} + \mathbf{b} + \mathbf{c}$?
   b) $\mathbf{a} = 3\angle 300^\circ$, $\mathbf{b} = 5\angle 120^\circ$, and $\mathbf{c} = 7\angle 30^\circ$, what is $\mathbf{d} = 2\mathbf{a} + \mathbf{b} - \mathbf{c}$?
   c) $\mathbf{a} = 4\angle -45^\circ$, $\mathbf{b} = 3\angle 150^\circ$, and $\mathbf{c} = 12\angle 60^\circ$, what is $\mathbf{d} = -\mathbf{a} - 3\mathbf{b} + 4\mathbf{c}$?

5. Given two vectors $\mathbf{a} = 5\angle 60^\circ$ and $\mathbf{b} = 8\angle 120^\circ$, determine $c = \mathbf{a} \bullet \mathbf{b}$ and $\mathbf{d} = \mathbf{a} \times \mathbf{b}$ using two different methods respectively.

6. Given $\mathbf{a} = 7\angle 45^\circ$ and $\mathbf{b} = 4\angle 310^\circ$, determine $c = \mathbf{a} \bullet \mathbf{b}$ and $\mathbf{d} = \mathbf{a} \times \mathbf{b}$ using two different methods respectively.

7. Assume $\mathbf{a} = -2\mathbf{i} + 3\mathbf{j}$, $\mathbf{b} = -B\mathbf{i} - A\mathbf{j}$, and $\mathbf{c} = A\mathbf{i} + B\mathbf{j}$. If $\mathbf{a} \bullet \mathbf{b} = 0$ and $\mathbf{a} \times \mathbf{b} = 13\mathbf{k}$, determine $\mathbf{b}$ and $\mathbf{c}$.

8. Given three vectors $\mathbf{a} = 3\mathbf{i} - \mathbf{j}$, $\mathbf{b} = -2\mathbf{i} + 3\mathbf{j}$, and $\mathbf{c} = -5\mathbf{i} + 2\mathbf{j}$, determine $\mathbf{a} \times \mathbf{c} \times \mathbf{b}$ and $\mathbf{a} \bullet (\mathbf{a} \times \mathbf{c} \times \mathbf{b})$.

8.2 Applications of vectors

Vectors are widely used in engineering. In theory, vectors in a 3D space can be manipulated commonly by matrix operations. Hence the examples presented below are mainly vectors on the same 2D plane. However, the principles of the 2D operations are applicable in 3D space.

Example 8.8: An airplane is heading to northeast (45°) with an air speed of 800 km/h. A wind of 60 km/h blows from southeast (–30°, from east). Find the ground speed of the airplane and the actual direction of flying. See Figure 8.6 for reference.

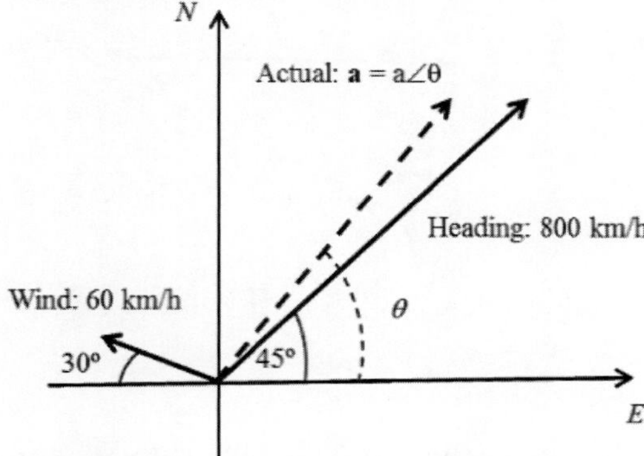

Figure 8.6: Sketch for Example 8.8

<u>**Solution**</u>

Decomposition of **h** = 800∠45° and **w** = 60∠150° (= 180° – 30°) to the x-axis and the y-axis is done as follows:

$$h_x = 800\cos 45° = 800\times\frac{\sqrt{2}}{2} = 400\sqrt{2},\text{ and } h_y = 800\sin 45° = 800\times\frac{\sqrt{2}}{2} = 400\sqrt{2};$$

$$w_x = 60\cos 150° = -60\cos 30° = -60\times\frac{\sqrt{3}}{2} = -30\sqrt{3},\text{ and } w_y = 60\sin 150° = 60\sin 30° = 60\times\frac{1}{2} = 30.$$

The actual speed and direction are determined by

$$X = h_x + w_x = 400\sqrt{2} - 30\sqrt{3},\text{ and } Y = h_y + w_y = 400\sqrt{2} + 30;$$

$$\mathbf{a} = X\mathbf{i} + Y\mathbf{j} = (400\sqrt{2} - 30\sqrt{3})\mathbf{i} + (400\sqrt{2} + 30)\mathbf{j};$$

$$\begin{aligned} V &= \sqrt{(400\sqrt{2} - 30\sqrt{3})^2 + (400\sqrt{2} + 30)^2} \\ &= \sqrt{(400\sqrt{2})^2 - 2\times 400\sqrt{2}\times 30\sqrt{3} + (30\sqrt{3})^2 + (400\sqrt{2})^2 + 2\times 400\sqrt{2}\times 30 + 30^2} \\ &= \sqrt{2\times 400^2 - 60\times 400\sqrt{6} + 2700 + 2\times 400^2 + 60\times 400\sqrt{2} + 900} \\ &\approx \sqrt{643600 - 24847} \approx 787\ (\text{km}/\text{h}). \end{aligned}$$

$$\theta = \arctan\frac{400\sqrt{2} + 30}{400\sqrt{2} - 30\sqrt{3}} \approx 49.2°.$$

Due to the wind, the actual flying is slightly slower and drifting about 4° further north.

Example 8.9: Two cables hold an object static vertically. Cable A has a pulling force of 50 N in a direction 30º with the x-axis and cable B has a pulling force of 40 N in a direction 60º with the opposite of the x-axis (Figure 8.7). How heavy is the object? [$g \approx 9.8$ m/s^2]

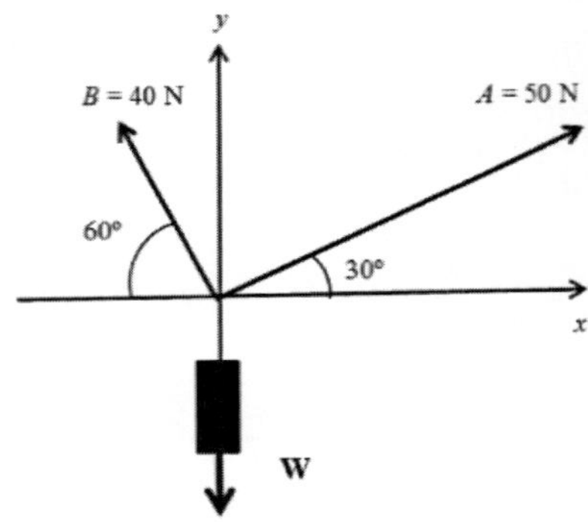

Figure 8.7: Sketch for Example 8.9

Solution

We decompose $\mathbf{A} = 50\angle 30°$, $\mathbf{B} = 40\angle 120°$ (=180º – 60º), and $\mathbf{W} = W\angle 270°$ to the x-axis and the y-axis as follows:

$$A_x = 50\cos 30° = 50\times\frac{\sqrt{3}}{2} = 25\sqrt{3}, \text{ and } A_y = 50\sin 30° = 50\times\frac{1}{2} = 25;$$
$$B_x = 40\cos 120° = -40\cos 60° = -40\times\frac{1}{2} = -20, \text{ and } B_y = 40\sin 120° = 40\times\frac{\sqrt{3}}{2} = 20\sqrt{3};$$
$$W_x = W\cos 270° = W\times 0 = 0, \text{ and } W_y = W\sin 270° = -W\sin 90° = -W.$$

As the object is static vertically, the sum of forces in the y-axis must be zero. Thus

$$Y\mathbf{j} = (A_y + B_y + W_y)\mathbf{j} = (25 + 20\sqrt{3} - W)\mathbf{j} \longrightarrow Y = 0$$
$$25 + 20\sqrt{3} - W = 0$$
$$W = 25 + 20\sqrt{3} \approx 59.64 \text{ (N)}$$
$$\therefore W = mg, \quad \therefore m = \frac{W}{g} = \frac{59.64}{9.8} \approx 6.1 \text{ kg}$$

The object is about 6 kg.

Example 8.10: A block of 51 kg is static on an inclined plane of 30° (Figure 8.8). How strong the friction is? [$g \approx 9.8$ m/s^2]

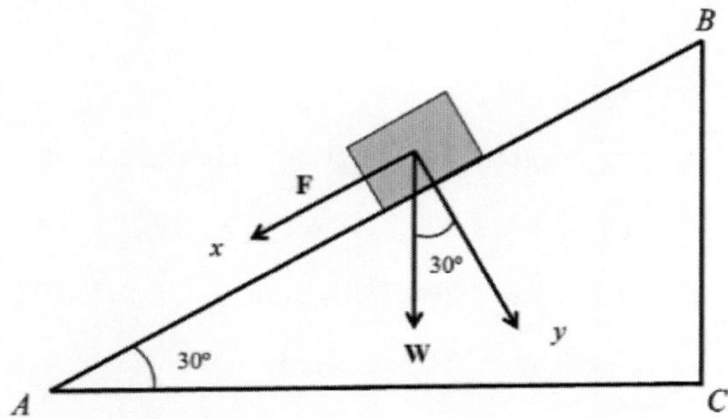

Figure 8.8: Sketch for Example 8.10

Solution

There are two forces exerted on the block: the weight **W** and the friction **F**. The weight is $W = mg = 51 \times 9.8 \approx 500$ N. We decompose **W** and **F** to the local rectangular x-axis and y-axis as follows:

$$W_x = 500\sin 30° = 500 \times \frac{1}{2} = 250, \text{ and } W_y = 500\cos 30° = 500 \times \frac{\sqrt{3}}{2} = 250\sqrt{3};$$
$$F_x = F\cos 0° = F, \text{ and } F_y = F\sin 0° = 0.$$

As the block is static on the inclined plane, the sum of forces in either axis must be zero. Thus

$$X\mathbf{i} = (W_x + F_x)\mathbf{i} = (250 + F)\mathbf{i} \longrightarrow X = 0$$
$$250 + F = 0 \longrightarrow F = -250 \text{ (N)}.$$

The friction is 250 N and parallel upwards to the inclined plane.

Example 8.11: In Example 8.10, if a horizontal force **P** of $400\sqrt{3}$ (N) is applied to the block, which moves the block on the inclined plane for 3 metres from point A to point B (Figure 8.9), how much work was done by force **P** for moving the block up to point B?

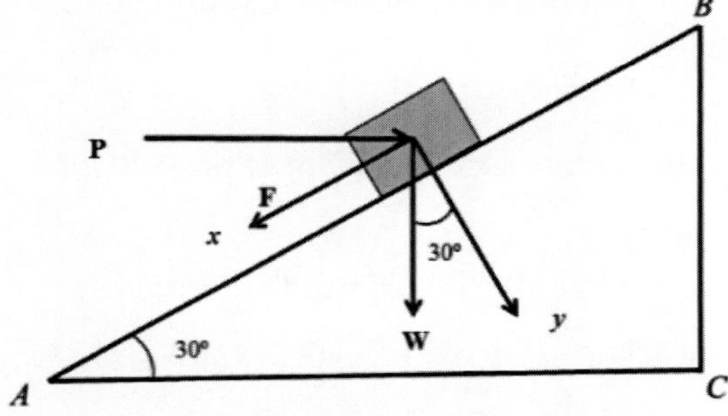

Figure 8.9: Sketch for Example 8.11

Solution
Still using the local rectangular x-axis and y-axis used in Example 8.10, all forces are

$$\mathbf{W} = W_x\mathbf{i} + W_y\mathbf{j} = 250\mathbf{i} + 250\sqrt{3}\mathbf{j},$$
$$\mathbf{F} = F_x\mathbf{i} + F_y\mathbf{j} = 250\mathbf{i} + 0\mathbf{j}, \longleftarrow \text{friction in the } x\text{-axis against the upward movement of the block}$$
$$\mathbf{P} = P_x\mathbf{i} + P_y\mathbf{j} = -400\sqrt{3}\cos 30°\mathbf{i} + 400\sqrt{3}\sin 30°\mathbf{j} = -600\mathbf{i} + 200\sqrt{3}\mathbf{j}. \text{ Thus the total force is}$$
$$\mathbf{T} = (W_x + F_x + P_x)\mathbf{i} + (W_y + F_y + P_y)\mathbf{j} = (250 + 250 - 600)\mathbf{i} + (250\sqrt{3} + 0 + 200\sqrt{3})\mathbf{j}$$
$$= -100\mathbf{i} + 450\sqrt{3}\mathbf{j}.$$

As the block was moved on the inclined plane for 3 metres, the displacement vector **D** in the local rectangular system is

$$\mathbf{D} = -3\mathbf{i} + 0\mathbf{j} = -3\mathbf{i}.$$

The work done by force **P** is

$$W_P = \mathbf{P}\bullet\mathbf{D} = (-600\mathbf{i} + 200\sqrt{3}\mathbf{j})\bullet(-3\mathbf{i}) = (-3)(-600) = 1800 \text{ (Nm)}.$$

Example 8.12: A plate in the xy-plane rotates around an axis perpendicular to the xy-plane at the origin (Figure 8.10). If a force $\mathbf{F}$ = 100N∠60° is applied at point $A(3, -2)$ on the xy-plane, find the torque vector $\boldsymbol{\tau} = \mathbf{OA} \times \mathbf{F}$ about the axis of rotation.

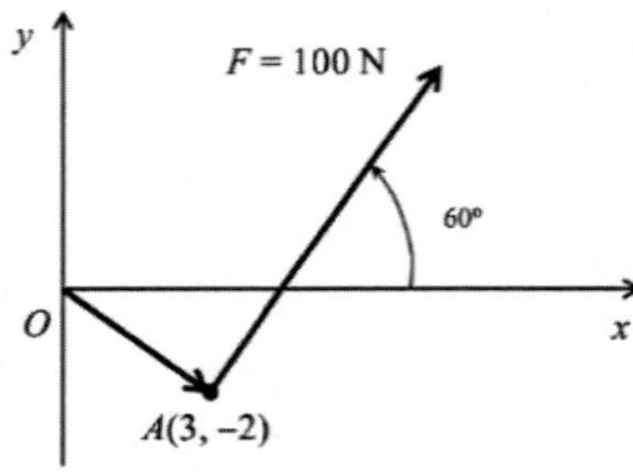

Figure 8.10: Sketch for Example 8.12

Solution
In the rectangular system, both vectors **OA** and **F** are expressed as

$$\mathbf{OA} = 3\mathbf{i} - 2\mathbf{j}, \text{ and } \mathbf{F} = 100\cos 60°\mathbf{i} + 100\sin 60°\mathbf{j} = 50\mathbf{i} + 50\sqrt{3}\mathbf{j}. \text{ Thus}$$
$$\boldsymbol{\tau} = \mathbf{OA} \times \mathbf{F} = (3\mathbf{i} - 2\mathbf{j}) \times (50\mathbf{i} + 50\sqrt{3}\mathbf{j}) = [3 \times 50\sqrt{3} - (-2) \times 50]\mathbf{k} = (150\sqrt{3} + 100)\mathbf{k}.$$

Exercises 8.2

1. An airplane is heading to northwest (135°) with an air speed of 880 km/h. A wind of 100 km/h blows from north (90°). Find the ground speed of the airplane and the actual direction of flying. [Hint: draw a diagram similar to Figure 8.6 for reference.]

2. Three cables hold an object static vertically. Cable A has a pulling force of 70 N in a direction 30° with the x-axis, cable B has a pulling force of 45 N in a direction 75° with the x-axis, and cable C has a pulling force of 100 N in a direction 45° with the opposite of the x-axis. How much does the object weigh? [Hint: draw a diagram similar to Figure 8.7 for reference. $g \approx 9.8$ m/s^2]

3. A block of 102 kg is static on an inclined plane of 20°. How strong the friction is? [Hint: draw a diagram similar to Figure 8.8. $g \approx 9.8$ m/s^2]

4. In Question 3, the block would slowly slide down to the bottom from the top of the inclined plane if its angle increases to 45°. How much work the gravity would have done on the block?

5. Three vehicles depart at the same point O. Vehicle A travels easterly with a speed of 100 km/h whereas vehicle B heads to the west with a speed of 60 km/h. Vehicle C travels with a speed of 80 km/h north-easterly with an angle of 50° from the east. Use vector operations to find how far vehicle C is from vehicles A and B respectively in half an hour? (keep 1 decimal place)

6. A radar speed camera is placed at point C off side of a road. It takes two readings of time in seconds at Points A and B respectively for a moving vehicle (See figure below). The speed limit of this road is 80 km/h. The time difference for vehicle X is 0.5 seconds between A and B whereas the time difference for vehicle Y is 0.4 seconds. Use vector operations to determine the speed of X and Y. Is any vehicle speeding?

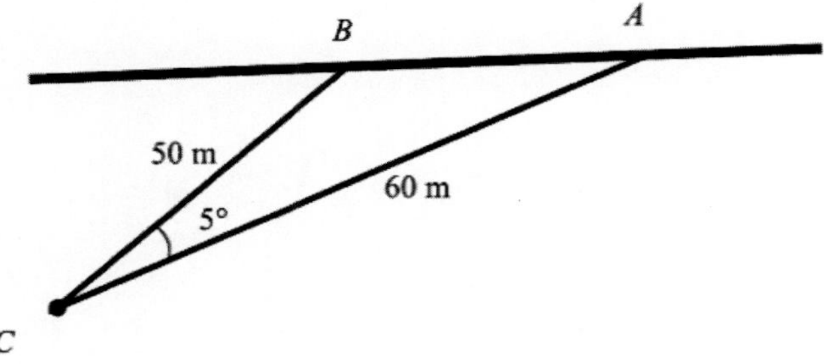

Chapter 8: Exercises Answers

Exercises 8.1

1. a) $\mathbf{V} = \frac{5}{2}\mathbf{i} + \frac{5\sqrt{3}}{2}\mathbf{j}$, b) $\mathbf{V} = 2\sqrt{2}\mathbf{i} - 2\sqrt{2}\mathbf{j}$, c) $\mathbf{V} = -3\sqrt{3}\mathbf{i} + 3\mathbf{j}$, d) $\mathbf{V} = \frac{7\sqrt{3}}{2}\mathbf{i} - \frac{7}{2}\mathbf{j}$
2. a) $\mathbf{a} = \sqrt{7}\angle 49.1°$, b) $\mathbf{b} = \sqrt{13}\angle 146.3°$, c) $\mathbf{c} = 10\angle -53.1° = 10\angle 306.9°$,
   d) $\mathbf{d} = 5\angle 216.9°$
3. a) $\mathbf{c} = 4.62\mathbf{i} + 6.45\mathbf{j},\ r \approx 7.9,\ \lambda \approx 54.4°$
   b) $\mathbf{c} = 9.76\mathbf{i} + 1.17\mathbf{j},\ r \approx 9.8,\ \lambda_c \approx 6.8°$, $\mathbf{d} = -4.1\mathbf{i} - 6.83\mathbf{j},\ r \approx 8.0,\ \lambda_d \approx 239°$
   c) $\mathbf{c} = -7.8\mathbf{i} + 1.5\mathbf{j},\ r \approx 7.94,\ \lambda \approx 169.1°$
   d) $\mathbf{c} = 6.06\mathbf{i} - 7.5\mathbf{j},\ r \approx 9.6,\ \lambda_c \approx 308.9°$, $\mathbf{d} = -6.06\mathbf{i} - 0.5\mathbf{j},\ r \approx 6.1,\ \lambda_d \approx 184.7°$
4. a) $\mathbf{d} = 1.86\mathbf{i} + 1.77\mathbf{j},\ r \approx 2.6,\ \theta \approx 43.6°$; b) $\mathbf{d} = -6.56\mathbf{i} - 4.37\mathbf{j},\ r \approx 7.9,\ \theta \approx 213.7°$
   c) $\mathbf{d} = 28.97\mathbf{i} + 39.89\mathbf{j},\ r \approx 49.3,\ \theta \approx 54°$
5. $c = \mathbf{a} \bullet \mathbf{b} = 20$, $\mathbf{d} = \mathbf{a} \times \mathbf{b} \approx 34.6\mathbf{k}$
6. $c = \mathbf{a} \bullet \mathbf{b} \approx -2.4$, $\mathbf{d} = \mathbf{a} \times \mathbf{b} \approx -27.9\mathbf{k}$
7. $A = 2,\ B = 3;\ \mathbf{b} = -3\mathbf{i} - 2\mathbf{j},\quad \mathbf{c} = 2\mathbf{i} + 3\mathbf{j}$
8. $\mathbf{a} \times \mathbf{c} \times \mathbf{b} = -3\mathbf{i} - 2\mathbf{j},\ \mathbf{a} \bullet (\mathbf{a} \times \mathbf{c} \times \mathbf{b}) = -7$

Exercises 8.2

1. $\mathbf{a} = -622.3\mathbf{i} + 522.3\mathbf{j}, V \approx 812$ (km/h), $\theta \approx 140°$
2. $m = 15.2$ (kg)
3. $F = -342$ (N)
4. $Work = 707L$ (Nm)
5. $AC \approx 39.1$ km, $\varphi_{AC} = \angle OAC \approx -51.6°$; $BC \approx 63.6$ km, $\varphi_{BC} = \angle OBC \approx 28.8°$.
6. $v_X \approx 79.9(km/h),\ v_Y \approx 99.7(km/h)$.
   By the speed limit of 80 km/h, vehicle *Y* is speeding, and vehicle *X* is within the speed limit.

CHAPTER 9

9 Essentials of Complex Numbers

CHAPTER OBJECTIVES

- Introduce complex numbers and representations
- Study operational rules of complex numbers
- Demonstrate applications of complex numbers

Essential statements on complex numbers:

- A complex number consists of a real part and an imaginary part.
- A complex number can be expressed by multiple forms.
- Complex numbers are widely used in electrical engineering.

Key topics:

- Complex numbers and representations
- Operations of complex numbers with different forms
- Complex numbers and vectors
- Quadratic equations with complex roots
- Ohm's law for alternating current by complex numbers

Flowchart of mathematical knowledge building

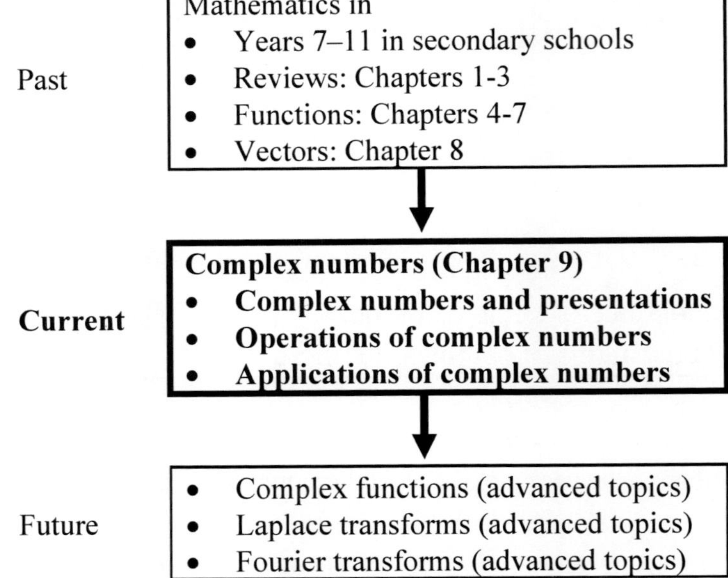

Chapter 9: Essentials of Complex Numbers

9.1 Complex Numbers in Rectangular System

9.1.1 The concept and representation in rectangular system

Using formula (1.16) in Chapter 1, the solution to quadratic equation $x^2+4x+8=0$ can be found as $x=2\pm i2$, in which $i=\sqrt{-1}$ (or $j=\sqrt{-1}$) represents the basic imaginary unit, contrary to real numbers. Complex numbers have the following format

$$\boxed{a+bi \text{ or } a+ib \quad (a+bj \text{ or } a+jb),} \tag{9.1}$$

where a and b are real numbers and are called values of the real part (*Re*) and imaginary part (*Im*) respectively. If b = 0, a complex number becomes a real number. If a = 0, a complex number has only the imaginary part and is called a *pure* imaginary number.

The two complex numbers, $a+ib$ and $a-ib$, are called ***conjugate*** complex numbers. Let $z=a+ib$, then $\bar{z}=a-ib$ is the conjugate of z.

Complex numbers $a+ib$ can be plotted as points onto a rectangular system, in which the horizontal axis represents the real part whereas the vertical axis represents the imaginary part. If we draw a straight line directed from the origin to a point, such diagram is called an ***Argand diagram*** (Figure 9.1a). The complex solution $z=2\pm 2i$ to $z^2+4z+8=0$ can be displayed as two directed points in the Argand diagram in Figure 9.1b. Note $z=2+2i$ and $\bar{z}=2-2i$ are conjugate numbers and symmetric about the real axis in the Argand diagram.

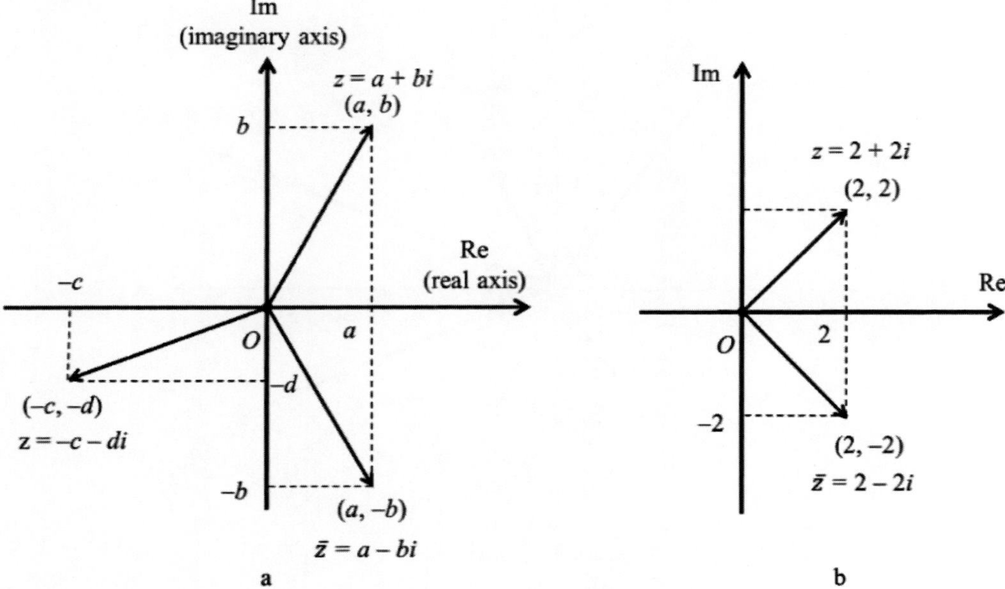

Figure 9.1: Argand diagrams

9.1.2 Operations in rectangular system

Addition and subtraction of complex numbers in rectangular system

If two or more complex numbers are for addition or subtraction, the outcome is a new complex number whose real part is the arithmetic sum of all real parts and whose imaginary part is the arithmetic sum of the imaginary parts of individual complex numbers, i.e.,

$$\boxed{(a_1+ib_1)\pm(a_2+ib_2)+\cdots\pm(a_n+ib_n)=(a_1\pm a_2\pm\cdots\pm a_n)+i(b_1\pm b_2\pm\cdots\pm b_n)}. \tag{9.2}$$

Example 9.1: Given $Z_1 = 5 + i2$, $Z_2 = -3 + i4$, $Z_3 = 1 - i2$, $Z_4 = -2 - i3$, $Z_5 = i$, find $A = Z_1 + Z_2 + Z_3 + Z_4 + Z_5$, and $B = Z_1 - Z_2 - Z_3 + Z_4 + Z_5$. Plot these complex numbers in an Argand diagram.

Solution

Using formula (9.2), A and B can be determined as follows:

$$\begin{aligned}
A &= Z_1+Z_2+Z_3+Z_4+Z_5=(5+i2)+(-3+i4)+(1-i2)+(-2-i3)+(i)\\
&=(5-3+1-2+0)+i(2+4-2-3+1)=1+i2\\
B &= Z_1-Z_2-Z_3+Z_4+Z_5=(5+i2)-(-3+i4)-(1-i2)+(-2-i3)+(i)\\
&=(5+3-1-2+0)+i(2-4+2-3+1)=5+i(-2)=5-i2
\end{aligned}$$

These complex numbers are plotted in Figure 9.2.

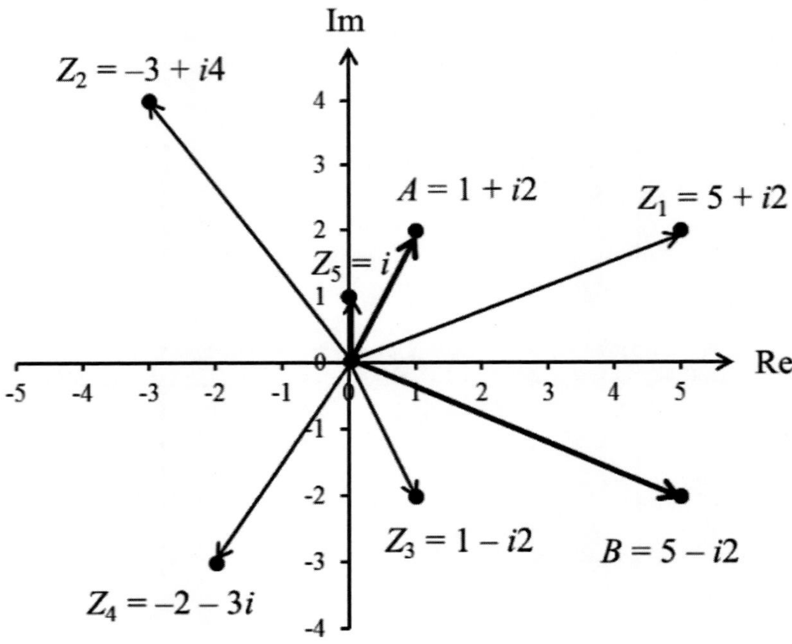

Figure 9.2: Argand diagram for Example 9.1

Multiplication of complex numbers in rectangular system

$$\begin{array}{l} i^n = (\sqrt{-1})^n \quad (n = 1, 2, \cdots) \\ n = 1 \;\rightarrow\; i^1 = i = (\sqrt{-1})^1 = \sqrt{-1} = i \\ n = 2 \;\rightarrow\; i^2 = (\sqrt{-1})^2 = -1 \\ n = 3 \;\rightarrow\; i^3 = (\sqrt{-1})^3 = (\sqrt{-1})^2\sqrt{-1} = -i \\ n = 4 \;\rightarrow\; i^4 = (\sqrt{-1})^2(\sqrt{-1})^2 = (-1)(-1) = 1 \\ n = 5 \;\rightarrow\; i^5 = (\sqrt{-1})^4(\sqrt{-1}) = 1 \times (\sqrt{-1}) = i \\ \cdots \\ n = 4k + m \;\rightarrow\; i^n = i^{4k+m} = i^{4k} i^m = [(\sqrt{-1})^4]^k i^m = 1^k \times i^m = i^m \\ (k = 1, 2 \ldots; 0 \le m \le 3),\ \textit{repeating if } n \ge 5 \end{array} \tag{9.3}$$

$$r(a \pm ib) = ra \pm irb \quad (r \text{ is a real number.}) \tag{9.4}$$

$$(a + ib)(c + id) = (c + id)(a + ib) = (ac - bd) + i(ad + bc) \tag{9.5}$$

$$(a + ib)(a - ib) = a^2 + b^2 \quad \text{(Conjugate multiplication is a positive real number!)} \tag{9.6}$$

Example 9.2: Given $Z_1 = 2 + i3$, $Z_2 = -i2$, and $Z_3 = i4$, find $A = 5Z_1$, $B = Z_1Z_3$, $C = Z_2Z_3$, and $D = Z_2Z_2$.

Solution

Using formulae (9.3) and (9.4), A, B, C and D can be determined as follows:

$$\begin{aligned} A &= 5Z_1 = 5(2 + i3) = 10 + i15 \\ B &= Z_1Z_3 = (2 + i3)(i4) = (i4)(2 + i3) = i8 + i^2 12 = i8 - 12 = -12 + i8 \\ C &= Z_2Z_3 = (-i2)(i4) = -8i^2 = -8 \times (-1) = 8 \\ D &= Z_2Z_2 = (-i2)(-i2) = 4i^2 = -4 \end{aligned}$$

Note D is different from the following wrong process:

$$(-i2)(-i2) \ne (-\sqrt{-4})(-\sqrt{-4}) = \sqrt{-4}\sqrt{-4} = \sqrt{(-4)(-4)} = \sqrt{16} = 4$$

Example 9.3: Let $Z_1 = 2 + i3$, $Z_2 = -1 + i2$, and $Z_3 = 1 - i4$. Find $A = Z_1Z_2$, $B = Z_1Z_2Z_3$, $C = Z_1^4$, and $D = Z_1\overline{Z}_1$.

Solution

Using formula (9.5), A, B and C can be determined as follows:

$$A = Z_1Z_2 = (2+i3)(-1+i2) = [2\times(-1)-3\times 2]+i[2\times 2+3\times(-1)] = -8+i$$
$$B = Z_1Z_2Z_3 = AZ_3 = (-8+i)(1-i4) = [-8\times 1-1\times(-4)]+i[1\times 1+(-8)\times(-4)] = -4+i33$$
$$C = Z_1^4 = (2+i3)^4 = [(2+i3)^2]^2 = [2^2+2\times 2\times i3+(i3)^2]^2 = (4+i12+9i^2)^2 = (4+i12-9)^2$$
$$= (-5+i12)^2 = (-5)^2+2\times(-5)\times i12+(i12)^2 = 25-i120+(-1)\times 144 = -119-i120$$

Use formula (9.6) to find D as follows:

$$D = Z_1\overline{Z}_1 = (2+i3)(2-i3) = 2^2+3^2 = 4+9 = 13$$

Division of complex numbers in rectangular system

Division between two complex numbers works by making the complex denominator the conjugate product. Hence,

$$\boxed{\frac{a+ib}{ic} = \frac{b}{c} - i\frac{a}{c}} \qquad (9.7)$$

$$\boxed{\frac{a+ib}{c+id} = \frac{ac+bd}{c^2+d^2} + i\frac{bc-ad}{c^2+d^2}} \qquad (9.8)$$

Example 9.4: Derive formulas (9.8).

Solution

Using formulas (9.3), (9.5) and (9.6), the denominator can be converted to a real number by using conjugate multiplication as follows:

$$\frac{a+ib}{c+id} = \frac{(a+ib)(c-id)}{(c+id)(c-id)} = \frac{ac-iad+ibc-(ib)(id)}{c^2+d^2} = \frac{ac+i(bc-ad)-i^2bd}{c^2+d^2} = \frac{ac+i(bc-ad)+bd}{c^2+d^2}$$
$$= \frac{ac+bd+i(bc-ad)}{c^2+d^2} = \frac{ac+bd}{c^2+d^2} + i\frac{bc-ad}{c^2+d^2}$$

Example 9.5: Given $Z_1 = 2 + i3$, $Z_2 = -i2$, and $Z_3 = 1 + i2$, find $A = Z_1/Z_2$, $B = 5Z_1/Z_3$, $C = Z_1Z_2/Z_3$ and $D = \dfrac{Z_1Z_2Z_3}{\overline{Z}_1\overline{Z}_2\overline{Z}_3}$.

Solution

Using formulas (9.7) and (9.8), A and B can be determined as follows:

$$A = Z_1 / Z_2 = \frac{2+i3}{-i2} = \frac{3}{-2} - i\frac{2}{-2} = -\frac{3}{2} + i$$

$$B = \frac{5Z_1}{Z_3} = 5(\frac{2+i3}{1+i2}) = 5(\frac{2\times 1+3\times 2}{1^2+2^2} + i\frac{3\times 1-2\times 2}{1^2+2^2}) = 5(\frac{8}{5} + i\frac{-1}{5}) = 8 - i$$

$$C = \frac{Z_1Z_2}{Z_3} = \frac{(2+i3)(-i2)}{1+i2} = \frac{(-i4+6)(1-i2)}{1^2+2^2} = \frac{(-i4-8+6-i12)}{5} = \frac{-2-i16}{5}$$

$$D = \frac{Z_1Z_2Z_3}{\overline{Z}_1\overline{Z}_2\overline{Z}_3} = \frac{(Z_1Z_2Z_3)(Z_1Z_2Z_3)}{(\overline{Z}_1\overline{Z}_2\overline{Z}_3)(Z_1Z_2Z_3)} = \frac{Z_1^2Z_2^2Z_3^2}{(Z_1\overline{Z}_1)(Z_2\overline{Z}_2)(Z_3\overline{Z}_3)} = \frac{(2+i3)^2(i2)^2(1+i2)^2}{(2^2+3^3)(2^2)(1^2+2^2)}$$

$$= \frac{(4+i12-9)(-4)(1+i4-4)}{(4+9)(4)(1+4)} = \frac{-(-5+i12)(4)(-3+i4)}{(13)(4)(5)} = \frac{(5-i12)(-3+i4)}{(13)(5)}$$

$$= \frac{-15+i20+i36+48}{65} = \frac{33+i56}{65}$$

Exercises 9.1

1. Combine and simplify the following complex numbers.

   a) $(2 + i) - (4 - i3)$
   b) $(a - i2) + (a - i5)$
   c) $(-3 + 7i) - (5 + i4)$

2. Evaluate the following operations.

   a) i^7 and i^{10}
   b) i^8 and i^{15}
   c) i^9 and i^{23}

3. Find the conjugate of the following complex numbers.

   a) $2 - i3$
   b) $2a + ib$
   c) $-3 + i4$

4. Multiply and simplify the following complex numbers.

   a) $(2 + i)(4 - i3)$
   b) $(a - i2)(a - i5)$
   c) $(-3 + 7i)(5 + i4)$

5. Divide and simplify the following complex numbers.

   a) $(2 + i) \div (4 - i3)$
   b) $(-3 + 7i) \div (5 + i4)$
   c) $(a - i2) \div (a - i5)$

6. Given $Z_1 = 1 + i3$, $Z_2 = 2 - i$, and $Z_3 = i4$, find $A = 3Z_1$, $B = 2Z_1Z_2$, and $D = Z_2Z_3$. Plot these complex numbers in an Argand diagram.

7. Given $Z_1 = 2 + i$, $Z_2 = -5 + i4$, and $Z_3 = -3 - i6$, find $A = Z_1 + 2Z_2 + 3Z_3$, $B = 4Z_1 - 2Z_2 - 3Z_3$, and $C = -3Z_1 - Z_2 + Z_3$. Plot these complex numbers in an Argand diagram.

8. Given $Z_1 = 2 + i3$, and $Z_2 = 1 - i4$, find $A = Z_1Z_2$, $B = Z_2^3$, and $D = Z_2\overline{Z}_2$.

9. Given $Z_1 = 3 + i2$, $Z_2 = i2$, and $Z_3 = -1 + i2$, find $A = 2Z_1/Z_2$ and $B = 3Z_3/Z_1$.

9.2 Complex Numbers in Polar System

9.2.1 Representation of complex numbers in polar system

Complex numbers in rectangular system can be converted to the polar form in a polar system shown in Figure 9.3. Since cosθ = a/r, sinθ = b/r, and tanθ = b/a, it is easy to derive the following equations between the two systems. If a polar angle is negative, it should be converted to positive first to have a unique conversation to the rectangular form. For example, if $\theta = -60°$, it needs to be converted to $\theta = 300°$ for a unique conversation to the rectangular form.

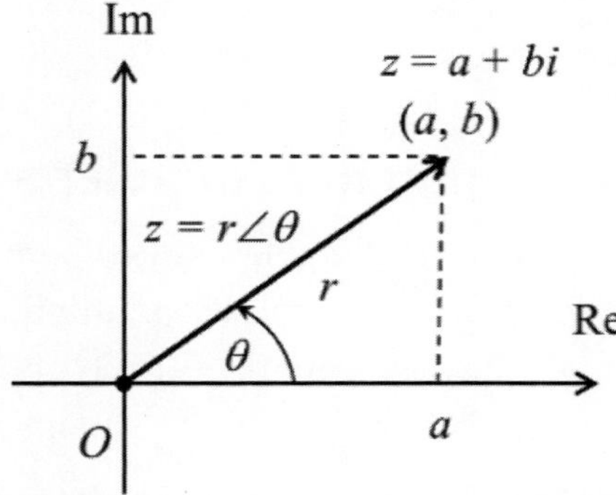

Figure 9.3: Polar form of a complex number

$$\begin{aligned} &a+ib = r\angle\theta \\ &a = r\cos\theta \quad \text{and} \quad b = r\sin\theta \\ &r = \sqrt{a^2+b^2} \quad \text{and} \quad \theta = \tan^{-1}(\frac{b}{a}) \end{aligned} \qquad (9.9)$$

$$a+ib = r(\cos\theta + i\sin\theta) \quad \text{(trigonometric form of complex number)} \qquad (9.10)$$

$$\begin{aligned} &i = 0 + i\cdot 1 = 1(\cos 90° + i\sin 90°) = 1\angle 90° \\ &-i = 0 + i\cdot(-1) = 1[\cos(-90°) + i\sin(-90°)] = 1\angle -90° \end{aligned} \quad \text{(the } i \text{ or } j \text{ operator)} \qquad (9.11)$$

Example 9.6: Express $Z = 4 + i3$ in polar and trigonometric forms respectively.

Solution

Using formulae (9.9) and (9.10), Z can be converted to the polar and trigonometric forms respectively.

$$r = \sqrt{a^2+b^2} = \sqrt{4^2+3^2} = \sqrt{16+9} = \sqrt{25} = 5$$

$$\theta = \tan^{-1}(\frac{b}{a}) = \tan^{-1}(\frac{3}{4}) = \tan^{-1} 0.75 = 36.9°$$

$$\textit{polar} \quad Z = 4 + i3 = 5\angle 36.9°$$

$$\textit{trigonometric} \quad Z = 4 + i3 = r(\cos\theta + i\sin\theta) = 5(\cos 36.9° + i\sin 36.9°)$$

Example 9.7: Express $Z = 5(\cos 50° + i\sin 50°)$ in polar and rectangular forms respectively.

Solution

Using formulas (9.9) and (9.10), Z can be converted to the polar and trigonometric forms respectively.

$$\textit{polar} \quad Z = 5\angle 50^\circ$$
$$a = r\cos\theta = 5\cos 50^\circ \approx 5\times 0.6428 = 3.214 \qquad b = r\sin\theta = 5\sin 50^\circ \approx 5\times 0.766 = 3.830$$
$$\textit{rectangular} \quad Z = a + ib = 3.214 + i3.830$$

9.2.2 Operations of complex numbers in polar system

Complex addition and subtraction in a rectangular system are fast but multiplication and division are much slower. In a polar system, complex multiplication and division are much faster but addition and subtraction are more complicated. Therefore, complex multiplication, division, and associated operations are preferred in a polar system as below.

$$\boxed{r_1\angle\theta_1 \bullet r_2\angle\theta_2 = r_1 r_2\angle(\theta_1 + \theta_2) \quad (\textit{product})} \tag{9.12}$$

$$\boxed{\begin{aligned} &i\bullet r\angle\theta = 1\angle 90°\bullet r\angle\theta = r\angle(\theta + 90°) \\ &-i\bullet r\angle\theta = 1\angle -90°\bullet r\angle\theta = r\angle(\theta - 90°) \end{aligned}} \tag{9.13}$$

$$\boxed{\frac{r_1\angle\theta_1}{r_2\angle\theta_2} = \frac{r_1}{r_2}\angle(\theta_1 - \theta_2) \quad (\textit{quotient})} \tag{9.14}$$

$$\boxed{(r\angle\theta)^n = r^n\angle n\theta \quad (\textit{power}\ \text{or}\ \textit{DeMoivre's theorem})} \tag{9.15}$$

$$\boxed{(r\angle\theta)^{1/n} = r^{1/n}\angle(\theta + 360k)/n \quad (k = 0, 1, \ldots n-1. \quad n\ \textit{numbers})} \tag{9.16}$$

Example 9.8: Given $Z_1 = 2\angle 50^\circ$ and $Z_2 = 3\angle 15^\circ$, find $Z_1 Z_2$, Z_1 / Z_2 and Z_2^4.

Solution

Using formulae (9.12), (9.14) and (9.15), we can work out the outcomes as follows.

$$Z_1 Z_2 = (2\angle 50°)(3\angle 15°) = (2\times 3)\angle(50° + 15°) = 6\angle 65°$$
$$\frac{Z_1}{Z_2} = \frac{2\angle 50°}{3\angle 15°} = \frac{2}{3}\angle(50° - 15°) = \frac{2}{3}\angle 35°$$
$$Z_2^4 = (3\angle 15°)^4 = 3^4\angle(4\times 15°) = 81\angle 60°$$

Example 9.9: Given $Z_1 = 4\angle 70°$ and $Z_2 = 5\angle 150°$, find iZ_1 and $-iZ_2$.

Solution

Using formula (9.13), we can work out the outcomes as follows.

$$iZ_1 = 4\angle(70° + 90°) = 4\angle 160°$$
$$-iZ_2 = 5\angle(150° - 90°) = 5\angle 60°$$

Note that if a complex number in polar form (e.g., Z_1) is multiplied by i, the product (iZ_1) is to rotate the complex number forwards by 90°. If a complex number in polar form (e.g., Z_2) is multiplied by $-i$, the product ($-iZ_2$) is to rotate the complex number backwards by 90°.

Example 9.10: Given $Z = 8\angle 60°$, find $Z^{1/3}$.

Solution

By formula (9.16),

$$Z^{1/3} = (8\angle 60^\circ)^{1/3} = 8^{1/3}\angle(60^\circ + 360^\circ k)/3 = 2\angle(60^\circ + 360^\circ k)/3$$
$$k = 0 \quad \rightarrow \quad Z^{1/3} = 2\angle 60^\circ/3 = 2\angle 20^\circ$$
$$k = 1 \quad \rightarrow \quad Z^{1/3} = 2\angle(60^\circ + 360^\circ)/3 = 2\angle 420^\circ/3 = 2\angle 140^\circ$$
$$k = 2 \quad \rightarrow \quad Z^{1/3} = 2\angle(60^\circ + 720^\circ)/3 = 2\angle 780^\circ/3 = 2\angle 260^\circ$$

These multiple solutions are resulted from the factor that cosine and sine functions are periodically repeating every 360º. These three are the basic angles within the first period of 360º and will repeatedly occur in sequence in the subsequent periods.

Exercises 9.2

1. Express the following complex numbers in polar and trigonometric forms respectively.

   a) $Z_1 = 2 + i$
   b) $Z_2 = 2 - i$
   c) $Z_3 = -3 - i6$

2. Express the following complex numbers in rectangular and polar forms respectively.

   a) $Z = 4(\cos 60° + i \sin 60°)$
   b) $Z = 6(\cos 150° + i \sin 150°)$
   c) $Z = 2(\cos 315° + i \sin 315°)$

3. Express the following complex numbers in rectangular and trigonometric forms respectively.

   a) $Z = 3\angle 60°$
   b) $Z = 5\angle 105°$
   c) $Z = 4\angle 280°$

4. Given $Z_1 = 3\angle 150^\circ$ and $Z_2 = 4\angle 45^\circ$, find Z_1Z_2, Z_1 / Z_2, and Z_2^4 in polar form respectively.

5. Given $Z = 16\angle 60^\circ$, find $Z^{1/4}$.

6. Given $Z_1 = 2(\cos 150° + i \sin 150°)$ and $Z_2 = 3(\cos 30° + i \sin 30°)$, find Z_1Z_2, Z_2 / Z_1, and Z_1^3 in polar form respectively.

7. Given $Z = 3(\cos 200° + i \sin 200°)$, find $Z^{1/3}$.

8. Given $Z_1 = 8\angle 45°$ and $Z_2 = 7\angle 109°$, find $-iZ_1Z_2$, and iZ_1 / Z_2.

9.3 Complex Numbers in Exponential Form

Complex numbers can also take the exponential form, commonly known as Euler's formula,

$$\boxed{re^{i\theta} = r(\cos\theta + i\sin\theta) \quad (\theta \text{ in } \textbf{radians} \text{ only})} \tag{9.17}$$

where e (≈ 2.71828) is the base of natural logarithm.

Complex multiplication, division, and associated operations in exponential form are similar to and maybe slightly easier than that in a polar system.

$$\boxed{r_1 e^{i\theta_1} \cdot r_2 e^{i\theta_2} = r_1 r_2 e^{i(\theta_1+\theta_2)} \quad (product)} \tag{9.18}$$

$$\boxed{\frac{r_1 e^{i\theta_1}}{r_2 e^{i\theta_2}} = \frac{r_1}{r_2} e^{i(\theta_1-\theta_2)} \quad (quotient)} \tag{9.19}$$

$$\boxed{(re^{i\theta})^n = r^n e^{in\theta} \quad (powers \text{ and } roots)} \tag{9.20}$$

Example 9.11: Covert $Z_1 = 3 + i4$ and $Z_2 = 3(\cos 60° + i\sin 60°)$ to their exponential forms, then find Z_2 / Z_1 and Z_2^3.

Solution

Convert Z_1 to polar form

$$r = \sqrt{3^2 + 4^2} = \sqrt{9+16} = \sqrt{25} = 5, \quad \theta_1 = \tan^{-1}\frac{4}{3} \approx 53.13° .$$

Then convert the angles in degrees in both complex numbers to radians

$$\theta_1 = \frac{53.13°}{180°}\pi \approx \frac{53.13°}{180°} \times 3.1416 \approx 0.9273$$

$$\theta_2 = 60° = \frac{\pi}{3} \approx 1.0472 .$$

In exponential form, both complex numbers and operations are

$$Z_1 = 5e^{i0.9273}, \quad Z_2 = 3e^{i\frac{\pi}{3}} = 3e^{i1.0472}$$

$$\frac{Z_2}{Z_1} = \frac{3}{5} e^{i(1.0472-0.9273)} = \frac{3}{5} e^{i0.1199} = 0.6e^{i0.1199}$$

$$Z_2^3 = (3e^{i\frac{\pi}{3}})^3 = 3^3 e^{i\frac{\pi}{3}\times 3} = 27e^{i\pi} = 27(\cos\pi + i\sin\pi) = 27(-1+i0) = -27.$$

Example 9.12: Covert $Z_1 = 2\angle 50^\circ$ and $Z_2 = 3\angle 15^\circ$ to their exponential forms, then find $Z_1 Z_2$, Z_1 / Z_2, Z_2^4 and $\sqrt[4]{Z_1}$.

Solution

We first convert the angles in degrees to radians by

$$\theta_1 = \frac{50^\circ}{180^\circ}\pi \approx \frac{50^\circ}{180^\circ} \times 3.1416 \approx 0.8727$$

$$\theta_2 = \frac{15^\circ}{180^\circ}\pi \approx \frac{15^\circ}{180^\circ} \times 3.1416 \approx 0.2618$$

Then using formulae (9.17) – (9.20), we can work out the required outcomes in exponential form.

$$Z_1 = r_1 e^{i\theta_1} = 2e^{i0.8727} \quad \text{and} \quad Z_2 = r_2 e^{i\theta_2} = 3e^{i0.2618}$$

$$Z_1 Z_2 = r_1 r_2 e^{i(\theta_1+\theta_2)} = (2\times 3)e^{i(0.8727+0.2618)} = 6e^{i1.1345}$$

$$\frac{Z_1}{Z_2} = \frac{r_1}{r_2} e^{i(\theta_1-\theta_2)} = \frac{2}{3} e^{i(0.8727-0.2618)} = \frac{2}{3} e^{i0.6109}$$

$$Z_2^4 = r_2^4 e^{i4\theta_2} = 3^4 e^{i4\times 0.2618} = 81e^{i1.0472}$$

$$\sqrt[4]{Z_1} = \sqrt[4]{r_1} e^{i\theta_1/4} = \sqrt[4]{2} e^{i0.8727/4} = \sqrt[4]{2} e^{i0.2182}$$

Exercises 9.3

1. Express the following complex numbers in their exponential forms respectively.

   a) $Z = 10\angle 40°$
   b) $Z = 8\angle 120°$
   c) $4(\cos 150° + i\sin 150°)$
   d) $8(\cos\frac{\pi}{6} + i\sin\frac{\pi}{6})$

2. Given $Z_1 = 1 + i3$, $Z_2 = 2 - i$, and $Z_3 = i4$, find $A = 3Z_1Z_3$, $B = 2Z_1Z_2$, and $D = Z_2Z_3$ in exponential form.

3. Given $Z_1 = 2 + i3$, and $Z_2 = 1 - i4$, find $A = Z_1/Z_2$, $B = Z_2^3$, and $C = Z_2\overline{Z}_2$ in exponential form.

4. Given $Z_1 = 3 + i2$, $Z_2 = i2$, and $Z_3 = -1 + i2$, find $A = 2Z_1Z_2Z_3$ and $B = 3\ Z_2Z_3/Z_1$ in exponential form.

5. Evaluate the following operations with complex numbers in exponential form respectively.

   a) $5e^{i\frac{\pi}{6}} \cdot 2e^{i\frac{\pi}{3}}$
   b) $1.5e^{i2} \cdot 2e^{i3}$
   c) $2e^{i1.5} \cdot 7e^{i2.9}$
   d) $\dfrac{6e^{i\pi}}{2e^{i2\pi}}$
   e) $\dfrac{3e^{i8}}{9e^{i2}}$
   f) $\dfrac{63e^{i3}}{7e^{i5}}$
   g) $(3e^{i2})^3$
   h) $(2e^{i\frac{\pi}{4}})^4$

6. Covert $Z_1 = 81\angle 40^\circ$ and $Z_2 = 2\angle 9^\circ$ to the exponential form, then find Z_1Z_2, Z_1 / Z_2, Z_2^4 and $\sqrt[4]{Z_1}$.

9.4 Applications of Complex Numbers

9.4.1 Quadratic equations with complex roots

One of the widely used applications of complex numbers is to present the solution of quadratic equations that do not have pure real roots. Reviewed in Chapter 1, the solution (roots) to the standard quadratic equation $ax^2+bx+c=0$ is generalized as

$$r_{1,2}=\frac{-b\pm\sqrt{b^2-4ac}}{2a}.$$

If $b^2-4ac\ge 0$, the solution is real numbers. If $b^2-4ac<0$, the solution is presented by a pair of conjugate complex numbers

$$r_{1,2}=\frac{-b\pm\sqrt{b^2-4ac}}{2a}=\frac{-b\pm\sqrt{-(4ac-b)^2}}{2a}=\frac{-b\pm\sqrt{-1}\sqrt{4ac-b^2}}{2a}=\frac{-b\pm i\sqrt{4ac-b^2}}{2a}.$$

Example 9.13: Solve $x^2-2x+5=0$.

Solution

This quadratic equation has $a = 1$, $b = -2$ and $c = 5$. Thus

$$r_{1,2}=\frac{-(-2)\pm\sqrt{(-2)^2-4\times1\times5}}{2\times1}=\frac{2\pm\sqrt{4-20}}{2}=\frac{2\pm\sqrt{-16}}{2}=\frac{2\pm\sqrt{(-1)(16)}}{2}=\frac{2\pm\sqrt{-1}\sqrt{16}}{2}$$

$$=\frac{2\pm i4}{2}=1\pm i2,\text{ or}$$

$$x_1=1+i2\text{ and }x_2=1-i2.$$

Example 9.14: Solve $2x^2+3x+2=0$.

Solution

This quadratic equation has $a = 2$, $b = 3$ and $c = 2$. Thus

$$r_{1,2}=\frac{-3\pm\sqrt{3^2-4\times2\times2}}{2\times2}=\frac{-3\pm\sqrt{9-16}}{4}=\frac{-3\pm\sqrt{-7}}{4}=\frac{-3\pm i\sqrt{7}}{4},\text{ or}$$

$$x_1=\frac{-3+i\sqrt{7}}{4}\text{ and }x_2=\frac{-3-i\sqrt{7}}{4}.$$

9.4.2 Complex numbers for operations of plane vectors

A plane vector can be directly mapped to the corresponding complex number in rectangular system shown in Figure 9.4, in which the x and y components of the vector correlate to the real and imaginary components of the complex number respectively. As a result, operations with several plane vectors can be carried out with the corresponding complex numbers.

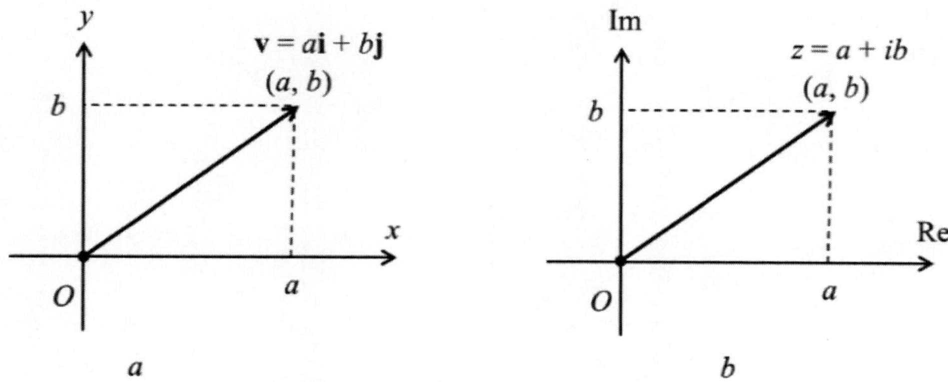

Figure 9.4: Mapping a vector to a complex number

However, although a plane vector in polar system has the same expression as the corresponding complex number represented in polar system, multiplication of plane vectors cannot be carried by the corresponding complex numbers in polar form. This is because vector multiplication creates either a scalar by the dot multiplication or a new vector on another plane by the cross multiplication.

Example 9.15: Given $\mathbf{A} = 5\mathbf{i} + 2\mathbf{j}$, $\mathbf{B} = -3\mathbf{i} + 4\mathbf{j}$, and $\mathbf{C} = \mathbf{i} - 2\mathbf{j}$, use complex numbers to find $\mathbf{D} = \mathbf{A} + \mathbf{B} + \mathbf{C}$, and $\mathbf{E} = \mathbf{A} - \mathbf{B} - \mathbf{C}$, respectively.

Solution

The given vectors in complex numbers are: $A = 5 + i2$, $B = -3 + i4$, $C = 1 - i2$. Thus

$D = A + B + C = (5+i2) + (-3+i4) + (1-i2) = (5-3+1) + i(2+4-2) = 3 + i4$,
$E = A - B - C = (5+i2) - (-3+i4) - (1-i2) = (5+3-1) + i(2-4+2) = 7 + i0 = 7$. Hence,
$\mathbf{D} = 3\mathbf{i} + 4\mathbf{j}$ and $\mathbf{E} = 7\mathbf{i} + 0\mathbf{j} = 7\mathbf{i}$.

This is the same as the direct operation with the vectors.

Example 9.16: Given $\mathbf{A} = 4\mathbf{i} + 3\mathbf{j}$ and $\mathbf{B} = -\mathbf{i} + 2\sqrt{2}\mathbf{j}$, verify that $C = \mathbf{A}\bullet\mathbf{B}$ and $\mathbf{D} = \mathbf{A}\times\mathbf{B}$ are different from multiplications by the corresponding complex numbers.

Solution

For vector multiplications,

$$C = \mathbf{A}\bullet\mathbf{B} = (4\mathbf{i}+3\mathbf{j})\bullet(-\mathbf{i}+2\sqrt{2}\mathbf{j}) = 4\times(-1)+3\times 2\sqrt{2} = -4+6\sqrt{2} \approx 4.4853$$
$$\mathbf{D} = \mathbf{A}\times\mathbf{B} = (4\mathbf{i}+3\mathbf{j})\times(-\mathbf{i}+2\sqrt{2}\mathbf{j}) = [4\times 2\sqrt{2}-3\times(-1)]\mathbf{k} = (8\sqrt{2}+3)\mathbf{k} \approx 14.3137\mathbf{k}$$

For complex number operations, the vectors are converted to complex numbers in polar forms first,

$$r_A = \sqrt{4^2+3^2} = \sqrt{16+9} = \sqrt{25} = 5, \quad \theta_A = \tan^{-1}(\frac{3}{4}) \approx 36.9° \longrightarrow \mathbf{A}: 5\angle 36.9°$$
$$r_B = \sqrt{(-1)^2+(2\sqrt{2})^2} = \sqrt{1+8} = \sqrt{9} = 3, \quad \theta_B = \tan^{-1}(\frac{2\sqrt{2}}{-1}) \approx -70.5° \longrightarrow \mathbf{B}: 3\angle -70.5°$$
$$(5\angle 36.9°)\cdot(3\angle -70.5°) = (5\times 3)\angle(36.9° - 70.5°) = 15\angle -33.6°$$
$$15\angle -33.6° = 15[\cos(-33.6°)+i\sin(-33.6°)] \approx 15(0.8329 - i0.5534) \approx 12.4935 - i8.3010$$

This converts back to the vector below

$$\mathbf{V} = 12.4935\mathbf{i} - 8.3010\mathbf{j}$$

which differs from either $C = \mathbf{A}\bullet\mathbf{B}$ or $\mathbf{D} = \mathbf{A}\times\mathbf{B}$.

9.4.3 Ohm's law for alternating current by complex numbers

The Ohm's law for direct current (DC) is

$$V = RI \tag{9.21}$$

where V is the direct voltage; R is the circuit resistance; I is the direct current. For alternating current (AC) $i = I_m\sin(\omega t+\varphi_i)$ and voltage $v = V_m\sin(\omega t+\varphi_v)$, both can be represented by complex numbers as

$$\tilde{I} = \frac{I_m}{\sqrt{2}}\angle\varphi_i \text{ and } \tilde{V} = \frac{V_m}{\sqrt{2}}\angle\varphi_v \tag{9.22}$$

where I_m and V_m are the maximum current and voltage respectively; φ_i and φ_v are the initial phases of the current and voltage respectively.

In AC circuits, the total resistance is from three sources: circuit resistance (R), capacitor known as capacitive reactance (X_C), inductor known as inductive reactance (X_L). These three sources together are called impedance and represented by the complex impedance $\tilde{Z}$

$$\boxed{\begin{array}{ll} \tilde{Z} = R + i(X_L - X_C) = R + iX = Z\angle\varphi = Ze^{i\varphi} & \\ X = (X_L - X_C) & \text{(circuit reactance)} \\ Z = \sqrt{R^2 + X^2} & \text{(magnitude of impedance)} \\ \varphi = \arctan\dfrac{X}{R} & \text{(phase angle).} \end{array}} \tag{9.23}$$

The Ohm's law in AC is represented as

$$\boxed{\tilde{V} = \tilde{Z}\tilde{I}}. \tag{9.24}$$

Example 9.17: An AC current is $i = 4.2426\sin(\omega t + 50°)$. Express this current in complex form.

Solution

$$\because I_m = 4.2426, \quad \varphi_i = 50° \qquad \therefore \tilde{I} = \frac{I_m}{\sqrt{2}}\angle\varphi_i = \frac{4.2426}{\sqrt{2}}\angle 50° \approx 3\angle 50°.$$

Example 9.18: An AC voltage in complex is $\tilde{V} = 51\angle -66°$. Convert this complex voltage back to the sinusoidal expression.

Solution

$$\because \frac{V_m}{\sqrt{2}} = 51 \text{ or } V_m = 51\sqrt{2}, \text{ and } \varphi_v = -66° \qquad \therefore v = V_m(\sin\omega t + \varphi_v) = 51\sqrt{2}(\sin\omega t - 66°).$$

Example 9.19: A circuit has a resistance of 8 Ω in series with an inductive reactance of 3 Ω and a capacitive reactance of 9 Ω. If a voltage $v = 28.2843\sin(\omega t + 60°)$ is applied to this circuit, what is the current in the circuit in sinusoidal form?

Solution

First to express both the voltage and impedance in complex forms

$$\because V_m = 28.2843 \text{ or } \frac{V_m}{\sqrt{2}} = \frac{28.2843}{\sqrt{2}} \approx 20, \text{ and } \varphi_v = 60° \qquad \therefore \tilde{V} = 20\angle 60°$$
$$\because R = 8\ \Omega,\ X_L = 3\ \Omega,\ X_C = 9\ \Omega, \text{ or } X = X_L - X_C = 3 - 9 = -6\ \Omega$$

$$\therefore Z = \sqrt{R^2 + X^2} = \sqrt{8^2 + (-6)^2} = \sqrt{64 + 36} = \sqrt{100} = 10,\ \varphi = \arctan\frac{-6}{8} \approx -36.9^\circ$$

$$\therefore \tilde{Z} = Z\angle\varphi = 10\angle -36.9^\circ$$

By the Ohm's law, the current is

$$\tilde{I} = \frac{\tilde{V}}{\tilde{Z}} = \frac{20\angle 60^\circ}{10\angle -36.9^\circ} = \frac{20}{10}\angle 60 - (-36.9^\circ) = 2\angle 96.9^\circ,\ \frac{I_m}{\sqrt{2}} = 2 \text{ or } I_m = 2\sqrt{2} \text{ and } \varphi_i = 96.9^\circ$$

$$\therefore i = I_m \sin(\omega t + \varphi_i) = 2\sqrt{2}\sin(\omega t + 96.9^\circ).$$

Note that the phase of current is about 37° more than that of the voltage, or the current is advanced about 37° ahead of the voltage.

Exercises 9.4

1. Solve the following equations.

   a) $x^2 + 9 = 0$
   b) $-3x^2 + 2x - 1 = 0$
   c) $\frac{1}{2}x^2 - 3x + 29 = 0$
   d) $\sqrt{x^2 - 1} = 2x + 1$
   e) $\frac{2x}{x+1} + \frac{3}{x-1} = 1$

2. Given **A** = –**i** + 3**j** and **B** = –3**i** – 7**j**, use complex numbers to find **C** = **A** + **B**, and **D** = **A** – **B** respectively.

3. Given **A** = –3**i** + 2**j**, **B** = 2**i** – 5**j**, and **C** = 3**i** + 4**j**, use complex numbers to find **D** = **A** – **B** + **C**, and **E** = –**A** + **B** + **C**, respectively.

4. Express the current or voltage below in complex form.

   a) $i = 122\sin(\omega t + 80°)$
   b) $i = 9\sin(\omega t - 150°)$
   c) $v = 25\sin(\omega t + 30°)$
   d) $v = 81\sin(\omega t - 45°)$

5. Express the current or voltage below in sinusoidal form.

   a) $\tilde{I} = 100\angle 70°$
   b) $\tilde{I} = 5\angle -55°$
   c) $\tilde{V} = 30\angle 110°$
   d) $\tilde{V} = 66\angle -25°$

6. Express the impedance below in complex form.

   a) $R = 5\ \Omega$, $X_L = 0\ \Omega$, $X_C = 4\ \Omega$
   b) $R = 10\ \Omega$, $X_L = 3\ \Omega$, $X_C = 0\ \Omega$
   c) $R = 7\ \Omega$, $X_L = 9\ \Omega$, $X_C = 2\ \Omega$

7. A circuit has a resistance of 9 Ω in series with an inductive reactance of 6 Ω and a capacitive reactance of 2 Ω. If a voltage $v = 120\sin(\omega t + 20°)$ is applied to this circuit, what is the current in the circuit in sinusoidal form?

Chapter 9: Exercises Answers

Exercises 9.1

1. a) $-2 + i4$, b) $2a - i7$, c) $-8 + i3$
2. a) $i^7 = -i,\ i^{10} = -1$; b) $i^8 = 1,\ i^{15} = -i$; c) $i^9 = i,\ i^{23} = -i$
3. a) $2 - i3 \rightarrow 2 + i3$; b) $2a + ib \rightarrow 2a - ib$; c) $-3 + i4 \rightarrow -3 - i4$
4. a) $11 - i2$; b) $(a^2 - 10) - i7a$; c) $-43 + i23$
5. a) $\dfrac{1+i2}{5}$, b) $\dfrac{13+i47}{41}$, c) $\dfrac{(a^2+10)+i3a}{a^2+25}$
6. $A = 3 + i9,\ B = 10 + i10,\ D = 4 + i8$
7. $A = -17 - i9,\ B = 27 + i14,\ C = -4 - i13$
8. $A = 14 - i5,\ B = -47 + i52,\ D = 17$
9. $A = 2 - i3,\ B = \dfrac{3}{13}(1 + i8)$

Exercises 9.2

1. a) $Z_1 = \sqrt{5}\angle 26.6° = \sqrt{5}(\cos 26.6° + i\sin 26.6°)$

   b) $Z_2 = \sqrt{5}\angle 333.4° = \sqrt{5}(\cos 333.4° + i\sin 333.4°)$

   c) $Z_3 = 3\sqrt{5}\angle 243.4° = 3\sqrt{5}(\cos 243.4° + i\sin 243.4°)$
2. a) $Z = 4\angle 60° = 2 + i2\sqrt{3}$, b) $Z = 6\angle 150° = -3\sqrt{3} + i3$, c) $Z = 2\angle 315° = \sqrt{2} - i\sqrt{2}$
3. a) $Z = 3(\cos 60° + i\sin 60°) = 1.5 + i2.5981$, b) $Z = 5(\cos 105° + i\sin 105°) = -1.2941 + i4.8296$,

   c) $Z = 4(\cos 280° + i\sin 280°) = 0.6946 - i3.9392$
4. $Z_1Z_2 = 12\angle 195°$, $\dfrac{Z_1}{Z_2} = \dfrac{3}{4}\angle 105°$, $Z_2^4 = 256\angle 180° = -256$
5. $Z^{1/4} = 2\angle 15°,\ 2\angle 105°,\ 2\angle 195°,\ 2\angle 285°$
6. $Z_1Z_2 = -6,\ \dfrac{Z_2}{Z_1} = \dfrac{3}{2}\angle -120°,\ Z_1^3 = i8$
7. $Z^{1/3} = \sqrt[3]{3}\angle 66.7°,\ \sqrt[3]{3}\angle 186.7°,\ \sqrt[3]{3}\angle 306.7°$
8. $-iZ_1Z_2 = 56\angle 64°,\ \dfrac{iZ_1}{Z_2} = \dfrac{8}{7}\angle 26°$

Exercises 9.3

1. a) $Z = 10e^{0.6981i}$, b) $Z = 8e^{2.0944i}$, c) $Z = 4e^{2.6180i}$, d) $Z = 8e^{0.5236i}$
2. $A = 12\sqrt{10}e^{2.8198i},\ B = 2\sqrt{50}e^{0.7854i},\ D = 4\sqrt{5}e^{1.1072i}$
3. $A = \sqrt{\dfrac{13}{17}}e^{2.3086i},\ B = 17\sqrt{17}e^{-3.9774i},\ C = 17$

4. $A = 4\sqrt{65}e^{1.0517i}$, $B = 6\sqrt{\frac{5}{13}}e^{-0.1243i}$

5. a) $10i$, b) $3e^{i5}$, c) $14e^{i4.4}$, d) -3, e) $\frac{1}{3}e^{i6}$, f) $9e^{-i2}$, g) $27e^{i6}$, h) -16

6. $Z_1 = 81e^{i0.6981}$, $Z_2 = r_2e^{i\theta_2} = 2e^{i0.1571}$, $Z_1Z_2 = 162e^{i0.8552}$, $\frac{Z_1}{Z_2} = 40.5e^{i0.541}$,

$Z_2^4 = 16e^{i0.6284}$, $\sqrt[4]{Z_1} = 3e^{i0.1745}$

Exercises 9.4

1. a) $x = \pm i3$, b) $x = \frac{1 \pm i\sqrt{2}}{3}$, c) $x = 3 \pm i7$, d) $x = \frac{-2 \pm i\sqrt{2}}{3}$, e) $x = \frac{-1 \pm i\sqrt{15}}{2}$
2. $\mathbf{C} = -4\mathbf{i} - 4\mathbf{j}$, $\mathbf{D} = 2\mathbf{i} + 10\mathbf{j}$
3. $\mathbf{D} = -2\mathbf{i} + 11\mathbf{j}$ and $\mathbf{E} = 8\mathbf{i} - 3\mathbf{j}$
4. a) $\tilde{I} \approx 86.267\angle 80°$, b) $\tilde{I} \approx 6.364\angle -150°$, c) $\tilde{V} \approx 17.6777\angle 30°$, d) $\tilde{V} \approx 57.2756\angle -45°$.
5. a) $i = 100\sqrt{2}(\sin \omega t + 70°)$, b) $i = 5\sqrt{2}(\sin \omega t - 55°)$, c) $v = 30\sqrt{2}(\sin \omega t + 110°)$,

   d) $v = 66\sqrt{2}(\sin \omega t - 25°)$.
6. a) $\tilde{Z} = \sqrt{41}\angle -26.6°$, b) $\tilde{Z} = \sqrt{109}\angle 16.7°$, c) $\tilde{Z} = 7\sqrt{2}\angle 45°$
7. $i \approx 12.1842\sin(\omega t - 4°)$.

CHAPTER 10

10 Essentials of Derivatives

CHAPTER OBJECTIVES

- Introduce concepts of limits, continuities, and derivatives of functions
- Explain the geometric and physical meanings of derivatives of functions
- Utilise the operational rules and advanced techniques to determine derivatives of functions
- Find higher-order derivatives of functions.

Essential statements on derivatives:

- The derivative of a function is the ***limit*** of rate of change of the function at a point.
- The act to find the derivative of a function is said to ***differentiate*** the function.
- The process of finding the derivative of a function is called ***differentiation***.

Key topics:

- Limits, continuities and derivatives of functions
- Geometric and physical interpretations of derivatives
- Derivatives of common functions
- Product, quotient, chain, and logarithmic rules
- Higher-order derivatives
- Derivatives of parametric and implicit functions

Flowchart of mathematical knowledge building

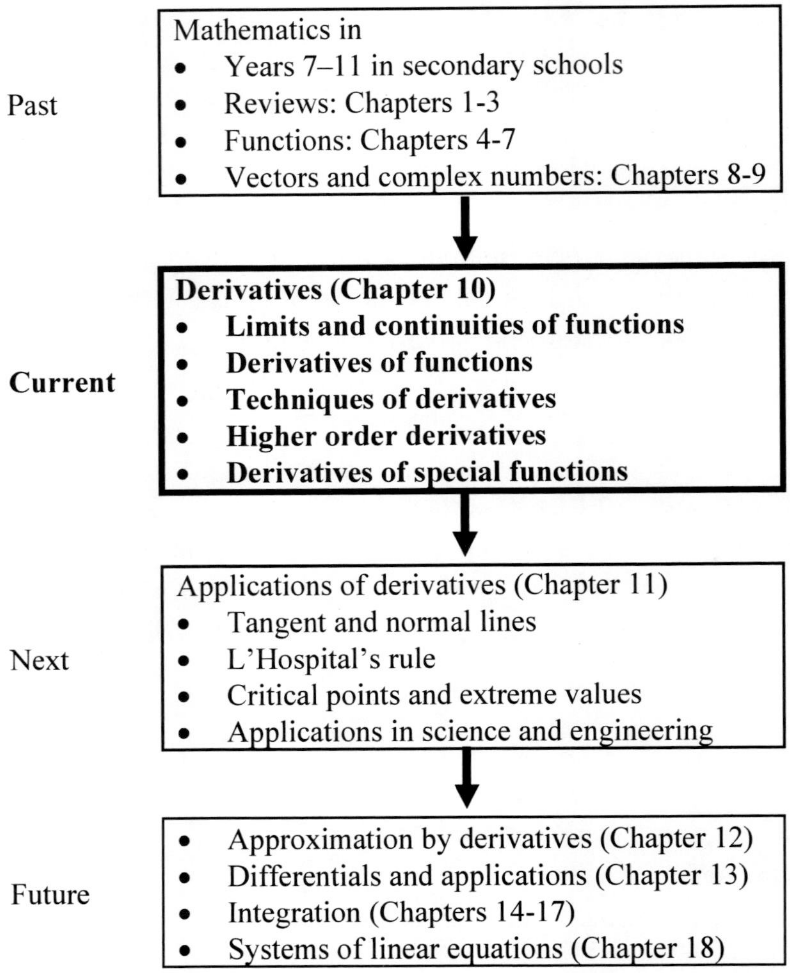

Past	Mathematics in • Years 7–11 in secondary schools • Reviews: Chapters 1-3 • Functions: Chapters 4-7 • Vectors and complex numbers: Chapters 8-9
	↓
Current	**Derivatives (Chapter 10)** • **Limits and continuities of functions** • **Derivatives of functions** • **Techniques of derivatives** • **Higher order derivatives** • **Derivatives of special functions**
	↓
Next	Applications of derivatives (Chapter 11) • Tangent and normal lines • L'Hospital's rule • Critical points and extreme values • Applications in science and engineering
	↓
Future	• Approximation by derivatives (Chapter 12) • Differentials and applications (Chapter 13) • Integration (Chapters 14-17) • Systems of linear equations (Chapter 18)

Chapter 10: Essentials of Derivatives

10.1 Limits and Continuities of Functions

10.1.1 Limits of functions

The concept of limits

Limit of a function at a point: Let $f(x)$ be a function defined in range R and a be a point within R. While x approaches a infinitely close but never reaches a, if $f(x)$ is infinitely close to a value L, then $\lim_{x \to a} f(x) = L$ is called the limit of $f(x)$ at point a. In general, $\lim_{x \to a} f(x) \neq f(a)$. However, for any smooth and continuous function, $\lim_{x \to a} f(x) = f(a)$ at any point within its range.

Limit of a function at infinity: Let $f(x)$ be a function defined in an open range R. While x extends to infinity ∞, if $f(x)$ is infinitely close to a value L, then $\lim_{x \to \infty} f(x) = L$ is called the limit of $f(x)$ at infinity. $\lim_{x \to \infty} f(x)$ must converge to a certain value or approach infinity (divergent). Otherwise $\lim_{x \to \infty} f(x)$ does not exist (or is not defined). Note that infinity is a symbol only and it does not represent any fixed number.

Example 10.1: Evaluate

a) $\lim_{x \to 1} \frac{2}{x}$;   b) $\lim_{x \to \infty} \frac{2}{x}$;   c) $\lim_{x \to 0} \frac{2}{x+1}$.

Solution

a) $\lim_{x \to 1} \frac{2}{x} = \frac{2}{1} = 2 = f(1) \longleftarrow f(x) = \frac{2}{x}$ is smooth and continuous everywhere except $x = 0$.

b) $\lim_{x \to \infty} \frac{2}{x} = \frac{2}{\infty} = 0 \overset{\neq}{\longleftarrow}$ This is not equal to $f(x)$ at ∞ since $f(\infty)$ is not defined.

c) $\lim_{x \to 0} \frac{2}{x+1} = \frac{2}{1} = 2 \longleftarrow$ This equals to $f(0) = 2$.

Example 10.2: Evaluate $\lim_{x \to \infty} \sin x$.

Solution

Since sin x oscillates between –1 and +1 while x increases infinitely, or in other words sin x does not converge to a certain value with increase in x, $\lim_{x \to \infty} \sin x$ does not exist.

Rules of operations on limits

If $\lim_{x\to a} f(x)$ and $\lim_{x\to a} g(x)$ exist, then

$$\begin{array}{|l|}\hline \lim_{x\to a}[f(x)\pm g(x)] = \lim_{x\to a} f(x) \pm \lim_{x\to a} g(x) \\ \lim_{x\to a} f(x)g(x) = \lim_{x\to a} f(x)\cdot \lim_{x\to a} g(x) \\ \lim_{x\to a}\dfrac{f(x)}{g(x)} = \dfrac{\lim_{x\to a} f(x)}{\lim_{x\to a} g(x)} \quad \text{if } \lim_{x\to a} g(x)\neq 0 \\ \lim_{x\to a}[kf(x)] = k\lim_{x\to a} f(x), \quad k\text{: constant} \\ \lim_{x\to a}[f(x)]^n = [\lim_{x\to a} f(x)]^n \;\&\; \lim_{x\to a}\sqrt[n]{f(x)} = \sqrt[n]{\lim_{x\to a} f(x)}, \; n\text{: positive integer} \\ \hline \end{array} \qquad (10.1)$$

$$\boxed{\text{If } f(x)\le g(x)\le h(x) \;\&\; \lim_{x\to a} f(x) = \lim_{x\to a} h(x) = L, \; \lim_{x\to a} g(x) = L.} \qquad (10.2)$$

This also applies to $\lim_{x\to\infty} f(x)$ and $\lim_{x\to\infty} g(x)$ if both exist.

Example 10.3: Evaluate

a) $\lim_{x\to 0}\dfrac{2x+1}{x-1}$;

b) $\lim_{x\to 1}\dfrac{2x+1}{x+1}$;

c) $\lim_{x\to\infty}\dfrac{2x+1}{x-1}$.

Solution

a) $\lim_{x\to 0}\dfrac{2x+1}{x-1} = \dfrac{\lim_{x\to 0}(2x+1)}{\lim_{x\to 0}(x-1)} = \dfrac{1}{-1} = -1$;

b) $\lim_{x\to 1}\dfrac{2x+1}{x+1} = \dfrac{\lim_{x\to 1}(2x+1)}{\lim_{x\to 1}(x+1)} = \dfrac{3}{2}$;

c) $\lim_{x\to\infty}\dfrac{2x+1}{x-1} = \lim_{x\to\infty}\dfrac{\frac{2x+1}{x}}{\frac{x-1}{x}} = \lim_{x\to\infty}\dfrac{2+\frac{1}{x}}{1-\frac{1}{x}} = \dfrac{\lim_{x\to\infty}(2+\frac{1}{x})}{\lim_{x\to\infty}(1-\frac{1}{x})} = \dfrac{2+\lim_{x\to\infty}\frac{1}{x}}{1-\lim_{x\to\infty}\frac{1}{x}} = \dfrac{2+0}{1-0} = 2.$

Example 10.4: Evaluate

a) $\lim_{x\to\infty} e^{-x}\sin x$; b) $\lim_{x\to\infty}(\sqrt{x^2+4}-x)$; c) $\lim_{x\to 2}\dfrac{2-x}{4-\sqrt{x^2+12}}$.

Solution

a) $\lim_{x\to\infty} e^{-x}\sin x = \lim_{x\to\infty}\dfrac{\sin x}{e^x} = \dfrac{\lim_{x\to\infty}\sin x}{\lim_{x\to\infty} e^x}$

Since $\dfrac{-1}{\lim_{x\to\infty} e^x} \le \dfrac{\lim_{x\to\infty}\sin x}{\lim_{x\to\infty} e^x} \le \dfrac{1}{\lim_{x\to\infty} e^x} \longrightarrow \dfrac{-1}{\infty} \le \dfrac{\lim_{x\to\infty}\sin x}{\lim_{x\to\infty} e^x} \le \dfrac{1}{\infty} \longrightarrow 0 \le \dfrac{\lim_{x\to\infty}\sin x}{\lim_{x\to\infty} e^x} \le 0,$

thus $\lim_{x\to\infty} e^{-x}\sin x = \dfrac{\lim_{x\to\infty}\sin x}{\lim_{x\to\infty} e^x} = 0;$ $\longleftarrow$ by limit (10.2)

b) Since $\sqrt{x^2+4}-x = \dfrac{(\sqrt{x^2+4}+x)(\sqrt{x^2+4}-x)}{\sqrt{x^2+4}+x} = \dfrac{(\sqrt{x^2+4})^2-x^2}{\sqrt{x^2+4}+x} = \dfrac{x^2+4-x^2}{\sqrt{x^2+4}+x} = \dfrac{4}{\sqrt{x^2+4}+x},$

thus $\lim_{x\to\infty}(\sqrt{x^2+4}-x) = \lim_{x\to\infty}\dfrac{4}{\sqrt{x^2+4}+x} = \dfrac{4}{\lim_{x\to\infty}\sqrt{x^2+4}+x} = \dfrac{4}{\infty} = 0.$

c) Since $\dfrac{2-x}{4-\sqrt{x^2+12}} = \dfrac{(2-x)(4+\sqrt{x^2+12})}{(4-\sqrt{x^2+12})(4+\sqrt{x^2+12})} = \dfrac{(2-x)(4+\sqrt{x^2+12})}{4^2-(x^2+12)}$

$= \dfrac{(2-x)(4+\sqrt{x^2+12})}{4-x^2} = \dfrac{(2-x)(4+\sqrt{x^2+12})}{(2-x)(2+x)} = \dfrac{4+\sqrt{x^2+12}}{2+x},$

thus $\lim_{x\to 2}\dfrac{2-x}{4-\sqrt{x^2+12}} = \lim_{x\to 2}\dfrac{4+\sqrt{x^2+12}}{2+x} = \dfrac{\lim_{x\to 2}(4+\sqrt{x^2+12})}{\lim_{x\to 2}(2+x)} = \dfrac{4+\sqrt{2^2+12}}{2+2} = \dfrac{4+4}{4} = 2.$

Two important limits and extensions

The two limits below are widely used in deriving other limits involving sinusoidal and/or exponential functions.

$$\lim_{x\to\infty}(1+\frac{1}{x})^x = e \rightleftharpoons \lim_{x\to 0}(1+x)^{\frac{1}{x}} = e \longrightarrow \lim_{x\to 0}\frac{e^x-1}{x} = 1 \tag{10.3}$$

$$\lim_{x\to 0}\frac{\sin x}{x} = 1 \rightleftharpoons \lim_{x\to 0}\frac{x}{\sin x} = 1 \tag{10.4}$$

Limit (10.3) can be extended to the following formulae

$$\boxed{\lim_{x\to\infty}(1-\frac{1}{x})^x=\frac{1}{e}\rightleftharpoons\lim_{x\to 0}(1-x)^{\frac{1}{x}}=\frac{1}{e}} \tag{10.5}$$

$$\boxed{\lim_{x\to\infty}(1+\frac{a}{x})^x=e^a\rightleftharpoons\lim_{x\to 0}(1+ax)^{\frac{1}{x}}=e^a} \tag{10.6}$$

Limit (10.4) can be extended to the following formulae

$$\boxed{\lim_{x\to 0}\frac{\tan x}{x}=1\rightleftharpoons\lim_{x\to 0}\frac{x}{\tan x}=1} \tag{10.7}$$

$$\boxed{\lim_{x\to 0}\frac{\sin ax}{x}=a\rightleftharpoons\lim_{x\to 0}\frac{x}{\sin ax}=\frac{1}{a}} \tag{10.8}$$

Example 10.5: Derive $\lim_{x\to 0}\frac{\sin ax}{x}=a$ and $\lim_{x\to 0}\frac{\tan x}{x}=1$ using limit (10.4) respectively.

Solution

$$\lim_{x\to 0}\frac{\sin ax}{x}=\lim_{x\to 0}a\bullet\frac{\sin ax}{ax}=a\lim_{x\to 0}\frac{\sin ax}{ax}=\xrightarrow[x\to 0,t\to 0]{t=ax}a\lim_{t\to 0}\frac{\sin t}{t}=a\times 1=a.$$

$$\lim_{x\to 0}\frac{\tan x}{x}=\lim_{x\to 0}\frac{\sin x}{x\cos x}=\lim_{x\to 0}(\frac{\sin x}{x})\lim_{x\to 0}(\frac{1}{\cos x})=1\times\frac{1}{\lim\limits_{x\to 0}\cos x}=\frac{1}{\cos 0}=\frac{1}{1}=1.$$

Example 10.6: Derive $\lim_{x\to\infty}(1-\frac{1}{x})^x=\frac{1}{e}$ and $\lim_{x\to 0}(1+ax)^{\frac{1}{x}}=e^a$ using limit (10.3) respectively.

Solution

$$\lim_{x\to\infty}(1-\frac{1}{x})^x=\xrightarrow[x\to\infty,t\to 0]{t=-\frac{1}{x},x=-\frac{1}{t}}\lim_{t\to 0}(1+t)^{-\frac{1}{t}}=\lim_{t\to 0}\frac{1}{(1+t)^{\frac{1}{t}}}=\frac{1}{\lim\limits_{t\to 0}(1+t)^{\frac{1}{t}}}=\frac{1}{e}.$$

$$\lim_{x\to 0}(1+ax)^{\frac{1}{x}}=\xrightarrow[x\to 0,t\to 0]{t=ax,x=\frac{t}{a}}\lim_{t\to 0}(1+t)^{\frac{1}{t/a}}=\lim_{t\to 0}(1+t)^{\frac{a}{t}}=[\lim_{t\to 0}(1+t)^{\frac{1}{t}}]^a=e^a.$$

Example 10.7: Evaluate

(a) $\lim\limits_{x\to 0}\dfrac{1-\cos 2x}{x^2}$; (b) $\lim\limits_{x\to 0}\dfrac{\sin 5x-\sin 3x}{x}$; (c) $\lim\limits_{x\to 1}\dfrac{\sin(x-1)}{x^2+x-2}$;

(d) $\lim\limits_{x\to\infty}(\dfrac{x-1}{x+1})^x$; (e) $\lim\limits_{x\to 0}(1+\tan x)^{\cot x}$; (f) $\lim\limits_{x\to 0}(1-\dfrac{x}{2})^{\frac{1}{x}}$.

<u>**Solution**</u>

(a) $\lim\limits_{x\to 0}\dfrac{1-\cos 2x}{x^2}=\lim\limits_{x\to 0}\dfrac{1-(1-2\sin^2 x)}{x^2}=\lim\limits_{x\to 0}\dfrac{1-1+2\sin^2 x}{x^2}=\lim\limits_{x\to 0}\dfrac{2\sin^2 x}{x^2}=2\lim\limits_{x\to 0}(\dfrac{\sin x}{x})^2=2;$

(b) $\lim\limits_{x\to 0}\dfrac{\sin 5x-\sin 3x}{x}=(\lim\limits_{x\to 0}\dfrac{\sin 5x}{x})-(\lim\limits_{x\to 0}\dfrac{\sin 3x}{x})=5-3=2;$ ← by limit (10.8)

(c) $\lim\limits_{x\to 1}\dfrac{\sin(x-1)}{x^2+x-2}=\lim\limits_{x\to 1}\dfrac{\sin(x-1)}{(x-1)(x+2)}=\lim\limits_{x\to 1}\dfrac{1}{x+2}\bullet\lim\limits_{x\to 1}\dfrac{\sin(x-1)}{x-1}=\dfrac{1}{3}\lim\limits_{x\to 1}\dfrac{\sin(x-1)}{x-1}=$

$\xrightarrow[x\to 1,t\to 0]{t=x-1}\dfrac{1}{3}\lim\limits_{t\to 0}\dfrac{\sin t}{t}=\dfrac{1}{3};$

(d) $\lim\limits_{x\to\infty}(\dfrac{x-1}{x+1})^x=\lim\limits_{x\to\infty}(\dfrac{x+1-2}{x+1})^x=\lim\limits_{x\to\infty}(\dfrac{x+1}{x+1}-\dfrac{2}{x+1})^x=\lim\limits_{x\to\infty}(1-\dfrac{2}{x+1})^x=$

$\xrightarrow[x\to\infty,t\to\infty]{t=x+1,x=t-1}\lim\limits_{t\to\infty}(1-\dfrac{2}{t})^{t-1}=\lim\limits_{t\to\infty}\dfrac{(1-\frac{2}{t})^t}{(1-\frac{2}{t})}=\dfrac{\lim\limits_{t\to\infty}(1-\frac{2}{t})^t}{\lim\limits_{t\to\infty}(1-\frac{2}{t})}=\dfrac{\lim\limits_{t\to\infty}(1+\frac{-2}{t})^t}{1}=e^{-2};$← by limit (10.6)

(e) $\lim\limits_{x\to 0}(1+\tan x)^{\cot x}=\xrightarrow[x\to 0,t\to 0]{t=\tan x,\cot x=\frac{1}{\tan x}=\frac{1}{t}}\lim\limits_{t\to 0}(1+t)^{\frac{1}{t}}=e;$ ← by limit (10.3)

(f) $\lim\limits_{x\to 0}(1-\dfrac{x}{2})^{\frac{1}{x}}=\lim\limits_{x\to 0}[1+(-\dfrac{1}{2})x]^{\frac{1}{x}}=e^{-\frac{1}{2}}.$ ← by limit (10.6)

10.1.2 Continuities of functions

Left-hand and right-hand limits

Existence of $\lim_{x\to a} f(x) = L$ ***at point a in range R***: While x approaches a infinitely close from both sides of a, $\lim_{x\to a} f(x) = L$ exists if and only if both the left-hand (x is smaller than a) and right-hand (x is greater than a) limits converge to the same value L, i.e., $\lim_{x\to a^-} f(x) = \lim_{x\to a^+} f(x) = L$.

Example 10.8: Evaluate

a) $\lim_{x\to 0}\dfrac{1}{1-e^{1/x}}$; b) $\lim_{x\to 2}\sqrt{4-x^2}$; c) $\lim_{x\to 0}|x|$.

Solution

a) $\lim_{x\to 0^-}\dfrac{1}{1-e^{1/x}} = \dfrac{1}{1-e^{-\infty}} = \dfrac{1}{1-\dfrac{1}{e^{\infty}}} = 1,\ \lim_{x\to 0^+}\dfrac{1}{1-e^{1/x}} = \dfrac{1}{1-e^{\infty}} = \dfrac{1}{1-\infty} = 0$ or $\lim_{x\to 0^-} f(x) \neq \lim_{x\to 0^+} f(x)$.

Thus, $\lim_{x\to 0}\dfrac{1}{1-e^{1/x}}$ does not exist.

b) $\lim_{x\to 2^-}\sqrt{4-x^2} = \sqrt{4-(1.9...)^2} = 0,\ \lim_{x\to 2^+}\sqrt{4-x^2} = \sqrt{4-(2.0...)^2} = \sqrt{4-4.0...} = \sqrt{-0.0...} \neq 0,$

or $\lim_{x\to 2^-} f(x) \neq \lim_{x\to 2^+} f(x)$. Thus, $\lim_{x\to 2}\sqrt{4-x^2}$ does not exist.

c) $\lim_{x\to 0^-}|x| = |-0| = 0,\ \lim_{x\to 0^+}|x| = |+0| = 0,$ thus $\lim_{x\to 0^-} f(x) = \lim_{x\to 0^+} f(x) = \lim_{x\to 0} f(x) = 0.$

Continuities of functions

Continuous function: Let $f(x)$ be a function defined in range R and a be a point within R. If $\lim_{x\to a} f(x) = f(a)$, $f(x)$ is said to be continuous at $x = a$. If this is met at any point within the range, $f(x)$ is a continuous function in the whole range. Note that $\lim_{x\to a} f(x) = f(a)$ equals $\lim_{h\to 0} f(a+h) = f(a)$.

Continuous function in an open interval: Function $f(x)$ is called a continuous function in an open interval (a, b) if it is continuous everywhere in (a, b).

Continuous function in a closed interval: Function $f(x)$ is called a continuous function in a closed interval $[a, b]$ if it is continuous everywhere in (a, b) and $\lim_{x\to a^+} f(x) = f(a)$ and $\lim_{x\to b^-} f(x) = f(b)$.

Operational rules of continuous functions: If $f(x)$ and $g(x)$ are continuous in (a, b), $f(x)\pm g(x)$, $f(x)g(x), \frac{f(x)}{g(x)}$ (if $g(x)\neq 0$) are continuous too.

Example 10.9: Evaluate if the following function is continuous at $x = 0$. If not, redefine $f(0)$ so as to make it continuous at $x = 0$.

$$f(x)=\begin{cases}\frac{1-\cos x}{x^2} & x\neq 0\\ 2 & x=0\end{cases}$$

Solution

$$\lim_{x\to 0}\frac{1-\cos x}{x^2}=\lim_{x\to 0}\frac{1-(\cos^2\frac{x}{2}-\sin^2\frac{x}{2})}{x^2}=\lim_{x\to 0}\frac{1-(1-2\sin^2\frac{x}{2})}{x^2}=\lim_{x\to 0}\frac{2\sin^2\frac{x}{2}}{x^2}=2\lim_{x\to 0}\frac{\sin^2\frac{x}{2}}{4\bullet\frac{x^2}{4}}$$

$$=\frac{2}{4}\lim_{x\to 0}\frac{\sin^2\frac{x}{2}}{(\frac{x}{2})^2}=\frac{1}{2}\lim_{x\to 0}(\frac{\sin\frac{x}{2}}{\frac{x}{2}})^2=\xrightarrow[x\to 0,t\to 0]{t=\frac{x}{2}}\frac{1}{2}\lim_{t\to 0}(\frac{\sin t}{t})^2=\frac{1}{2}.$$

Currently $\lim_{x\to 0}f(x)=L=\frac{1}{2}\neq f(0)=2$. If we define $f(0)=\frac{1}{2}$, this makes $\lim_{x\to 0}f(x)=f(0)=\frac{1}{2}$. Hence $x = 0$ is called a *point of removable discontinuity* for $f(x)$, which can be removed by redefining

$$f(x)=\begin{cases}\frac{1-\cos x}{x^2} & x\neq 0\\ \frac{1}{2} & x=0\end{cases}.$$

Example 10.10: Evaluate if sin$x$, cos$x$, tan$x$ and cot$x$ are continuous functions in the open interval $(0, \pi/2)$, or in the closed interval $[0, \pi/2]$, or in a mixed interval from 0 to $\pi/2$.

Solution

Since sin$x$ is continuous from 0 to $\pi/2$ and $\lim_{x\to 0^+}\sin x=0=\sin 0$ & $\lim_{x\to\frac{\pi}{2}^-}\sin x=1=\sin\frac{\pi}{2}$, sin$x$ is continuous in the closed interval $[0, \pi/2]$.

Since cos$x$ is continuous from 0 to $\pi/2$ and $\lim_{x\to 0^+}\cos x=1=\cos 0$ & $\lim_{x\to\frac{\pi}{2}^-}\cos x=0=\cos\frac{\pi}{2}$, cos$x$ is continuous in the closed interval $[0, \pi/2]$.

Since tan$x$ is continuous in (0, π/2) and $\lim_{x\to 0^+} \tan x = 0 = \tan 0$, but $\lim_{x\to \frac{\pi}{2}^-} \tan x \to \infty$, or undefined, tan$x$ is continuous in the mixed interval [0, π/2).

Since cot$x$ is continuous in (0, π/2) and $\lim_{x\to \frac{\pi}{2}^-} \cot x = 0$, but $\lim_{x\to 0^+} \cot x \to \infty$, or undefined, thus cot$x$ is continuous in the mixed interval (0, π/2].

Exercises 10.1

1. Evaluate

   (a) $\lim_{x\to 1}(2x-1)$; (b) $\lim_{x\to 2}(x^2-2x+5)$; (c) $\lim_{x\to 3}\frac{x-2}{x+2}$;

   (d) $\lim_{x\to -2}\frac{3x+1}{x^2+6}$; (e) $\lim_{x\to 6}\sqrt{x^2-2x-8}$; (f) $\lim_{x\to 5}\frac{\sqrt{2x-1}}{2x}$;

   g) $\lim_{x\to -1}(x-|x|)$.

2. Evaluate

   (a) $\lim_{x\to\infty}\frac{x+2}{x-1}$; (b) $\lim_{x\to\infty}\frac{x+3}{3x^2+2x-3}$; (c) $\lim_{x\to\infty}\frac{-x+3}{2x-5}$;

   (d) $\lim_{x\to\infty}\frac{x^3+3x^2-7}{x^2-1}$; (e) $\lim_{x\to 0}\frac{x+2}{x}$; (f) $\lim_{x\to -4}\frac{x+4}{\sqrt{x^2+2x-8}}$;

   (g) $\lim_{x\to 1}\frac{x^2-1}{x-1}$; (h) $\lim_{x\to 2}\frac{x-2}{\sqrt{x^2+5}-3}$; (i) $\lim_{x\to 2}\frac{x(x-2)}{\sqrt{x+2}-2}$;

3. Derive $\lim_{x\to 0}\frac{e^x-1}{x}=1$ and $\lim_{x\to 0}\frac{x}{\sin ax}=\frac{1}{a}$ respectively.

4. Evaluate

   (a) $\lim_{x\to 0}\frac{\sin x\tan x}{x^2}$; (b) $\lim_{x\to 0}\frac{\sin\frac{x}{2}-\sin\frac{x}{3}}{x}$; (c) $\lim_{x\to 2}\frac{\sin\frac{x-2}{3}}{x-2}$;

   (d) $\lim_{x\to\infty}(\frac{x-2}{x+1})^x$; (e) $\lim_{x\to 0}(1+2\tan x)^{2\cot x}$; (f) $\lim_{x\to 0}(1-\frac{x}{3})^{\frac{2}{x}}$.

5. Evaluate

   a) $\lim_{x\to 0}(1-e^{1/x})$; b) $\lim_{x\to 3}\sqrt{x^2-9}$; c) $\lim_{x\to 0}(|x|-x)$.

6. Evaluate if sec$x$ and csc$x$ are continuous functions in the open interval (0, π/2), or in the closed interval [0, π/2], or in a mixed interval from 0 to π/2.

10.2 Derivatives of Continuous Functions

10.2.1 The concept and meanings of derivatives of functions

Concepts and definitions

Increments: The increment δx of an independent variable x is the change in x as it increases (or decreases) from one point x_i to another point x_j in its range. Hence $\delta x_{ij} = x_j - x_i$ or $x_j = x_i + \delta x_{ij}$. For a continuous function $y = f(x)$ defined in range R, an increment in independent variable x from x_i to x_j leads to an increment in dependent variable $\delta y_{ij} = f(x_j) - f(x_i) = f(x_i + \delta x_{ij}) - f(x_i)$. Both δy_{ij} and δx_{ij} are instantaneous increments of function $y = f(x)$ whilst x changes from x_i to x_j in the range.

Average rate of change: The ratio of the instantaneous increments δy_{ij} to δx_{ij} is defined as the average rate of change of function $y = f(x)$ at point x_i in the interval $\delta x_{ij} = x_j - x_i$, i.e.,

$$\boxed{\text{Average rate of change at } x_i:\ \frac{\delta y_{ij}}{\delta x_{ij}} = \frac{f(x_j) - f(x_i)}{x_j - x_i} = \frac{f(x_i + \delta x_{ij}) - f(x_i)}{x_j - x_i}}. \tag{10.9}$$

Derivative: The derivative of a function $y = f(x)$ with respect to x at point $x = x_i$ is defined as the limit of the average rate of change of $y = f(x)$ at point x_i whilst $\delta x \to 0$ (or $x_j \to x_i$), i.e.,

$$\text{Derivative of } y_i = f(x_i) = \lim_{\delta x \to 0} \frac{\delta y}{\delta x} = \lim_{\delta x \to 0} \frac{f(x_i + \delta x) - f(x_i)}{\delta x}, \tag{10.10}$$

provided the limit exists. In practice, the subscript i is often omitted and the derivative of $y = f(x)$ with respect to x is commonly noted as $\frac{dy}{dx}$, $\frac{df(x)}{dx}$, y', or $f'(x)$, i.e.,

$$\boxed{\frac{dy}{dx} = y' = f'(x) = \lim_{\delta x \to 0} \frac{\delta y}{\delta x} = \lim_{\delta x \to 0} \frac{f(x + \delta x) - f(x)}{\delta x}}. \tag{10.11}$$

Example 10.11: Find the derivative for $y = f(x) = x^2$. Determine the derivative of $y = f(x) = x^2$ at $x = 1$.

Solution

We need to find $f(x + \delta x)$ in order to use formula (10.11) for the derivative of $y = f(x) = x^2$.

$$f(x + \delta x) = (x + \delta x)^2 = x^2 + 2x\delta x + (\delta x)^2.$$

By formula (10.11),

$$\frac{dy}{dx} = \lim_{\delta x \to 0} \frac{f(x+\delta x) - f(x)}{\delta x} = \lim_{\delta x \to 0} \frac{[x^2 + 2x\delta x + (\delta x)^2] - x^2}{\delta x} = \lim_{\delta x \to 0} \frac{2x\delta x + (\delta x)^2}{\delta x}$$
$$= \lim_{\delta x \to 0} (2x + \delta x) = 2x + \lim_{\delta x \to 0} (\delta x) = 2x.$$

At $x = 1$, the derivative of $y = x^2$ is $f'(1) = 2 \times 1 = 2$.

Example 10.12: Find the derivative for $y = \sin x$.

Solution

For $y = \sin x$,

$$y(x+\delta x) = \sin(x+\delta x);$$
$$\delta y = y(x+\delta x) - y(x) = \sin(x+\delta x) - \sin x$$
$$= 2\sin(\frac{x+\delta x - x}{2})\cos(\frac{x+\delta x + x}{2}) \qquad \leftarrow \sin A - \sin B = 2\sin(\frac{A-B}{2})\cos(\frac{A+B}{2})$$
$$= 2\sin(\frac{\delta x}{2})\cos(x+\frac{\delta x}{2}).$$

By formula (10.11),

$$\frac{dy}{dx} = \lim_{\delta x \to 0} \frac{\delta y}{\delta x} = \lim_{\delta x \to 0} \frac{2\sin(\frac{\delta x}{2})\cos(x+\frac{\delta x}{2})}{\delta x} = \lim_{\delta x \to 0} \frac{2\sin(\frac{\delta x}{2})\cos(x+\frac{\delta x}{2})}{2 \times \frac{\delta x}{2}}$$

$$= \lim_{\delta x \to 0} \frac{\sin(\frac{\delta x}{2})\cos(x+\frac{\delta x}{2})}{\frac{\delta x}{2}} = \lim_{t \to 0} \frac{\sin t \cos(x+t)}{t} \qquad \leftarrow (\text{let } t = \frac{\delta x}{2},\ t \to 0 \text{ if } \delta x \to 0)$$

$$= \lim_{t \to 0} \frac{\sin t}{t} \lim_{t \to 0} \cos(x+t) = \cos x \qquad \leftarrow \text{by limit (10.4)}$$

Example 10.13: Find the derivative for $y = e^x$, $y = \dfrac{1}{x+3}$ and $y = \dfrac{1}{\sqrt{1+2x}}$ respectively.

Solution

For $y = e^x$

$$y(x+\delta x) = e^{x+\delta x} = e^x e^{\delta x}; \qquad \delta y = y(x+\delta x) - y(x) = e^x e^{\delta x} - e^x = e^x(e^{\delta x} - 1).$$

By formula (10.11),

$$\frac{dy}{dx} = \lim_{\delta x \to 0} \frac{\delta y}{\delta x} = \lim_{\delta x \to 0} \frac{e^x(e^{\delta x} - 1)}{\delta x} = e^x \lim_{\delta x \to 0} \frac{e^{\delta x} - 1}{\delta x} = e^x \qquad \leftarrow \text{by limit (10.3)}$$

For $y = \frac{1}{x+3}$,

$$\delta y = f(x+\delta x) - f(x) = \frac{1}{x+\delta x+3} - \frac{1}{x+3} = \frac{x+3}{(x+\delta x+3)(x+3)} - \frac{x+\delta x+3}{(x+\delta x+3)(x+3)}$$

$$= \frac{x+3-x-\delta x-3}{(x+3)(x+\delta x+3)} = \frac{-\delta x}{(x+3)(x+\delta x+3)}$$

$$\frac{dy}{dx} = \lim_{\delta x\to 0}\frac{\delta y}{\delta x} = \lim_{\delta x\to 0}\frac{\frac{-\delta x}{(x+3)(x+\delta x+3)}}{\delta x} = \lim_{\delta x\to 0}\frac{-1}{(x+3)(x+\delta x+3)} = \frac{-1}{(x+3)^2}$$

For $y = \frac{1}{\sqrt{1+2x}}$,

$$y(x+\delta x) = \frac{1}{\sqrt{1+2(x+\delta x)}}$$

$$\delta y = y(x+\delta x) - y(x) = \frac{1}{\sqrt{1+2(x+\delta x)}} - \frac{1}{\sqrt{1+2x}} = \frac{\sqrt{1+2x} - \sqrt{1+2(x+\delta x)}}{[\sqrt{1+2(x+\delta x)}](\sqrt{1+2x})}$$

$$= \frac{[\sqrt{1+2x} - \sqrt{1+2(x+\delta x)}][\sqrt{1+2x} + \sqrt{1+2(x+\delta x)}]}{\{[\sqrt{1+2(x+\delta x)}](\sqrt{1+2x})\}[\sqrt{1+2x} + \sqrt{1+2(x+\delta x)}]}$$

$$= \frac{(\sqrt{1+2x})^2 - [\sqrt{1+2(x+\delta x)}]^2}{\{[\sqrt{1+2(x+\delta x)}](\sqrt{1+2x})\}[\sqrt{1+2x} + \sqrt{1+2(x+\delta x)}]}$$

$$= \frac{1+2x-[1+2x+2(\delta x)]}{\{[\sqrt{1+2(x+\delta x)}](\sqrt{1+2x})\}[\sqrt{1+2x} + \sqrt{1+2(x+\delta x)}]}$$

$$= \frac{1+2x-1-2x-2(\delta x)}{\{[\sqrt{1+2(x+\delta x)}](\sqrt{1+2x})\}[\sqrt{1+2x} + \sqrt{1+2(x+\delta x)}]}$$

$$= \frac{-2(\delta x)}{\{[\sqrt{1+2(x+\delta x)}](\sqrt{1+2x})\}[\sqrt{1+2x} + \sqrt{1+2(x+\delta x)}]}$$

$$\frac{dy}{dx} = \lim_{\delta x\to 0}\frac{\delta y}{\delta x} = \lim_{\delta x\to 0}\frac{\frac{-2(\delta x)}{\{[\sqrt{1+2(x+\delta x)}](\sqrt{1+2x})\}[\sqrt{1+2x} + \sqrt{1+2(x+\delta x)}]}}{\delta x}$$

$$= \lim_{\delta x\to 0}\frac{-2}{\{[\sqrt{1+2(x+\delta x)}](\sqrt{1+2x})\}[\sqrt{1+2x} + \sqrt{1+2(x+\delta x)}]}$$

$$= \frac{-2}{\{[\sqrt{1+2(x+0)}](\sqrt{1+2x})\}[\sqrt{1+2x} + \sqrt{1+2(x+0)}]}$$

$$= \frac{-2}{[\sqrt{1+2x}\sqrt{1+2x}][\sqrt{1+2x}+\sqrt{1+2x}]} = \frac{-2}{[\sqrt{1+2x}\sqrt{1+2x}][2\sqrt{1+2x}]}$$

$$= \frac{-1}{(\sqrt{1+2x})^3}$$

Geometric interpretation of derivatives

If we treat function $y = f(x)$ as a continuous curve in a defined range, the derivative of $y = f(x)$ at point x on the curve is the gradient (or slope) of the tangent line to the curve at that point (Figure 10.1).

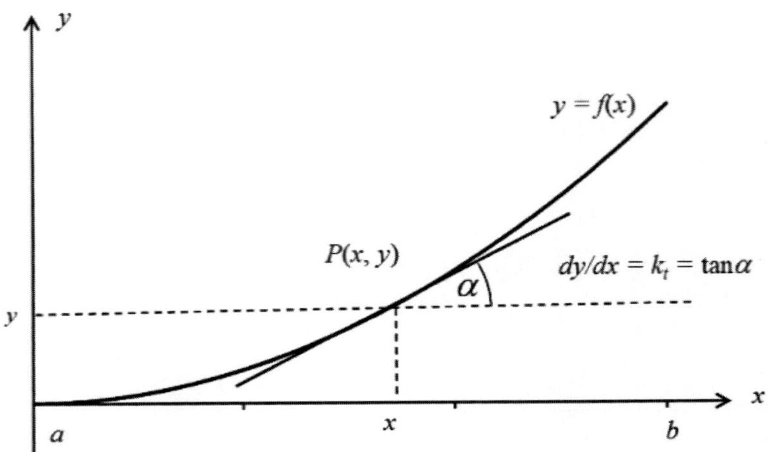

Figure 10.1: Geometric interpretation of derivatives

Example 10.14: Determine the tangent line of $y = e^x$ at $x = 1$.

Solution

The derivative of function $y = e^x$ was obtained in Example 10.13 as $y' = \frac{dy}{dx} = e^x$. At $x = 1$, the gradient of the tangent line is $k = y'(1) = e^1 = e.$ Let this tangent line be

$$y = kx + c = ex + c. \text{(} c \text{ is an unknown constant)}$$

Since the tangent line crosses $y = e^x$ at point $(1, e)$, substitute $x = 1$ and $y = e$ into the linear function to determine the value for c, i.e.,

$$e = e \times 1 + c \quad \rightarrow \quad c = 0.$$

Hence the tangent line of $y = e^x$ at $x = 1$ is $y = ex$.

The physical interpretation of derivatives

If we treat function $y = f(t)$ as the displacement of an object travelling with time t in a defined range, the derivative of $y = f(t)$ at time t is the velocity of the travelling object at that time.

Example 10.15: An object with mass m travels horizontally. The distance (s) travelled can be determined by

$$s(t) = v_0 t + \frac{1}{2}at^2,$$

where v_0 is the initial speed the object has at $t = 0$; a is the constant acceleration the object maintains during the travel. This movement is illustrated in Figure 10.2.

(a) Find the derivative of $s(t)$ at any time of $t > 0$;

(b) Assume $v_0 = 10\ m/s$ and $a = -0.1\ m/s^2$, how long has the object travelled before stopping?

v_0 ■ m →

$t = 0$ s

Figure 10.2: Horizontal movement with a constant acceleration (a) and an initial speed (v_0)

Solution

(a) As $s(t) = v_0 t + \frac{1}{2}at^2$,

$$s(t+\delta t) = v_0(t+\delta t) + \frac{1}{2}a(t+\delta t)^2 = v_0 t + v_0(\delta t) + \frac{1}{2}a[(t^2 + 2t(\delta t) + (\delta t)^2]$$

$$= v_0 t + v_0(\delta t) + \frac{1}{2}at^2 + at(\delta t) + \frac{1}{2}a(\delta t)^2 = (v_0 t + \frac{1}{2}at^2) + v_0(\delta t) + at(\delta t) + \frac{1}{2}a(\delta t)^2$$

$$= s(t) + v_0(\delta t) + at(\delta t) + \frac{1}{2}a(\delta t)^2.$$

By formula (10.11),

$$\frac{ds}{dt} = \lim_{\delta t \to 0} \frac{s(t+\delta t) - s(t)}{\delta t} = \lim_{\delta t \to 0} \frac{[s(t) + v_0(\delta t) + at(\delta t) + \frac{1}{2}a(\delta t)^2] - s(t)}{\delta t}$$

$$= \lim_{\delta t \to 0} \frac{v_0(\delta t) + at(\delta t) + \frac{1}{2}a(\delta t)^2}{\delta t} = \lim_{\delta t \to 0}[v_0 + at + \frac{1}{2}a(\delta t)] = v_0 + at + \lim_{\delta t \to 0}[\frac{1}{2}a(\delta t)] = v_0 + at.$$

This is the speed of this travelling object at any time. In the beginning ($t = 0$), it travels with an initial speed v_0. If acceleration a is negative, the object will slow down ($v_0 + at$ decreases with time) till it stops (when $v_0 + at = 0$).

(b) Let $v_0 + at = 0$; then $t = -v_0/a$. Substitute $v_0 = 10\ m/s$ and $a = -0.1\ m/s^2$ to get

$t = -v_0/a = -10/(-0.1) = 100$ (seconds).

The object stops moving in 100 seconds.

Derivative and continuity of a function

- If function $f(x)$ is differentiable everywhere in a defined range, $f(x)$ must be a continuous function in the same range.
- However, a continuous function in a defined range is not necessarily differentiable everywhere in the defined range.
- It is such function $f(x)$ *both continuous and smooth* in a defined range that is differentiable everywhere in the defined range.

Example 10.16: Evaluate if $y = 1 - x^3$ is differentiable at $x = 0$.

Solution

Function $y = 1 - x^3$ is continuous and smooth everywhere; hence it is differentiable at any point, including $x = 0$. This can be verified as below.

$$y(x+\delta x) = 1-(x+\delta x)^3 = 1-[x^3+3x^2\delta x+3x(\delta x)^2+(\delta x)^3] = 1-x^3-3x^2\delta x-3x(\delta x)^2-(\delta x)^3$$

$$\delta y = y(x+\delta x)-y(x) = 1-x^3-3x^2\delta x-3x(\delta x)^2-(\delta x)^3-(1-x^3) = -3x^2\delta x-3x(\delta x)^2-(\delta x)^3$$

$$\therefore \frac{\delta y}{\delta x} = \frac{-3x^2\delta x-3x(\delta x)^2-(\delta x)^3}{\delta x} = -[3x^2+3x\delta x+(\delta x)^2]$$

$$\lim_{\delta x\to 0^-}\frac{\delta y}{\delta x} = \lim_{\delta x\to 0^-}\{-[3x^2+3x\delta x+(\delta x)^2]\} = -3x^2;\ x=0,\quad \lim_{\delta x\to 0^-}\frac{\delta y}{\delta x} = -3\times 0^2 = 0$$

$$\lim_{\delta x\to 0^+}\frac{\delta y}{\delta x} = \lim_{\delta x\to 0^+}\{-[3x^2+3x\delta x+(\delta x)^2]\} = -3x^2;\ x=0,\quad \lim_{\delta x\to 0^+}\frac{\delta y}{\delta x} = -3\times 0^2 = 0$$

Since $\lim\limits_{\delta x\to 0^-}\frac{\delta y}{\delta x} = \lim\limits_{\delta x\to 0^+}\frac{\delta y}{\delta x}$, $y = 1 - x^3$ is differentiable at any point, including $x = 0$.

Example 10.17: Evaluate if $y = x$ and $y = |x|$ are differentiable at $x = 0$.

Solution

Function $y = x$ is continuous and smooth around $x = 0$ (Figure 10.3).

$$y(x+\delta x) = x+\delta x \text{ and } \delta y = y(x+\delta x)-y(x) = x+\delta x-x = \delta x.$$

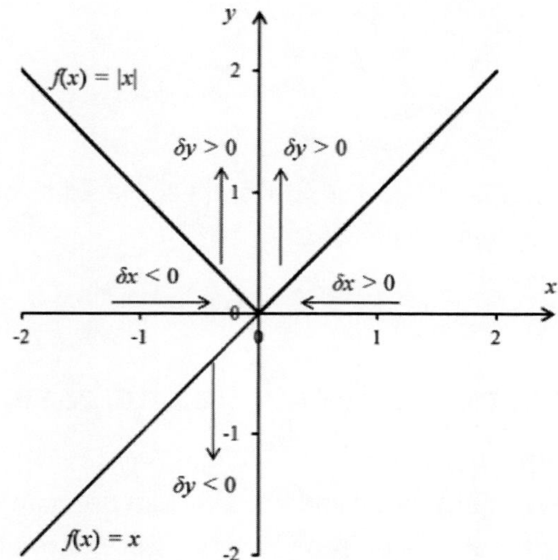

Figure 10.3: Graphs of $y = x$ and $y = |x|$ (thicker lines)

By formula (10.11),

$$\frac{dy}{dx} = \lim_{\delta x \to 0} \frac{\delta y}{\delta x} = \lim_{\delta x \to 0} \frac{\delta x}{\delta x} = 1.$$

This is actually true for all x, including $x = 0$, for linear function $y = x$.

Function $y = |x|$ is equivalent to

$$y = \begin{cases} -x & x < 0 \\ 0 & x = 0 \\ x & x > 0 \end{cases},$$

which is continuous and smooth everywhere except at $x = 0$ where $y = |x|$ has a vertex (Figure 10.3). Its derivative must be evaluated by examining the left-hand and right-hand limits of $y = |x|$ at $x = 0$.

For the left-hand limit, $y = -x$

$$y(x + \delta x) = |x + \delta x| = -(x + \delta x);$$

$$\delta y = y(x + \delta x) - y(x) = |x + \delta x| - |x| = -(x + \delta x) - (-x) = -x - \delta x + x = -\delta x.$$

$$\frac{dy}{dx} = \lim_{\delta x \to 0^-} \frac{\delta y}{\delta x} = \lim_{\delta x \to 0} \frac{-\delta x}{\delta x} = -1.$$

For the right-hand limit, $y = x$

$$y(x + \delta x) = |x + \delta x| = x + \delta x;$$

$$\delta y = y(x+\delta x) - y(x) = |x+\delta x| - |x| = x + \delta x - x = \delta x\,.$$

$$\frac{dy}{dx} = \lim_{\delta x \to 0^+} \frac{\delta y}{\delta x} = \lim_{\delta x \to 0} \frac{\delta x}{\delta x} = 1.$$

The left-hand and right-hand limits at $x = 0$ are different so $y = |x|$ has no derivative at $x = 0$ even though it is continuous at $x = 0$, *but not smooth.*

10.2.2 Derivatives of common functions and basic rules

Derivatives of common functions

In theory, the derivative of any continuous and smooth function can be obtained using formula (10.11). In practice, the derivatives of common functions are normally compiled as a table for easy use as shown in Table 10.1.

Table 10.1: Derivatives of common functions

$f(x)$	$f'(x)$	$f(x)$	$f'(x)$
c (constant)	0	$\sin ax$	$a\cos ax$
x^r (r: real number)	rx^{r-1}	$\cos ax$	$-a\sin ax$
$x\ (r=1)$	$1 \equiv \frac{d(x)}{dx} = (x)'$	$\tan ax$	$a\sec^2 ax = \frac{a}{\cos^2 ax}$
$\frac{1}{x}\ (r=-1)$	$\frac{-1}{x^2}$	$\cot ax$	$-a\csc^2 ax = \frac{-a}{\sin^2 ax}$
$\sqrt{x}\ (r=\frac{1}{2})$	$\frac{1}{2\sqrt{x}}$	$\sec ax$	$\frac{a\sin ax}{\cos^2 ax}$
e^{rx}	re^{rx}	$\csc ax$	$\frac{-a\cos ax}{\sin^2 ax}$
$e^x\ (r=1)$	e^x	$\arcsin ax$	$\frac{a}{\sqrt{1-(ax)^2}}$
$e^{-x}\ (r=-1)$	$-e^{-x}$	$\arccos ax$	$\frac{-a}{\sqrt{1-(ax)^2}}$
$a^{rx}\ (a>0)$	$ra^{rx}\ln a$	$\arctan ax$	$\frac{a}{1+(ax)^2}$
$x^{ax}\ (x>0)$	$ax^{ax}(1+\ln x)$	$\sinh ax$	$a\cosh ax$
$\ln x$	$\frac{1}{x}$	$\cosh ax$	$a\sinh ax$
$\log_a x$	$\frac{1}{x\ln a}$	$\tanh ax$	$\frac{a}{\cosh^2 ax}$

Example 10.18: Find the derivative for $y = x^3$, $y = \frac{1}{x^4}$, and $y = \sqrt[3]{x}$ respectively by Table 10.1.

Solution

In Table 10.1, $y = f(x) = x^r \longrightarrow y' = f'(x) = rx^{r-1}$. For $y = x^3$ where $r = 3$,

$$y' = \frac{dy}{dx} = \frac{d(x^3)}{dx} = 3x^{3-1} = 3x^2 .$$

For $y = \frac{1}{x^4} = x^{-4}$ where $r = -4$,

$$y' = \frac{dy}{dx} = \frac{d(x^{-4})}{dx} = -4x^{-4-1} = -4x^{-5} = -\frac{4}{x^5}.$$

For $y = \sqrt[3]{x} = x^{\frac{1}{3}}$ where $r = \frac{1}{3}$,

$$y' = \frac{dy}{dx} = \frac{d(\sqrt[3]{x})}{dx} = \frac{1}{3}x^{\frac{1}{3}-1} = \frac{1}{3}x^{\frac{1}{3}-\frac{3}{3}} = \frac{1}{3}x^{\frac{-2}{3}} = \frac{1}{3x^{\frac{2}{3}}} = \frac{1}{3\sqrt[3]{x^2}}.$$

Operational rules of derivatives

If $u = f(x)$ and $v = g(x)$ are differentiable functions within a shared range, and $u' = f'(x)$ and $v' = g'(x)$,

$$\begin{array}{|l|}\hline \frac{d(au)}{dx} = a\frac{du}{dx} \Longleftrightarrow (au)' = au', \quad a\text{: constant} \\ \frac{d(u \pm v)}{dx} = \frac{du}{dx} \pm \frac{dv}{dx} \Longleftrightarrow (u \pm v)' = u' \pm v' \\ \frac{d(au \pm bv)}{dx} = a\frac{du}{dx} \pm b\frac{dv}{dx} \Longleftrightarrow (au \pm bv)' = au' \pm bv', \quad a, b\text{: constants} \\ \frac{d(uv)}{dx} = \frac{du}{dx}v + u\frac{dv}{dx} \Longleftrightarrow (uv)' = u'v + uv' \quad \longleftarrow \text{product rule} \\ \frac{d(\frac{u}{v})}{dx} = \frac{\frac{du}{dx}v - u\frac{dv}{dx}}{v^2} \Longleftrightarrow (\frac{u}{v})' = \frac{u'v - uv'}{v^2} \quad \longleftarrow \text{quotient rule} \\ \hline \end{array} \qquad (10.12)$$

Derivatives of many combinative functions from common functions can be obtained by combining these rules with the derivatives of common functions listed in Table 10.1.

Example 10.19: Given $y = \sinh x$, prove $y' = \cosh x$.

Solution

As $y = \sinh x = \dfrac{e^x - e^{-x}}{2} = \dfrac{1}{2}(e^x - e^{-x})$,

$$y' = (\sinh x)' = \frac{1}{2}(e^x - e^{-x})' = \frac{1}{2}[(e^x)' - (e^{-x})'] = \frac{1}{2}[1\bullet e^x - (-1\bullet e^{-x})] = \frac{e^x + e^{-x}}{2} = \cosh x.$$

Example 10.20: Find the derivative for $y = 2x^3 + 3\sqrt[3]{x}$.

Solution

Let $u = x^3 \longrightarrow u' = 3x^2$ and $v = \sqrt[3]{x} \longrightarrow v' = \dfrac{1}{3\sqrt[3]{x^2}}$, which were obtained in Example 10.18.

$$y' = (2x^3 + 3\sqrt[3]{x})' = (2u + 3v)' = 2u' + 3v' = 2 \times 3x^2 + 3 \times \frac{1}{3\sqrt[3]{x^2}} = 6x^2 + \frac{1}{\sqrt[3]{x^2}}.$$

Example 10.21: Find the derivative for $y = 2x^2 \ln\sqrt{x}$.

Solution

Let $u = 2x^2 \longrightarrow u' = 2 \times 2x^{2-1} = 4x$ and $v = \ln\sqrt{x} = \dfrac{1}{2}\ln x \longrightarrow v' = \dfrac{1}{2}(\ln x)' = \dfrac{1}{2x}$. Apply product rule to $y = 2x^2 \ln\sqrt{x} = uv$,

$$y' = (2x^2 \ln\sqrt{x})' = (uv)' = u'v + uv' = 4x \times \frac{1}{2}\ln x + 2x^2 \times \frac{1}{2x} = 2x\ln x + x = x(2\ln x + 1).$$

Example 10.22: Determine the derivative for $y = e^x \sin x$.

Solution

Let $u = e^x \longrightarrow u' = (e^x)' = e^x$ and $v = \sin x \longrightarrow v' = (\sin x)' = \cos x$. By product rule for $y = uv$:

$$y' = (e^x \sin x)' = (uv)' = u'v + uv' = e^x \sin x + e^x \cos x = e^x(\sin x + \cos x).$$

Example 10.23: Given $y = \tan x$, use quotient rule to prove $y' = \sec^2 x = \dfrac{1}{\cos^2 x}$.

Solution

Let $u = \sin x \longrightarrow u' = (\sin x)' = \cos x$ and $v = \cos x \longrightarrow v' = (\cos x)' = -\sin x$.

As $y = \tan x = \dfrac{\sin x}{\cos x} = \dfrac{u}{v}$, apply quotient rule to $y = \tan x = \dfrac{\sin x}{\cos x} = \dfrac{u}{v}$:

$$y' = (\tan x)' = (\frac{\sin x}{\cos x})' = (\frac{u}{v})' = \frac{u'v - uv'}{v^2} = \frac{\cos x \cos x - \sin x(-\sin x)}{\cos^2 x} = \frac{\cos^2 x + \sin^2 x}{\cos^2 x}$$
$$= \frac{1}{\cos^2 x} = \sec^2 x.$$

Example 10.24: Determine the derivative for $y = \frac{2x^2 \ln\sqrt{x}}{e^x \sin x}$.

Solution

Let $u = 2x^2 \ln\sqrt{x} = x^2 \ln x \longrightarrow u' = x(2\ln x + 1)$ that was obtained in Example 10.21 using product rule; $v = e^x \sin x \longrightarrow y' = e^x(\sin x + \cos x)$, which was obtained in Example 10.22. By quotient rule:

$$y' = (\frac{2x^2 \ln\sqrt{x}}{e^x \sin x})' = (\frac{u}{v})' = \frac{u'v - uv'}{v^2} = \frac{x(2\ln x + 1)(e^x \sin x) - (2x^2 \ln\sqrt{x})e^x(\sin x + \cos x)}{(e^x \sin x)^2}$$
$$= \frac{xe^x[(2\ln x + 1)\sin x - 2x(\frac{1}{2}\ln x)(\sin x + \cos x)]}{(e^x \sin x)^2} = \frac{x[(2\ln x + 1)\sin x - x\ln x(\sin x + \cos x)]}{e^x \sin^2 x}.$$

Example 10.25: Determine the derivative for $y = \frac{x^2 - e^{-2x}}{2\sin x}$.

Solution

Let $u = x^2 - e^{-2x} \longrightarrow u' = (x^2 - e^{-2x})' = 2x - (-2e^{-x}) = 2x + 2e^{-x}$; $v = 2\sin x \longrightarrow y' = 2\cos x$.
By quotient rule:

$$y' = (\frac{x^2 - e^{-2x}}{2\sin x})' = (\frac{u}{v})' = \frac{u'v - uv'}{v^2} = \frac{(2x + 2e^{-2x})(2\sin x) - (x^2 - e^{-2x})(2\cos x)}{4\sin^2 x}$$
$$= \frac{4x\sin x + 4e^{-2x}\sin x - 2x^2\cos x + 2e^{-2x}\cos x}{4\sin^2 x} = \frac{x(2\sin x - x\cos x) + e^{-2x}(2\sin x + \cos x)}{2\sin^2 x}.$$

Exercises 10.2

1. Find derivative for $y=\sqrt{2x+3}$ by means of the limit operations on the average rate of change.

2. Find derivatives for the following functions respectively using the rules of differentiation.

(a) $y=x^4-2x^3-x^2+5x+3$; (b) $y=x^2-2x+5$; (c) $y=\sqrt[3]{8x^2}$;

(d) $y=\frac{3}{\sqrt{x^3}}$; (e) $y=e^{2x}$; (f) $y=e^{-3x}$; (g) $y=\ln x^{\frac{2}{3}}$;

(h) $y=\log x^{-2}$; (i) $y=\log_2\sqrt{x}$; (j) $y=\sin 3x+2\cos 2x-\cos 3x$;

(k) $y=\tan x+\sin 2x$; (l) $y=3e^{-2x}-2e^{3x}$; (m) $y=\ln\frac{x}{\sqrt{2x}}$; (n) $y=-2x+\ln\frac{1}{x}$.

3. Find derivatives for the following functions respectively using the rules of differentiation.

(a) $y=x^3e^{2x}$; (b) $y=2x^2\ln\sqrt{x}$; (c) $y=2\sqrt[3]{x^2}\sin x$;

(d) $y=e^{-3x}\ln\sqrt[3]{x}$; (e) $y=e^{2x}\cos 2x$; (f) $y=e^{-3x}\sin\frac{x}{3}$; (g) $y=\frac{\ln x^2}{2x}$;

(h) $y=\frac{x^2-e^{2x}}{\ln x}$; (i) $y=\frac{4\sqrt{x}}{e^{2x}}$; (j) $y=\sin 3x\cos 2x$;

(k) $y=\frac{\tan x}{\cos 2x}$; (l) $y=\frac{3e^{-2x}-2e^{3x}}{3x^2}$; (m) $y=\frac{\ln x}{\sqrt{2x}}$; (n) $y=-2x\ln\frac{1}{x}$.

4. Find the slope for the following functions at point $x=1$ or $t=\pi/4$.

(a) $y=x^2+3x-4$; (b) $y=-2\sqrt{x}$; (c) $y=e^{-3x}$;

(d) $y=2x^2\ln\sqrt{x}$; (e) $y=\frac{4\sqrt{x}}{e^{2x}}$; (f) $y=\sin 3t+2\cos 2t$; (g) $y=\tan t+\sin t$;

10.3 Advanced Techniques of Differentiation

10.3.1 The extended product rule

The product rule of two functions $u(x)$ and $v(x)$ can be extended to the product of more functions $p(x)$, $q(x)$, …, $w(x)$ if their derivatives exist, i.e.,

$$\boxed{y = u(x)v(x)w(x) \longrightarrow y' = u'(x)v(x)w(x) + u(x)v'(x)w(x) + u(x)v(x)w'(x)} \tag{10.13}$$

Example 10.26: Determine the derivative for $y = x^2 e^x \cos x$.

Solution

$u = x^2 \longrightarrow u' = 2x$; $v = e^x \longrightarrow v' = e^x$; $w = \cos x \longrightarrow w' = -\sin x$. Apply (10.13) to $y = uvw$:

$$\begin{aligned} y' &= (x^2 e^x \cos x)' = u'vw + uv'w + uvw' = 2xe^x \cos x + x^2 e^x \cos x + x^2 e^x (-\sin x) \\ &= 2xe^x \cos x + x^2 e^x \cos x - x^2 e^x \sin x = xe^x (2\cos x + x\cos x - x\sin x). \end{aligned}$$

Example 10.27: Determine the derivative for $y = \dfrac{e^{-x}\ln x + x\sin x}{\sqrt{x}}$.

Solution

$$y = \frac{e^{-x}\ln x + x\sin x}{\sqrt{x}} = x^{-\frac{1}{2}}(e^{-x}\ln x + x\sin x) = x^{-\frac{1}{2}}e^{-x}\ln x + x^{1-\frac{1}{2}}\sin x = x^{-\frac{1}{2}}e^{-x}\ln x + x^{\frac{1}{2}}\sin x,$$

$$y = x^{-\frac{1}{2}}e^{-x}\ln x + x^{\frac{1}{2}}\sin x = uvw + pq \quad \text{where}$$

$$u = x^{-\frac{1}{2}} \longrightarrow u' = (x^{-\frac{1}{2}})' = -\frac{1}{2}x^{-\frac{1}{2}-1} = -\frac{1}{2}x^{-\frac{3}{2}};\quad v = e^{-x} \longrightarrow v' = (e^{-x}) = -e^{-x};$$

$$w = \ln x \longrightarrow w' = (\ln x)' = \frac{1}{x};\quad p = x^{\frac{1}{2}} \longrightarrow p' = (x^{\frac{1}{2}})' = \frac{1}{2}x^{\frac{1}{2}-1} = \frac{1}{2}x^{-\frac{1}{2}};$$

$$q = \sin x \longrightarrow q' = (\sin x)' = \cos x.$$

$$\begin{aligned} \therefore y' &= (uvw + pq)' = (uvw)' + (pq)' = u'vw + uv'w + uvw' + p'q + pq' \\ &= -\frac{1}{2}x^{-\frac{3}{2}}e^{-x}\ln x + x^{-\frac{1}{2}}(-e^{-x})\ln x + x^{-\frac{1}{2}}e^{-x}\frac{1}{x} + \frac{1}{2}x^{-\frac{1}{2}}\sin x + x^{\frac{1}{2}}\cos x \\ &= -\frac{1}{2}x^{-\frac{3}{2}}e^{-x}\ln x - x^{-\frac{1}{2}}e^{-x}\ln x + x^{-\frac{3}{2}}e^{-x} + \frac{1}{2}x^{-\frac{1}{2}}\sin x + x^{\frac{1}{2}}\cos x \\ &= \frac{1}{2}x^{-\frac{3}{2}}[-e^{-x}\ln x - 2xe^{-x}\ln x + 2e^{-x} + x\sin x + 2x^2\cos x] \\ &= \frac{e^{-x}(2 - \ln x - 2x\ln x) + x(\sin x + 2x\cos x).}{2\sqrt{x^3}} \end{aligned}$$

10.3.2 The chain rule

The Standard chain rule

If a continuous function $y = y(x)$ is convertible to an equivalent chained (or nested) function $y = y(u)$ through substitution of $u = u(x)$, the derivative of y with respect to x can be obtained by the product of the two nested derivatives dy/du and du/dx if both derivatives exist, i.e.,

$$y = y(x) = y(u(x)) \longrightarrow \begin{cases} \dfrac{dy}{dx} = \dfrac{dy}{du}\dfrac{du}{dx} = \dfrac{dy(u)}{du}\dfrac{du(x)}{dx} \quad \text{or} \\ \dfrac{dy}{dx} = y' = y'_u u'_x \end{cases} \tag{10.14}$$

This is called the chain rule of derivatives for nested functions.

Example 10.28: Determine the derivative for $y = \sin(ax+b)$ where a and b are constants.

Solution

Let $u = ax+b$. Then

$$y = \sin(ax+b) = \sin u \longrightarrow y'_u = \frac{dy}{du} = (\sin u)' = \cos u = \cos(ax+b);$$

$$u = ax+b \longrightarrow u'_x = \frac{du}{dx} = (ax+b)' = (ax)' + (5)' = a + 0 = a.$$

Apply chain rule (10.14) to $y = \sin(ax+b) = \sin u$:

$$y' = y'_u u'_x = \cos u \times a = a\cos(ax+b).$$

Example 10.29: Determine the derivative for $y = \ln(x^2+5x+1)$.

Solution

Let $u = x^2+5x+1$. Then

$$y = \ln(x^2+5x+1) = \ln u \longrightarrow y'_u = \frac{dy}{du} = (\ln u)' = \frac{1}{u} = \frac{1}{x^2+5x+1};$$

$$u = x^2+5x+1 \longrightarrow u'_x = \frac{du}{dx} = (x^2+5x+1)' = 2x+5.$$

Apply chain rule (10.14) to $y = \ln(x^2+5x+1) = \ln u$:

$$y' = y'_u u'_x = \frac{1}{u}(2x+5) = \frac{2x+5}{x^2+5x+1}.$$

Example 10.30: Determine the derivative for $y = \ln(x^2 + 5x + 1)\sin(ax + b)$ where a & b are constants.

Solution

Function $y = \ln(x^2 + 5x + 1)\sin(ax + b)$ is the product of those two functions in Examples 10.28-29. We can combine the product rule and chain rule together to obtain its derivative.

Let $g = \ln(x^2 + 5x + 1)$ and $h = \sin(ax + b)$. Then

$$y = gh \rightarrow y' = g'h + gh'.$$

As known from Examples 10.28-29,

$$g' = \frac{2x+5}{x^2+5x+1} \text{ and } h' = a\cos(ax+b).$$

Hence

$$\begin{aligned} y' = g'h + gh' &= \frac{2x+5}{x^2+5x+1}\sin(ax+b) + \ln(x^2+5x+1)a\cos(ax+b) \\ &= \frac{(2x+5)\sin(ax+b)}{x^2+5x+1} + a\ln(x^2+5x+1)\cos(ax+b). \end{aligned}$$

The extended chain rule

If a continuous function $y = y(x)$ is convertible to an equivalent chained (or nested) function through multiple substitutions, for example, $y = y(u)$, $u = u(v)$, $v = v(w)$, and $w = w(x)$, the derivative of y with respect to x can be obtained by the product of all the nested derivatives if these derivatives exist, i.e.,

$$y = y(x) = y(u(v(w(x)))) \longrightarrow \begin{cases} \dfrac{dy}{dx} = \dfrac{dy}{du}\dfrac{du}{dv}\dfrac{dv}{dw}\dfrac{dw}{dx} \text{ or} \\ \dfrac{dy}{dx} = y' = y'_u u'_v v'_w w'_x \end{cases} \tag{10.15}$$

Example 10.31: Determine the derivatives for $y = e^{\sin(x^2+3)}$ and $y = \ln[\cos(3-x)^2]$ respectively.

Solution

For $y = e^{\sin(x^2+3)}$, let $v = x^2+3$, $u = \sin(x^2+3) = \sin v$, then $y = e^{\sin(x^2+3)} = e^u$.

$$v = x^2+3 \longrightarrow v'_x = \frac{dv}{dx} = (x^2+3)' = 2x;$$

$$u = \sin v \longrightarrow u'_v = \frac{du}{dv} = (\sin v)' = \cos v = \cos(x^2+3);$$

$$y = e^u \longrightarrow y'_u = (e^u)' = e^u = e^{\sin(x^2+3)}$$

By the extended chain rule (10.15),

$$y' = y'_u u'_v v'_x = [e^{\sin(x^2+3)}][\cos(x^2+3)][2x] = 2x\cos(x^2+3)e^{\sin(x^2+3)}.$$

For $y = \ln[\cos(3-x)^2]$, let $w = 3-x$, $v = (3-x)^2 = w^2$, $u = \cos(3-x)^2 = \cos v$, $y = \ln u$.

$$w = 3-x \to w'_x = \frac{dw}{dx} = (3-x)' = (3)' - (x)' = 0-1 = -1;$$

$$v = w^2 \to v'_w = \frac{dv}{dw} = (w^2)' = 2w = 2(3-x);$$

$$u = \cos v \to u'_v = \frac{du}{dv} = (\cos v)' = -\sin v = -\sin(3-x)^2;$$

$$y = \ln u \to y'_u = (\ln u)' = \frac{1}{u} = \frac{1}{\cos(3-x)^2}.$$

By the extended chain rule (10.15),

$$y' = y'_u u'_v v'_w w'_x = [\frac{1}{\cos(3-x)^2}][-\sin(3-x)^2][2(3-x)](-1) = \frac{2(3-x)\sin(3-x)^2}{\cos(3-x)^2} = 2(3-x)\tan(3-x)^2.$$

10.3.3 The logarithmic rule

The logarithmic rule is an extension of the chain rule to functions with form $y = \ln f(x)$ as if using $u = f(x)$ for substitution, which results in the following logarithmic rule:

$$\boxed{y = \ln f(x) \longrightarrow y' = \frac{d[\ln f(x)]}{dx} = \frac{f'(x)}{f(x)}} \qquad (10.16)$$

The logarithmic rule is useful in deriving derivatives for combined functions made up by multiple terms of functions clustered together by multiplications and/or divisions.

Example 10.32: Determine the derivative for $y = \ln[\cos(3-x)^2]$ using the logarithmic rule.

Solution

This is the same function in Example 10.31. Let $f(x) = \cos(3-x)^2$. Then $f'(x)$ can be obtained using the chain rule.

$$v = 3-x \longrightarrow v'_x = (3-x)' = -1; \quad u = (3-x)^2 = v^2 \longrightarrow u'_v = (v^2)' = 2v = 2(3-x);$$
$$f(u) = \cos u \longrightarrow f'_u = (\cos u)' = -\sin u = -\sin(3-x)^2.$$

Thus by formula (10.15),

$$f'(x) = f'_u u'_v v'_x = -\sin(3-x)^2 2(3-x)(-1) = 2(3-x)\sin(3-x)^2.$$

Then by formula (10.16),

$$y' = \ln[\cos(3-x)^2]' = \ln[f(x)]' = \frac{f'(x)}{f(x)} = \frac{2(3-x)\sin(3-x)^2}{\cos(3-x)^2} = 2(3-x)\tan(3-x)^2.$$

This is the same outcome as obtained in Example 10.31.

Example 10.33: Determine the derivative for $y = e^{2x+1}(3x-1)$ using the logarithmic rule.

Solution

Apply natural logarithmic operation to both sides of the function:

$$\ln y = \ln[e^{2x+1}(3x-1)] = \ln e^{2x+1} + \ln(3x-1) = (2x+1) + \ln(3x-1), \text{ or}$$
$$\ln y = (2x+1) + \ln(3x-1).$$

Apply the logarithmic rule (10.16) to both sides of the above logarithmic equation:

$$(\ln y)' = [2x+1+\ln(3x-1)]' \longrightarrow \frac{y'}{y} = (2x)' + (1)' + [\ln(3x-1)]'$$
$$\frac{y'}{y} = 2+0+\frac{(3x-1)'}{3x-1} \longrightarrow \frac{y'}{y} = 2+\frac{3}{3x-1}$$

This can be reorganised as

$$y' = y(2+\frac{3}{3x-1}) = e^{2x+1}(3x-1)(2+\frac{3}{3x-1}) = e^{2x+1}[2(3x-1)+3] = e^{2x+1}[6x-2+3]$$
$$= e^{2x+1}(6x+1).$$

Example 10.34: Determine the derivative for $y=(x^2-1)e^{-2x}\tan x$ using the logarithmic rule.

Solution

Apply natural logarithmic operation to both sides of the function:

$$\ln y=\ln[(x^2-1)e^{-2x}\tan x]=\ln(x^2-1)+\ln e^{-2x}+\ln\tan x=\ln(x^2-1)+(-2x)+\ln\frac{\sin x}{\cos x},\ \text{or}$$

$$\ln y=\ln(x^2-1)-2x+\ln\sin x-\ln\cos x.$$

Apply the logarithmic rule (10.16) to both sides of the above logarithmic equation:

$$(\ln y)'=[\ln(x^2-1)-2x+\ln\sin x-\ln\cos x]'\longrightarrow\frac{y'}{y}=[\ln(x^2-1)]'-(2x)'+(\ln\sin x)'-(\ln\cos x)'$$

$$\frac{y'}{y}=\frac{(x^2-1)'}{x^2-1}-2+\frac{(\sin x)'}{\sin x}-\frac{(\cos x)'}{\cos x}\longrightarrow\frac{y'}{y}=\frac{2x}{x^2-1}-2+\frac{\cos x}{\sin x}-\frac{-\sin x}{\cos x}$$

$$\frac{y'}{y}=\frac{2x}{x^2-1}+\cot x+\tan x-2$$

This can be reorganised as

$$\begin{aligned}y'&=y(\frac{2x}{x^2-1}+\cot x+\tan x-2)=(x^2-1)e^{-2x}\tan x(\frac{2x}{x^2-1}+\cot x+\tan x-2)\\&=e^{-2x}[2x\tan x+(x^2-1)\tan x\cot x+(x^2-1)\tan x\tan x-2(x^2-1)\tan x]\quad\leftarrow\tan x\cot x=1\\&=e^{-2x}[2x\tan x+(x^2-1)+(x^2-1)\tan^2 x-2(x^2-1)\tan x]\\&=e^{-2x}\{[2x-2(x^2-1)]\tan x+(x^2-1)(1+\tan^2 x)\}\quad\leftarrow 1+\tan^2 x=\sec^2 x\\&=e^{-2x}[(2x-2x^2+2)\tan x+(x^2-1)\sec^2 x]=e^{-2x}[(x^2-1)\sec^2 x-2(x^2-x-1)\tan x].\end{aligned}$$

Example 10.35: Determine the derivative for $y=\frac{\sqrt{2x+1}}{\sqrt[3]{3x^2+3x+2}}$ using the logarithmic rule.

Solution

Apply natural logarithmic operation to both sides of the function:

$$\ln y=\ln\frac{\sqrt{2x+1}}{\sqrt[3]{3x^2+3x+2}}.$$

$$\ln\frac{\sqrt{2x+1}}{\sqrt[3]{3x^2+3x+2}}=\ln\sqrt{2x+1}-\ln\sqrt[3]{3x^2+3x+2}=\frac{1}{2}\ln(2x+1)-\frac{1}{3}\ln(3x^2+3x+2),\ \text{or}$$

$$\ln y=\frac{1}{2}\ln(2x+1)-\frac{1}{3}\ln(3x^2+3x+2).$$

Apply the logarithmic rule (10.16) to both sides of the above logarithmic equation:

$$(\ln y)' = [\frac{1}{2}\ln(2x+1) - \frac{1}{3}\ln(3x^2+3x+2)]' \longrightarrow \frac{y'}{y} = \frac{1}{2}[\ln(2x+1)]' - \frac{1}{3}[\ln(3x^2+3x+2)]'$$

$$\frac{y'}{y} = \frac{1}{2}\frac{(2x+1)'}{2x+1} - \frac{1}{3}\frac{(3x^2+3x+2)'}{3x^2+3x+2} \longrightarrow \frac{y'}{y} = \frac{1}{2}\frac{2}{2x+1} - \frac{1}{3}\frac{(6x+3)}{3x^2+3x+2}$$

$$\frac{y'}{y} = \frac{1}{2x+1} - \frac{2x+1}{3x^2+3x+2}$$

This can be reorganised as

$$y' = y(\frac{1}{2x+1} - \frac{2x+1}{3x^2+3x+2}) = \frac{\sqrt{2x+1}}{\sqrt[3]{3x^2+3x+2}}(\frac{1}{2x+1} - \frac{2x+1}{3x^2+3x+2}).$$

Exercises 10.3

1. Find derivatives of the following functions respectively.

(a) $y = 2x^3e^x\ln x$; (b) $y = 4\sqrt{x}\ln\sqrt{x}e^{-x}$; (c) $y = xe^{-x}\ln x\sin 2x$;

(d) $y = \tan x\sin 2x$; (e) $y = 3e^{-2x}\cos 2x$; (f) $y = \ln x\sin^2 x$; (g) $y = x^2\tan x\ln\frac{1}{x}$

2. Find derivatives of the following functions respectively.

(a) $y = (x-2)^3$; (b) $y = \ln\sqrt{2x+5}$; (c) $y = 3\sqrt[3]{(x+2)^2}$;

(d) $y = e^{-x^2+2x}$; (e) $y = e^{\sqrt{2x+3}}$; (f) $y = \sin(3x-2)$; (g) $y = \cos(x^2-2x+1)$;

(h) $y = \cos(3x+2) + 2\cos[2(x-1)]$; (i) $y = \ln(x^2+2x+5)$; (j) $y = e^{\sin(x+2)}$.

3. Find derivatives of the following functions respectively.

(a) $y = x^3e^{2x}\cos^2 x$; (b) $y = \frac{4\sqrt{x^2+3}}{e^{2x}}$; (c) $y = 6\sqrt{x^2+3x}e^{x^2+3x}$;

(d) $y = 2(x+2)^2\ln\sqrt{x+2}$; (e) $y = 6\sin(3x-1)\cos(x^2+2)$; (f) $y = x^{-2}e^{\frac{1}{4}x}\cos 2x$.

10.4 Higher Order Derivatives

The derivative of $y = f(x)$ is a new function, i.e., $y' = f'(x) = g(x)$. If $g(x)$ is differentiable, a further differentiation on $g(x)$ creates another function $g'(x) = y'' = f''(x)$, called the second-order derivative of $y = f(x)$. Repeating such process will create the third-order derivative y''', fourth-order derivative $y^{(4)}$, and so forth.

The notations for higher-order derivatives are slightly different for the first-order to third-order from the fourth-order and above. These are listed as follows for reference:

$$\begin{aligned}
&y' = \frac{dy}{dx} = Dy \longleftarrow \text{first-order} \\
&y'' = \frac{dy'}{dx} = \frac{d^2y}{dx^2} = D^2y \longleftarrow \text{second-order} \\
&y''' = \frac{dy''}{dx} = \frac{d^3y}{dx^3} = D^3y \longleftarrow \text{third-order} \\
&y^{(4)} = \frac{dy'''}{dx} = \frac{d^4y}{dx^4} = D^4y \longleftarrow \text{fourth-order} \\
&\cdots \\
&y^{(n)} = \frac{dy^{(n-1)}}{dx} = \frac{d^ny}{dx^n} = D^ny \longleftarrow n\text{th-order.}
\end{aligned} \tag{10.17}$$

In these formulae, $D = \frac{d}{dx}$ is called the derivative operator that has wide applications in advanced mathematics.

Example 10.36: Given $y = \sqrt{x}$, find its first-order, second-order, third-order and fourth-order derivatives.

Solution

$$y' = (\sqrt{x})' = \frac{1}{2}x^{\frac{1}{2}-1} = \frac{1}{2}x^{-\frac{1}{2}} = \frac{1}{2\sqrt{x}}$$

$$y'' = (y')' = (\frac{1}{2}x^{-\frac{1}{2}})' = \frac{1}{2}(-\frac{1}{2})x^{-\frac{1}{2}-1} = -\frac{1}{4}x^{-\frac{3}{2}} = -\frac{1}{4\sqrt{x^3}}$$

$$y''' = (y'')' = (-\frac{1}{4}x^{-\frac{3}{2}})' = -\frac{1}{4}(-\frac{3}{2})x^{-\frac{3}{2}-1} = \frac{3}{8}x^{-\frac{5}{2}} = \frac{3}{8\sqrt{x^5}}$$

$$y^{(4)} = (y''')' = (\frac{3}{8}x^{-\frac{5}{2}})' = \frac{3}{8}(-\frac{5}{2})x^{-\frac{5}{2}-1} = -\frac{15}{16}x^{-\frac{7}{2}} = -\frac{15}{16\sqrt{x^7}}.$$

Example 10.37: Given $y = e^{2x}$, find its first-order, second-order, third-order and fourth-order derivatives. Infer the pattern of the *n*th-order derivative.

Solution

$$y' = (e^{2x})' = e^{2x}(2x)' = 2e^{2x} \leftarrow \text{chain rule, } u = 2x,\ y = e^u$$
$$y'' = (y')' = (2e^{2x})' = 2e^{2x}(2x)' = 4e^{2x} = 2^2 e^{2x}$$
$$y''' = (y'')' = (4e^{2x})' = 4e^{2x}(2x)' = 8e^{2x} = 2^3 e^{2x}$$
$$y^{(4)} = (y''')' = (8e^{2x})' = 8e^{2x}(2x)' = 16e^{2x} = 2^4 e^{2x}.$$

This pattern indicates the *n*th-order derivative is $y^{(n)} = 2^n e^{2x}$.

Example 10.38: Given $y = \sin ax$, find its first-order, second-order, third-order and fourth-order derivatives. Infer the pattern of the *n*th-order derivative.

Solution

$$y' = (\sin ax)' = a\cos ax = a\sin(\frac{\pi}{2} + ax)$$
$$y'' = [a\cos ax]' = -a^2 \sin ax = a^2 \sin(\pi + ax) = a^2 \sin(\frac{2\pi}{2} + ax)$$
$$y''' = [-a^2 \sin ax]' = -a^3 \cos ax = a^3 \sin(\frac{3\pi}{2} + ax)$$
$$y^{(4)} = [-a^3 \cos ax]' = a^4 \sin ax = a^4 \sin(2\pi + ax) = a^4 \sin(\frac{4\pi}{2} + ax)$$

This pattern indicates the *n*th-order derivative is $y^{(n)} = a^n \sin(\frac{n\pi}{2} + ax)$.

Exercises 10.4

1. Find the first-order to the third-order derivatives of the following functions respectively.

   (a) $y = e^{2x}$; (b) $y = x^4 - 2x^3 - 3x^2 - 4x^4 + 5$; (c) $y = \ln x$;

   (d) $y = \sin 2x$; (e) $y = \cos(2x - 1)$; (f) $y = e^{x^2}$.

10.5 Derivatives of Special Functions

10.5.1 Derivatives of parametric functions

In a parametric function, a relationship between x and y is defined through another parameter (t) commonly shared by both x and y as

$$\begin{cases} x = f(t) \\ y = g(t) \end{cases}.$$

The derivative for a parametric function is determined through the ratio of the derivatives of y and x separately with respect to parameter (t) as follows:

$$\boxed{\begin{cases} x = f(t) \\ y = g(t) \end{cases} \longrightarrow \frac{dy}{dx} = \frac{\frac{dy}{dt}}{\frac{dx}{dt}} = \frac{y'_t}{x'_t} \longleftarrow \begin{cases} x'_t = f'(t) \\ y'_t = g'(t) \end{cases}}. \tag{10.18}$$

Example 10.39: A parametric function is defined by $x = t - 2$ and $y = t^2 + 5$. Determine the derivative dy/dx for this parametric function.

Solution

$$\begin{cases} x = t - 2 \\ y = t^2 + 5 \end{cases} \longrightarrow \begin{cases} \dfrac{dx}{dt} = (t-2)' = 1 \\ \dfrac{dy}{dt} = (t^2+5)' = 2t \end{cases},$$

thus, by formula (10.18), the derivative dy/dx for this parametric function is

$$\frac{dy}{dx} = \frac{\frac{dy}{dt}}{\frac{dx}{dt}} = \frac{2t}{1} = 2t.$$

Example 10.40: A parametric function is defined by $x = 3\sin t$ and $y = 4\cos t$. Find the derivative dy/dx for this parametric function at $t = \pi/6$.

Solution

$$\begin{cases} x = 3\sin t \\ y = 4\cos t \end{cases} \longrightarrow \begin{cases} \dfrac{dx}{dt} = (3\sin t)' = 3\cos t \\ \dfrac{dy}{dt} = (4\cos t)' = -4\sin t \end{cases}.$$

Thus, by formula (10.18), the derivative *dy*/*dx* for this parametric function is

$$\frac{dy}{dx}=\frac{\frac{dy}{dt}}{\frac{dx}{dt}}=\frac{-4\sin t}{3\cos t}=-\frac{4}{3}\tan t.$$

At $t=\pi/6$, $\frac{dy}{dx}=-\frac{4}{3}\tan\frac{\pi}{6}=-\frac{4}{3}\frac{\sqrt{3}}{3}=-\frac{4\sqrt{3}}{9}$.

10.5.2 Derivatives of implicit functions

If a function between *x* and *y* is given as a general equation $f(x,y)=c$, rather than an explicit function $y=f(x)$, the derivative of such an implicit function between *x* and *y* can be obtained by applying the chain rule to both sides of $f(x,y)=c$ and then separating y' from the resultant equation, which is likely to be expressed by *x* and *y* together as a new implicit function.

Example 10.41: Given $x^2+y^2=r^2$, find the derivative *dy*/*dx* for this implicit function (*r* is constant).

Solution

Differentiate both sides of the equation with respect to *x*,

$$\frac{d(x^2+y^2)}{dx}=\frac{dr^2}{dx}\longrightarrow\frac{d(x^2)}{dx}+\frac{d(y^2)}{dx}=0\longrightarrow 2x+\frac{d(y^2)}{dy}\frac{dy}{dx}=0 \quad\leftarrow \text{chain rule to } y(x)$$

$$2x+2y\frac{dy}{dx}=0\longrightarrow 2y\frac{dy}{dx}=-2x\longrightarrow y'=\frac{dy}{dx}=\frac{-2x}{2y}=-\frac{x}{y}.$$

Example 10.42: Given $3\sin x+\cos y=\ln y-2x^2$, find the derivative *dy*/*dx*.

Solution

Differentiate both sides of the equation with respect to *x*,

$$(3\sin x+\cos y)'=(\ln y-2x^2)'\longrightarrow 3(\sin x)'+(\cos y)'=(\ln y)'-2(x^2)'$$

$$3\cos x-\sin y\bullet y'=\frac{y'}{y}-4x \quad\leftarrow \text{chain rule to } y(x)$$

$$3y\cos x-y\sin y\bullet y'=y'-4xy \quad\leftarrow \text{multiply both sides by } y$$

$$3y\cos x+4xy=y'+y\sin y\bullet y'\longrightarrow y'(1+y\sin y)=(3\cos x+4x)y$$

$$y'=\frac{dy}{dx}=\frac{(3\cos x+4x)y}{1+y\sin y}.$$

Example 10.43: Given $e^{\sin y}e^{\cos x} = x^2$, find the derivative *dy*/*dx* for this implicit function.

Solution

Apply natural logarithm to both sides of the equation,

$$\ln(e^{\sin y}e^{\cos x}) = \ln x^2 \longrightarrow \ln e^{\sin y} + \ln e^{\cos x} = 2\ln x \longrightarrow \sin y + \cos x = 2\ln x$$

Differentiate both sides of the above equation with respect to *x*,

$$(\sin y + \cos x)' = (2\ln x)' \longrightarrow (\sin y)' + (\cos x)' = 2(\ln x)' \longrightarrow \cos y \bullet y' - \sin x = \frac{2}{x}$$

$$\cos y \bullet y' = \frac{2}{x} + \sin x \longrightarrow y' = \frac{dy}{dx} = \frac{2}{x\cos y} + \frac{\sin x}{\cos y} = \frac{2}{x\cos y} + \frac{x\sin x}{x\cos y} = \frac{2 + x\sin x}{x\cos y}.$$

Example 10.44: Find the second-order derivative for implicit function $x^2 + y^2 = r^2$ that was differentiated in Example 10.41.

Solution

The first-order derivative of $x^2 + y^2 = r^2$ was found in Example 10.41 as

$$y' = -\frac{x}{y} \quad \text{or} \quad yy' = -x.$$

The 2nd-order derivative can be obtained using either of the two expressions. By $y' = -\frac{x}{y}$:

$$\begin{aligned} y'' &= (y')' = (-\frac{x}{y})' = (-xy^{-1})' = -[(x)'y^{-1} + x(y^{-1})'] \quad \leftarrow \text{product rule} \\ &= -[1 \bullet y^{-1} + x(-1y^{-2}y')] = -y^{-1} + xy^{-2}y' \quad \leftarrow \text{chain rule } y = u,\ u = u(x) \\ &= -\frac{1}{y} + \frac{x}{y^2}y' = -\frac{1}{y} + \frac{x}{y^2}(-\frac{x}{y}) \quad \leftarrow y' = -\frac{x}{y} \\ &= -\frac{y^2}{y^3} - \frac{x^2}{y^3} = -\frac{x^2 + y^2}{y^3} = -\frac{r^2}{y^3}. \quad \leftarrow x^2 + y^2 = r^2 \end{aligned}$$

By $yy' = -x$:

$$(yy')' = (-x)' \longrightarrow (y)' \bullet y' + y(y')' = -1 \longrightarrow (y')^2 + yy'' = -1 \longrightarrow yy'' = -1 - (y')^2$$

$$y'' = -\frac{1 + (y')^2}{y} = -\frac{1 + (-\frac{x}{y})^2}{y} = -\frac{1 + \frac{x^2}{y^2}}{y} = -\frac{\frac{y^2}{y^2} + \frac{x^2}{y^2}}{y} = -\frac{\frac{y^2 + x^2}{y^2}}{y} = -\frac{x^2 + y^2}{y^3} = -\frac{r^2}{y^3}.$$

Exercises 10.5

1. Find derivatives of the following functions respectively.

(a) $\begin{cases} x = 2t - 1 \\ y = t^2 + 3 \end{cases}$; (b) $\begin{cases} x = \sqrt{t^2 + 4} \\ y = \dfrac{3}{\sqrt{t+2}} \end{cases}$; (c) $\begin{cases} x = 2\cos t + 1 \\ y = 2\sin t + 2 \end{cases}$;

(d) $\begin{cases} x = 4\sin 2t \\ y = 3\cos 2t \end{cases}$; (e) $\begin{cases} x = e^t + 2 \\ y = e^{-t} + 3 \end{cases}$; (f) $\begin{cases} x = \ln(2t^2 + 1) \\ y = 2t^2 + 3 \end{cases}$.

2. Find derivatives of the following functions respectively.

(a) $(xy - 2)^3 - x^2 = 0$; (b) $y + x = e^{x^2 + 2y}$; (c) $\sqrt{y + x} = e^{\sqrt{2x+3}}$;

(d) $2x^2 + 3y^2 = 5$; (e) $\cos(3y + x) = 2\sin(2x - y)$; (f) $y = \ln(y^2 + 2xy + 4x^2)$;

(g) $y = e^{\sin(x+y)}$.

3. Find the second-order derivatives of the following functions respectively.

(a) $(xy - 2)^3 - x^2 = 0$; (b) $y + x = e^{x^2 + 2y}$; (c) $2x^2 + 3y^2 = 5$.

Chapter 10: Exercises Answers

Exercises 10.1

1. a) 1, b) 5, c) 1/5, d) −1/2, e) 4, f) 3/10, g) −2
2. a) 1, b) 0, c) −1/2, d) not exist, e) not exist, f) 0, g) 2, h) 3/2, i) 8
4. a) 1, b) 1/6, c) 1/3, d) e^{-3}, e) e^{4}, f) $e^{-2/3}$
5. a) not exist, b) not exist, c) 0
6. sec*x* is continuous in [0, π/2); csc*x* is continuous in (0, π/2].

Exercises 10.2

1. $\frac{dy}{dx}=\frac{1}{\sqrt{2x+3}}$

2. a) $y'=4x^3-6x^2-2x+5$, b) $y'=2x-2$, c) $y'=\frac{4}{3\sqrt[3]{x}}$, d) $y'=-\frac{9}{2\sqrt{x^5}}$, e) $y'=2e^{2x}$,

   f) $y'=-3e^{-3x}$, g) $y'=\frac{2}{3x}$, h) $y'=\frac{-2}{x\ln 10}$, i) $y'=\frac{1}{2x\ln 2}$,

   j) $y'=3\cos 3x-4\sin 2x+3\sin 3x$, k) $y'=\frac{1}{\cos^2 x}+2\cos 2x$, l) $y'=-6e^{-2x}-6e^{3x}$,

   m) $y'=\frac{1}{2x}$, n) $y'=-2-\frac{1}{x}$

3. a) $y'=3x^2e^{2x}+2x^3e^{2x}$, b) $y'=2x\ln x+x$, c) $y'=\frac{4\sin x}{3\sqrt[3]{x}}+2x^{\frac{2}{3}}\cos x$,

   d) $y'=e^{-3x}(\frac{1}{3x}-\ln x)$, e) $y'=2e^{2x}(\cos x-\sin 2x)$, f) $y'=e^{-3x}(\frac{1}{3}\cos\frac{x}{3}-3\sin\frac{x}{3})$,

   g) $y'=\frac{1-\ln x}{x^2}$, h) $y'=\frac{x^2(2\ln x-1)+e^{2x}(1-2x\ln x)}{x(\ln x)^2}$, i) $y'=\frac{1-4x}{e^{2x}\sqrt{x}}$,

   j) $y'=3\cos 3x\cos 2x-2\sin 3x\sin 2x$, k) $y'=\frac{1-\tan^2 x+4\sin^2 x}{\cos^2 2x}$

   l) $y'=\frac{-6e^{-2x}(1+x)+2e^{3x}(2-3x)}{3x^3}$, m) $y'=\frac{2-\ln x}{2\sqrt{2x^3}}$, n) $y'=2(\ln x+1)$

4. a) 5, b) −1, c) $-3e^{-3}$, d) 1, e) $-6e^{-2}$, f) $-\frac{3\sqrt{2}}{2}-4$, g) $\frac{4+\sqrt{2}}{2}$

Exercises 10.3

1. a) $y'=6x^2e^x\ln x+2x^3e^x\ln x+2x^2e^x$; b) $y'=e^{-x}(\frac{\ln x+2-2x\ln x}{\sqrt{x}})$,

   c) $y'=e^{-x}(\ln x\sin 2x-x\ln x\sin 2x+\sin 2x+2x\ln x\cos 2x)$, d) $y'=2\sin 2x$,

   e) $y'=-6e^{-2x}(\cos 2x+\sin 2x)$, f) $y'=\frac{\sin^2 x}{x}+\sin 2x\ln x$,

   g) $y'=-2x\tan x\ln x-x^2\sec^2 x\ln x-x\tan x$

2. a) $y'=3(x-2)^2$, b) $y'=\frac{1}{2x+5}$, c) $y'=\frac{2}{\sqrt[3]{x+2}}$, d) $y'=-2(x-1)e^{-x^2+2x}$,

e) $y' = \dfrac{e^{\sqrt{2x+3}}}{\sqrt{2x+3}}$, f) $y' = 3\cos(3x-2)$, g) $y' = -(2x-2)\sin(x^2-2x+1)$,

h) $y' = -3\sin(3x+2) - 4\sin 2(x-1)$, i) $y' = \dfrac{2x+2}{x^2+2x+5}$, j) $y' = \cos(x+2)e^{\sin(x+2)}$

3. a) $y' = x^2 e^{2x}\cos^2 x(3+2x+2x\tan x)$, b) $y' = \dfrac{-4(2x^2-x+6)}{e^{2x}\sqrt{x^2+3}}$,

c) $y' = \dfrac{3(2x+3)(2x^2+6x+1)e^{x^2+3x}}{\sqrt{x^2+3x}}$, d) $y' = (x+2)[\ln(x+3)+1]$,

e) $y' = 18\cos(3x-1)\cos(x^2+2) - 12x\sin(3x-1)\sin(x^2+2)$,

f) $y' = 4x^{-3}e^{\frac{1}{4}x}(x\cos 2x - 8\cos 2x - 8x\sin 2x)$

Exercises 10.4

1. a) $y' = 2e^{2x}, y'' = 4e^{2x}, y''' = 8e^{2x}, y^{(4)} = 16e^{2x}$

b) $y' = -12x^3 - 6x^2 - 6x,\ y'' = -36x^2 - 12x - 6,\ y''' = -72x - 12,\ y^{(4)} = -72$

c) $y' = \dfrac{1}{x},\ y'' = -\dfrac{1}{x^2},\ y''' = \dfrac{2}{x^3},\ y^{(4)} = \dfrac{-6}{x^4}$;

d) $y' = 2\cos 2x,\ y'' = -4\sin 2x,\ y''' = -8\cos 2x,\ y^{(4)} = 16\sin 2x$

e) $y' = -2\sin(2x-1),\ y'' = -4\cos(2x-1),\ y''' = 8\sin(2x-1),\ y^{(4)} = 16\cos(2x-1)$

f) $y' = 2xe^{x^2},\ y'' = 2(2x^2+1)e^{x^2},\ y''' = 4x(2x^2+3)e^{x^2},\ y^{(4)} = 4(4x^4+12x^2+3)e^{x^2}$

Exercises 10.5

1. (a) $\dfrac{dy}{dx} = t$ (b) $\dfrac{dy}{dx} = -\dfrac{3}{2t}\sqrt{\dfrac{t^2+4}{(t+2)^3}}$ (c) $\dfrac{dy}{dx} = -\cot t$ (d) $\dfrac{dy}{dx} = -\dfrac{3}{4}\tan 2t$

(e) $\dfrac{dy}{dx} = -e^{-2t}$ (f) $\dfrac{dy}{dx} = 2t^2+1$

2. (a) $y' = \dfrac{2}{3(xy-2)^2} - \dfrac{y}{x}$ (b) $y' = \dfrac{1-2x^2-2xy}{2x+2y-1}$ (c) $y' = \dfrac{2(x+y)}{\sqrt{2x+3}} - 1$

(d) $y' = \dfrac{-4x}{6y} = -\dfrac{2x}{3y}$ (e) $y' = \dfrac{4\cos(2x-y)+\sin(3y+x)}{2\cos(2x-y)-3\sin(3y+x)}$

(f) $y' = \dfrac{8x+2y}{4x^2+y^2+2xy-2x-2y}$ (g) $y' = \dfrac{y\cos(x+y)}{1-y\cos(x+y)}$

3. (a) $y' = \dfrac{-4-xy}{3x^2},\quad y'' = \dfrac{28}{9x^3} + \dfrac{4y}{9x^2} = \dfrac{4(7+xy)}{9x^3}$ (b) $\begin{cases} y'' = \dfrac{-4x-2y-(2x+2+2y')y'}{2x+2y-1} \\ y' = \dfrac{1-2x^2-2xy}{2x+2y-1} \end{cases}$

(c) $y' = \dfrac{-4x}{6y} = -\dfrac{2x}{3y},\quad y'' = -\dfrac{2(2x^2+3y^2)}{9y^3}$

CHAPTER 11

11 Applications of Derivatives

CHAPTER OBJECTIVES

- Determine the tangent and normal lines at a point of a function using derivatives
- Determine the limits of indeterminate forms by L'Hospital's rule
- Explain the critical points and extreme values of functions
- Find critical points and determine their natures using derivatives
- Solve rates and optimization problems in science and engineering by derivatives.

Essential statements on applications of derivatives:

- Derivatives are useful in determining the tangent and normal lines at a given point in a function.
- L'Hospital's rule is to determine the limit of some indeterminate forms.
- Derivatives are widely used to deal with optimization problems in science and engineering.

Key topics:

- Tangent and normal lines of a function at a point
- L'Hospital's rule
- Critical points and extreme values of functions
- The first-order and second-order derivative tests
- Rate of changes by derivatives
- Optimization using derivatives

Flowchart of mathematical knowledge building

Past	Mathematics in • Years 7–11 in secondary schools • Reviews: Chapters 1-3 • Functions: Chapters 4-7 • Derivatives: Chapter 10
	↓
Current	**Applications of derivatives (Chapter 11)** • **Tangent and normal lines** • **L'Hospital's rule** • **Critical points and extreme values** • **Applications in science and engineering**
	↓
Next	Approximation by derivatives (Chapter 12) • Solve nonlinear equations by Newton's method • Taylor polynomials • Taylor and Maclaurin series
	↓
Future	• Differentials and applications (Chapter 13) • Integration (Chapters 14-17) • Systems of linear equations (Chapter 18)

Chapter: 11 Applications of Derivatives

11.1 Tangent and Normal Lines

The tangent line at a point

Known from the geometric meaning, the derivative of function $f(x)$ at point $P(x_1, y_1)$ is the slope or gradient of the tangent line crossing the same point $P(x_1, y_1)$, i.e., $k_t = y' = f'(x_1)$. By the point-slope form of linear functions, the tangent line at point $P(x_1, y_1)$ on $y = f(x)$ is

$$\boxed{y - y_1 = f'(x_1)(x - x_1) = y'_{x=x_1}(x - x_1)}. \tag{11.1}$$

The normal line at a point

Since the slope of the normal line that crosses the same point $P(x_1, y_1)$ on $y = f(x)$ is the negative reciprocal of the slope of the tangent line, i.e., $k_n k_t = -1$ or $k_n = -1/k_t$, by the point-slope form of linear functions, the normal line at point $P(x_1, y_1)$ on $y = f(x)$ is

$$\boxed{y - y_1 = -\frac{1}{f'(x_1)}(x - x_1) = -\frac{1}{y'_{x=x_1}}(x - x_1)}. \tag{11.2}$$

Example 11.1: Given $y = x^3 - 3x^2 + 6x - 5$, determine the tangent and normal functions that cross point (1, –1).

<u>**Solution**</u>

For the tangent line, its slope is

$$y'_{x=1} = (x^3 - 3x^2 + 6x - 5)' = 3x^2 - 6x + 6 = 3 \times 1^2 - 6 \times 1 + 6 = 3.$$

By formula (11.1), the tangent function is

$$y - y_1 = y'_{x=1}(x - x_1) \longrightarrow y - (-1) = 3(x - 1) \longrightarrow y + 1 = 3x - 3 \longrightarrow y = 3x - 4.$$

By formula (11.2), the normal function is

$$y - y_1 = -\frac{1}{y'_{x=1}}(x - x_1) \longrightarrow y - (-1) = -\frac{1}{3}(x - 1) \longrightarrow y + 1 = -\frac{1}{3}x + \frac{1}{3}$$

$$y = -\frac{1}{3}x + \frac{1}{3} - 1 = -\frac{1}{3}x + \frac{1}{3} - \frac{3}{3} = -\frac{1}{3}x - \frac{2}{3} = -\frac{1}{3}(x + 2).$$

Example 11.2: Given $y = 2\sin x - 3x + 2$, determine the tangent and normal functions that cross point (0, 2).

Solution

For the tangent line, its slope is

$$y'_{x=0} = (2\sin x - 3x + 2)' = 2(\sin x)' - (3x)' + (2)' = 2\cos x - 3 + 0 = 2\cos 0 - 3 = 2 - 3 = -1.$$

By formula (11.1), the tangent function is

$$y - y_1 = y'_{x=0}(x - x_1) \longrightarrow y - 2 = -1(x - 0) \longrightarrow y - 2 = -x \longrightarrow y = -x + 2.$$

By formula (11.2), the normal function is

$$y - y_1 = -\frac{1}{y'_{x=0}}(x - x_1) \longrightarrow y - 2 = -\frac{1}{-1}(x - 0) \longrightarrow y - 2 = x \longrightarrow y = x + 2.$$

Example 11.3: A section of curved railway is formed as $y = 0.05x^2 - 10$. Find the tangent and normal equations at point (20, 10).

Solution

The slope of the tangent line of $y = 0.05x^2 - 10$ is

$$y' = (0.05x^2 - 10)' = 2 \times 0.05x = 0.1x.$$

At point (20, 10), the slope is $k = y'(20) = 0.1 \times 20 = 2.$ Thus, the tangent line is

$$y - 10 = 2(x - 20) \longrightarrow y = 2x - 30.$$

At point (20, 10), the normal line is

$$y - 10 = -\frac{1}{2}(x - 20) \longrightarrow y = -\frac{1}{2}x + 20.$$

Example 11.4: Given $x = 2\sin t$ and $y = 2\sqrt{3}\cos t$ defined in quadrant I, determine the tangent and normal functions that cross point (1, 3).

Solution

Since this parametric function is defined in quadrant I, we can find the angle t at the given point from $x = 1$ by $x = 2\sin t$ as follows

$$2\sin t = 1 \longrightarrow \sin t = \frac{1}{2} \longrightarrow t = \arcsin\frac{1}{2} = \frac{\pi}{6}.$$

$$x' = (2\sin t)' = 2\cos t \text{ and } y' = (2\sqrt{3}\cos t)' = -2\sqrt{3}\sin t.$$

For the tangent line, its slope at $t = \frac{\pi}{6}$ is

$$y'_{t=1} = \frac{y'}{x'} = \frac{-2\sqrt{3}\sin t}{2\cos t} = -\sqrt{3}\tan t = -\sqrt{3}\tan\frac{\pi}{6} = -\sqrt{3}\times\frac{\sqrt{3}}{3} = -1.$$

By formula (11.1), the tangent function is

$$y - y_1 = y'_{t=1}(x - x_1) \longrightarrow y - 3 = -1(x-1) \longrightarrow y - 3 = -x + 1 \longrightarrow y = -x + 4.$$

By formula (11.2), the normal function is

$$y - y_1 = -\frac{1}{y'_{t=1}}(x - x_1) \longrightarrow y - 3 = -\frac{1}{-1}(x-1) \longrightarrow y - 3 = x - 1 \longrightarrow y = x + 2.$$

Example 11.5: Given $5y^2 - 3xy + 2x^2 = 4$, determine the tangent and normal functions that cross point (–1, –1).

Solution

This is an implicit function. We can find y' by applying differentiation to both sides of the equation:

$$(5y^2 - 3xy + 2x^2)' = (4)' \longrightarrow 10yy' - 3(x'y + xy') + 4x = 0 \longrightarrow 10yy' - 3y - 3xy' + 4x = 0$$

$$(10y - 3x)y' = 3y - 4x \longrightarrow y' = \frac{3y - 4x}{10y - 3x}.$$

For the tangent line at point (–1, –1), its slope is

$$y'_{\substack{x=-1\\y=-1}} = \frac{3y-4x}{10y-3x} = \frac{3\times(-1) - 4\times(-1)}{10\times(-1) - 3\times(-1)} = \frac{-3+4}{-10+3} = -\frac{1}{7}.$$

By formula (11.1), the tangent function is

$$y - y_1 = y'_{\substack{x=-1\\y=-1}}(x - x_1)$$

$$y-(-1) = -\frac{1}{7}[x-(-1)] \longrightarrow y+1 = -\frac{1}{7}(x+1) \longrightarrow y = -\frac{1}{7}x - \frac{1}{7} - 1 = -\frac{1}{7}(x+8).$$

By formula (11.2), the normal function is

$$y - y_1 = -\frac{1}{y'_{\substack{x=-1\\y=-1}}}(x - x_1)$$

$$y-(-1) = 7[x-(-1)] \longrightarrow y+1 = 7(x+1) \longrightarrow y = 7x+7-1 = 7x+6.$$

Exercises 11.1

1. Given $y = 3x^2 + 2x - 6$, determine the tangent and normal functions that cross point (–1, –5).

2. Given $y = \ln x + 6x - 5$, determine the tangent and normal functions that cross point (1, 1).

3. Given $y = \sin 2x - \cos x$, determine the tangent and normal functions that cross point (π/6, 0).

4. Given $y = 2e^x - 3e^{-x}$, determine the tangent and normal functions that cross point (0, –1).

5. Given $y = \sqrt{5 + x^2}$, determine the tangent and normal functions that cross point (2, 3).

6. Given $x = 4\cos t$ and $y = 4\sqrt{3}\sin t$ defined in quadrant I, determine the tangent and normal functions that cross point (2, 6).

7. Given $x^3 + y^2 - xy = x^2$, determine the tangent and normal functions that cross point (1, 1).

8. Given $ay^2 + bxy = 4 - x^2$ crossing point (3, 1) with $y'(3,1) = 3$, determine the constants *a* and *b* in this implicit equation.

11.2 Limits of Indeterminate Forms (L'Hospital's Rule)

Indeterminate forms refer to expressions that approach $\frac{0}{0}$ or $\frac{\infty}{\infty}$. These indeterminate forms do not guarantee the existence of a limit. Suppose $f(x)$ and $g(x)$ are differentiable and $g'(c) \neq 0$ at $x = c$ (or ∞) in $[a, b]$ or (a, ∞). If $\lim_{x \to c} f(x) = 0$ AND $\lim_{x \to c} g(x) = 0$ or $\lim_{x \to c} f(x) = \infty$ AND $\lim_{x \to c} g(x) = \infty$, then the limit is (if it exists),

$$\boxed{\lim_{x \to c} \frac{f(x)}{g(x)} = \frac{\lim_{x \to c} f(x)}{\lim_{x \to c} g(x)} = \frac{\lim_{x \to c} f'(x)}{\lim_{x \to c} g'(x)}. \longleftarrow \text{L'Hospital's rule}} \qquad (11.3)$$

This is known as the L'Hospital's rule. Other indeterminate forms may be converted to $\frac{0}{0}$ or $\frac{\infty}{\infty}$, for example,

$$y = 0 \bullet \infty \text{ or } \infty \bullet 0 \longrightarrow \frac{0}{\frac{1}{\infty}} \underset{\leftrightarrow}{\rightleftarrows} \frac{0}{0} \text{ or } 0 \bullet \infty \longrightarrow \frac{\infty}{\frac{1}{0}} \underset{\leftrightarrow}{\rightleftarrows} \frac{\infty}{\infty}$$

$$y = 1^{\infty} \text{ or } \infty^{0} \longrightarrow \ln y = \infty \bullet \ln 1 \underset{\leftrightarrow}{\rightleftarrows} \infty \bullet 0 \text{ or } \ln y = 0 \bullet \ln \infty \underset{\leftrightarrow}{\rightleftarrows} 0 \bullet \infty$$

$$y = \infty - \infty = \frac{a}{0} - \frac{b}{0} \longrightarrow \frac{0 \bullet a - 0 \bullet b}{0 \bullet 0} \underset{\leftrightarrow}{\rightleftarrows} \frac{0}{0}.$$

Example 11.6: Verify $\lim_{x \to 0} \frac{\sin x}{x} = 1$ using rule (11.3).

Solution

Since $\lim_{x \to 0} \frac{\sin x}{x}$ is the type of $\frac{0}{0}$, rule (11.3) can be applied to

$$\lim_{x \to 0} \frac{\sin x}{x} = \frac{\lim_{x \to 0} \sin x}{\lim_{x \to 0} x} = \frac{\lim_{x \to 0} (\sin x)'}{\lim_{x \to 0} (x)'} = \frac{\lim_{x \to 0} \cos x}{\lim_{x \to 0} 1} = \frac{1}{1} = 1.$$

Example 11.7: Verify $\lim_{x \to 0} (1 + x)^{\frac{1}{x}} = e$ using rule (11.3).

Solution

This is the type of 1^{∞}. Let $F(x) = (1 + x)^{\frac{1}{x}}$. Then $\ln F(x) = \ln[(1 + x)^{\frac{1}{x}}] = \frac{1}{x} \ln(1 + x) = \frac{\ln(1 + x)}{x}$.

$\lim_{x \to 0} \ln F(x) = \lim_{x \to 0} \frac{\ln(1 + x)}{x}$ is the type of $\frac{0}{0}$ so rule (11.13) can be applied to:

$$\lim_{x\to 0}\ln F(x)=\lim_{x\to 0}\frac{\ln(1+x)}{x}=\frac{\lim\limits_{x\to 0}\ln(1+x)}{\lim\limits_{x\to 0}x}=\frac{\lim\limits_{x\to 0}[(\ln(1+x)]'}{\lim\limits_{x\to 0}(x)'}=\frac{\lim\limits_{x\to 0}\frac{(1+x)'}{1+x}}{\lim\limits_{x\to 0}1}=\lim_{x\to 0}\frac{1}{1+x}=1=\ln e.$$

Thus $\lim\limits_{x\to 0}\ln F(x)=\ln e$ or $\ln\lim\limits_{x\to 0}F(x)=\ln e$. This means

$$\lim_{x\to 0}F(x)=e\longrightarrow\lim_{x\to 0}(1+x)^{\frac{1}{x}}=e.$$

Example 11.8: Evaluate $\lim\limits_{x\to\infty}e^{-x}x^2$.

Solution

This is the type of $0\bullet\infty$. Since $e^{-x}x^2=\frac{x^2}{e^x}$, $\lim\limits_{x\to\infty}e^{-x}x^2=\lim\limits_{x\to\infty}\frac{x^2}{e^x}\to\frac{\infty}{\infty}$. Thus rule (11.3) can be applied to:

$$\lim_{x\to\infty}e^{-x}x^2=\lim_{x\to\infty}\frac{x^2}{e^x}=\frac{\lim\limits_{x\to\infty}x^2}{\lim\limits_{x\to\infty}e^x}=\frac{\lim\limits_{x\to\infty}(x^2)'}{\lim\limits_{x\to\infty}(e^x)'}=\frac{\lim\limits_{x\to\infty}(2x)}{\lim\limits_{x\to\infty}e^x}\qquad\longleftarrow\frac{\infty}{\infty}\text{ again}$$

$$=\frac{\lim\limits_{x\to\infty}(2x)'}{\lim\limits_{x\to\infty}(e^x)'}=\frac{\lim\limits_{x\to\infty}2}{\lim\limits_{x\to\infty}e^x}=\frac{2}{\infty}=0.\qquad\longleftarrow\text{apply L'Hospital's rule again}$$

Example 11.9: Evaluate $\lim\limits_{x\to 0}(\frac{\cos x}{x}-\frac{1}{\sin x})$.

Solution

This is the type of $\infty-\infty$.

$$\lim_{x\to 0}(\frac{\cos x}{x}-\frac{1}{\sin x})=\lim_{x\to 0}(\frac{\cos x\sin x}{x\sin x}-\frac{x}{x\sin x})=\lim_{x\to 0}\frac{\cos x\sin x-x}{x\sin x}=\lim_{x\to 0}\frac{\frac{1}{2}\sin 2x-x}{x\sin x}=\lim_{x\to 0}\frac{\sin 2x-2x}{2x\sin x}$$

$$\therefore\lim_{x\to 0}(\frac{\cos x}{x}-\frac{1}{\sin x})=\lim_{x\to 0}\frac{\sin 2x-2x}{2x\sin x}\xrightarrow{0/0}\lim_{x\to 0}\frac{(\sin 2x-2x)'}{(2x\sin x)'}=\lim_{x\to 0}\frac{2\cos 2x-2}{2(\sin x+x\cos x)}$$

$$=\lim_{x\to 0}\frac{\cos 2x-1}{\sin x+x\cos x}\xrightarrow{0/0}\lim_{x\to 0}\frac{(\cos 2x-1)'}{(\sin x+x\cos x)'}=\lim_{x\to 0}\frac{-2\sin 2x}{\cos x+\cos x-x\sin x}$$

$$=\lim_{x\to 0}\frac{-2\sin 2x}{2\cos x-x\sin x}=\frac{-2\sin 0}{2\cos 0-0\sin 0}=\frac{0}{2-0}=0.$$

Example 11.10: Evaluate $\lim_{x \to 0^+} (\cot x)^x$.

Solution

This is the type of ∞^0.

$$(\cot x)^x \longrightarrow \ln y = \ln(\cot x)^x = x \ln(\frac{\cos x}{\sin x}) = x \ln \cos x - x \ln \sin x,$$

$$\lim_{x \to 0^+} \ln y = \lim_{x \to 0^+} (x \ln \cos x - x \ln \sin x) = 0 \times \ln 1 - \lim_{x \to 0^+} (x \ln \sin x) = \lim_{x \to 0^+} (\frac{-\ln \sin x}{\frac{1}{x}}) \to \frac{\infty}{\infty}$$

$$\lim_{x \to 0^+} \ln y = \lim_{x \to 0^+} (\frac{-\ln \sin x}{\frac{1}{x}}) = \lim_{x \to 0^+} \frac{(-\ln \sin x)'}{(\frac{1}{x})'} = \lim_{x \to 0^+} \frac{-\frac{\cos x}{\sin x}}{-\frac{1}{x^2}} = \lim_{x \to 0^+} \frac{x^2 \cos x}{\sin x}$$

$$= \lim_{x \to 0^+} \frac{x}{\sin x} \bullet \lim_{x \to 0^+} x \cos x = 1 \bullet 0 \bullet 1 = 0$$

$$\therefore \lim_{x \to 0^+} y = \lim_{x \to 0^+} (\cot x)^x = e^{\ln y} = e^0 = 1.$$

Exercises 11.2

1. Evaluate

a) $\lim_{x \to \infty} \frac{x^2 - 2x + 9}{x \ln x}$ b) $\lim_{x \to 1} \frac{2x^2 - 3x + 1}{x \ln x}$ c) $\lim_{x \to 0} \frac{2x^2 - x + \cos x - 1}{3x}$ d) $\lim_{x \to \infty} \frac{2x}{e^x}$

e) $\lim_{x \to 0} \frac{\tan x - x}{x - \sin x}$ f) $\lim_{x \to 0^+} \frac{\ln \tan 5x}{\ln \tan 3x}$ g) $\lim_{x \to 2} \frac{\sin x - \sin 2}{x - 2}$ h) $\lim_{x \to 0} \frac{e^x - e^{-x}}{\sin x}$

2. Evaluate

a) $\lim_{x \to \infty} e^{-2x} x^3$ b) $\lim_{x \to 1} (\frac{x}{x-1} - \frac{1}{\ln x})$ c) $\lim_{x \to 0} (\frac{1}{x} - \frac{1}{e^x - 1})$ d) $\lim_{x \to 0^+} x^{\sin x}$

11.3 Critical Points and Extreme Values of Functions

11.3.1 Concepts of critical points and extreme values

Interior critical points and extreme values of a continuous function

Let $f(x)$ be a function defined in range $[a, b]$. Any point x within $[a, b]$ that makes $f(x)$ a local extreme value (either maximum or minimum) in its vicinity is called an interior critical point of function $f(x)$. For example, points at c, d, e, f and g in Figure 11.1 are all critical points of $f(x)$ since $f(c)$, $f(e)$ and $f(g)$ are local maximum values in the vicinities around these points; $f(d)$ and $f(f)$ are local minimum values in the vicinities around these points. However, point $x = h$ is not a critical point as $f(h)$ is not an extreme value around point h.

Boundary critical points and extreme values of a continuous function

Let $f(x)$ be a function defined in range $[a, b]$. If $f(a)$ or $f(b)$ is the largest (or smallest) value in the whole range, it is a boundary critical point of function $f(x)$ in range $[a, b]$. In Figure 11.1, $f(a)$ is the smallest value (or the global minimum) of $f(x)$ in the whole range so point a is a boundary critical point of $f(x)$. However, point b is not a boundary critical point as $f(b)$ is neither the smallest value nor the largest value (or the global maximum) of $f(x)$ in the whole range.

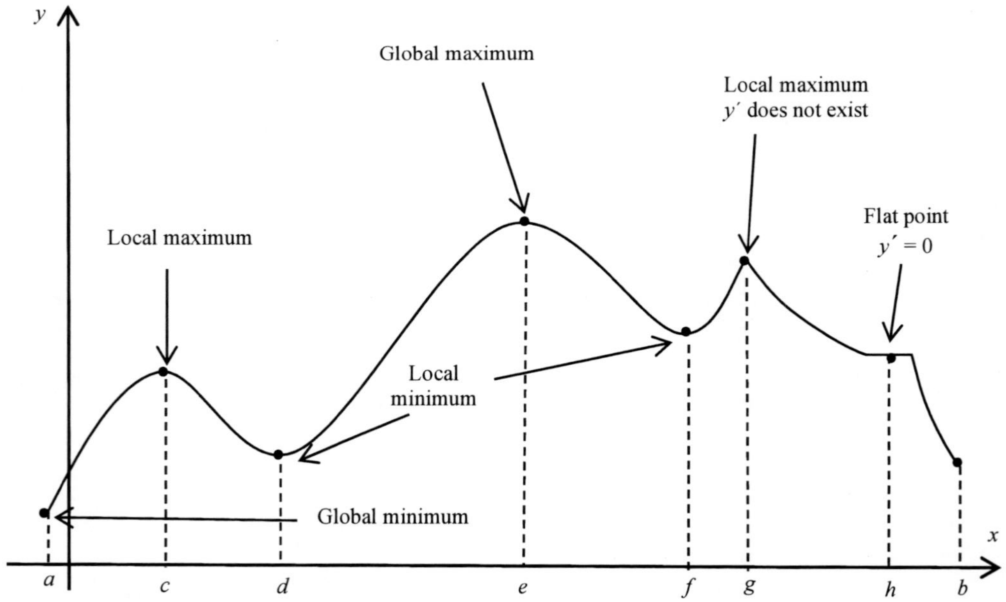

Figure 11.1: Critical points on the curve $y = f(x)$ in range $[a, b]$

The global and local extreme values of a continuous function

- An interior critical point correlates to either a local extreme value or a global extreme value;
- A boundary point can only be either a point of global extreme value or an ordinary point;
- The global extreme values (the global maximum and global minimum) are among these extreme values at all interior critical points and the two values at the two endpoints of range $[a, b]$.

If not specifically stated for the global extreme values, determining the extreme values of a function $f(x)$ in range $[a, b]$ means to find the **interior critical points**, and determine the corresponding extreme value at each of these points and its nature/status of either a maximum or minimum point/value in $[a, b]$.

Determining the interior critical points and extreme values of continuous functions

An interior critical point c for function $f(x)$ defined in $[a, b]$ must meet one of these two conditions:

$$\boxed{f'(c)=0 \text{ or } f'(c) \text{ is not defined at } x=c.} \qquad (11.4)$$

Note a point at $f'(c)=0$ is not necessarily an interior critical point, like $f(h)$ in Figure 11.1.

Example 11.11: Find the interior critical point(s) for $y=f(x)=x^2-4x+3$ in $[-3, 3]$ and determine its nature.

Solution

This function is continuous and smooth in the given range. The interior critical points (if any) must meet condition (11.4).

$$y'=f'(x)=(x^2-4x+3)'=2x-4=0 \longrightarrow 2x=4 \longrightarrow x=2.$$

The extreme value at $x = 2$ is $y=f(2)=2^2-4\times2+3=4-8+3=-1$.

This extreme value must be either greater or smaller than any value on both sides of $x = 2$. Therefore we choose $x_1 = 1$ (on the left of 2) and $x_2 = 3$ (on the right of 2) to test its status.

$$y_1=f(1)=1^2-4\times1+3=1-4+3=0>f(2)=-1;$$
$$y_2=f(3)=3^2-4\times3+3=9-12+3=0>f(2)=-1.$$

Both are great than $f(2)$; thus there is a minimum point at $x = 2$ and its value is $y_{\min}=f(2)=-1$.

Example 11.12: Find the interior critical points for $y = |\sin x|$ in $[-\pi, \pi]$ and plot these critical points in a graph in the curve. By observation, define the nature of each of these critical points.

Solution

The given function $y = |\sin x|$ can be expressed as

$$y = |\sin x| = \begin{cases} -\sin x & -\pi \le x < 0 \\ \sin x & 0 \le x \le \pi \end{cases}.$$

In range $[-\pi, 0)$, $y' = (-\sin x)' = -\cos x = 0 \longrightarrow x = -\frac{\pi}{2}$.

In range $(0, \pi]$, $y' = (\sin x)' = \cos x = 0 \longrightarrow x = \frac{\pi}{2}$.

Since $\sin x$ is continuous and smooth in these sub-ranges, both $x = -\frac{\pi}{2}$ and $x = \frac{\pi}{2}$ must be the interior critical points. At $x = 0$,

$$y'(0^-) = \lim_{x \to 0^-}(-\cos x) = -1 \text{ and } y'(0^+) = \lim_{x \to 0^+} \cos x = 1.$$

As both the left-hand and right-hand limits are different, the derivative at point $x = 0$ does not exist. According to condition (11.4), there is another interior critical point at point $x = 0$.

In Figure 11.2, it is clearly seen that the maximum values are at both $x = -\frac{\pi}{2}$ and $x = \frac{\pi}{2}$ whereas a minimum value is at $x = 0$.

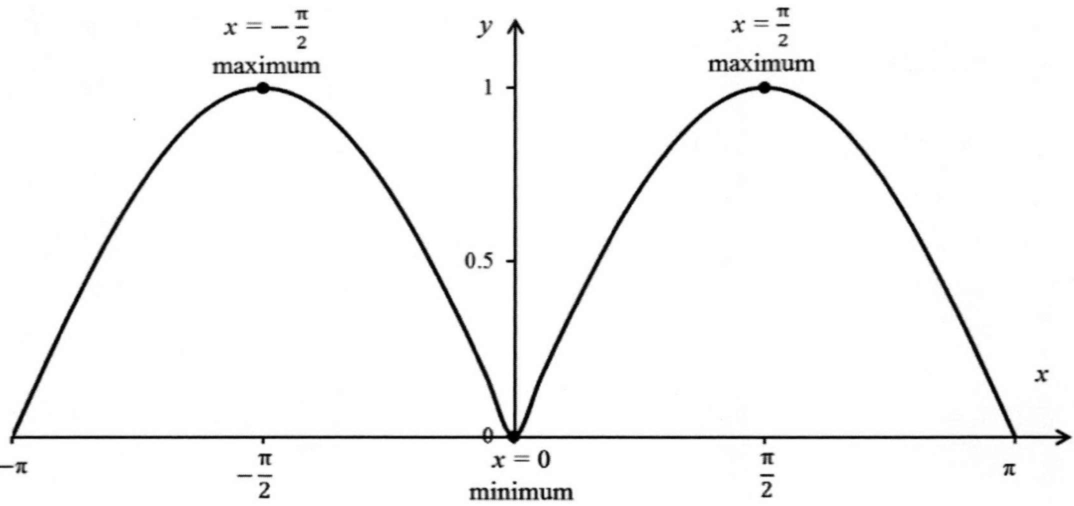

Figure 11.2: Critical points on the curve of $y = |\sin x|$ in $[-\pi, \pi]$

11.3.2 Determining the nature of extreme values

The interior critical points can be found by using condition (11.4). However, determining the nature of these critical points by the natural test used in Example 11.11 is rather tedious. Fortunately there are two more effective ways to make this task easier.

The first derivative test (FDT)

At an interior critical point c for continuous function $f(x)$,

1) Moving x (near c) from the left of c to the right of c, if $f'(x)$ changes sign from positive (+) to negative (–), then $f(c)$ is a local maximum value at $x = c$; such as the case of $g(-2)$ in Figure 11.3.
2) Moving x (near c) from the left of c to the right of c, if $f'(x)$ changes sign from negative (–) to positive (+), then $f(c)$ is a local minimum value at $x = c$; such as the case of $f(2)$ in Figure 11.3.
3) Moving x (near c) from the left of c to the right of c, if $f'(x)$ does not change sign, then there is no extreme value at $x = c$.

Note the first derivative test (FDT) applies to both cases in condition (11.4).

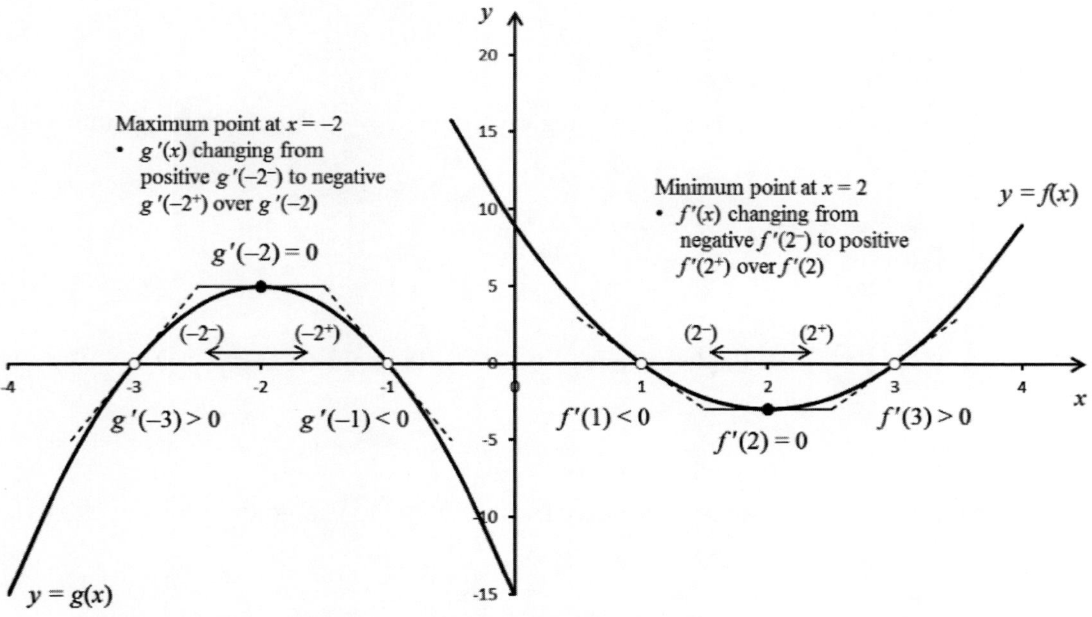

Figure 11.3: Essential points about the first derivative test (FDT)

Example 11.13: Use the first derivative test to determine the nature of each of those critical points found in Example 11.12 for $y = |\sin x|$.

Solution

The following three interior critical points were found for $y = |\sin x|$ in Example 11.12:

- Point at $x = 0$ at which y' does not exist;
- Points at $x = -\dfrac{\pi}{2}$ and $x = \dfrac{\pi}{2}$ where $y' = 0$.

In $[-\pi, 0)$, $y' = -\cos x$. For $x = -\dfrac{\pi}{2}$, $x_L = -\dfrac{\pi}{2} - \dfrac{\pi}{6} = -\dfrac{2\pi}{3}$ is on the left and $x_R = -\dfrac{\pi}{2} + \dfrac{\pi}{6} = -\dfrac{\pi}{3}$ is on the right.

$$y'(x_L) = -\cos(-\frac{2\pi}{3}) = -\cos\frac{2\pi}{3} = -(-\frac{1}{2}) = \frac{1}{2};\ \ y'(x_R) = -\cos(-\frac{\pi}{3}) = -\cos\frac{\pi}{3} = -\frac{1}{2}.$$

As $y'(x)$ changes sign from positive to negative over $x = -\dfrac{\pi}{2}$, there is a maximum value at $x = -\dfrac{\pi}{2}$,

$$y_{\max} = \left|\sin(-\frac{\pi}{2})\right| = -\sin(-\frac{\pi}{2}) = \sin\frac{\pi}{2} = 1.$$

In $(0, \pi]$, $y' = \cos x$. For $x = \dfrac{\pi}{2}$, $x_L = \dfrac{\pi}{2} - \dfrac{\pi}{6} = \dfrac{\pi}{3}$ is on the left and $x_R = \dfrac{\pi}{2} + \dfrac{\pi}{6} = \dfrac{2\pi}{3}$ is on the right.

$$y'(x_L) = \cos\frac{\pi}{3} = \frac{1}{2};\ \ y'(x_R) = \cos\frac{2\pi}{3} = -\frac{1}{2}.$$

As $y'(x)$ also changes sign from positive to negative over $x = \dfrac{\pi}{2}$, there is a maximum value at $x = \dfrac{\pi}{2}$.

$$y_{\max} = \left|\sin\frac{\pi}{2}\right| = \sin\frac{\pi}{2} = 1.$$

For $x = 0$, $x_L = 0 - \dfrac{\pi}{6} = -\dfrac{\pi}{6}$ is on the left and $x_R = 0 + \dfrac{\pi}{6} = \dfrac{\pi}{6}$ is on the right of $x = 0$.

$$y'(x_L) = -\cos(-\frac{\pi}{6}) = -\cos\frac{\pi}{6} = -\frac{\sqrt{3}}{2};\ \ y'(x_R) = \cos\frac{\pi}{6} = \frac{\sqrt{3}}{2}.$$

As $y'(x)$ changes sign from negative to positive over $x = 0$, there is a minimum value at $x = 0$,

$$y_{\min} = |\sin 0| = 0.$$

The second derivative test (SDT)

At an interior critical point c for continuous function $f(x)$,

1) If $f''(c) < 0$, then $f(c)$ is a local maximum value at $x = c$; such as the case of $g(-2)$ in Figure 11.4.
2) If $f''(c) > 0$, then $f(c)$ is a local minimum value at $x = c$; such as the case of $f(2)$ in Figure 11.4.
3) If $f''(c) = 0$ or $f''(c)$ does not exist, then the second derivative test is not applicable.

The second derivative test (SDT) is relatively simpler in use but not applicable if $f''(c) = 0$ or $f''(c)$ does not exist. The first derivative test is relatively more tedious in use but is applicable to all situations.

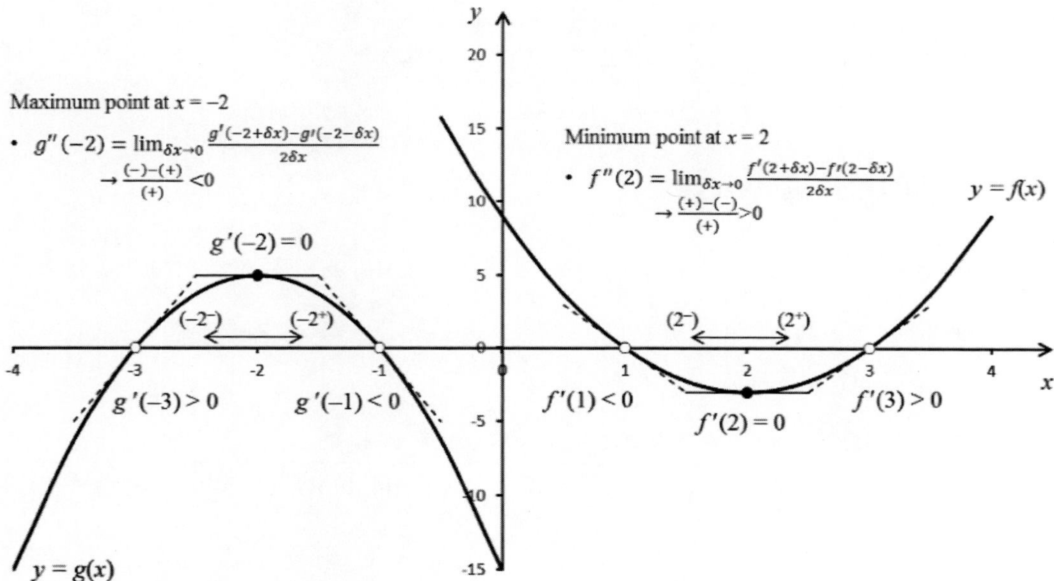

Figure 11.4: Essential points about the second derivative test (SDT)

Example 11.14: Use the second derivative test to determine the nature of each of those critical points found in Example 11.12 for $y = |\sin x|$.

Solution

In the three interior critical points found for $y = |\sin x|$ in Example 11.12, $f''(x)$ does not exist at $x = 0$; thus the second derivative test is not applicable at $x = 0$.

At point $x = -\frac{\pi}{2}$, $y'' = (y')' = (-\cos x)' = \sin x \longrightarrow y''(-\frac{\pi}{2}) = \sin(-\frac{\pi}{2}) = -1 < 0.$

Thus, there is a maximum value at $x=-\frac{\pi}{2}$, i.e., $y_{max}=\left|\sin(-\frac{\pi}{2})\right|=-\sin(-\frac{\pi}{2})=\sin\frac{\pi}{2}=1$.

At point $x=\frac{\pi}{2}$, $y''=(y')'=(\cos x)'=-\sin x \longrightarrow y''(\frac{\pi}{2})=-\sin\frac{\pi}{2}=-1<0$.

Thus, there is a maximum value at $x=\frac{\pi}{2}$, i.e., $y_{max}=\left|\sin\frac{\pi}{2}\right|=\sin\frac{\pi}{2}=1$.

Example 11.15: Use the second derivative test to find the interior extreme value(s) for $y=f(x)=x^2-4x+3$ in [–3, 3] presented in Example 11.11.

Solution

The interior critical points (if any) must meet condition (11.4).

$$y'=f'(x)=(x^2-4x+3)'=2x-4=0 \longrightarrow 2x=4 \longrightarrow x=2$$
$$y''=(y')'=(2x-4)'=2>0 \text{ everywhere in } [-3,3].$$

Thus there is a minimum value at $x = 2$:

$$y_{min}=f(2)=2^2-4\times2+3=4-8+3=-1.$$

Classifying all possible critical points and extreme values

For all interior critical points (c_1, c_2, c_3,...) and the two endpoints in a closed range [a, b] for continuous function $f(x)$,

1) Sort all $f(c_i)$, $f(a)$ and $f(b)$ in ascending order;
2) If $f(a)$ or $f(b)$ is the smallest among all the sorted values, $f(a)$ or $f(b)$ is the global minimum value in [a, b];
3) If $f(a)$ or $f(b)$ is the largest among all the sorted values, $f(a)$ or $f(b)$ is the global maximum value in [a, b];
4) If $f(a)$ or $f(b)$ is in the middle of the sorted values, $f(a)$ or $f(b)$ is not an extreme value in [a, b];
5) All interior extreme values $f(c_i)$ are either the local or global extreme values in [a, b] depending on the nature of $f(a)$ and $f(b)$.

Example 11.16: Find ALL extreme points and values for $y = x^3 - \frac{3}{2}x^2 - 6x + 5$ in [–3, 3].

Solution

The interior critical points must meet condition (11.4). Since $y = x^3 - \frac{3}{2}x^2 - 6x + 5$ is continuous and smooth everywhere in [–3, 3], the critical points are those interior critical points and the two endpoints in the range. Hence the second derivative test is preferred.

$$y' = (x^3 - \frac{3}{2}x^2 - 6x + 5)' = 3x^2 - 3x - 6 = 0 \longrightarrow x^2 - x - 2 = 0$$
$$(x+1)(x-2) = 0 \longrightarrow x = -1,\ x = 2 \text{ in } [-3,3]$$
$$y'' = (y')' = (3x^2 - 3x - 6)' = 6x - 3,$$
$$y''(-1) = 6(-1) - 3 = -6 - 3 = -9 < 0,$$
$$y''(2) = 6 \times 2 - 3 = 12 - 3 = 9 > 0.$$

Thus there is a minimum value at $x = 2$ and a maximum value at $x = -1$ respectively:

$$y_{\min} = y(2) = 2^3 - \frac{3}{2} \times 2^2 - 6 \times 2 + 5 = 8 - 6 - 12 + 5 = -5$$
$$y_{\max} = y(-1) = (-1)^3 - \frac{3}{2} \times (-1)^2 - 6 \times (-1) + 5 = -1 - \frac{3}{2} + 6 + 5 = -2.5 + 11 = 8.5.$$

At the two endpoints $x = -3$ and $x = 3$,

$$y(3) = 3^3 - \frac{3}{2} \times 3^2 - 6 \times 3 + 5 = 27 - \frac{27}{2} - 18 + 5 = \frac{27}{2} - 13 = 13.5 - 13 = 0.5$$
$$y(-3) = (-3)^3 - \frac{3}{2} \times (-3)^2 - 6 \times (-3) + 5 = -27 - \frac{27}{2} + 18 + 5 = -40.5 + 23 = -17.5.$$

As $y(-3) < y(2) < y(3) < y(-1)$, $y(-3) = -17.5$ is the global minimum value; $y(2) = -5$ is a local minimum value; $y(-1) = 8.5$ is the global maximum value for $y = x^3 - \frac{3}{2}x^2 - 6x + 5$ in [–3, 3]. The endpoint $x = 3$ is an ordinary point in [–3, 3].

Example 11.17: Find ALL extreme points and values for $y = 2\sqrt{x} - x$ in [0, 3].

Solution

The interior critical points must meet condition (11.4). Since $y = 2\sqrt{x} - x$ is continuous and smooth everywhere in [0, 3], the critical points are those interior critical points and the two endpoints in the range. Hence the second derivative test is preferred.

$$y' = (2\sqrt{x} - x)' = 2 \times \frac{1}{2} x^{\frac{1}{2}-1} - 1 = x^{-\frac{1}{2}} - 1 = \frac{1}{\sqrt{x}} - 1.$$

$$y'' = (y')' = (x^{-\frac{1}{2}} - 1)' = -\frac{1}{2} x^{-\frac{1}{2}-1} - 0 = -\frac{1}{2} x^{-\frac{3}{2}} = -\frac{1}{2\sqrt{x^3}}.$$

Note $y'(x)$ does not exist at $x = 0$ so $x = 0$ is both an endpoint and a critical point. Let $y'(x) = 0$; then

$$\frac{1}{\sqrt{x}} - 1 = 0 \longrightarrow \frac{1}{\sqrt{x}} = 1 \longrightarrow \sqrt{x} = 1 \xrightarrow{\text{square both sides}} x = 1 \text{ in } [0,3].$$

$$y''(1) = -\frac{1}{2\sqrt{1^3}} = -\frac{1}{2} < 0.$$

Thus there is a maximum value at $x = 1$:

$$y_{\max} = y(1) = 2\sqrt{1} - 1 = 1.$$

At the two endpoints $x = 0$ and $x = 3$,

$$y(0) = 2\sqrt{0} - 0 = 0$$
$$y(3) = 2\sqrt{3} - 3 \approx 3.4641 - 3 = 0.4641.$$

As $y(0) < y(3) < y(1)$, $y(0) = 0$ is the global minimum value; $y(1) = 1$ is the global maximum value in [0, 3]. The endpoint $x = 3$ is an ordinary point in [0, 3]. These points are shown in Figure 11.5.

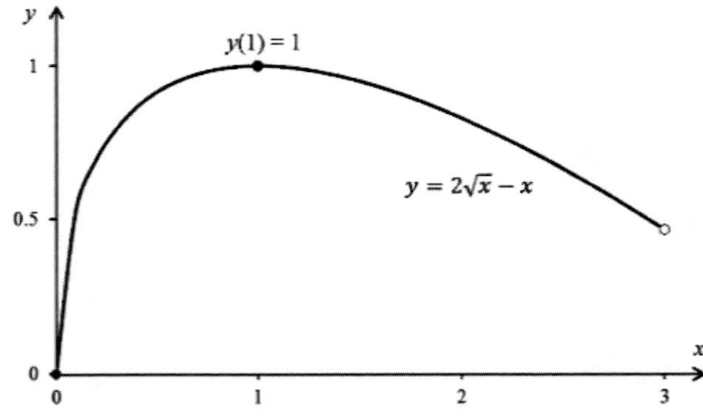

Figure 11.5: The extreme points (solid dots) for $y = 2\sqrt{x} - x$ in [0, 3]

Exercises 11.3

1. Use both the first and second derivative tests to find the interior extreme value(s) for the following functions.

   a) $y = 2x^2 + 4x - 1$
   b) $y = x \ln x$
   c) $y = x + \cos 2x, \quad x \in [0, \pi]$
   d) $y = -2x^3 + 7x^2 - 8x - 2$
   e) $y = 2 - x^{\frac{2}{3}}$
   f) $y = 2x + \sin 2x, \quad x \in [-\pi, \pi]$

2. Find ALL extreme points and values for $y = x^3 - x^2 - 8x - 3$ in [–4, 3].

3. Find ALL extreme points and values for $y = \sin 2x + 2\cos x$ in [0, $\pi/2$].

4. Find ALL extreme points and values for $y = 2\sqrt{x} + \dfrac{1}{x}$ in [1/4, 4].

11.4 Applications of Derivatives in Science and Engineering

This section presents some applications of derivatives mainly in science and engineering. More applications are provided in the exercises of this section.

Example 11.18: In the two-gear system shown below, the angular speed of gear A is known as a constant ω rad/s. The radius of gear A is R and that of gear B is r. First find the angular speed of gear B with respect to that of gear A. Then find the angular speed of gear B assuming the size of gear A is the twice of the size of gear B.

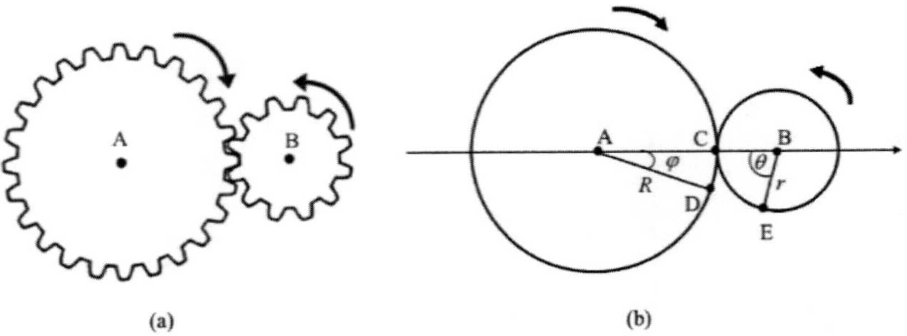

Solution

In the simplified diagram of this two-gear system (b), assume initially the two gears are contacted at point C. After time t, the initial contact moved to point D in gear A and point E in gear B. During this time, the arc lengths moved on both gears must be the same even though their angular speeds are different. This gives the following relationship:

$$\overset{\frown}{CD} = \overset{\frown}{CE} \longrightarrow R\varphi = r\theta.$$

Differentiate both sides with respect to time t, noting R and r are constants and $\dfrac{d\varphi}{dt} = \omega$,

$$\frac{d(R\varphi)}{dt} = \frac{d(r\theta)}{dt} \longrightarrow R\frac{d\varphi}{dt} = r\frac{d\theta}{dt} \longrightarrow r\frac{d\theta}{dt} = R\omega \longrightarrow \frac{d\theta}{dt} = \frac{R}{r}\omega.$$

This indicates that the angular speed of gear B is proportional to the angular speed of gear A. If the angular speed of gear A is fixed, the angular speed of gear B is proportional to the radius of gear A and inversely proportional to the radius of gear B. This means that the larger gear always rotates slower than the smaller gear, or in other words, the smaller gear always rotates faster than the larger gear. If the size of gear A is twice of that of gear B, or $R = 2r$, the angular speed of gear B will be

$$\frac{d\theta}{dt} = \frac{R}{r}\omega = \frac{2r}{r}\omega = 2\omega.$$

Gear B rotates two times faster than gear A does.

Example 11.19: A ladder of 15 m leans against one wall of an alley 9 m wide (see figure below). The ladder slips whilst the top of the ladder slides down. What is the speed of the top of the ladder by the time when the foot reaches the opposite wall at a speed of 2 m/s?

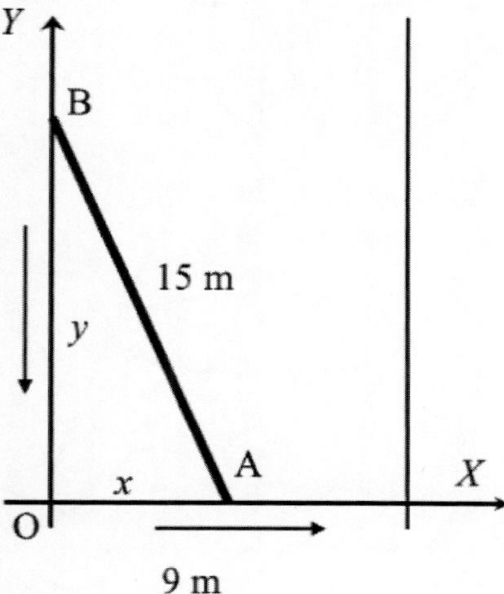

Solution

Assume at time t the top of the ladder is y m above ground and the foot of the ladder is x m from the wall. By Pythagorean theorem,

$$x^2 + y^2 = 15^2.$$

Differentiate both sides of the equation with respect to time t:

$$\frac{d(x^2)}{dt} + \frac{d(y^2)}{dt} = \frac{d(15^2)}{dt} \longrightarrow 2x\frac{dx}{dt} + 2y\frac{dy}{dt} = 0 \longrightarrow 2y\frac{dy}{dt} = -2x\frac{dx}{dt} \longrightarrow \frac{dy}{dt} = -\frac{x}{y}\frac{dx}{dt}.$$

When the foot reaches the opposite wall, $\frac{dx}{dt} = 2$ m/s, $x = 9$ m and

$$y = \sqrt{15^2 - 9^2} = \sqrt{225 - 81} = \sqrt{144} = 12 \text{ m}.$$

$$\frac{dy}{dt} = -\frac{x}{y}\frac{dx}{dt} = -\frac{9}{12} \times 2 = -\frac{9}{6} = -\frac{3}{2} = -1.5\ m/s.$$

Hence, the speed of the top of the ladder by the time when the foot reaches the opposite wall is 1.5 m/s. The negative sign indicates the ladder is sliding down.

Example 11.20: Referring to the figure below, water is flowing into a conical cylinder from the top at a rate of 5 m^3/h. If the radius of the base of this cylinder is 4 metres and the height is 10 metres, find the rate at which the water level is rising when it is 4 metres from the bottom.

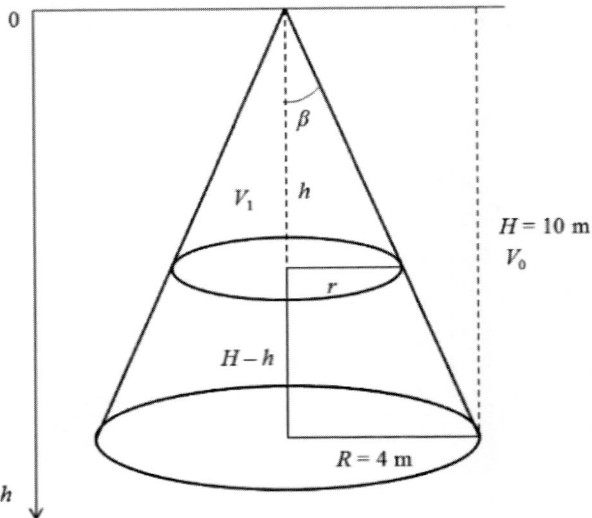

Solution

The whole volume of this conical cylinder is a constant $V_0 = \frac{1}{3}\pi R^2 H$. The volume of the top conical cylinder that is still unfilled is $V_1 = \frac{1}{3}\pi r^2 h$. The volume of the bottom part that is filled with water is

$$V = V_0 - V_1 = V_0 - \frac{1}{3}\pi r^2 h.$$

As $\tan\beta = \frac{r}{h} = \frac{R}{H}$, thus $r = \frac{R}{H}h$.

$$V = V_0 - V_1 = V_0 - \frac{1}{3}\pi(\frac{R}{H}h)^2 h = V_0 - \frac{1}{3}\pi(\frac{R}{H})^2 h^3.$$

Differentiate both sides of the equation with respect to time t (h is a function of t):

$$\frac{dV}{dt} = (V_0 - \frac{1}{3}\pi(\frac{R}{H})^2 h^3)' = \frac{dV_0}{dt} - \frac{1}{3}\pi(\frac{R}{H})^2 \frac{d(h^3)}{dt} = 0 - \frac{1}{3}\pi(\frac{R}{H})^2 3h^2 \frac{dh}{dt} = -\pi(\frac{Rh}{H})^2 \frac{dh}{dt}$$

$$\frac{dh}{dt} = \frac{-H^2}{\pi R^2 h^2}\frac{dV}{dt}.$$

Substitute $H=10,\ R=4,\ h=10-4=6,\ \dfrac{dV}{dt}=5$ into above equation:

$$\frac{dh}{dt}=\frac{-10^2}{\pi\times 4^2\times 6^2}\times 5\approx\frac{-500}{3.14159\times 16\times 36}\approx -0.2763\ \text{m/h}.$$

The rate at which the water level is rising when it is 4 metres from the bottom is 0.2763 m/h.

Example 11.21: A conical cylinder has height H and base radius R. If its volume is set to 10 m^3, find the H and R that make the surface area of such conical cylinder minimum.

Solution

The surface area A of such conical cylinder is the sum of the base area A_b and the side area A_s:

$$A=A_b+A_s=\pi R^2+\pi(R^2+H^2)=2\pi R^2+\pi H^2.$$

Since the volume of this conical cylinder is fixed to 10 m^3, thus

$$\frac{1}{3}\pi R^2H=10\longrightarrow R^2=\frac{30}{\pi H};\quad A=2\pi R^2+\pi H^2=2\pi\frac{30}{\pi H}+\pi H^2=\frac{60}{H}+\pi H^2.$$

Differentiate both sides of the equation with respect to H,

$$\frac{dA}{dH}=\left(\frac{60}{H}\right)'+(\pi H^2)'=\frac{-60}{H^2}+2\pi H.\ \text{ Let } A'=0,$$

$$\frac{-60}{H^2}+2\pi H=0\longrightarrow 2\pi H=\frac{60}{H^2}\longrightarrow \pi H=\frac{30}{H^2}$$

$$\pi H^3=30\longrightarrow H^3=\frac{30}{\pi}\longrightarrow H=\sqrt[3]{\frac{30}{\pi}}\approx\sqrt[3]{\frac{30}{3.1416}}\approx 2.1216\ \text{m}.$$

$$R^2=\frac{30}{\pi H}\longrightarrow R=\sqrt{\frac{30}{\pi H}}\approx\sqrt{\frac{30}{3.1416\times 2.1216}}\approx 2.1216\ \text{m}.$$

This means that the height and the base radius should take the same size of about 2.1 m. We still need checking if this makes the surface area minimum.

$$\frac{d^2A}{dH^2}=\left(\frac{-60}{H^2}\right)'+(2\pi H)'=(-2)\left(\frac{-60}{H^3}\right)+2\pi=\frac{120}{H^3}+2\pi>0.$$

Therefore, the surface area is indeed the minimum if the height and the base radius take the same size of about 2.1 m.

Example 11.22: The cross-section of a water canal is shown below. Find the angle x that makes the area of cross-section ABCD maximum.

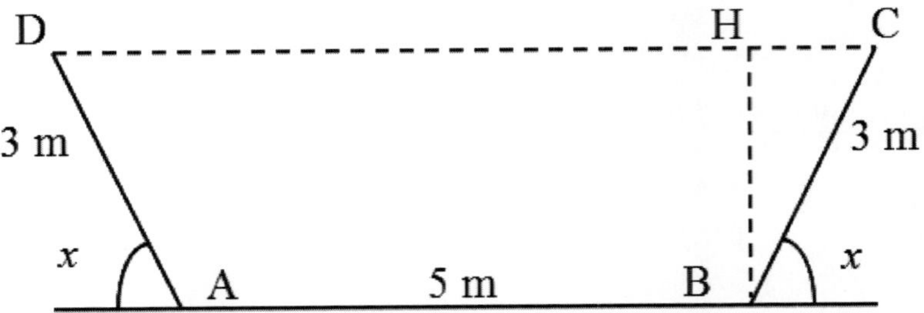

Solution

The area of cross-section ABCD is

$$A = \frac{1}{2}(\overline{AB} + \overline{CD})\overline{BH} = \frac{1}{2}(\overline{AB} + \overline{AB} + 2\overline{CH})\overline{BC}\cdot\sin x = \frac{1}{2}(2\overline{AB} + 2\overline{CH})\overline{BC}\cdot\sin x$$
$$= (\overline{AB} + \overline{CH})\overline{BC}\cdot\sin x = (\overline{AB} + \overline{BC}\cdot\cos x)\overline{BC}\cdot\sin x = 3\sin x(5 + 3\cos x)$$
$$= 15\sin x + 9\sin x\cos x = 15\sin x + \frac{9}{2}\sin 2x.$$

Differentiate both sides of the equation with respect to x,

$$A' = 15(\sin x)' + \frac{9}{2}(\sin 2x)' = 15\cos x + \frac{9}{2}\cos 2x(2x)' \quad \leftarrow \text{chain rule}$$
$$= 15\cos x + 9\cos 2x = 15\cos x + 9(2\cos^2 x - 1) = 15\cos x + 18\cos^2 x - 9.$$

Let $A' = 0$ and $t = \cos x$,

$$15t + 18t^2 - 9 = 0 \longrightarrow 6t^2 + 5t - 3 = 0\,;\ t = \frac{-5 \pm \sqrt{5^2 - 4\times 6\times(-3)}}{2\times 6} = \frac{-5 \pm \sqrt{25+72}}{12} = \frac{-5 \pm \sqrt{97}}{12}.$$

As $t = \cos x$ must be positive,

$$\cos x = \frac{-5 + \sqrt{97}}{12} \approx \frac{-5 + 9.84886}{12} = \frac{4.84886}{12} \approx 0.40407$$
$$x = \arccos(0.40407) \approx 66.17°.$$

To verify that this angle will give the maximum area for the cross-section, we check A''.

$$A'' = (A')' = (15\cos x + 9\cos 2x)' = -15\sin x - 18\sin 2x < 0$$
$$A_{\max} = 3\sin x(5 + 3\cos x) = 3\sin 66.17°(5 + 3\cos 66.17°) \approx 17.05\ m^2$$

Example 11.23: Given 720-metre long fencing materials, how to choose the dimensions so that the complex of six same sized blocks (see figure below) has the maximum area (i.e., the total area of the six blocks).

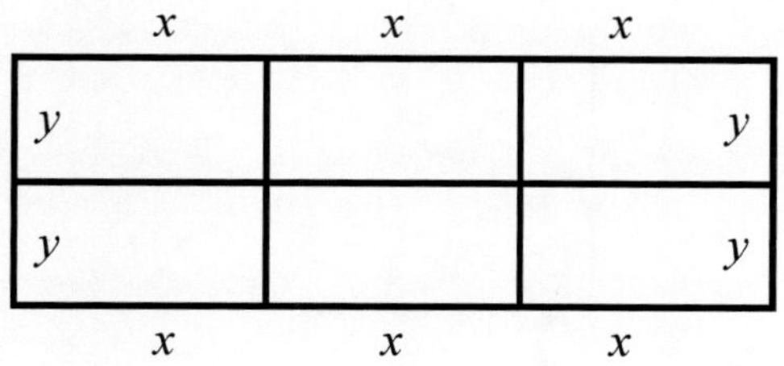

Solution

The sum of all sections (both x and y) of the six blocks must be 720 metres, i.e., $9x+8y=720$. The total area of these six blocks is $A=(3x)(2y)=6xy$. From $9x+8y=720$,

$$8y=720-9x \longrightarrow y=\frac{720-9x}{8}=90-\frac{9}{8}x.$$

Substitute y into $A=6xy$,

$$A=6xy=6x(90-\frac{9}{8}x)=540x-\frac{27}{4}x^2.$$

$$\frac{dA}{dx}=(540x-\frac{27}{4}x^2)'=540-\frac{27}{2}x.$$

Let $\frac{dA}{dx}=0$,

$$540-\frac{27}{2}x=0 \longrightarrow \frac{27}{2}x=540 \longrightarrow x=\frac{1080}{27}=40\text{ m}.$$

$$y=90-\frac{9}{8}x=90-\frac{9}{8}\times 40=90-45=45\text{ m}.$$

$$\frac{d^2A}{dx^2}=(540-\frac{27}{2}x)'=-\frac{27}{2}<0.$$

Therefore, there is a maximum area if $x=40$ m and $y=45$ m for

$$A_{\max}=6xy=6\times 40\times 45=10800\text{ m}^2.$$

Exercises 11.4

1. A man 1.8 m tall is walking at a rate of 2 m/s toward a light 18 m above the ground. How fast is the length of his shadow changing? How fast is the tip of his shadow moving?

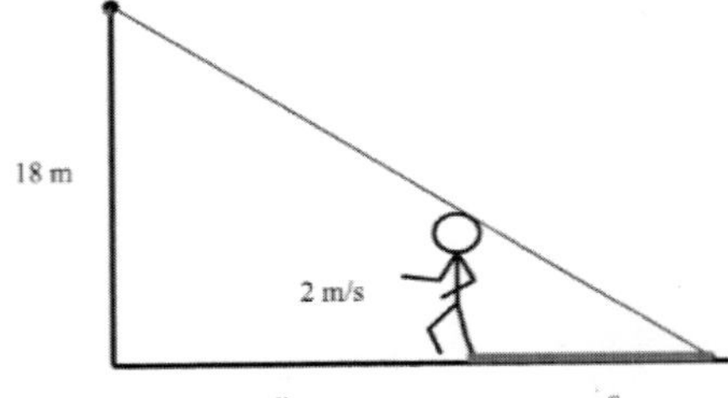

2. A balloon (regarded as a ball) is being inflated by pumped air at the rate of 10 cm^3 per second. How fast is the diameter of the balloon increasing when the radius is 5 cm?

3. A TV manufacturing company wants to market a new TV. In order to sell x TV sets, the price per set must be $p = 200 - 0.45x$. If the total cost of producing x TV sets is $T(x) = 4500 + 0.55x^2$. How many TV sets should the company sell for maximising profit? What is the maximum profit?

4. An open box is to be made from a 20 cm by 30 cm cardboard by cutting squares out of the four corners and folding up the sides. What dimensions of the box will have the largest volume? What is the maximum volume of the box?

5. A cylindrical can is to hold 2000 cm^3 of water. Find the dimensions (surface area) that use the least material.

6. The sum of two positive numbers is 100. Find the numbers that makes the sum of their squares minimum.

7. If 30 orange trees are planted in a fixed area of land, each tree yields 400 oranges on average. The yield will decrease about 10 oranges per tree for each additional tree planted in the land. How many trees should be planted for the maximum total yield? Is your result justified?

8. An east-west road meets a north-south road at point C (see the figure below). A diagonal road from A to B is designed so that this road must go through point E that is 4 km west of C and 2 km north of C. Find where points A and B should be located so that the area of ABC is the minimum.

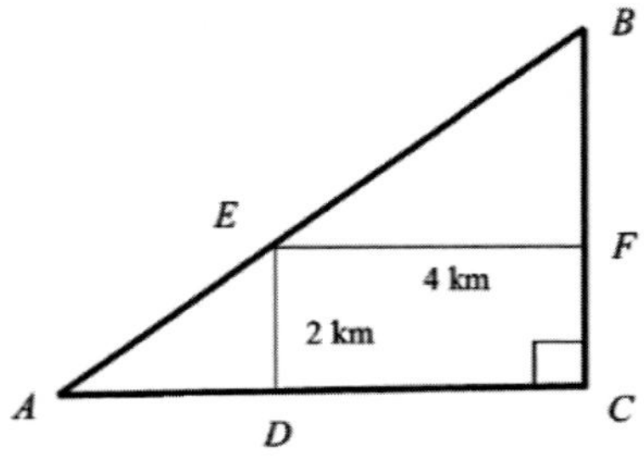

Chapter 11: Exercises Answers

Exercises 11.1

1. The tangent function: $y=-4x-9$; the normal function: $y=\frac{1}{4}(x-19)$
2. The tangent function: $y=7x-6$; the normal function: $y=-\frac{1}{7}(x-8)$
3. The tangent function: $y=\frac{3}{2}x-\frac{3\pi}{12}$; the normal function: $y=-\frac{2}{3}x+\frac{\pi}{9}$.
4. The tangent function: $y=5x-1$ the normal function: $y=-\frac{1}{5}x-1$
5. The tangent function: $y=\frac{2}{3}x-\frac{2}{3}$ the normal function: $y=-\frac{3}{2}x+6$.
6. The tangent function: $y=-x+8$; the normal function: $y=x+4$.
7. The tangent function: $y = 1$; the normal function: $x = 1$.
8. $a=4,\ b=-3 \rightarrow 4y^2-3xy=4-x^2$.

Exercises 11.2

1. a) ∞, b) 1, c) –1/3, d) 0, e) 2, f) 1, g) cos2, h) 2
2. a) 0, b) ½, c) ½, d) 1

Exercises 11.3

1. a) There is a local minimum at $x=-1$ with $y_{min}=-3$.
   b) There is a local minimum at $x=e^{-1}$ with $y_{min}=-e^{-1}$.
   c) $x=\frac{\pi}{12}\rightarrow y_{max}=\frac{\pi+6\sqrt{3}}{12};\ x=\frac{6\pi}{12}\rightarrow y_{min}=\frac{5\pi-6\sqrt{3}}{12}$
   d) $x=\frac{4}{3}\rightarrow y_{max}=\frac{-134}{27};\ x=1\rightarrow y_{min}=-5$
   e) $FDT: x=0\rightarrow y_{max}=2$
   f) $FDT: x=\pm\frac{\pi}{2}$ are not critical points.
2. $x = -4$ the global minimum; $x = 2$ is a local minimum; $x = -4/3$ a local and the global maximum but $x = 3$ is an ordinary point.
3. $x = \pi/2$ is the global minimum; $x = \pi/6$ is a local and the global maximum but $x = 0$ is an ordinary point.
4. Therefore, $x = 1$ is the global minimum; $x = 1/4$ is the global maximum but $x = 4$ is an ordinary point.

Exercises 11.4

1. 20/9 m/s ≈ 2.22 m/s

2. $0.2/\pi$ m/s $\approx$ 0.064 m/s
3. Selling 100 TV sets to gain the maximum profit of \$5500.
4. If the side of squares is about 3.9 cm, the volume of the box is the maximum of 1056 cm^3.
5. Taking the same diameter and the height of 13.66 cm uses the least material of 879 cm^2.
6. An even split of 50 each makes the sum of their squares minimum at 5000.
7. Extra 5 trees for a total of 35 trees for the total of 12250 oranges
8. *A* should be located at 8 km west of *C* and *B* should be located at 4 km north of *C*.

CHAPTER 12

12 Derivatives for Approximation

CHAPTER OBJECTIVES

- Solve nonlinear equations by Newton's method
- Introduce Taylor polynomials and approximation
- Introduce Taylor and Maclaurin series and approximation

Essential statements on approximation by derivatives:

- Newton's method is useful in solving some nonlinear equations.
- Taylor (and Maclaurin) polynomials and series are behind many approximations in scientific computation.

Key topics:

- Newton's method
- Taylor polynomials
- Taylor and Maclaurin series

Flowchart of mathematical knowledge building

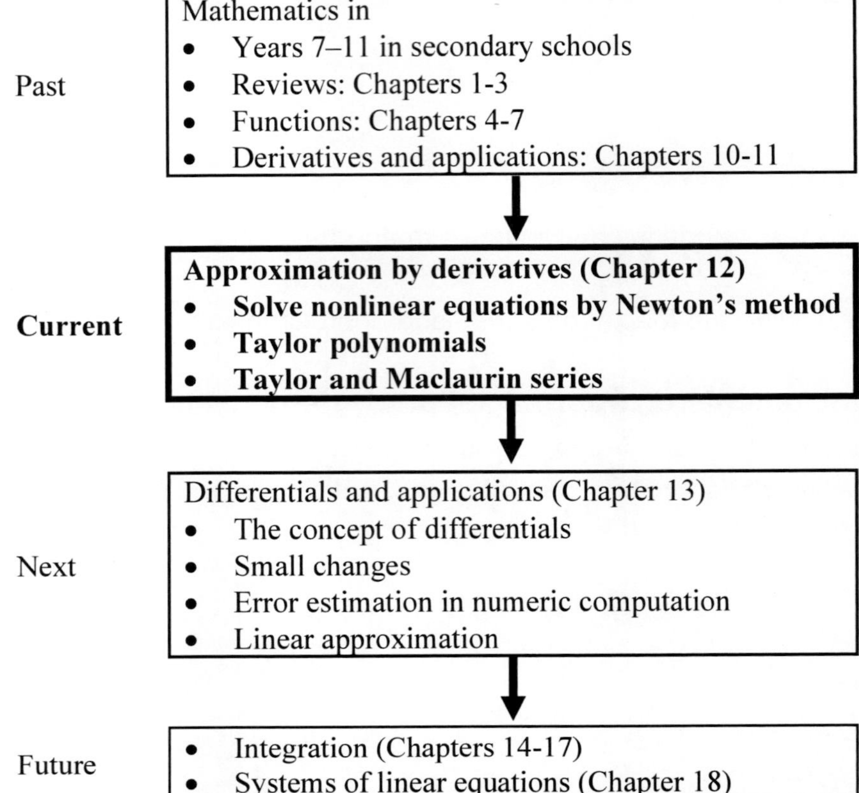

Chapter 12: Derivatives for Approximation

12.1 Solving Nonlinear Equations by Newton's Method

In real-world applications, a great number of nonlinear equations cannot be solved perfectly with exact solutions. The common strategy to solve such complicated problem is to obtain its approximate solution using an appropriate procedure known as '***numeric approach***' or '***numeric computation***'. There are many different established numeric methods for solving various types of equations. Netwton's method is a popular approach to solve some nonlinear equations under special conditions.

Solving an equation $f(x) = 0$ within $[a, b]$ is equivalent to finding the point $x = x_r$ ($\in [a, b]$) where $f(x)$ intersects the x-axis (Figure 12.1). Such a point $x = x_r$ is called *the root* of $f(x) = 0$ if it exists. It is not easy to judge if any root exists for a general equation $f(x) = 0$, but in a situation where $f(a)$ and $f(b)$ have opposite signs, such point indeed exists. This is because $f(x)$ must cross the x-axis at least once within $[a, b]$ so that $f(a)$ and $f(b)$ have opposite signs.

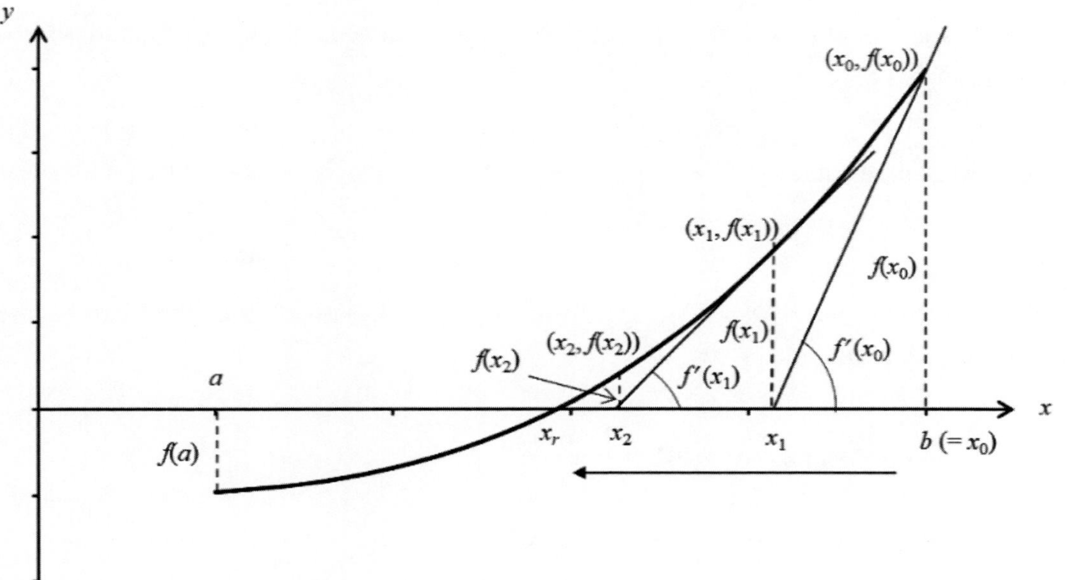

Figure 12.1: Process of approaching the root $x = x_r$ for equation $f(x) = 0$ by Newton's method

Newton's method uses a sequence of x-intercepts of the tangent lines from an initial point x_0 (usually either a or b) to approximate the root of equation $f(x) = 0$ within $[a, b]$. Suppose the initial point to be $x_0 = b$ (Figure 12.1). The tangent line on $f(x)$ at $(x_0, f(x_0))$ crosses the x-axis at x_1 and the tangent of this line is the derivative at $x_0 = b$, i.e.,

$$f'(x_0) = \frac{f(x_0)}{x_0 - x_1} \longrightarrow x_0 - x_1 = \frac{f(x_0)}{f'(x_0)} \longrightarrow x_1 = x_0 - \frac{f(x_0)}{f'(x_0)}.$$

This means that x_1 can be determined by the values of x, y and y' at the initial point x_0. The new point x_1, closer to x_r compared to x_0, can be used to find a new x-intercept x_2 in a similar way as follows

$$f'(x_1) = \frac{f(x_1)}{x_1 - x_2} \longrightarrow x_1 - x_2 = \frac{f(x_1)}{f'(x_1)} \longrightarrow x_2 = x_1 - \frac{f(x_1)}{f'(x_1)}.$$

Obviously x_2 is even closer to x_r compared to x_1. Repeating this process will take the consecutive x-intercepts approaching the root of the equation. This iterative procedure is captured by the following recurrent relation, called *Newton's method*:

$$\boxed{x_{i+1} = x_i - \frac{f(x_i)}{f'(x_i)}, \quad i = 0,1,2,3,\ldots} \tag{12.1}$$

The iteration terminates if the difference of two consecutive x-intercepts is smaller than a pre-set error limit, i.e., $|x_i - x_{i-1}| < \varepsilon$, which indicates how close the approximate value of x_i is to the root x_r. If such condition is met after the ith iteration, an acceptable approximate solution is found for $f(x) = 0$, i.e., $x_r \approx x_i$.

Example 12.1: Solve equation $f(x) = \frac{1}{x} - \ln x = 0$ in [1, 3] using Newton's method with $\varepsilon = 0.01$.

Solution

Since $f(a) = f(1) = 1$ and $f(b) = f(3) = 1/3 - \ln 3 \approx -0.7653$, Newton's method can be used to solve this nonlinear equation.

$$f(x) = \frac{1}{x} - \ln x \longrightarrow f(x_i) = \frac{1}{x_i} - \ln x_i = \frac{1 - x_i \ln x_i}{x_i}$$

$$f'(x) = (\frac{1}{x} - \ln x)' = -\frac{1}{x^2} - \frac{1}{x} = -\frac{1+x}{x^2} \longrightarrow f'(x_i) = -\frac{1+x_i}{x_i^2}$$

Formula (12.1) becomes

$$x_{i+1} = x_i - \frac{f(x_i)}{f'(x_i)} = x_i - \frac{\dfrac{1 - x_i \ln x_i}{x_i}}{-\dfrac{1+x_i}{x_i^2}} = x_i + \frac{x_i(1 - x_i \ln x_i)}{1+x_i}.$$

Let $x_0 = a = 1$,

$$x_{0+1} = x_1 = x_0 + \frac{x_0(1 - x_0 \ln x_0)}{1 + x_0} = 1 + \frac{1 \times (1 - 1 \times \ln 1)}{1 + 1} = 1 + \frac{1}{2} = 1.5.$$

Since $|x_1 - x_0| = |1.5 - 1| = 0.5 > \varepsilon = 0.01$, let $x_1 = 1.5$,

$$x_2 = x_1 + \frac{x_1(1 - x_1 \ln x_1)}{1 + x_1} = 1.5 + \frac{1.5 \times (1 - 1.5 \times \ln 1.5)}{1 + 1.5} \approx 1.7351.$$

Since $|x_2 - x_1| = |1.7351 - 1.5| = 0.2351 > \varepsilon = 0.01$, let $x_2 = 1.7351$,

$$x_3 = x_2 + \frac{x_2(1 - x_2 \ln x_2)}{1 + x_2} = 1.7351 + \frac{1.7351 \times (1 - 1.7351 \times \ln 1.7351)}{1 + 1.7351} \approx 1.7629.$$

Since $|x_3 - x_2| = |1.7629 - 1.7351| = 0.0278 > \varepsilon = 0.01$, let $x_3 = 1.7629$,

$$x_4 = x_3 + \frac{x_3(1 - x_3 \ln x_3)}{1 + x_3} = 1.7629 + \frac{1.7629 \times (1 - 1.7629 \times \ln 1.7629)}{1 + 1.7629} \approx 1.7631.$$

Since $|x_4 - x_3| = |1.7631 - 1.7629| = 0.0002 < \varepsilon = 0.01$, the approximate solution is $x \approx x_4 = 1.7631$.

The iterative process above can also be tabulated in the following table for simplicity. This can be implemented using Excel too.

i	x_i	x_{i+1}	$\lvert x_{i+1} - x_i \rvert$	$\lvert x_{i+1} - x_i \rvert < \varepsilon$
0	1	1.5	0.5	N
1	1.5	1.7351	0.2351	N
2	1.7351	1.7629	0.0278	N
3	1.7629	**1.7631**	**0.0002**	**Y**
$x \approx x_4 = 1.7631$				

Example 12.2: Solve equation $f(x) = x^2 - \sin x = 0$ in [0.5, 1] using Newton's method with $\varepsilon =$ 0.0001.

Solution

Since $f(a) = f(0.5) = 0.25 - \sin 0.5 \approx 0.25 - 0.4794 = -0.2294$ and $f(b) = f(1) = 1 - \sin 1 \approx 1 - 0.8415 = 0.1585$, Newton's method can be used to solve this nonlinear equation.

$$f(x) = x^2 - \sin x \longrightarrow f(x_i) = x_i^2 - \sin x_i$$

$$f'(x) = (x^2 - \sin x)' = 2x - \cos x \longrightarrow f'(x_i) = 2x_i - \cos x_i$$

Formula (12.1) becomes

$$x_{i+1} = x_i - \frac{f(x_i)}{f'(x_i)} = x_i - \frac{x_i^2 - \sin x_i}{2x_i - \cos x_i}.$$

$$x_0 = b = 1: \quad x_1 = x_0 - \frac{x_0^2 - \sin x_0}{2x_0 - \cos x_0} = 1 - \frac{1^2 - \sin(1)_{rad}}{2 \times 1 - \cos(1)_{rad}} = 0.89140$$

$$|x_1 - x_0| = 0.10860 > 0.0001$$

$$x_1 = 0.89140: \quad x_2 = x_1 - \frac{x_1^2 - \sin x_1}{2x_1 - \cos x_1} = 0.89140 - \frac{0.89140^2 - \sin(0.89140)_{rad}}{2 \times 0.89140 - \cos(0.89140)_{rad}} = 0.87698$$

$$|x_2 - x_1| = 0.01441 > 0.0001$$

$$x_2 = 0.87698: \quad x_3 = x_2 - \frac{x_2^2 - \sin x_2}{2x_2 - \cos x_2} = 0.87698 - \frac{0.87698^2 - \sin(0.87698)_{rad}}{2 \times 0.87698 - \cos(0.87698)_{rad}} = 0.87673$$

$$|x_3 - x_2| = 0.00026 > 0.0001$$

$$x_3 = 0.87673: \quad x_4 = x_3 - \frac{x_3^2 - \sin x_3}{2x_3 - \cos x_3} = 0.87698 - \frac{0.87673^2 - \sin(0.87673)_{rad}}{2 \times 0.87673 - \cos(0.87673)_{rad}} = 0.87673$$

$$|x_4 - x_3| = 0.00000 < 0.0001$$

The approximate solution is $x \approx x_4 = 0.8767$.

Note: 1) The value for sine and cosine is in radians. 2) Five decimal places are kept during the iteration as the error limit is in four decimal places. Keeping one more decimal place than that of the error limit ensures the accuracy of the solution.

Example 12.3: Solve equation $x^3 - 2x + 2 = \sqrt{x+2}$ in [0, 0.5] using Newton's method with $\varepsilon = 0.001$.

Solution

We first rearrange $x^3 - 2x + 2 = \sqrt{x+2}$ to $f(x) = x^3 - 2x + 2 - \sqrt{x+2}$. Solving the equation becomes solving equation $f(x) = x^3 - 2x + 2 - \sqrt{x+2} = 0$. Since $f(a) = f(0) = 2 - \sqrt{2} \approx 0.5858$ and $f(b) = f(0.5) = 0.5^3 - 2 \times 0.5 + 2 - \sqrt{2.5} \approx -0.4561$, Newton's method can be used to solve this nonlinear equation.

$$f(x) = x^3 - 2x + 2 - \sqrt{x+2}. \longrightarrow f(x_i) = x_i^3 - 2x_i + 2 - \sqrt{x_i + 2}.$$

$$f'(x) = (x^3 - 2x + 2 - \sqrt{x+2})' = 3x^2 - 2 - \frac{1}{2\sqrt{x+2}} \longrightarrow f'(x_i) = 3x_i^2 - 2 - \frac{1}{2\sqrt{x_i + 2}}$$

We can use formula (12.1) directly to conduct the iterative process as follows.

$$x_0 = a = 0:$$

$$f(x_0) = x_0^3 - 2x_0 + 2 - \sqrt{x_0 + 2} = 0^3 - 2 \times 0 + 2 - \sqrt{0+2} = 0.58579$$

$$f'(x_0) = 3x_0^2 - 2 - \frac{1}{2\sqrt{x_0 + 2}} = 3 \times 0^2 - 2 - \frac{1}{2\sqrt{0+2}} = -2.35355$$

$$x_1 = x_0 - \frac{f(x_0)}{f'(x_0)} = 0 - \frac{0.58579}{-2.35355} = 0.24889 \longrightarrow |x_1 - x_0| = 0.24889 > 0.001$$

$x_1 = 0.24889:$

$f(x_1) = x_1^3 - 2x_1 + 2 - \sqrt{x_1 + 2} = 0.24889^3 - 2\times 0.24889 + 2 - \sqrt{0.24889 + 2} = 0.01800$

$f'(x_1) = 3x_1^2 - 2 - \dfrac{1}{2\sqrt{x_1 + 2}} = 3\times 0.24889^2 - 2 - \dfrac{1}{2\sqrt{0.24889 + 2}} = -2.14757$

$x_2 = x_1 - \dfrac{f(x_1)}{f'(x_1)} = 0.24889 - \dfrac{0.01800}{-2.14757} = 0.25728 \longrightarrow |x_2 - x_1| = 0.00838 > 0.001$

$x_2 = 0.25728:$

$f(x_2) = x_2^3 - 2x_2 + 2 - \sqrt{x_2 + 2} = 0.25728^3 - 2\times 0.25728 + 2 - \sqrt{0.25728 + 2} = 0.00006$

$f'(x_2) = 3x_2^2 - 2 - \dfrac{1}{2\sqrt{x_2 + 2}} = 3\times 0.25728^2 - 2 - \dfrac{1}{2\sqrt{0.25728 + 2}} = -2.13422$

$x_3 = x_2 - \dfrac{f(x_2)}{f'(x_2)} = 0.25728 - \dfrac{0.00006}{-2.13422} = 0.25730 \longrightarrow |x_3 - x_2| = 0.00003 < 0.001$

The approximate solution is $x \approx x_3 = 0.2573$.

Exercises 12.1

1. Solve equation $e^x - x - 3 = 0$ in [1, 2] with $\varepsilon = 0.001$.

2. Solve equation $x^3 - 4x^2 + x - 7 = 0$ in [3.5, 4.5] with $\varepsilon = 0.001$.

3. Solve equation $\cos x - \sqrt{x} = 0$ in [0, 1] with $\varepsilon = 0.001$.

4. Solve equation $\sin x - \dfrac{1}{x+1} = 0$ in [0, 1] with $\varepsilon = 0.001$.

5. Solve equation $x^2 = \ln x + 5$ in [2, 3] with $\varepsilon = 0.0001$.

6. Solve equation $\sqrt[3]{x} - \dfrac{1}{\sqrt{x+1}} = 0$ in [0, 1] with $\varepsilon = 0.0001$.

7. Solve equation $2\sin x + 1 = 2\cos x$ in [0, π/4] with $\varepsilon = 0.001$.

8. Solve equation $xe^x = x^2 + 1$ in [0, 1] with $\varepsilon = 0.0001$.

9. Solve equation $y = x^4 - 3x^3 + 2x^2 - 5x + 7 = 0$ in [1, 2] with $\varepsilon = 0.001$.

12.2 Taylor Polynomials

General Taylor polynomials

For $y = f(x)$, if its first-order, second-order,…, and nth-order derivatives exist at $x = a$, an nth-order polynomial $p_n(x)$ can be created using these known values at point $x = a$, i.e., $f(a)$, $f'(a), f''(a),..., f^{(n)}(a)$, to approximate function $f(x)$. Such an nth-order polynomial $p_n(x)$ is called the <u>*nth-order Taylor polynomial*</u> or the <u>*general Taylor polynomial*</u> written as

$$\boxed{p_n(x)=f(a)+f'(a)(x-a)+f''(a)\frac{(x-a)^2}{2!}+f'''(a)\frac{(x-a)^3}{3!}+...+f^{(n)}(a)\frac{(x-a)^n}{n!}} \quad (12.2)$$

Commonly used lower-order Taylor polynomials

For $n = 1$, the *general Taylor polynomial* becomes a linear function:

$$\boxed{p_1(x)=f(a)+f'(a)(x-a)} \quad (12.3)$$

For $n = 2$, the *general Taylor polynomial* becomes a quadratic function:

$$\boxed{p_2(x)=f(a)+f'(a)(x-a)+f''(a)\frac{(x-a)^2}{2!}=p_1(x)+f''(a)\frac{(x-a)^2}{2!}} \quad (12.4)$$

For $n = 3$, the *general Taylor polynomial* becomes a cubic function:

$$\boxed{p_3(x)=f(a)+f'(a)(x-a)+f''(a)\frac{(x-a)^2}{2!}+f'''(a)\frac{(x-a)^3}{3!}=p_2(x)+f'''(a)\frac{(x-a)^3}{3!}} \quad (12.5)$$

Example 12.4: Given $f(x)=\ln x$, create a fifth-order Taylor polynomial at $x = 1$ to approximate $f(x)$. Use this polynomial to estimate ln1.5, assuming the exact value of ln1.5 = 0.4055.

<u>Solution</u>

We firstly find the first to fifth derivatives of $f(x)=\ln x$.

$$f'(x)=(\ln x)'=\frac{1}{x},\quad f''(x)=[\frac{1}{x}]'=-\frac{1}{x^2},\quad f'''(x)=(-\frac{1}{x^2})'=\frac{2}{x^3},$$
$$f^{(4)}(x)=[\frac{2}{x^3}]'=-\frac{6}{x^4},\quad f^{(5)}(x)=[\frac{-6}{x^4}]'=\frac{24}{x^5}.$$

At $x = a = 1$:

$$f(a) = f(1) = \ln 1 = 0, \quad f'(1) = \frac{1}{1} = 1, \quad f''(1) = -\frac{1}{1^2} = -1,$$

$$f'''(1) = \frac{2}{1^3} = 2, \quad f^{(4)}(1) = -\frac{6}{1^4} = -6, \quad f^{(5)}(1) = \frac{24}{1^5} = 24.$$

Create the fifth-order Taylor polynomial at $a = 1$ using formula (12.2).

$$\begin{aligned} p_5(x) &= f(1) + f'(1)(x-1) + f''(1)\frac{(x-1)^2}{2!} + f'''(1)\frac{(x-1)^3}{3!} + f^{(4)}(1)\frac{(x-1)^4}{4!} + f^{(5)}(1)\frac{(x-1)^5}{5!} \\ &= 0 + (x-1) - \frac{(x-1)^2}{2!} + \frac{2(x-1)^3}{3!} - \frac{6(x-1)^4}{4!} + \frac{24(x-1)^5}{5!} \\ &= (x-1) - \frac{(x-1)^2}{2} + \frac{(x-1)^3}{3} - \frac{(x-1)^4}{4} + \frac{(x-1)^5}{5}. \end{aligned}$$

Substitute $x = 1.5$ into this polynomial to approximate $\ln 1.5$ as follows:

$$\begin{aligned} \ln 1.5 \approx p_5(1.5) &= (1.5-1) - \frac{(1.5-1)^2}{2} + \frac{(1.5-1)^3}{3} - \frac{(1.5-1)^4}{4} + \frac{(1.5-1)^5}{5} \\ &= 0.5 - \frac{0.5^2}{2} + \frac{0.5^3}{3} - \frac{0.5^4}{4} + \frac{0.5^5}{5} = 0.5 - \frac{0.25}{2} + \frac{0.125}{3} - \frac{0.0625}{4} + \frac{0.03125}{5} \approx 0.4073 \end{aligned}$$

As the 'exact value' is given as $\ln 1.5 = 0.4055$, the error between this true value and the approximate value from the Taylor polynomial is

$$|f(1.5) - p_5(1.5)| = |0.4055 - 0.4073| = 0.0018.$$

Example 12.5: Given $f(x) = \sqrt{2x+1}$, use linear, quadratic, and cubic Taylor polynomials to approximate $f(4.5) = \sqrt{2 \times 4.5 + 1} = \sqrt{10}$ and estimate the errors respectively assuming $\sqrt{10} = 3.1623$.

Solution

We firstly find the first, second and the third-order derivatives of $f(x) = \sqrt{2x+1}$.

$$f'(x) = (\sqrt{2x+1})' = \frac{1}{2}(2x+1)^{-\frac{1}{2}}(2x+1)' = \frac{1}{2}(2x+1)^{-\frac{1}{2}}(2) = (2x+1)^{-\frac{1}{2}} = \frac{1}{\sqrt{2x+1}},$$

$$f''(x) = [(2x+1)^{-\frac{1}{2}}]' = -\frac{1}{2}(2x+1)^{-\frac{3}{2}}(2x+1)' = -\frac{1}{2}(2x+1)^{-\frac{3}{2}}(2) = -(2x+1)^{-\frac{3}{2}} = \frac{-1}{\sqrt{(2x+1)^3}},$$

$$f'''(x) = [-(2x+1)^{-\frac{3}{2}}]' = \frac{3}{2}(2x+1)^{-\frac{5}{2}}(2x+1)' = \frac{3}{2}(2x+1)^{-\frac{5}{2}}(2) = 3(2x+1)^{-\frac{5}{2}} = \frac{3}{\sqrt{(2x+1)^5}}.$$

Let $a = 4$ to obtain all the required values at $x = a = 4$:

$$f(a) = f(4) = \sqrt{2\times 4+1} = \sqrt{9} = 3,$$
$$f'(a) = f'(4) = \frac{1}{\sqrt{2\times 4+1}} = \frac{1}{\sqrt{9}} = \frac{1}{3},$$
$$f''(a) = f''(4) = \frac{-1}{\sqrt{(2\times 4+1)^3}} = \frac{-1}{\sqrt{9^3}} = \frac{-1}{\sqrt{3^6}} = \frac{-1}{3^3} = -\frac{1}{27},$$
$$f'''(a) = f'''(4) = \frac{3}{\sqrt{(2\times 4+1)^5}} = \frac{3}{\sqrt{9^5}} = \frac{3}{\sqrt{3^{10}}} = \frac{3}{3^5} = \frac{1}{3^4} = \frac{1}{81}.$$

Create the linear, quadratic, and cubic Taylor polynomials at $a = 4$ using formulae (12.3-12.5).

$$p_1(x) = f(a) + f'(a)(x-a) = f(4) + f'(4)(x-4) = 3 + \frac{1}{3}(x-4),$$
$$p_2(x) = p_1(x) + f''(a)\frac{(x-a)^2}{2!} = 3 + \frac{1}{3}(x-4) - \frac{1}{27}\times\frac{1}{2!}(x-4)^2 = 3 + \frac{1}{3}(x-4) - \frac{1}{54}(x-4)^2,$$
$$p_3(x) = p_2(x) + f'''(a)\frac{(x-a)^3}{3!} = 3 + \frac{1}{3}(x-4) - \frac{1}{54}(x-4)^2 + \frac{1}{81}\times\frac{1}{3!}(x-4)^3$$
$$= 3 + \frac{1}{3}(x-4) - \frac{1}{54}(x-4)^2 + \frac{1}{486}(x-4)^3.$$

Substitute $x = 4.5$ into these polynomials to approximate $f(4.5) = \sqrt{2\times 4.5+1} = \sqrt{10}$ as follows:

$$\sqrt{10} \approx p_1(4.5) = 3 + \frac{1}{3}(4.5-4) = 3 + \frac{0.5}{3} \approx 3.1667$$
$$\sqrt{10} \approx p_2(4.5) = 3.1667 - \frac{1}{54}(4.5-4)^2 = 3.1667 - \frac{0.5^2}{54} = 3.1667 - \frac{0.25}{54} \approx 3.1621$$
$$\sqrt{10} \approx p_3(4.5) = 3.1621 + \frac{1}{486}(4.5-4)^3 = 3.1621 + \frac{0.5^3}{486} = 3.1621 + \frac{0.125}{486} \approx 3.1624.$$

As the 'true value' is given as $\sqrt{10} = 3.1623,$ the errors between this true value and those from the Taylor polynomials are as follows. The quadratic and cubic polynomials produce fine approximates for $\sqrt{10}$.

$$|f(4.5) - p_1(4.5)| = |3.1623 - 3.1667| = 0.0044,$$
$$|f(4.5) - p_2(4.5)| = |3.1623 - 3.1621| = 0.0002,$$
$$|f(4.5) - p_3(4.5)| = |3.1623 - 3.1624| = 0.0001.$$

Example 12.6: Given $f(x)=\sin x$ and $f(30°)=f(\frac{\pi}{6})=\frac{1}{2}$, use a cubic Taylor polynomial to approximate $f(35°)=\sin 35°$ and estimate the error assuming $\sin 35°=0.5736$.

Solution

We firstly find the first, second and the third-order derivatives of $f(x)=\sin x$.

$$f'(x)=(\sin x)'=\cos x,\quad f''(x)=(\cos x)'=-\sin x,\quad f'''(x)=(-\sin x)'=-\cos x.$$

At a = 30° = π/6, all the required values are:

$$f'(a)=f'(\frac{\pi}{6})=\cos\frac{\pi}{6}=\frac{\sqrt{3}}{2},\quad f''(a)=f''(\frac{\pi}{6})=-\sin\frac{\pi}{6}=-\frac{1}{2},\quad f'''(a)=f'''(\frac{\pi}{6})=-\cos\frac{\pi}{6}=-\frac{\sqrt{3}}{2}.$$

Create the cubic Taylor polynomial at a = 30° = π/6 using formula (12.5). Note x and a must be in radians in the Taylor polynomial.

$$\begin{aligned}p_3(x)&=f(\frac{\pi}{6})+f'(\frac{\pi}{6})(x-\frac{\pi}{6})+f''(\frac{\pi}{6})\frac{(x-\frac{\pi}{6})^2}{2!}+f'''(\frac{\pi}{6})\frac{(x-\frac{\pi}{6})^3}{3!}\\&=\frac{1}{2}+\frac{\sqrt{3}}{2}(x-\frac{\pi}{6})-\frac{1}{2}\times\frac{1}{2!}(x-\frac{\pi}{6})^2-\frac{\sqrt{3}}{2}\times\frac{1}{3!}(x-\frac{\pi}{6})^3=\frac{1}{2}+\frac{\sqrt{3}}{2}(x-\frac{\pi}{6})-\frac{1}{4}(x-\frac{\pi}{6})^2-\frac{\sqrt{3}}{12}(x-\frac{\pi}{6})^3\end{aligned}$$

Convert x = 35° to radians and then substitute it into the cubic polynomial to approximate $f(35°)=\sin 35°$ as follows:

$$\begin{aligned}&x=35°=\frac{35°\pi}{180°}\approx 0.6109,\\&\sin 35°=\sin 0.6109\approx p_3(0.6109)\\&\quad=\frac{1}{2}+\frac{\sqrt{3}}{2}(0.6109-\frac{\pi}{6})-\frac{1}{4}(0.6109-\frac{\pi}{6})^2-\frac{\sqrt{3}}{12}(0.6109-\frac{\pi}{6})^3\\&\quad\approx\frac{1}{2}+\frac{\sqrt{3}}{2}(0.6109-0.5236)-\frac{1}{4}(0.6109-0.5236)^2-\frac{\sqrt{3}}{12}(0.6109-0.5236)^3\\&\quad\approx\frac{1}{2}+\frac{\sqrt{3}}{2}(0.0873)-\frac{1}{4}(0.0873)^2-\frac{\sqrt{3}}{12}(0.0873)^3\approx 0.5736.\end{aligned}$$

As the 'true value' is given as $\sin 35°=0.5736$, there is no error between this true value and the value approximated by the cubic Taylor polynomial.

Estimating errors of Taylor polynomials in approximation

By using an nth-order Taylor polynomial $p_n(x)$ to approximate $f(x)$, if the error between them is expressed as $E_n(x)$, then the following formula must stand:

$$\boxed{f(x) = p_n(x) + E_n(x).} \tag{12.6}$$

The error term, also called the **remainder of order *n*** $R_n(x)$, can be estimated by

$$\boxed{E_n(x) = R_n(x) = f^{(n+1)}(c)\frac{(x-a)^{n+1}}{(n+1)!} \quad c \text{ is between } x \text{ and } a.} \tag{12.7}$$

Example 12.7: In example 12.5, estimate the error of the quadratic Taylor polynomial in approximating $\sqrt{10}$.

Solution

By formula (12.7), the error is about

$$E_2(x) = f^{(2+1)}(c)\frac{(4.5-4)^{2+1}}{(2+1)!} = f'''(c)\frac{0.5^3}{3!} = \frac{3}{\sqrt{(2c+1)^5}}\frac{0.125}{6} = \frac{0.0625}{\sqrt{(2c+1)^5}},$$

where c is a value between $a = 4$ and $x = 4.5$. If $c = 4$, the error would be the largest, i.e.,

$$E_2(x) < \frac{0.0625}{\sqrt{(2\times 4+1)^5}} = \frac{0.0625}{\sqrt{9^5}} = \frac{0.0625}{\sqrt{3^{10}}} = \frac{0.0625}{3^5} = \frac{0.0625}{243} < 0.0003.$$

Example 12.8: For $f(x) = e^x$, assuming $f(1) = e = 2.71828$, to make sure that the approximate to $f(x) = e^x$ within [1, 1.5] using a Taylor polynomial is within an error limit $\varepsilon = 0.0001$, estimate the order of the Taylor polynomial that would be sufficient to satisfy the error limit.

Solution

As all orders of derivatives of $f(x) = e^x$ are the function itself, we can use formula (12.7) to estimate the required order of the Taylor polynomial as follows.

$E_n(x) = f^{(n+1)}(c)\dfrac{(x-a)^{n+1}}{(n+1)!} < \varepsilon$, in which $f^{(n+1)}(c) = e^c$, $\varepsilon = 0.0001$, $x = 1.5$, $a = 1$. Thus,

$$e^c\frac{(1.5-1)^{n+1}}{(n+1)!} < 0.0001 \longrightarrow \frac{0.5^{n+1}}{(n+1)!} < \frac{0.0001}{e^c} \longrightarrow \frac{0.5^n}{(n+1)!} < \frac{0.0001}{0.5e^c} \longrightarrow \frac{0.5^n}{(n+1)!} < \frac{0.0002}{e^c}$$

As c is within (1, 1.5), the largest value for c is 1.5. Therefore, a safer condition would be

$$\frac{0.5^n}{(n+1)!} < \frac{0.0002}{e^{1.5}} \longrightarrow \frac{0.5^n}{(n+1)!} < 0.00005$$

Let $F(n) = \dfrac{0.5^n}{(n+1)!}$. By trying n = 1, 2, …, we find that $F(4)$ < 0.00053 and $F(5)$ < 0.00005. Therefore, a 5th-order Taylor polynomial will be sufficient to guarantee that the approximate to $f(x) = e^x$ within [1, 1.5] is within the error limit $\varepsilon = 0.0001$.

Exercises 12.2

1. Given $f(x) = \dfrac{5}{\sqrt{x+1}}$, use linear, quadratic, and cubic Taylor polynomials to approximate $f(4)$ and estimate the errors respectively assuming $\sqrt{5} = 2.2361$.

2. Given $f(x) = \cos x$, use linear, quadratic, and cubic Taylor polynomials to approximate $f(62°) = \cos 62°$. Estimate the error of the quadratic Taylor polynomial in approximating $\cos 62°$, assuming $\cos 62° = 0.4695$.

3. For $f(x) = \ln x$, assuming $f(2) = \ln 2 = 0.69315$, if we want to make sure that the approximate to $f(x) = \ln x$ within [1.5, 2.5] using a Taylor polynomial is within an error limit $\varepsilon = 0.001$, estimate the order of the Taylor polynomial that would be sufficient to meet the error limit.

4. For $f(x) = \cos x$, if we want to make sure that the approximate to $f(x) = \cos x$ within [$\pi/6$, $\pi/3$] using a Taylor polynomial is within an error limit $\varepsilon = 0.0001$, estimate the order of the Taylor polynomial that would be sufficient to meet the error limit.

12.3 Taylor and Maclaurin Series

Taylor series

If $f(x)$ has infinitely many orders of derivatives, the nth-order Taylor polynomial can be extended to an infinite series as

$$p(x)=f(a)+f'(a)(x-a)+f''(a)\frac{(x-a)^2}{2!}+...+f^{(n)}(a)\frac{(x-a)^n}{n!}+... \tag{12.8}$$

Particularly, if $p(x)$ is convergent in a range within the domain where $f(x)$ is defined, $p(x)$ and $f(x)$ will be equivalent in that range where $p(x)$ is convergent, i.e.,

$$f(x)=p(x)=f(a)+f'(a)(x-a)+f''(a)\frac{(x-a)^2}{2!}+...+f^{(n)}(a)\frac{(x-a)^n}{n!}+... \tag{12.9}$$

This is called the Taylor series for function $f(x)$.

It is easy to understand that the nth-order Taylor polynomial of a function is just a special case of truncating the Taylor series by cutting its terms of the (n+1)th and higher orders. The error from such truncation can be estimated using formula (12.7).

Maclaurin series

If $f(x)$ and its infinitely many orders of derivatives exist at $x = a = 0$, the Taylor series defined in formula (12.9) becomes the Maclaurin series for function $f(x)$, subject to convergence of $p(x)$ in a range within the domain where $f(x)$ is defined, i.e.,

$$f(x)=p(x)=f(0)+f'(0)x+f''(0)\frac{x^2}{2!}+...+f^{(n)}(0)\frac{x^n}{n!}+... \tag{12.10}$$

Example 12.9: Find the Maclaurin series for $f(x)=\frac{1}{1-x}$ in (–1, 1).

Solution

We firstly find the first-order, the second-order and the third-order derivatives of $f(x)=\frac{1}{1-x}$.

$$f'(x)=(\frac{1}{1-x})'=[(1-x)^{-1}]'=-1(1-x)^{-2}(1-x)'=-1(1-x)^{-2}(-1)=(1-x)^{-2},$$
$$f''(x)=[(1-x)^{-2}]'=-2(1-x)^{-3}(1-x)'=-2(1-x)^{-3}(-1)=2\times 1(1-x)^{-3}=2!(1-x)^{-3},$$
$$f'''(x)=[2!(1-x)^{-3}]'=-3\times 2!(1-x)^{-4}(1-x)'=-3!(1-x)^{-4}(-1)=3!(1-x)^{-4}$$

This indicates that the nth-order of derivatives of $f(x)=\frac{1}{1-x}$ has the following pattern:

$$f^{(n)}(x)=n!(1-x)^{-(n+1)}$$

Let x = 0 to obtain all the required values:

$$f(0)=\frac{1}{1-x}=1,$$
$$f'(0)=(1-0)^{-2}=1,$$
$$f''(0)=2!(1-0)^{-3}=2!,$$
$$f'''(0)=3!(1-0)^{-4}=3!,$$
$$\ldots$$
$$f^{(n)}(0)=n!(1-0)^{-(n+1)}=n!.$$

Create the Maclaurin series using formulae (12.10).

$$f(x)=\frac{1}{1-x}=1+x+2!\frac{x^2}{2!}+3!\frac{x^3}{3!}\ldots+n!\frac{x^n}{n!}+\ldots=1+x+x^2+x^3\ldots+x^n+\ldots=\sum_{n=0}^{\infty}x^n.$$

Although $f(x)=\frac{1}{1-x}$ is defined in the whole domain except x = 1, its Maclaurin series is only convergent in (–1, 1). This is because $\sum_{n=0}^{\infty}x^n$ is simply an infinite geometric series of ratio of x, which by formula $S=S_\infty=\frac{a_1}{1-r}$ $(|r|<1)$ has a finite value only within $|x|<1$, or in (–1, 1).

Example 12.10: Find the Maclaurin series for $f(x)=e^x$.

Solution

As all orders of derivatives of $f(x)=e^x$ are the function itself, let x = 0 to obtain all the required values:

$$f(0)=f'(0)=f''(0)=\ldots=f^{(n)}(0)=\ldots=e^0=1.$$

Create the Maclaurin series using formulae (12.10).

$$f(x)=e^x=1+x+1\times\frac{x^2}{2!}+1\times\frac{x^3}{3!}\ldots+1\times\frac{x^n}{n!}+\ldots=1+x+\frac{x^2}{2!}+\frac{x^3}{3!}\ldots+\frac{x^n}{n!}+\ldots=\sum_{n=0}^{\infty}\frac{x^n}{n!}.$$

This series is convergent to $f(x)=e^x$ in the whole range.

It seems the Maclaurin series is simpler than the Taylor series for function $f(x)$. However, the Taylor series always exists for a well-defined function $f(x)$ in its range of convergence; such is not guaranteed for the Maclaurin series. For example, for $f(x) = \ln x$ in Example 12.4, its Maclaurin series does not exist as $f(x)$ and its derivatives at $x = 0$ do not exist.

Maclaurin series of common functions

Following the same process, we can find the Maclaurin series for some commonly used functions as follows.

$$\begin{aligned}
&e^x = 1 + x + \frac{x^2}{2!} + \frac{x^3}{3!} \ldots + \frac{x^n}{n!} + \ldots = \sum_{n=0}^{\infty} \frac{x^n}{n!} && -\infty < x < \infty \\
&\sin x = x - \frac{x^3}{3!} + \frac{x^5}{5!} - \frac{x^7}{7!} \ldots = \sum_{n=1}^{\infty} \frac{(-1)^{n-1} x^{2n-1}}{(2n-1)!} && -\infty < x < \infty \\
&\cos x = 1 - \frac{x^2}{2!} + \frac{x^4}{4!} - \frac{x^6}{6!} \ldots = \sum_{n=0}^{\infty} \frac{(-1)^n x^{2n}}{(2n)!} && -\infty < x < \infty \\
&\frac{1}{1-x} = 1 + x + x^2 + x^3 \ldots + x^n + \ldots = \sum_{n=0}^{\infty} x^n && -1 < x < 1 \\
&\frac{1}{1+x} = 1 - x + x^2 - x^3 \ldots + x^n + \ldots = \sum_{n=0}^{\infty} (-1)^n x^n && -1 < x < 1 \\
&\ln(1+x) = x - \frac{x^2}{2} + \frac{x^3}{3} - \frac{x^4}{4} \ldots = \sum_{n=1}^{\infty} (-1)^{n-1} \frac{x^n}{n} && -1 < x \le 1
\end{aligned} \tag{12.11}$$

Operations with Maclaurin series

If two or more Maclaurin series are defined in the same range, addition and/or subtraction among these Maclaurin series follows the term-wise addition and/or subtraction and the resultant series is also defined in the same range, i.e.,

$$\sum_{n}^{\infty} a_n(x) \pm \sum_{n}^{\infty} b_n(x) = \sum_{n}^{\infty} [a_n(x) \pm b_n(x)] \tag{12.12}$$

If a Maclaurin series is convergent to $f(x)$, differentiation (or integration) can be applied to both $f(x)$ and the Maclaurin series simultaneously, subject to the existence of the derivatives (or integrals) of all involved terms, i.e.,

$$f(x) = \sum_{n}^{\infty} a_n(x) \longrightarrow f'(x) = \sum_{n}^{\infty} a_n'(x) \tag{12.13}$$

Example 12.11: Given $\sin x = x - \frac{x^3}{3!} + \frac{x^5}{5!} - \frac{x^7}{7!} ...$, find the Maclaurin series for $f(x) = \cos x$.

Solution

As $\sin x = x - \frac{x^3}{3!} + \frac{x^5}{5!} - \frac{x^7}{7!} ...$, differentiate both sides of this series with respect to x:

$$(\sin x)' = (x - \frac{x^3}{3!} + \frac{x^5}{5!} - \frac{x^7}{7!} ...)'$$

$$\cos x = (x)' - (\frac{x^3}{3!})' + (\frac{x^5}{5!})' - (\frac{x^7}{7!})' + ... = 1 - 3\times\frac{x^2}{3!} + 5\times\frac{x^4}{5!} - 7\times\frac{x^6}{7!} + ...$$

$$= 1 - \frac{x^2}{2!} + \frac{x^4}{4!} - \frac{x^6}{6!} + ... = \sum_{n=0}^{\infty} \frac{(-1)^n x^{2n}}{(2n)!}.$$

Example 12.12: Find the Maclaurin series for $f(x) = \frac{2}{1-x^2}$.

Solution

Method 1

As $f(x) = \frac{2}{1-x^2} = \frac{1}{1-x} + \frac{1}{1+x} = a(x) + b(x)$ where $a(x) = \frac{1}{1-x}$ and $b(x) = \frac{1}{1+x}$, by formulae (12.11) the Maclaurin series for $a(x)$ and $b(x)$ can be found as:

$$a(x) = \frac{1}{1-x} = 1 + x + x^2 + x^3 ... + x^n + ...; \qquad b(x) = \frac{1}{1+x} = 1 - x + x^2 - x^3 ... + x^n + ...$$

By formula (12.12),

$$f(x) = \frac{2}{1-x^2} = a(x) + b(x) = \frac{1}{1-x} + \frac{1}{1+x}$$

$$= (1 + x + x^2 + x^3 ... + x^n + ...) + (1 - x + x^2 - x^3 ... + x^n + ...) = 2 + 2x^2 + 2x^4 ... + 2x^{2n} + ...$$

$$= 2(1 + x^2 + x^4 ... + x^{2n} + ...) = 2\sum_{n=0}^{\infty} x^{2n}.$$

Method 2

As $f(x) = \frac{2}{1-x^2} \xrightarrow{t=x^2} g(t) = \frac{2}{1-t}$, by formulae in (12.11) the Maclaurin series for $g(t)$ can be found as:

$$g(t) = \frac{2}{1-t} = 2(1 + t + t^2 + t^3 ... + t^n + ...).$$

Thus, by backward substitution,

$$f(x)=\frac{2}{1-x^2}\xleftrightarrow{t=x^2}g(t)=2(1+t+t^2+t^3...+t^n+...)=2(1+x^2+x^4...+x^{2n}+...)=2\sum_{n=0}^{\infty}x^{2n}.$$

Example 12.13: Given $\sin x = x-\frac{x^3}{3!}+\frac{x^5}{5!}-\frac{x^7}{7!}...$, approximate sin30° using the first three terms of this Maclaurin series. Assess the error of this approximation.

__Solution__

We first convert the angle in degrees into radians:

$$30° = \frac{\pi}{6} \approx 0.52360 .$$

Then use the first three terms of this Maclaurin series to approximate sin30° as follows:

$$\sin 30° \approx x-\frac{x^3}{3!}+\frac{x^5}{5!} = 0.52360-\frac{0.52360^3}{3!}+\frac{0.52360^5}{5!} \approx 0.52360-0.02392+0.00033 = 0.50001$$

Since sin30° = 0.5, the error is only about 0.00001, a very accurate approximate. This means other terms higher than x^5 all together contributed at most |0.00001| to the true value.

Exercises 12.3

1. Find the Maclaurin series for $f(x)=\frac{1-x}{1+x}$.

2. Find the Maclaurin series for $f(x)=\cos x$.

3. Given $\sin x = x-\frac{x^3}{3!}+\frac{x^5}{5!}-\frac{x^7}{7!}...$ and $\cos x = 1-\frac{x^2}{2!}+\frac{x^4}{4!}-\frac{x^6}{6!}...$, find the Maclaurin series for $f(x)=\sin x+\cos x$.

4. Given $\cos x = 1-\frac{x^2}{2!}+\frac{x^4}{4!}-\frac{x^6}{6!}...$, find the Maclaurin series for $f(x)=\sin x$.

5. Given $\frac{1}{1-x}=1+x+x^2+x^3...+x^n+...$, find the Maclaurin series for $f(x)=\frac{3}{x-3}$.

6. Approximate cos45° using the first three terms of Maclaurin series $\cos x = 1-\frac{x^2}{2!}+\frac{x^4}{4!}-\frac{x^6}{6!}...$. Assess the error of this approximation.

Chapter 12: Exercises Answers

Exercises 12.1

1. $x_4 \approx 1.5052$
2. $x_4 \approx 4.1636$
3. $x_5 \approx 0.6419$
4. $x_4 \approx 0.6508$
5. $x_4 \approx 2.4262$
6. $x_4 \approx 0.5289$
7. $x_3 \approx 0.4240$
8. $x_5 \approx 0.7384$.
9. $x_3 \approx 1.3234$

Exercises 12.2

1. $\sqrt{5} \approx p_1(4) \approx 2.1875, \ \sqrt{5} \approx p_2(4) \approx 2.2461, \ \sqrt{5} \approx p_3(4) \approx 2.2339.$
2. $p_1(62°) \approx 0.46977, \ p_2(62°) \approx 0.46947, \ p_3(62°) \approx 0.46951.$
3. $F(n) = (n+1)3^{n+1} \longrightarrow n \geq 4$
4. $F(n) = \dfrac{(\frac{\pi}{6})^{n+1}}{(n+1)!} \longrightarrow n \geq 5$

Exercises 12.3

1. $f(x) = \dfrac{1-x}{1+x} = 1 - 2x + 2x^2 - 2x^3 + 2x^4 - 2x^5 ... - 2x^{n+1} + ...$
2. $\cos x = 1 - \dfrac{x^2}{2!} + \dfrac{x^4}{4!} - \dfrac{x^6}{6!} + ...$
3. $f(x) = \sin x + \cos x = 1 + x - \dfrac{x^2}{2!} - \dfrac{x^3}{3!} + \dfrac{x^4}{4!} + \dfrac{x^5}{5!} - \dfrac{x^6}{6!} - \dfrac{x^7}{7!} ...$
4. $\sin x = -(\cos x)' = x - \dfrac{x^3}{3!} + \dfrac{x^5}{5!} ...$
5. $f(x) = -\sum_{n=0}^{\infty} (\frac{x}{3})^n$
6. $\cos 45° \approx 1 - \frac{1}{2!}(\frac{\pi}{4})^2 + \frac{1}{4!}(\frac{\pi}{4})^4 = 1 - \frac{1}{2}(\frac{\pi}{4})^2 + \frac{1}{24}(\frac{\pi}{4})^4 \approx 0.7074, \qquad e = |0.7071 - 0.7074| = 0.0003$

CHAPTER 13

13 Differentials and Applications

CHAPTER OBJECTIVES

- Introduce the concept of differentials
- Introduce error analysis using differentials
- Use differentials for approximate computation

Essential statements on differentials:

- The differential of a function is the small change in the dependent variable with respect to the small change in the independent variable of the function at a point.
- The small change in the dependent variable is proportional to the small change in the independent variable by the derivative of the function at a point.
- The derivative and the differential of the same function are related but have different meanings, with different applications.

Key topics:

- Differential of a function at a point
- Small changes
- Absolute and relative errors and error bounds
- Linear approximation

Flowchart of mathematical knowledge building

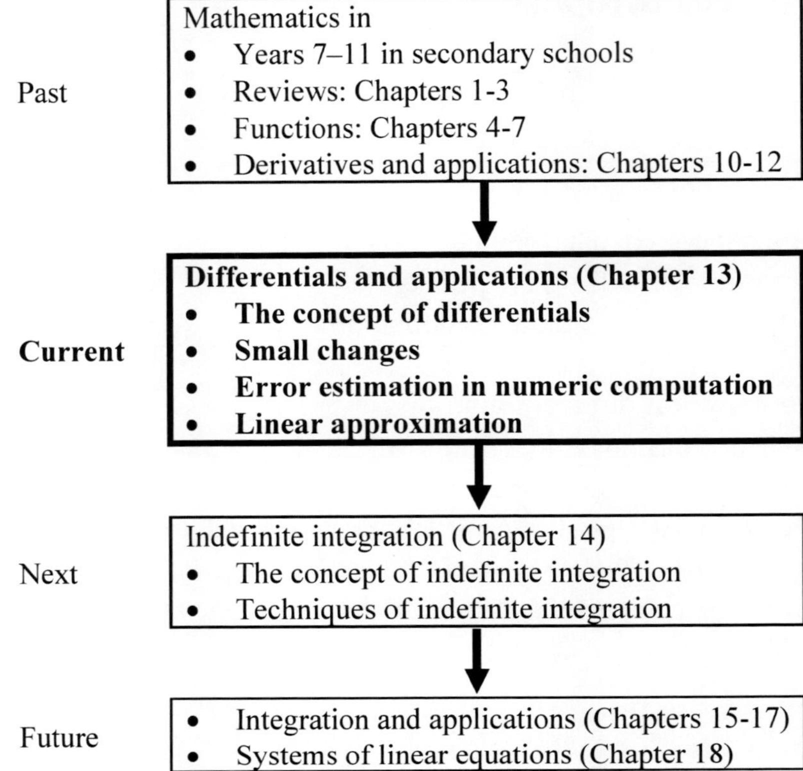

Chapter 13: Differentials and Applications

13.1 Differentials of Functions

13.1.1 Small changes and differentials of functions

The derivative of function $y = f(x)$ with respect to x is the limit of the average *rate of change*, i.e.,

$$\frac{dy}{dx} = y' = f'(x) = \lim_{\delta x \to 0} \frac{\delta y}{\delta x} = \lim_{\delta x \to 0} \frac{f(x+\delta x) - f(x)}{\delta x}.$$

The derivative $f'(x)$ measures *how fast* $f(x)$ changes if x has a change at a given point. Suppose dx is a *small change* in x at a point, dy should be the corresponding *instantaneous small change* in $f(x)$ at the same point. By derivative, the relationship between these two small changes is:

$$\boxed{dy = y'dx = f'(x)dx = \frac{d[f(x)]}{dx}dx = d[f(x)].} \qquad (13.1)$$

This relationship between the two *instantaneous small changes dy* and dx at point x is called the *differential of function* $y = f(x)$ at point x. For $y = f(x)$, its differential dy is simply its derivative y' or $f'(x)$ appended by dx. The differentials of common functions are listed in Table 13.1.

Table 13.1: Differentials of common functions

$f(x)$	$d[f(x)]$	$f'(x)dx$	$f(x)$	$d[f(x)]$	$f'(x)dx$
c (constant)	$d[c]$	0	$\cot ax$	$d[\cot ax]$	$-a\csc^2 axdx$
x^r (r: real number)	$d[x^r]$	$rx^{r-1}dx$	$\sec ax$	$d[\sec ax]$	$\frac{a\sin ax}{\cos^2 ax}dx$
e^{rx}	$d[e^{rx}]$	$re^{rx}dx$	$\csc ax$	$d[\csc ax]$	$\frac{-a\cos ax}{\sin^2 ax}dx$
a^{rx} $(a>0)$	$d[a^{rx}]$	$ra^{rx}\ln adx$	$\arcsin ax$	$d[\arcsin ax]$	$\frac{a}{\sqrt{1-(ax)^2}}dx$
x^{ax} $(x>0)$	$d[x^{ax}]$	$ax^{ax}(1+\ln x)dx$	$\arccos ax$	$d[\arccos ax]$	$\frac{-a}{\sqrt{1-(ax)^2}}dx$
$\ln x$	$d[\ln x]$	$\frac{dx}{x}$	$\arctan ax$	$d[\arctan ax]$	$\frac{a}{1+(ax)^2}dx$
$\log_a x$	$d[\log_a x]$	$\frac{dx}{x\ln a}$	$\operatorname{arc\,cot} ax$	$d[\operatorname{arc\,cot} ax]$	$\frac{-a}{1+(ax)^2}dx$
$\sin ax$	$d[\sin ax]$	$a\cos axdx$	$\sinh ax$	$d[\sinh ax]$	$a\cosh axdx$
$\cos ax$	$d[\cos ax]$	$-a\sin axdx$	$\cosh ax$	$d[\cosh ax]$	$a\sinh axdx$
$\tan ax$	$d[\tan ax]$	$a\sec^2 axdx$	$\tanh ax$	$d[\tanh ax]$	$\frac{a}{\cosh^2 ax}dx$

Example 13.1: Determine the differential for $y = \sin x + \cos x$ and $y = 3\ln x$ respectively.

Solution

For $y = \sin x + \cos x$,

$$dy = y'dx = (\sin x + \cos x)'dx = (\cos x - \sin x)dx = \cos x dx - \sin x dx.$$

For $y = 3\ln x$,

$$dy = y'dx = (3\ln x)'dx = \frac{3}{x}dx = \frac{3dx}{x}.$$

Example 13.2: Determine the differential for $y = e^x(x^2+1)\cos 2x$.

Solution

For $y = e^x(x^2+1)\cos 2x$,

$$\begin{aligned} y' &= [e^x(x^2+1)\cos 2x]' = (e^x)'(x^2+1)\cos 2x + e^x(x^2+1)'\cos 2x + e^x(x^2+1)(\cos 2x)' \\ &= e^x(x^2+1)\cos 2x + e^x(2x)\cos 2x + e^x(x^2+1)(-2\sin 2x) \\ &= e^x[(x^2+1)\cos 2x + 2x\cos 2x - 2(x^2+1)\sin 2x] = e^x[(x^2+2x+1)\cos 2x - 2(x^2+1)\sin 2x] \\ &= e^x[(x+1)^2\cos 2x - 2(x^2+1)\sin 2x] \end{aligned}$$

$$\therefore dy = y'dx = e^x[(x+1)^2\cos 2x - 2(x^2+1)\sin 2x]dx.$$

Example 13.3: Determine the differential for $y = (x+1)\ln(x^2-1)$ at $x = 2$.

Solution

For $y = (x+1)\ln(x^2-1)$,

$$\begin{aligned} y' &= [(x+1)\ln(x^2-1)]' = (x+1)'\ln(x^2-1) + (x+1)[\ln(x^2-1)]' \\ &= \ln(x^2-1) + (x+1)\frac{2x}{x^2-1} = \ln(x^2-1) + (x+1)\frac{2x}{(x-1)(x+1)} = \ln(x^2-1) + \frac{2x}{x-1}. \end{aligned}$$

$$\therefore dy = y'dx = [\ln(x^2-1) + \frac{2x}{x-1}]dx.$$

At $x = 2$,

$$dy\big|_{x=2} = [\ln(2^2-1) + \frac{2\times 2}{2-1}] = (\ln 3 + 4)dx.$$

Example 13.4: Determine the differential for $y = \dfrac{e^{-x} - e^{x}}{\sin x - \cos x}$.

Solution

For $y = \dfrac{e^{-x} - e^{x}}{\sin x - \cos x}$,

$$y' = \left(\frac{e^{-x} - e^{x}}{\sin x - \cos x}\right)' = \frac{(e^{-x} - e^{x})'(\sin x - \cos x) - (e^{-x} - e^{x})(\sin x - \cos x)'}{(\sin x - \cos x)^2}$$

$$= \frac{(-e^{-x} - e^{x})(\sin x - \cos x) - (e^{-x} - e^{x})(\cos x + \sin x)}{\sin^2 x - 2\sin x \cos x + \cos^2 x}$$

$$= \frac{-e^{-x}\sin x + \cancel{e^{-x}\cos x} - \cancel{e^{x}\sin x} + e^{x}\cos x - \cancel{e^{-x}\cos x} - e^{-x}\sin x + e^{x}\cos x + \cancel{e^{x}\sin x}}{1 - \sin 2x}$$

$$= \frac{2e^{x}\cos x - 2e^{-x}\sin x}{1 - \sin 2x} = \frac{2(e^{x}\cos x - e^{-x}\sin x)}{1 - \sin 2x}.$$

$$\therefore dy = y'dx = \frac{2(e^{x}\cos x - e^{-x}\sin x)}{1 - \sin 2x}dx.$$

Example 13.5: Determine the differential for $y^2 + e^{-x} + x^2 = xy + 2$ at (0, 1).

Solution

For $y^2 + e^{-x} + x^2 = xy + 2$,

$$(y^2 + e^{-x} + x^2)' = (xy + 2)' \longrightarrow 2yy' - e^{-x} + 2x = y + xy' \longrightarrow 2yy' - xy' = y - 2x + e^{-x}$$

$$(2y - x)y' = y - 2x + e^{-x} \longrightarrow y' = \frac{y - 2x + e^{-x}}{2y - x}$$

$$\therefore dy = y'dx = \frac{y - 2x + e^{-x}}{2y - x}dx.$$

At (0, 1),

$$dy\big|_{(0,1)} = \frac{1 - 0 + e^{0}}{2 - 0} = \frac{2}{2}dx = dx.$$

13.1.2 Estimation of small changes of a function at a given point

As both differentials dy and dx are abstract terms of *small changes* in theory, in practice both dy and dx are replaced by *real small changes* of δy and δx respectively. The relationship (13.1) then becomes a formula for approximating real small changes as

$$\boxed{\delta y \approx y'\delta x = f'(x)\delta x.} \qquad (13.2)$$

Note if the given point is x_1, y' or $f'(x)$ must be $y'(x_1)$ or $f'(x_1)$.

Example 13.6: Determine δy for $y = (x+1)\ln(x^2-1)$ at $x = 2$ given $\delta x = 0.01$.

Solution

In Example 13.3, the differential for $y = (x+1)\ln(x^2-1)$ at $x = 2$ is found as

$$dy\big|_{x=2} = [\ln(2^2-1) + \frac{2\times 2}{2-1}] = (\ln 3 + 4)dx.$$

By formula (13.2), given $\delta x = 0.01$, the corresponding change δy for $y = (x+1)\ln(x^2-1)$ at $x = 2$ is

$$\delta y\big|_{x=2} \approx (\ln 3 + 4)\times 0.01 \approx 0.05.$$

Example 13.7: A radioactive substance decays by $N = N_0 e^{-\lambda t}$, in which N_0 is the initial amount and λ (per day) is the decay constant. If $N_0 = 100$ g and $\lambda = 0.02$ per day, estimate the daily change of the substance around the third day and the 100th day respectively.

Solution

For $N = N_0 e^{-\lambda t}$, its derivative with respect to time t is

$$\frac{dN}{dt} = (N_0 e^{-\lambda t})' = -\lambda N_0 e^{-\lambda t}. \text{ Hence } \quad \delta N \approx N'\delta t = -\lambda N_0 e^{-\lambda t}\delta t.$$

Substitute $N_0 = 100$ g, $\lambda = 0.02$ per day, and $\delta t = 1$ (daily) into δN:

$$\delta N \approx -\lambda N_0 e^{-\lambda t}\delta t = -0.02\times 100 e^{-0.02t}\times 1 = -2e^{-0.02t}.$$

$$t = 3rd\ day: \quad \delta N \approx -2e^{-0.02t} = -2e^{-0.02\times 3} = -2e^{-0.06} \approx -1.8835 \text{ g/day}$$

$$t = 100th\ day: \quad \delta N \approx -2e^{-0.02t} = -2e^{-0.02\times 100} = -2e^{-2} \approx -0.2707 \text{ g/day}.$$

The substance decays faster in the earlier period and slower in the later stage. The negative sign indicates the amount of substance decreases with time.

Example 13.8: Given two spherical containers with different radiuses of 1 metre and 20 metres respectively, if the radius has a small change of 0.05 metres due to the daily change in temperature, estimate the changes in volume for the two spherical containers respectively.

Solution

The volume of a sphere is $V(r)=\frac{4\pi}{3}r^3$ and its derivative with respect to radius r is

$$V'(r)=(\frac{4\pi}{3}r^3)'=4\pi r^2. \text{ Hence}$$

$$\delta V \approx V'(r)\delta r = 4\pi r^2 \delta r.$$

For the spherical container with radius $r = 1$ metre, a small change in the radius $\delta r = 0.05$ metres would cause a change in volume as follow:

$$\delta V(1) \approx V'(1)\delta r = 4\pi 1^2(0.05) \approx 4\times 3.141593\times 0.05 \approx 0.6283 \text{ m}^3.$$

For the spherical container with radius $r = 20$ metres, the change in its volume is

$$\delta V(20) \approx V'(20)\delta r = 4\pi 20^2(0.05) \approx 4\times 3.141593\times 400\times 0.05 \approx 251.3274 \text{ m}^3.$$

Example 13.9: Determine δy for $y^2+yx^2-x^3=1$ at (1, 1) given $\delta x = 0.03$.

Solution

For $y^2+yx^2-x^3=1$,

$$(y^2+yx^2-x^3)'=(1)' \longrightarrow 2yy'+y'x^2+2xy-3x^2=0 \longrightarrow 2yy'+x^2y'=3x^2-2xy$$

$$(2y+x^2)y'=x(3x-2y) \longrightarrow y'=\frac{x(3x-2y)}{2y+x^2}$$

$$\therefore dy = y'dx = \frac{x(3x-2y)}{2y+x^2}dx.$$

At (1, 1),

$$dy\big|_{(1,1)} = \frac{x(3x-2y)}{2y+x^2} = \frac{1\times(3\times 1-2\times 1)}{2\times 1+1}dx = \frac{1}{3}dx.$$

By formula (13.2), given $\delta x = 0.03$, the corresponding change δy at (1, 1) is

$$\delta y\big|_{(1,1)} \approx \frac{1}{3}\delta x = \frac{1}{3}\times 0.03 = 0.01.$$

Exercises 13.1

1. Determine the differentials of following functions.

(a) $y = x^4 - 2x^3 - x^2 + 5x + 3$; (b) $y = x^2 - 2x + 5$; (c) $y = \sqrt[3]{8x^2}$;

(d) $y = \dfrac{3}{\sqrt{x^3}}$; (e) $y = e^{2x}$; (f) $y = e^{-3x}$; (g) $y = \ln x^{\frac{2}{3}}$;

(h) $y = \dfrac{x^2 - e^{2x}}{\ln x}$; (i) $y = \dfrac{4\sqrt{x}}{e^{2x}}$; (j) $y = \sin 3x \cos 2x$;

(k) $y = \dfrac{\tan x}{\cos 2x}$; (l) $y = \dfrac{3e^{-2x} - 2e^{3x}}{3x^2}$; (m) $y = \dfrac{\ln x}{\sqrt{2x}}$; (n) $y = -2x \ln \dfrac{1}{x}$.

2. In the following questions, assume $x = 0$ and $\delta x = 0.05$. Estimate the change in y with respect to the change in x respectively.

(a) $y = x^2 + 3x - 4$; (b) $y = -2\sqrt{x + 4}$; (c) $y = e^{-3x}$;

(d) $y = 2x^2 \ln \sqrt{1 + x}$; (e) $y = \sin 3x + 2\cos 2x$.

13.2 Applications of Differentials

13.2.1 Error estimation in numeric computation

Absolute error and error bound

The result of numeric computation is an approximate to the exact value of a function. Therefore, we must assess how close the approximate value is to the exact value. The difference between the exact value (y^*) and the approximate value (y) is called the ***error*** (e), i.e.,

$$e = y^* - y. \tag{13.3}$$

In practice, the exact value is often unknown; hence we usually regard an approximate value that falls within a pre-set limit of error as an acceptable solution to a given problem. Such a pre-set limit of error is called the ***error bound*** denoted as ε. Apply this definition to formula (13.3),

$$|e| = |y^* - y| \le \varepsilon. \tag{13.4}$$

Relative error and error bound

The relative error of the approximate value y to the exact value y^* is defined as

$$e_r = \frac{e}{y^*} = \frac{y^* - y}{y^*} \approx \frac{e}{y} \quad (e << y). \tag{13.5}$$

Since the exact value is often unknown, the relative error is estimated by using the approximate value if it is far greater than the absolute error. Similarly, the relative error bound ε_r is defined as

$$|e_r| = \left|\frac{y^* - y}{y^*}\right| \approx \left|\frac{e}{y}\right| \le \varepsilon_r. \tag{13.6}$$

Errors of elementary functions

Errors can be accumulated and propagated with multiple operations during numeric computation. Generally, for a function with only one independent variable $y = f(x)$, an error (δx) in the independent variable x will cause an error (δy) for the dependent variable y as follows

$$\delta y \approx |f'(x)\delta x|. \tag{13.7}$$

The correspondent relative error is

$$e_r = \left|\frac{\delta y}{y}\right| \approx \left|\frac{f'(x)}{f(x)}\delta x\right|. \tag{13.8}$$

Example 13.10: In calculating the volume of a cylinder $V = \pi R^2 H$, if you take $\pi = 3.14$ to approximate $\pi = 3.141593$, an error in calculating the volume will occur. Assuming the radius of the base and the height of the cylinder share the same size of 1 metre, determine the absolute error and relative error in volume caused by the error (difference) of using $\pi = 3.14$ rather than $\pi = 3.141593$. If the radius and the height of the cylinder are 10 metres and 20 metres respectively, reassess the absolute error and relative error in volume caused by the same error in π.

Solution

Although the volume of a cylinder $V = \pi R^2 H$ is a function of three parameters (π, R, H), this question implies that there is no error in both the radius and the height of the cylinder. Therefore, the only source of error in this case is *variable* π, which has an error $\delta\pi = 3.141593 - 3.14 = 0.001593$. The corresponding absolute error and relative error in volume are

$$\delta V \approx \left|V'\delta\pi\right| = \left|\frac{dV}{d\pi}\delta\pi\right| = \left|R^2 H\delta\pi\right|,$$

$$e_V = \left|\frac{\delta V}{V}\right| \approx \left|\frac{R^2 H\delta\pi}{\pi R^2 H}\right| = \left|\frac{\delta\pi}{\pi}\right|.$$

For $R = H = 1$ metre,

$$\delta V \approx \left|R^2 H\delta\pi\right| = 1^2 \times 1 \times 0.001593 = 0.001593\ m^3,$$

$$e_V \approx \left|\frac{\delta\pi}{\pi}\right| = \frac{0.001593}{3.14} = 0.000507 = 0.05\%.$$

For $R = 10$ metres and $H = 20$ metres,

$$\delta V \approx \left|R^2 H\delta\pi\right| = 10^2 \times 20 \times 0.001593 = 3.186\ m^3,$$

$$e_V \approx \left|\frac{\delta\pi}{\pi}\right| = \frac{0.001593}{3.14} = 0.000507 = 0.05\%.$$

Where the size (radius and height) of a cylinder is relatively small, a difference of 0.001593 for π does not result in significant difference in calculating the volume for the cylinder. Where the size of a cylinder is relatively large, the same difference in π will produce noticeable difference in calculating the volume for the cylinder. However, the change in the size of a cylinder does not affect the *relative error* in the volume as no error is supposed in both the radius and height.

Example 13.11: Use the outcomes in Example 13.8 to estimate the relative changes (similar to relative errors) in the volume of the two spherical containers caused by the small change $\delta r = 0.05$ metres in the radius respectively.

Solution

The relative error in volume of a sphere is

$$e_r \approx \frac{\delta V}{V} = \frac{4\pi r^2 \delta r}{\frac{4\pi}{3} r^3} = \frac{3\delta r}{r}.$$

For the spherical container with radius $r = 1$ metre,

$$e_r \approx \frac{3\delta r}{r} = \frac{3 \times 0.05}{1} = 0.15 = 15\%.$$

For the spherical container with radius $r = 20$ metres,

$$e_r \approx \frac{3\delta r}{r} = \frac{3 \times 0.05}{20} = 0.0075 = 0.75\%.$$

This indicates that the absolute error of the spherical container of 1-metre radius is 0.6283 m^3 but this is about 15% of the volume of this sphere. The absolute error of the spherical container with a radius of 20 metres is as large as 251 m^3 but it is only about 0.75% of the volume of this bigger sphere.

Example 13.12: An elliptical sport field is constructed with a semi-major axis of 75 m and a semi-minor axis of 50 m. If the measurement on the semi-major and semi-minor axes has an error of 0.1 m, estimate the absolute error and relative error in its area caused by the error in measurement.

Solution

The area of an elliptical shape is $A = \pi ab$ where a and b are the semi-major and semi-minor axes respectively. Since the error in measurement of the semi-major axis and the semi-minor axis occurs independently, the total error is roughly the sum of the errors occurred along these two directions. Note when measuring a, the length of b has nothing to do with a (or b is constant); when measuring b, the length of a has nothing to do with b (or a is constant). Therefore,

$$\delta A_a \approx \frac{dA}{da}\delta a = \frac{d(\pi ab)}{da}\delta a = \pi b \delta a, \quad \delta A_b \approx \frac{dA}{db}\delta b = \frac{d(\pi ab)}{db}\delta b = \pi a \delta b$$

$$\delta A \approx \delta A_a + \delta A_b \approx \pi b \delta a + \pi a \delta b = \pi(a\delta b + b\delta a) = \pi(a+b)\delta a \longleftarrow \delta a = \delta b$$

$$e_A = \left|\frac{\delta A}{A}\right| \approx \left|\frac{\pi(a+b)\delta a}{\pi ab}\right| = \frac{(a+b)\delta a}{ab}$$

For $a = 75$ m, $b = 50$ m and $\delta a = \delta b = 0.1\,\text{m}$,

$$\delta A \approx \pi(a+b)\delta a = \pi(75+50)\times 0.1 = 12.5\pi \approx 39.27\ m^2$$

$$e_A \approx \frac{(a+b)\delta a}{ab} = \frac{(75+50)\times 0.1}{75\times 50} = \frac{12.5}{3750} \approx 0.0033 = 0.33\%.$$

13.2.2 Linear approximation

Formula (13.2) for estimating small changes can be reorganised as

$$\boxed{\begin{array}{l} y_2 - y_1 \approx f'(x_1)(x_2 - x_1) \text{ or} \\ y_2 \approx y_1 + f'(x_1)(x_2 - x_1), \end{array}} \qquad (13.9)$$

where y_1 is known at x_1 and y_2 is unknown at $\boldsymbol{x_2}$ **that is near** $\boldsymbol{x_1}$. This is called the formula of linear approximation because it is similar to a linear equation in the point-slope form of $y_2 - y_1 = k(x_2 - x_1)$. Linear approximation is to use the value at the known point (x_1, y_1) and the tangent at x_1 to estimate the value y_2 at the near point (x_2, y_2) shown in Figure 13.1.

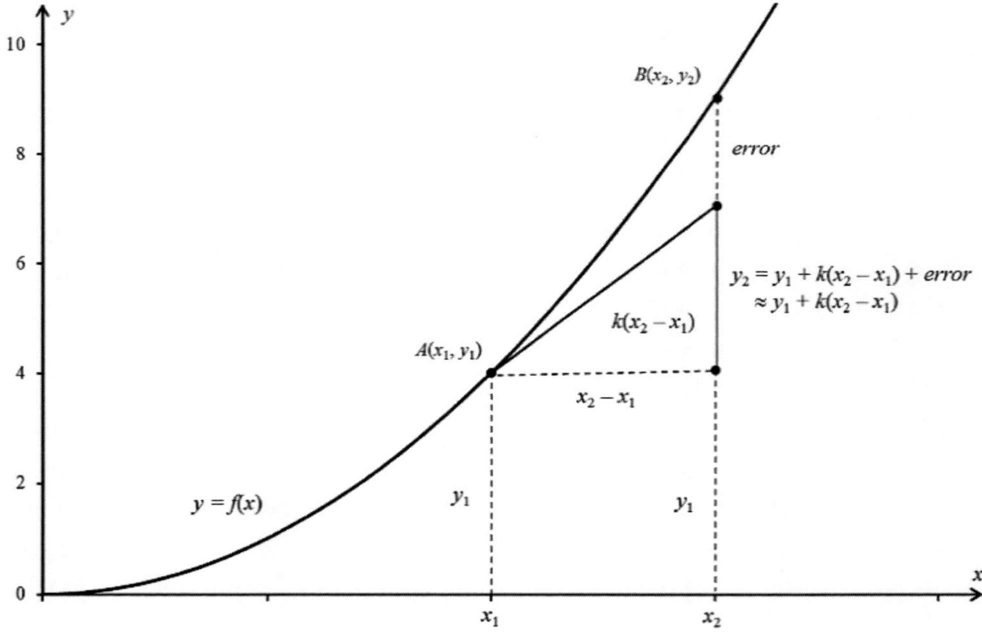

Figure 13.1: The concept of linear approximation

There are two key factors when choosing x_1 and x_2:

- The two points x_1 and x_2 must be close to each other, i.e., $\delta x = x_2 - x_1$ should be small to reduce the error of approximation shown in Figure 13.1.
- $y_1 = f(x_1)$ must be a value that is given or easily obtained without error. This avoids passing any existing error from y_1 onto y_2 before the actual approximation takes place.

Example 13.13: Approximate $y = \sqrt{3x+1}$ at $x = 0.99$ and $x = 1.004$ respectively.

Solution

Let $x_1 = 1$. Then $y_1 = \sqrt{3x+1} = \sqrt{4} = 2$, without any error. For $y = \sqrt{3x+1}$,

$$y' = (\sqrt{3x+1})' = \frac{1}{2}(3x+1)^{\frac{1}{2}-1}(3x+1)' = \frac{3}{2}(3x+1)^{-\frac{1}{2}} = \frac{3}{2\sqrt{3x+1}}.$$

At $x_1 = 1$,

$$y'(1) = \frac{3}{2\sqrt{3\times 1+1}} = \frac{3}{2\sqrt{4}} = \frac{3}{4}.$$

For $x = 0.99$ and $x = 1.004$, by formula (13.9),

$$y(0.99) \approx y(1) + y'(1)(0.99-1) = 2 - \frac{3}{4}\times 0.01 = 2 - 0.0075 = 1.9925,$$

$$y(1.004) \approx y(1) + y'(1)(1.004-1) = 2 + \frac{3}{4}\times 0.004 = 2 + 0.003 = 2.003.$$

These values are almost the same as the exact values.

Example 13.14: Approximate $y = \sin 25°$.

Solution

Since $25° = 30° - 5° = \frac{\pi}{6} - \frac{5\pi}{180} = \frac{\pi}{6} - 0.08727$, then $y = \sin 25° = \sin(\frac{\pi}{6} - 0.08727)$.

Let $x_1 = \frac{\pi}{6}$. Then $y_1 = \sin\frac{\pi}{6} = \frac{1}{2}$, without any error. For $y = \sin x$, $y' = (\sin x)' = \cos x$.

At $x_1 = \frac{\pi}{6}$, $y'(\frac{\pi}{6}) = \cos\frac{\pi}{6} = \frac{\sqrt{3}}{2}$.

For $x = 25° = \frac{\pi}{6} - 0.08727$, by formula (13.9),

$$\sin 25° \approx \sin\frac{\pi}{6} + y'(\frac{\pi}{6})(\frac{\pi}{6} - 0.08727 - \frac{\pi}{6}) = \frac{1}{2} - \frac{\sqrt{3}}{2}\times 0.08727 \approx 0.4244.$$

The exact value of $\sin 25°$ is 0.4226. Hence the error is 0.0018.

Example 13.15: Approximate the value of the following function near the given point.

$$y = e^x\sqrt{x^2 - 3x + 2}, \quad x_1 = 0 \text{ and } \delta x = 0.05$$

Solution

Since $x_1 = 0$, then $y_1 = e^0\sqrt{0-0+2} = \sqrt{2}$, $x_2 = x_1 + \delta x = 0 + 0.05 = 0.05$. Note $y_1 = \sqrt{2}$ is an irrational number and will be kept as it is until the final calculation to reduce error accumulation.

For $y = e^x\sqrt{x^2 - 3x + 2}$,

$$y' = (e^x\sqrt{x^2-3x+2})' = e^x\sqrt{x^2-3x+2} + e^x(\sqrt{x^2-3x+2})' = e^x[\sqrt{x^2-3x+2} + \frac{(x^2-3x+2)'}{2\sqrt{x^2-3x+2}}]$$

$$= e^x[\sqrt{x^2-3x+2} + \frac{2x-3}{2\sqrt{x^2-3x+2}}]$$

At $x_1 = 0$,

$$y'(0) = e^0[\sqrt{0-0+2} + \frac{0-3}{2\sqrt{0-0+2}}] = \sqrt{2} - \frac{3}{2\sqrt{2}} = \sqrt{2} - \frac{3\sqrt{2}}{4} = \frac{4\sqrt{2}-3\sqrt{2}}{4} = \frac{\sqrt{2}}{4}.$$

For $x_2 = 0.05$, by formula (13.9),

$$y(0.05) \approx y(0) + y'(0)(0.05-0) = \sqrt{2} + \frac{\sqrt{2}}{4}\times 0.05 = \sqrt{2}(1+0.0125) = 1.025\sqrt{2} \approx 1.4319$$

By using calculator directly, $y(0.05) \approx 1.4308$, very close to the approximated value.

Exercises 13.2

1. In the following questions, assume $x = 1$ and $\delta x = 0.01$. Estimate the absolute and relative errors in y with respect to the error in x respectively.

   (a) $y = x^2 + 3x - 4$; (b) $y = -2\sqrt{x}$; (c) $y = e^{-3x}$; (d) $y = 2x^2 \ln\sqrt{1+x}$.

2. Approximate the values of the following functions near the given point.

   (a) $y = x^3 - 2x^2 + 3,\ \ x_1 = 1$ and $\delta x = 0.01$;

   (b) $y = \ln\sqrt{x^2 + 3},\ \ x_1 = 1$ and $\delta x = 0.02$;

   (c) $y = 2\sin x + 3\cos x,\ \ x_1 = \pi/3$ and $\delta x = 0.01$;

   (d) $y = (x+2)^2 e^x,\ \ x_1 = 0$ and $\delta x = 0.01$.

Chapter 13: Exercises Answers

Exercises 13.1

1. (a) $dy = (4x^3 - 6x^2 - 2x + 5)dx$ (b) $dy = (2x - 2)dx$ (c) $dy = \frac{4}{3\sqrt[3]{x}}dx$

(d) $dy = -\frac{9}{2\sqrt{x^5}}dx$ (e) $dy = 2e^{2x}dx$ (f) $dy = -3e^{-3x}dx$ (g) $y = \frac{2}{3x}dx$

(h) $dy = \frac{x(2x - 2e^{2x})\ln x - x^2 + e^{2x}}{x(\ln x)^2}dx$ (i) $dy = 4e^{-2x}[-2\sqrt{x} + \frac{1}{2\sqrt{x}}]dx$

(j) $dy = (3\cos 3x \cos 2x - 2\sin 3x \sin 2x)dx$ (k) $dy = \frac{\sec^2 x + 2\tan x \tan 2x}{\cos 2x}dx$

(l) $dy = \frac{2e^{3x}(2-3x) - 6e^{-2x}(x+1)}{3x^3}dx$ (m) $dy = \frac{\sqrt{2x}(2 - \ln x)}{4x^2}dx$

(n) $dy = 2(\ln x + 1)dx$

2. (a) $\delta y \approx 0.15$ (b) $\delta y \approx -0.025$ (c) $\delta y \approx -0.15$ (d) $\delta y \approx 0$ (e) $\delta y \approx 0.15$

Exercises 13.2

1. (a) $\delta y \approx 0.05,\ e_r \to \infty$ (b) $\delta y \approx 0.01,\ e_r \approx 0.5\%$ (c) $\delta y \approx 0.0015,\ e_r \approx 3\%$

(d) $\delta y \approx 0.0189,\ e_r \approx 2.72\%$

2. (a) $y \approx 1.99$ (b) $y \approx 0.6981$ (c) $y \approx 3.2161$ (d) $y \approx 4.08$

CHAPTER 14

14 Indefinite Integration

CHAPTER OBJECTIVES

- Introduce the concept of indefinite integration
- Find integrals of simple functions by mapping and basic rules of integration
- Explain the principles of advanced techniques of integration
- Use advanced techniques to obtain integrals of complicated functions

Essential statements on indefinite integration:

- The integral $F(x)$ of function $f(x)$ is the inverse of the differential $d[F(x)]$ or $f(x)dx$ of function $F(x)$.
- The act to find the integral of function $f(x)$ is said to integrate the function.
- The process of finding the integral of function $f(x)$ is called integration.

Key topics:

- The concept of indefinite integral or integration
- Basic rules of indefinite integration
- Integration by substitutions
- Integration by parts
- Integration by complete differentials
- Integration by partial fractions

Flowchart of mathematical knowledge building

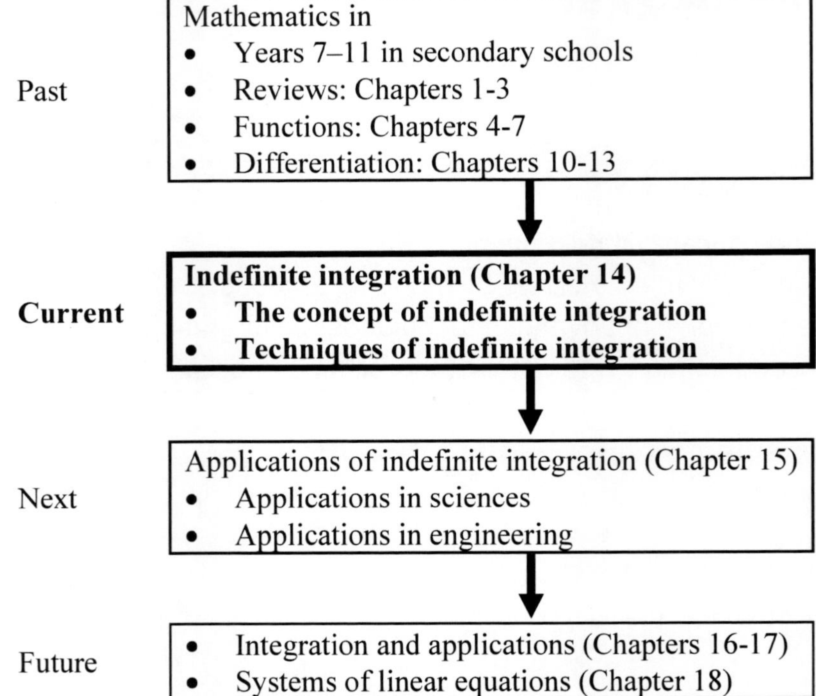

Chapter 14: Indefinite Integration

14.1 Fundamentals of Indefinite Integration

14.1.1 The concept of indefinite integrals

Let $f(x)$ be the derivative of function $F(x)$ or $f(x)=F'(x)$ and both be continuous functions defined in the same or different ranges. Given $f(x)$, the original function $F(x)$ determined by inversing $f(x)$ through its differential $f(x)dx = F'(x)dx = d[F(x)]$ is called the integral of function $f(x)$, denoted as:

$$\boxed{F(x)=\int f(x)dx=\int F'(x)dx=\int d[F(x)].} \qquad (14.1)$$

$f(x)=F'(x)$ is commonly called the *integrand* of the integral.

If $G(x)$ is an inverse function of $f(x)$, $F(x)$ = $G(x)$ + *constant* must be another inverse function of $f(x)$ too. This is because the derivative of any standalone constant term in a multi-term function is zero, thus $F'(x)=G'(x)+c'=f(x)+0=f(x)$. As c can be any constant, therefore formula (14.1) is also expressed as

$$\boxed{F(x)=\int f(x)dx+c.} \qquad (14.2)$$

This is called the *indefinite integral* of $f(x)$ because c can be any unknown constant.

As indefinite integral is the inverse of the corresponding differential of the original function, a backward mapping process from $f'(x)$ to $f(x)$ for common functions listed in Table 10.1 or Table 13.1 is the general approach to determine the integrals of common functions. Note that in formulae (14.1) and (14.2) $F(x)$, rather than $f(x)$, is used for the original function, and $f(x)$, rather than $f'(x)$, is used for the derivative.

Mappings for the indefinite integrals of some common functions are listed in Table 14.1. Given $f(x)$ for $F(x)$, the following three points are helpful:

- $F(x)=\int f(x)dx$ is to directly map the integrand $f(x)$ to $F(x)$, like the cases in the third column in Table 14.1. This is commonly used in mapping a single-term integrand to its original function.
- $F(x)=\int d[F(x)]$ is to firstly find $d[F(x)]$, 'the unexpanded differential of $F(x)$', and then obtain $F(x)$ as the part within $d[F(x)]$, like the cases in the fourth column in Table 14.1. This is very useful in determining the original function of a multi-term integrand.
- A correct inverse of integrand $f(x)$ to $F(x)$ should see $F'(x)=f(x)$.

Table 14.1: Indefinite integrals of some common functions

	$f(x)$	$f(x)dx = F'(x)dx$	$d[F(x)]$	$F(x) = \int f(x)dx = \int d[F(x)]$
1	0	0	$d[c]$	c (constant)
2	a (constant)	$ax^0 dx$	$d[ax]$	$ax + c$
3	$x^r\ (r \neq -1)$	$x^r dx$	$d[\frac{1}{r+1}x^{r+1}]$	$\frac{1}{r+1}x^{r+1} + c$
4	$\frac{1}{\sqrt{x}}\ (r = -\frac{1}{2})$	$x^{-\frac{1}{2}}dx$	$d[2\sqrt{x}]$	$2\sqrt{x} + c$
5	$\frac{1}{x^2}\ (r = -2)$	$x^{-2}dx$	$d[-\frac{1}{x}]$	$-\frac{1}{x} + c$
6	e^{rx}	$e^{rx}dx$	$d[\frac{1}{r}e^{rx}]$	$\frac{1}{r}e^{rx} + c$
7	e^x	$e^x dx$	$d[e^x]$	$e^x + c$
8	a^{rx}	$a^{rx}dx$	$d[\frac{1}{r\ln a}a^{rx}]$	$\frac{1}{r\ln a}e^{rx} + c\ (a > 0,\ a \neq 1)$
9	e^{-x}	$e^{-x}dx$	$d[-e^{-x}]$	$-e^{-x} + c$
10	$\frac{1}{x}\ (x > 0)$	$\frac{dx}{x}$	$d[\ln x]$	$\ln x + c$
11	$\sin ax$	$\sin ax dx$	$d[-\frac{1}{a}\cos ax]$	$-\frac{1}{a}\cos ax + c$
12	$\cos ax$	$\cos ax dx$	$d[\frac{1}{a}\sin ax]$	$\frac{1}{a}\sin ax + c$
13	$\tan ax$	$\tan ax dx$	$d[-\frac{1}{a}\ln\|\cos ax\|]$	$-\frac{1}{a}\ln\|\cos ax\| + c$
14	$\cot ax$	$\cot ax dx$	$d[\frac{1}{a}\ln\|\sin ax\|]$	$\frac{1}{a}\ln\|\sin ax\| + c$
15	$\sec^2 ax$	$\sec^2 ax dx$	$d[\frac{1}{a}\tan ax]$	$\frac{1}{a}\tan ax + c$
16	$\csc^2 ax$	$\csc^2 ax dx$	$d[-\frac{1}{a}\cot ax]$	$-\frac{1}{a}\cot ax + c$
17	$\frac{1}{\sqrt{a^2 - x^2}}$	$\frac{1}{\sqrt{a^2 - x^2}}dx$	$d[\arcsin\frac{x}{a}]$	$\arcsin\frac{x}{a} + c$
18	$\frac{1}{a^2 + x^2}$	$\frac{1}{a^2 + x^2}dx$	$d[\frac{1}{a}\arctan\frac{x}{a}]$	$\frac{1}{a}\arctan\frac{x}{a} + c$
19	$\sinh ax$	$\sinh ax dx$	$d[\frac{1}{a}\cosh ax]$	$\frac{1}{a}\cosh ax + c$
20	$\cosh ax$	$\cosh ax dx$	$d[\frac{1}{a}\sinh ax]$	$\frac{1}{a}\sinh ax + c$

Example 14.1: Find $\int x^3 dx$ and $\int \sqrt{x}dx$, and verify the results respectively.

Solution

Given $f(x)=x^3$ where r = 3, by the #3 rule in Table 14.1,

$$F(x)=\int x^3 dx=\frac{1}{3+1}x^{3+1}+c=\frac{1}{4}x^4+c.$$

To verify the correctness of this outcome, check if $F'(x)=f(x)$.

$$F'(x)=(\frac{1}{4}x^4+c)'=(\frac{1}{4}x^4)'+(c)'=4\times\frac{1}{4}x^{4-1}+0=x^3=f(x).$$

Given $f(x)=\sqrt{x}$ where r = 1/2, by the #3 rule in Table 14.1,

$$F(x)=\int\sqrt{x}dx=\int x^{\frac{1}{2}}dx=\frac{1}{\frac{1}{2}+1}x^{\frac{1}{2}+1}+c=\frac{1}{\frac{1}{2}+\frac{2}{2}}x^{\frac{1}{2}+\frac{2}{2}}+c=\frac{1}{\frac{3}{2}}x^{\frac{3}{2}}+c=\frac{2}{3}x^{\frac{3}{2}}+c.$$

To verify the correctness of this outcome, check if $F'(x)=f(x)$.

$$F'(x)=(\frac{2}{3}x^{\frac{3}{2}}+c)'=(\frac{2}{3}x^{\frac{3}{2}})'+(c)'=\frac{3}{2}\times\frac{2}{3}x^{\frac{3}{2}-1}+0=x^{\frac{3}{2}-\frac{2}{2}}=x^{\frac{1}{2}}=\sqrt{x}=f(x).$$

Example 14.2: Find $\int e^{3x}dx$ and $\int e^{-2x}dx$, and verify the results respectively.

Solution

Given $f(x)=e^{3x}$ where r = 3, by the #6 rule in Table 14.1,

$$F(x)=\int e^{3x}dx=\frac{1}{3}e^{3x}+c.$$

To verify the correctness of this outcome, check if $F'(x)=f(x)$.

$$F'(x)=(\frac{1}{3}e^{3x}+c)'=(\frac{1}{3}e^{3x})'+(c)'=3\times\frac{1}{3}e^{3x}+0=e^{3x}=f(x).$$

Given $f(x)=e^{-2x}$ where r = –2, by the #6 rule in Table 14.1,

$$F(x)=\int e^{-2x}dx=\frac{1}{-2}e^{-2x}+c=-\frac{1}{2}e^{-2x}+c\longrightarrow F'(x)=(-\frac{1}{2}e^{-2x}+c)'=e^{-2x}=f(x)$$

Example 14.3: Evaluate $\int \sin 2x dx$ and $\int \cos\frac{x}{5} dx$, and verify the results respectively.

Solution

Given $f(x) = \sin 2x$ where $a = 2$, by the #11 rule in Table 14.1,

$$F(x) = \int \sin 2x dx = -\frac{1}{2}\cos 2x + c.$$

To verify the correctness of this outcome, check if $F'(x) = f(x)$.

$$F'(x) = (-\frac{1}{2}\cos 2x + c)' = (-\frac{1}{2}\cos 2x)' + (c)' = (-2)\times(-\frac{1}{2}\sin 2x) + 0 = \sin 2x = f(x).$$

Given $f(x) = \cos\frac{x}{5}$ where $a = 1/5$, by the #12 rule in Table 14.1,

$$F(x) = \int \cos\frac{x}{5} dx = \frac{1}{\frac{1}{5}}\sin\frac{x}{5} + c = 5\sin\frac{x}{5} + c.$$

To verify the correctness of this outcome, check if $F'(x) = f(x)$.

$$F'(x) = (5\sin\frac{x}{5} + c)' = (5\sin\frac{x}{5})' + (c)' = \frac{1}{5}\times(5\cos\frac{x}{5}) + 0 = \cos\frac{x}{5} = f(x).$$

14.1.2 Basic rules for indefinite integrals

Operational rules

Let $F(x)$ be the integral or original function of integrand $f(x)$ and $G(x)$ be the integral of integrand $g(x)$. Let a and b be constants; then the following operations are applicable to the indefinite integrals:

$$\begin{aligned}&\int af(x)dx = a\int f(x)dx = aF(x) + c\\&\int [af(x) \pm bg(x)]dx = a\int f(x)dx \pm b\int g(x)dx = aF(x) \pm bG(x) + c.\end{aligned} \tag{14.3}$$

Example 14.4: Evaluate $\int \frac{3}{x}dx$ and $\int \frac{1}{2\sqrt{x}}dx$, and verify the results respectively.

Solution

Given $f(x) = \frac{3}{x}$, by the #10 rule in Table 14.1,

$$F(x) = \int \frac{3}{x}dx = 3\int \frac{1}{x}dx = 3\ln x + c.$$

To verify the correctness of this outcome, check if $F'(x) = f(x)$.

$$F'(x) = (3\ln x + c)' = (3\ln x)' + (c)' = 3(\ln x)' + 0 = \frac{3}{x} = f(x).$$

Given $f(x) = \frac{1}{2\sqrt{x}} = \frac{1}{2}\frac{1}{\sqrt{x}}$, by the #4 rule in Table 14.1,

$$F(x) = \int \frac{1}{2\sqrt{x}}dx = \frac{1}{2}\int \frac{1}{\sqrt{x}}dx = \frac{1}{2}(2\sqrt{x}) + c = \sqrt{x} + c$$

To verify the correctness of this outcome, check if $F'(x) = f(x)$.

$$F'(x) = (\sqrt{x} + c)' = (\sqrt{x})' + (c)' = \frac{1}{2}x^{\frac{1}{2}-1} + 0 = \frac{1}{2}x^{-\frac{1}{2}} = \frac{1}{2\sqrt{x}} = f(x).$$

Example 14.5: Evaluate $\int e^x(3e^{2x} - 2e^{-3x})dx$.

Solution

$$\int e^x(3e^{2x} - 2e^{-3x})dx = \int (3e^x e^{2x} - 2e^x e^{-3x})dx = \int (3e^{3x} - 2e^{-2x})dx$$
$$= \int 3e^{3x}dx - \int 2e^{-2x}dx = \frac{3}{3}e^{3x} - \frac{2}{-2}e^{-2x} + c = e^{3x} + e^{-2x} + c.$$

Example 14.6: Evaluate $\int x^3(3x-2)^2 dx$.

Solution

$$\int x^3(3x-2)^2 dx = \int x^3(9x^2 - 12x + 4)dx = \int (9x^5 - 12x^4 + 4x^3)dx$$
$$= \int 9x^5 dx - \int 12x^4 dx + \int 4x^3 dx = \frac{9}{6}x^6 - \frac{12}{5}x^5 + \frac{4}{4}x^4 + c = \frac{3}{2}x^6 - \frac{12}{5}x^5 + x^4 + c.$$

Example 14.7: Prove $\int(\cos 2x+\cos 4x)dx = 2\sin x\cos^3 x + c$.

Solution

$$\int(\cos 2x+\cos 4x)dx = \int \cos 2xdx + \int \cos 4xdx = \frac{1}{2}\sin 2x + c_1 + \frac{1}{4}\sin 4x + c_2 \quad (c = c_1 + c_2 \text{ is constant.})$$

$$= \frac{1}{2}\sin 2x + \frac{1}{4}\sin 4x + c = \frac{1}{2}\sin 2x + \frac{1}{4}(2\sin 2x\cos 2x) + c \quad \longleftarrow (\sin 2\theta = 2\sin\theta\cos\theta)$$

$$= \frac{1}{2}\sin 2x + \frac{1}{2}\sin 2x\cos 2x + c = \frac{1}{2}\sin 2x(1+\cos 2x) + c$$

$$= \frac{1}{2}(2\sin x\cos x)(1+\cos 2x) + c = \sin x\cos x(1+2\cos^2 x - 1) + c \quad \longleftarrow (\cos 2\theta = 2\cos^2\theta - 1)$$

$$= \sin x\cos x(2\cos^2 x) + c = 2\sin x\cos^3 x + c.$$

Example 14.8: Prove $\int \frac{1}{2}e^x dx + \int \frac{1}{2}e^{-x}dx = \sinh x + c$.

Solution

$$\int \frac{1}{2}e^x dx + \int \frac{1}{2}e^{-x}dx = \frac{1}{2}\int e^x dx + \frac{1}{2}\int e^{-x}dx = \frac{1}{2}e^x + c_1 - \frac{1}{2}e^{-x} + c_2 = \frac{e^x - e^{-x}}{2} + c \quad (c = c_1 + c_2 \text{ is constant.})$$

$$= \sinh x + c. \quad \longleftarrow (\sinh x = \frac{e^x - e^{-x}}{2})$$

Determining constant *c* by known conditions

The constant c associated with the integral or original function $F(x)$ can be uniquely defined if sufficient conditions are provided. If there is one constant in $F(x)$, one known condition is enough to determine the constant. If $F(x)$ has n constants (this happens when $F(x)$ is resulted from n subsequent integrations), n independent known conditions must be provided to define all n constants.

Example 14.9: Known in Example 14.4 that $F(x) = \sqrt{x} + c$ is the solution to $\int \frac{1}{2\sqrt{x}}dx$. If $F(1) = 2$, determine constant c for $F(x)$.

Solution

Substitute $x = 1$, $F(1) = 2$ into $F(x) = \sqrt{x} + c$,

$$2 = \sqrt{1} + c \longrightarrow 2 = 1 + c \longrightarrow c = 1.$$

The solution satisfying $F(1) = 2$ is $F(x) = \sqrt{x} + 1$.

Example 14.10: Known in Example 14.7 that $F(x)=2\sin x\cos^3 x+c$. Given $F(\frac{\pi}{4})=1$, determine constant c for $F(x)$.

Solution

Substitute $F(\frac{\pi}{4})=1$ into $F(x)=2\sin x\cos^3 x+c$,

$$1=2\sin\frac{\pi}{4}\cos^3\frac{\pi}{4}+c \longrightarrow 1=2\times\frac{\sqrt{2}}{2}\times(\frac{\sqrt{2}}{2})^3+c \longrightarrow 1=\frac{(\sqrt{2})^4}{2^3}+c$$

$$1=\frac{2^2}{2^3}+c \longrightarrow 1=\frac{1}{2}+c \longrightarrow c=1-\frac{1}{2}=\frac{1}{2}.$$

The solution satisfying $F(\frac{\pi}{4})=1$ is $F(x)=2\sin x\cos^3 x+\frac{1}{2}$.

Exercises 14.1

1. Find the integrals for the following integrands.

   a) $\int 5x^4 dx$ and $\int \sqrt[3]{x^2}dx$

   b) $\int 4x^{-2}dx$ and $\int \frac{2}{\sqrt{x}}dx$

   c) $\int e^{\frac{1}{3}x}dx$ and $\int e^{-2x}dx$

   d) $\int 2\sin^2 x dx$ and $\int \sin\frac{x}{3}dx$

   e) $\int \frac{-1}{x}dx$ and $\int \frac{1}{2x}dx$

2. Find the integrals for the following integrands.

   a) $\int (2x^3-x^2+4x-3)dx$ and $\int (\sqrt{x}-\frac{1}{2\sqrt{x}})dx$

   b) $\int (4x+\frac{3}{x}+\frac{1}{x^2}+\frac{2}{x^3})dx$ and $\int (\frac{-1}{2x}+e^{-2x}+e^{2x})dx$

   c) $\int (e^{\frac{1}{2}x}-e^{\frac{1}{3}x})dx$ and $\int (2\sin 2x-\cos 2x)dx$

   d) $\int (3x^2-4x+5)dx$ given $x=1, y=2$.

   e) $\int (1-\frac{1}{x}-\frac{1}{x^2})dx$ given $x=1, y=-2$.

   f) $\int (2\cos^2 x-2\sin x\cos x)dx$ given $x=0, y=1$.

   g) $\int (3\sqrt{x}-\frac{1}{\sqrt{x}})dx$ given $x=2, y=10$.

14.2 Advanced Techniques for Indefinite Integration

14.2.1 Integration by substitution

The concept of integration by substitution

Integration by substitution is basically the inverse of the chain rule of differentiation. If $G(x)$ can be rewritten as $G(u)$ by substituting a combined term of x with $u = u(x)$, its derivative with respect to x can be obtained by $G'(x)=\frac{dG(x)}{dx}=\frac{dG(u)}{du}\frac{du}{dx}=g(x)$. The integral of $g(x)$ can be determined by

$$\int g(x)dx=\int \frac{dG(u)}{du}\frac{du}{\cancel{dx}}\cancel{dx}=\int G'(u)du=\int g(u)du=G(u)+c.$$

$G(u)$ can be mapped by $g(u)$ using Table 14.1 by regarding u as x. The integral $G(x)$ is determined by substituting $u = u(x)$ back into $G(u)$. A few common substitutions are introduced as follows. Note that not every substitution results in a successful integration. It is more a *guided trial* for individual cases.

Type of $u=ax\pm b$ *for* $g(ax\pm b)dx$

$$g(ax\pm b)dx \xrightarrow{\text{from } g(x)dx} \begin{cases} u=ax\pm b \\ du=adx \\ dx=\frac{1}{a}du \end{cases} \xrightarrow{\text{to } g(u)du} g(u)du \xrightarrow{\text{by } du} G(u) \xrightarrow{u=ax\pm b} G(x)$$

Example 14.11: Evaluate $\int 6(x+2)^2dx$ and $\int \sqrt{5-3x}dx$ respectively.

Solution

For $\int 6(x+2)^2dx$, let $u=x+2 \to du=dx$. Substitute $u=x+2$ and $du=dx$ into the integral:

$$\int 6(x+2)^2dx=\int 6u^2du=\frac{6}{3}u^3+c=2(x+2)^3+c.$$

For $\int \sqrt{5-3x}dx$, let $u=5-3x \to du=-3dx \to dx=-\frac{1}{3}du$. Put $u=5-3x$ and $dx=-\frac{1}{3}du$ into the integral:

$$\int \sqrt{5-3x}dx=\int \sqrt{u}(-\frac{1}{3}du)=-\frac{1}{3}\int \sqrt{u}du=-\frac{1}{3}(\frac{2}{3}u^{\frac{3}{2}})+c=-\frac{2}{9}(5-3x)^{\frac{3}{2}}+c.$$

Example 14.12: Evaluate $\int e^{3x+1}dx$ and $\int \frac{1}{2x+3}dx$ respectively.

Solution

For $\int e^{3x+1}dx$, let $u = 3x+1 \rightarrow du = 3dx \rightarrow dx = \frac{1}{3}du$. Substitute $u = 3x+1$ and $dx = \frac{1}{3}du$ into the integral:

$$\int e^{3x+1}dx = \int e^u(\frac{1}{3}du) = \frac{1}{3}\int e^u du = \frac{1}{3}e^u + c = \frac{1}{3}e^{3x+1} + c.$$

For $\int \frac{1}{2x+3}dx$, let $u = 2x+3 \rightarrow du = 2dx \rightarrow dx = \frac{1}{2}du$. Put $u = 2x+3$ and $dx = \frac{1}{2}du$ into the integral:

$$\int \frac{1}{2x+3}dx = \int \frac{1}{u}(\frac{1}{2}du) = \frac{1}{2}\int \frac{du}{u} = \frac{1}{2}\ln u + c = \frac{1}{2}\ln(2x+3) + c.$$

Type of $u = ax^2 \pm b$ *for* $g(ax^2 \pm b)(xdx)$

$$g(ax^2 \pm b)(xdx) \xrightarrow{\text{from } g(x)dx} \begin{cases} u = ax^2 \pm b \\ du = 2axdx \\ xdx = \frac{1}{2a}du \end{cases} \xrightarrow{\text{to } g(u)du} g(u)du \xrightarrow{\text{by } du} G(u) \xrightarrow{u=ax^2 \pm b} G(x)$$

Example 14.13: Evaluate $\int \frac{3xdx}{2x^2+1}$.

Solution

For $\int \frac{3xdx}{2x^2+1}$, let $u = 2x^2+1 \rightarrow du = 4xdx \rightarrow xdx = \frac{1}{4}du$. Substitute $u = 2x^2+1$ and $xdx = \frac{1}{4}du$ into the integral:

$$\int \frac{3xdx}{2x^2+1} = 3\int \frac{\frac{1}{4}du}{u} = \frac{3}{4}\int \frac{du}{u} = \frac{3}{4}\ln u + c = \frac{3}{4}\ln(2x^2+1) + c.$$

Example 14.14: Evaluate $\int x\sqrt{4-x^2}dx$.

Solution

For $\int x\sqrt{4-x^2}dx = \int x\sqrt{-x^2+4}dx,$ let $u=-x^2+4 \to du=-2xdx \to xdx=-\frac{1}{2}du.$ Substitute $u=-x^2+4$ and $xdx=-\frac{1}{2}du$ into the integral:

$$\int x\sqrt{4-x^2}dx = \int \sqrt{-x^2+4}(xdx) = \int \sqrt{u}(-\frac{1}{2}du) = -\frac{1}{2}\int \sqrt{u}du = -\frac{1}{2}(\frac{2}{3}u^{\frac{3}{2}})+c = -\frac{1}{3}(4-x^2)^{\frac{3}{2}}+c.$$

Type of $x = a\sin u$ *for* $g(\sqrt{a^2-x^2})dx$

$$g(\sqrt{a^2-x^2})dx \xrightarrow{\text{from } g(x)dx} \begin{cases} x = a\sin u \\ u = \arcsin\frac{x}{a} \\ dx = a\cos u du \end{cases} \xrightarrow{\text{to } g(u)du} g(u)du \xrightarrow{\text{by } du} G(u) \xrightarrow{u=\arcsin\frac{x}{a}} G(x)$$

Note $x=a\sin u$ is to transfer $\sqrt{a^2-x^2} = \sqrt{a^2-a^2\sin^2 u} = \sqrt{a^2(1-\sin^2 u)} = \sqrt{a^2\cos^2 u} = a\cos u.$

Example 14.15: Evaluate $\int \frac{1}{\sqrt{1-x^2}}dx$.

Solution

For $\int \frac{1}{\sqrt{1-x^2}}dx,$ let $x=\sin u \to dx=\cos u du.$ Substitute $x=\sin u$ and $dx=\cos u du$ into the integral:

$$\int \frac{1}{\sqrt{1-x^2}}dx = \int \frac{1}{\sqrt{1-\sin^2 u}}(\cos u du) = \int \frac{\cos u du}{\sqrt{\cos^2 u}} = \int \frac{\cos u du}{\cos u} = \int du = u+c = \arcsin x + c.$$

Example 14.16: Evaluate $\int \sqrt{4-x^2}dx$.

Solution

For $\int \sqrt{4-x^2}dx = \int \sqrt{2^2-x^2}dx,$ let $x=2\sin u \to dx=2\cos u du.$ Substitute $x=2\sin u$ and $dx=2\cos u du$ into the integral:

$$\int \sqrt{4-x^2}dx = \int \sqrt{2^2-x^2}dx = \int \sqrt{2^2-(2\sin u)^2}(2\cos u)du = 2\int 2\sqrt{1-\sin^2 u}\cos udu$$
$$= 4\int \sqrt{\cos^2 u}\cos udu = 4\int \cos u \cos udu = 4\int \cos^2 udu$$
$$= 4\int \frac{1+\cos 2u}{2}du \longleftarrow \cos^2\theta = \frac{1+\cos 2\theta}{2}$$
$$= 2\int (1+\cos 2u)du = 2\int du + 2\int \cos 2udu = 2u + \frac{2}{2}\sin 2u + c = 2u + \sin 2u + c$$
$$= 2u + 2\sin u\cos u + c \xleftarrow[x=2\sin u]{u=\arcsin\frac{x}{2}} \cos u = \sqrt{\cos^2 u} = \sqrt{1-\sin^2 u} = \sqrt{1-(\frac{x}{2})^2}$$
$$= 2\arcsin\frac{x}{2} + x\sqrt{1-(\frac{x}{2})^2} + c = 2\arcsin\frac{x}{2} + x\sqrt{\frac{4-x^2}{4}} + c = 2\arcsin\frac{x}{2} + \frac{x}{2}\sqrt{4-x^2} + c.$$

Type of $u = \sin x$ *or* $u = \cos x$ *for* $g(\sin x \bullet \div \cos x)dx$

$$g(\sin x \bullet \div \cos x)dx \xrightarrow{\text{from } g(x)dx} \begin{cases} u = \sin x \\ du = \cos xdx \end{cases} \xrightarrow{\text{to } g(u)du} g(u)du \xrightarrow{\text{by } du} G(u) \xrightarrow{u=\sin x} G(x)$$
$$g(\sin x \bullet \div \cos x)dx \xrightarrow{\text{from } g(x)dx} \begin{cases} u = \cos x \\ du = -\sin xdx \end{cases} \xrightarrow{\text{to } g(u)du} g(u)du \xrightarrow{\text{by } du} G(u) \xrightarrow{u=\cos x} G(x)$$

Example 14.17: Evaluate $\int \sin x \cos xdx$.

Solution

Let $u = \sin x \to du = \cos xdx$. Substitute $u = \sin x$ and $\cos xdx = du$ into the integral:

$$\int \underbrace{\sin x}_{u}\underbrace{\cos xdx}_{du} = \int udu = \frac{1}{2}u^2 + c = \frac{1}{2}\sin^2 x + c.$$

This can be verified by

$$F'(x) = (\frac{1}{2}\sin^2 x + c)' = (\frac{1}{2}\sin^2 x)' + (c)' = 2\times\frac{1}{2}\sin x(\sin x)' + 0 = \sin x\cos x = f(x).$$

Example 14.18: Evaluate $\int \tan xdx$.

Solution

Let $u = \cos x \to du = -\sin xdx$. Substitute $u = \cos x$ and $\sin xdx = -du$ into the integral:

$$\int \tan xdx = \int \frac{\sin xdx}{\cos x} = \int \frac{-du}{u} = -\ln u + c = -\ln\cos x + c.$$

Example 14.19: Evaluate $\int \cos 2x \cos x dx$.

Solution

Let $u = \sin x \rightarrow du = \cos x dx$. Substitute $u = \sin x$ and $\cos x dx = du$ into the integral:

$$\int \cos 2x \cos x dx = \int (1 - 2\sin^2 x)(\cos x dx) \longleftarrow \cos 2\theta = 1 - 2\sin^2 \theta$$
$$= \int (1 - 2u^2)(du) = \int du - 2\int u^2 du = u - \frac{2}{3}u^3 + c = \sin x - \frac{2}{3}\sin^3 x + c.$$

14.2.2 Integration by parts

The concept of integration by parts

Integration by pars is basically the inverse of the product rule of differentiation. If $u = f(x)$ and $v = g(x)$ are differentiable functions, from formulae (10.12) in Chapter 10, the product rule of differentiation is

$$\frac{d(uv)}{dx} = \frac{du}{dx}v + u\frac{dv}{dx}.$$

Multiply dx to both sides of above equation to transfer it to its differential form:

$$\cancel{dx} \times \frac{d(uv)}{\cancel{dx}} = (\frac{du}{\cancel{dx}}v + u\frac{dv}{\cancel{dx}}) \times \cancel{dx}$$
$$d(uv) = vdu + udv.$$

Apply integral to both sides of above differential equation and remember $F(x) = \int d[F(x)]$,

$$\int d(uv) = \int vdu + \int udv$$
$$uv = \int vdu + \int udv \quad \text{or}$$

$$\boxed{\int vdu = uv - \int udv \rightleftarrows \int udv = uv - \int vdu} \qquad (14.4)$$

Formula (14.4) is commonly known as *integration by parts*. It converts one combinative part of an integral to the other combinative part of the integral if the integrand can be treated as a product of two differentiable functions. Note again that only some combined integrands can be integrated by integration by parts.

In practice, once v (or u) and du (or dv) on the left of formula (14.4) are identified, u (or v) and dv (du) can be obtained through:

$$\boxed{\begin{array}{l}\int vdu = uv - \int udv \longleftarrow u = \int du, \quad dv = v'dx \\ \int udv = uv - \int vdu \longleftarrow v = \int dv, \quad du = u'dx\end{array}} \qquad (14.5)$$

The LIATE guide for integration by parts

Kasube (1983) proposed a *rule of thumb* commonly known as LIATE to choose the u function that comes first in $\int udv = uv - \int vdu$. LIATE stands for

- L – logarithmic functions, for example, ln$x$, log$_b x$
- I – inverse trigonometric functions, arcsin$x$, arctan$x$
- A – algebraic functions, for example, $2x^3$, $3x - 5$
- T – trigonometric functions, for example, sin$x$, cos$2x$, 4tan$3x$
- E – exponential functions, for example, e^x, e^{-3x}, 5^{2x}.

LIATE indicates the order of the selection of the first function (first part) u (or v) from the combined integrand in integration by parts. Once u (or v) is selected, the remainder of the combined integrand become the differential dv (or du).

For example, in $\int xe^{-x}dx$, the combined integrand is $xe^{-x}dx$. By LIATE, x is classified as the A and is ahead of e^{-3x} (the E), so the selection should be $u = x$ and $dv = e^{-x}dx$.

LIATE works for most cases where integration by parts can be applied.

Direct application of integration by parts

For simple integrands, one round of integration by parts may be sufficient in obtaining the solution.

Example 14.20: Evaluate $\int xe^{-x}dx$.

Solution

For $\int xe^{-x}dx$, x is an algebraic function (A) and e^{-x} is an exponential function (E). By formula (14.5) and LIATE, let $v = x$ and $du = e^{-x}dx$.

$$\int \underset{v}{\underline{x}}\,\underset{du}{\underline{e^{-x}dx}} = (\underset{v}{\underline{x}})(\underset{u}{\underline{-e^{-x}}}) - \int (\underset{u}{\underline{-e^{-x}}})(\underset{dv}{\underline{dx}}) \longleftarrow u = \int du = \int e^{-x}dx = -e^{-x}, \quad dv = v'dx = (x)'dx = dx$$

$$= -xe^{-x} + \int e^{-x}dx = -xe^{-x} + (-e^{-x}) + c = -xe^{-x} - e^{-x} + c = -e^{-x}(x+1) + c.$$

Example 14.21: Evaluate $\int 3\ln x dx$.

Solution

For $\int 3\ln x dx$, 3 is an algebraic function (A) of zero-order and ln$x$ is a logarithmic function (L). By formula (14.5) and LIATE, let $v = \ln x$ and $du = 3dx$.

$$\int \underbrace{\ln x}_{v}\underbrace{(3dx)}_{du} = \underbrace{(\ln x)}_{v}\underbrace{(3x)}_{u} - \int \underbrace{(3x)}_{u}\underbrace{(\frac{dx}{x})}_{dv} \longleftarrow u = \int du = \int 3dx = 3x,\ \ dv = v'dx = (\ln x)'dx = \frac{dx}{x}$$

$$= 3x\ln x - \int 3dx = 3x\ln x - 3x + c = 3x(\ln x - 1) + c.$$

Example 14.22: Evaluate $\int x^3 \ln x dx$.

Solution

For $\int x^3 \ln x dx$, x^3 is an algebraic function (A) and ln$x$ is a logarithmic function (L). By formula (14.5) and LIATE, let $v = \ln x$ and $du = x^3 dx$.

$$\int \underbrace{\ln x}_{v}\underbrace{x^3 dx}_{du} = \underbrace{(\ln x)}_{v}\underbrace{(\frac{1}{4}x^4)}_{u} - \int \underbrace{(\frac{1}{4}x^4)}_{u}\underbrace{(\frac{dx}{x})}_{dv} \longleftarrow u = \int x^3 dx = \frac{1}{4}x^4,\ \ dv = (\ln x)'dx = \frac{dx}{x}$$

$$= \frac{1}{4}x^4 \ln x - \frac{1}{4}\int x^3 dx = \frac{1}{4}x^4 \ln x - \frac{1}{16}x^4 + c = \frac{1}{4}x^4(\ln x - \frac{1}{4}) + c = \frac{1}{16}x^4(4\ln x - 1) + c.$$

Example 14.23: Evaluate $\int 2x\cos x dx$.

Solution

For $\int 2x\cos x dx$, $2x$ is an algebraic function (A) and cos$x$ is a trigonometric function (T). By formula (14.5) and LIATE, let $v = 2x$ and $du = \cos x dx$. By formula (14.5),

$$\int \underbrace{2x}_{v}\underbrace{\cos x dx}_{du} = \underbrace{(2x)}_{v}\underbrace{(\sin x)}_{u} - \int \underbrace{(\sin x)}_{u}\underbrace{(2dx)}_{dv} \longleftarrow u = \int \cos x dx = \sin x,\ \ dv = v'dx = (2x)'dx = 2dx$$

$$= 2x\sin x - \int 2\sin x dx = 2x\sin x - 2(-\cos x) + c = 2x\sin x + 2\cos x + c = 2(x\sin x + \cos x) + c.$$

Example 14.24: Evaluate $\int 2\arcsin x dx$.

Solution

For $\int 2\arcsin x dx$, 2 is a constant (A) and arcsin$x$ is an inverse trigonometric function (I). By formula (14.5) and LIATE, let $v = \arcsin x$ and $du = 2dx$.

$$\int \underbrace{\arcsin x}_{v}\underbrace{(2dx)}_{du} = (\underbrace{\arcsin x}_{v})(\underbrace{2x}_{u}) - \int (\underbrace{2x}_{u})(\underbrace{\frac{dx}{\sqrt{1-x^2}}}_{dv}) \longleftarrow u = 2x,\ dv = (\arcsin x)'dx = \frac{dx}{\sqrt{1-x^2}}$$

$$= 2x\arcsin x - \int \frac{2xdx}{\sqrt{1-x^2}} \longleftarrow u = 1-x^2, du = -2xdx \leftrightarrow 2xdx = -du$$

$$= 2x\arcsin x - \int \frac{-du}{\sqrt{u}} = 2x\arcsin x + 2\sqrt{u} + c = 2x\arcsin x + 2\sqrt{1-x^2} + c.$$

Example 14.25: Evaluate $\int x^2 \arctan x dx$.

Solution

For $\int x^2 \arctan x dx$, x^2 is an algebraic function (A) and arctan$x$ is an inverse trigonometric function (I). By formula (14.5) and LIATE, let $v = \arctan x$ and $du = x^2dx$.

$$\int \underbrace{\arctan x}_{v}\underbrace{(x^2dx)}_{du} = (\underbrace{\arctan x}_{v})(\underbrace{\frac{1}{3}x^3}_{u}) - \int (\underbrace{\frac{1}{3}x^3}_{u})(\underbrace{\frac{dx}{1+x^2}}_{dv}) \longleftarrow u = \frac{1}{3}x^3,\ dv = (\arctan x)'dx = \frac{dx}{1+x^2}$$

$$= \frac{1}{3}x^3 \arctan x - \frac{1}{3}\int \frac{x^3}{1+x^2}dx \longleftarrow \frac{x^3}{1+x^2} = x - \frac{x}{1+x^2} \textit{ by long division}$$

$$= \frac{1}{3}x^3 \arctan x - \frac{1}{3}\int (x - \frac{x}{1+x^2})dx = \frac{1}{3}x^3 \arctan x - \frac{1}{3}(\frac{1}{2}x^2 - \int \frac{x}{1+x^2}dx)$$

$$= \frac{1}{3}x^3 \arctan x - \frac{1}{6}x^2 + \frac{1}{3}\int \frac{x}{1+x^2}dx \longleftarrow u = 1+x^2, du = 2xdx \leftrightarrow xdx = \frac{1}{2}du$$

$$= \frac{1}{3}x^3 \arctan x - \frac{1}{6}x^2 + \frac{1}{3}\int \frac{\frac{1}{2}du}{u} = \frac{1}{3}x^3 \arctan x - \frac{1}{6}x^2 + \frac{1}{6}\ln u + c$$

$$= \frac{1}{3}x^3 \arctan x - \frac{1}{6}x^2 + \frac{1}{6}\ln(1+x^2) + c = \frac{1}{6}[2x^3 \arctan x - x^2 + \ln(1+x^2)] + c.$$

Example 14.26: Evaluate $\int x\cosh xdx$.

Solution

For $\int x\cosh xdx$, x is an algebraic function (A) and cosh$x$ is a hyperbolic function, similar to trigonometric functions (T). By formula (14.5) and LIATE, let $v=x$ and $du=\cosh xdx$.

$$\int \underset{v}{\underline{x}}\underset{du}{\underline{(\cosh xdx)}} = \underset{v}{(\underline{x})}\underset{u}{(\underline{\sinh x})} - \int \underset{u}{(\underline{\sinh x})}\underset{dv}{(\underline{dx})} \longleftarrow u = \int \cosh xdx = \sinh x,\quad dv = dx$$
$$= x\sinh x - \int \sinh xdx = x\sinh x - \cosh x + c.$$

Nested applications of integration by parts

For more complicated integrands, the process of integration by parts may be repeated in the subsequent integral(s) nested within a complicated integration. Add constant c to the solution when no further integration is needed.

Example 14.27: Evaluate $\int x^2e^xdx$.

Solution

For $\int x^2e^xdx$, let $v=x^2$ and $du=e^xdx$. By formula (14.5),

$$\int \underset{v}{\underline{x^2}}\,\underset{du}{\underline{e^xdx}} = \underset{v}{(\underline{x^2})}\underset{u}{(\underline{e^x})} - \int \underset{u}{(\underline{e^x})}\underset{dv}{(\underline{2xdx})} \longleftarrow u = \int e^xdx = e^x,\quad dv = (x^2)'dx = 2xdx$$
$$= x^2e^x - \int e^x(2xdx) = x^2e^x - 2\int xe^xdx = x^2e^x - 2I \longleftarrow I = \int xe^xdx \qquad (1)$$

Repeat integration by parts to $I=\int xe^xdx$:

$$I = \int xe^xdx = (x)(e^x) - \int e^xdx = xe^x - e^x.$$

Substitute I into (1):

$$\int x^2e^xdx = x^2e^x - 2(xe^x - e^x) + c = x^2e^x - 2xe^x + 2e^x + c = e^x(x^2 - 2x + 2) + c.$$

Example 14.28: Evaluate $\int (\ln x)^2 x dx$.

Solution

For $\int (\ln x)^2 x dx$, let $v = (\ln x)^2$ and $du = x dx$. By formula (14.5),

$$\int \underbrace{(\ln x)^2}_{v} \underbrace{x dx}_{du} \longleftarrow u = \int x dx = \frac{1}{2}x^2, \quad dv = [(\ln x)^2]' dx = 2\ln x (\ln x)' dx = \frac{2\ln x dx}{x}$$

$$= \underbrace{(\ln x)^2}_{v}\underbrace{(\frac{1}{2}x^2)}_{u} - \int \underbrace{(\frac{1}{2}x^2)}_{u}\underbrace{(\frac{2\ln x dx}{x})}_{dv} \longleftarrow \textit{1st - round of integration by parts}$$

$$= \frac{1}{2}x^2(\ln x)^2 - \int (\ln x)(x dx)$$

$$= \frac{1}{2}x^2(\ln x)^2 - [(\ln x)(\frac{1}{2}x^2) - \int (\frac{1}{2}x^2)(\frac{dx}{x})] \longleftarrow \textit{2nd - round of integration by parts}$$

$$= \frac{1}{2}x^2(\ln x)^2 - (\frac{1}{2}x^2 \ln x - \frac{1}{2}\int x dx) = \frac{1}{2}x^2(\ln x)^2 - \frac{1}{2}x^2 \ln x + \frac{1}{4}x^2 + c$$

$$= \frac{1}{2}x^2[(\ln x)^2 - \ln x + \frac{1}{2}] + c = \frac{1}{4}x^2[2(\ln x)^2 - 2\ln x + 1] + c.$$

Example 14.29: Evaluate $\int 4x^2 \cos 2x dx$.

Solution

For $\int 4x^2 \cos 2x dx$, let $v = 4x^2$ and $du = \cos 2x dx$. By formula (14.5),

$$\int \underbrace{4x^2}_{v} \underbrace{\cos 2x dx}_{du} \longleftarrow u = \int \cos 2x dx = \frac{1}{2}\sin 2x, \quad dv = (4x^2)' dx = 8x dx$$

$$= \underbrace{4x^2}_{v}\underbrace{(\frac{1}{2}\sin 2x)}_{u} - \int \underbrace{(\frac{1}{2}\sin 2x)}_{u}\underbrace{(8x dx)}_{dv} \longleftarrow \textit{1st - round of integration by parts}$$

$$= 2x^2 \sin 2x - 4\int (x)(\sin 2x dx)$$

$$= 2x^2 \sin 2x - 4[(x)(-\frac{1}{2}\cos 2x) - \int (-\frac{1}{2}\cos 2x)(dx)] \longleftarrow \textit{2nd - round of integration by parts}$$

$$= 2x^2 \sin 2x - 4[-\frac{1}{2}x\cos 2x + \frac{1}{2}\int \cos 2x dx] = 2x^2 \sin 2x - 4[-\frac{1}{2}x\cos 2x + \frac{1}{4}\sin 2x] + c$$

$$= 2x^2 \sin 2x + 2x\cos 2x - \sin 2x + c.$$

Cyclic applications of integration by parts

For some special integrands involving combinations of logarithmic, exponential, and trigonometric functions, or higher-order functions, the process of integration by parts may be repeated in the subsequent integral(s) until the recurrence of the original integral or its lower-order form. If the original integral reoccurs, the integral can be obtained by means of solving algebraic equations from that point. If the lower-order integral occurs, the integral can be obtained as a *reduction formula*.

Example 14.30: Evaluate $\int \frac{\ln x}{x} dx$.

Solution

For $I = \int \frac{\ln x}{x} dx$, let $v = \ln x$ and $du = \frac{dx}{x}$. By formula (14.5),

$$I = \int \frac{\ln x}{x} dx = \int \underbrace{\ln x}_{v} \underbrace{(\frac{dx}{x})}_{du} \longleftarrow u = \int \frac{dx}{x} = \ln x, \;\; dv = (\ln x)' dx = \frac{dx}{x}$$

$$= \underbrace{(\ln x)}_{v} \underbrace{(\ln x)}_{u} - \int \underbrace{(\ln x)}_{u} \underbrace{(\frac{dx}{x})}_{dv} = (\ln x)^2 - \underbrace{\int \frac{\ln x}{x} dx}_{I} = (\ln x)^2 - I \;\; or$$

$$I = (\ln x)^2 - I \longrightarrow 2I = (\ln x)^2, \text{ thus } I = \int \frac{\ln x}{x} dx = \frac{1}{2} (\ln x)^2 + c.$$

Example 14.31: Evaluate $\int e^{-2x} \sin x dx$.

Solution

For $I = \int e^{-2x} \sin x dx$, let $v = \sin x$ and $du = e^{-2x} dx$. By formula (14.5),

$$I = \int \underbrace{\sin x}_{v} \underbrace{e^{-2x} dx}_{du} \longleftarrow u = \int e^{-2x} dx = -\frac{1}{2} e^{-2x}, \;\; dv = (\sin x)' dx = \cos x dx$$

$$= \underbrace{(\sin x)}_{v} \underbrace{(-\frac{1}{2} e^{-2x})}_{u} - \int \underbrace{(-\frac{1}{2} e^{-2x})}_{u} \underbrace{(\cos x dx)}_{dv} \longleftarrow 1st\text{-}round\ of\ integration\ by\ parts$$

$$= -\frac{1}{2} e^{-2x} \sin x + \frac{1}{2} \int (\cos x)(e^{-2x} dx) \longleftarrow u = \int e^{-2x} dx = -\frac{1}{2} e^{-2x}, \;\; dv = (\cos x)' dx = -\sin x dx$$

$$= -\frac{1}{2} e^{-2x} \sin x + \frac{1}{2} [-\frac{1}{2} e^{-2x} \cos x - \int \frac{1}{2} e^{-2x} \sin x dx] \longleftarrow 2nd\text{-}round\ of\ integration\ by\ parts$$

$$= -\frac{1}{2} e^{-2x} \sin x + \frac{1}{2} [-\frac{1}{2} e^{-2x} \cos x - \frac{1}{2} \int e^{-2x} \sin x dx]$$

$$= -\frac{1}{2} e^{-2x} \sin x - \frac{1}{4} e^{-2x} \cos x - \frac{1}{4} \underbrace{\int e^{-2x} \sin x dx}_{I} = -\frac{1}{2} e^{-2x} \sin x - \frac{1}{4} e^{-2x} \cos x - \frac{1}{4} I \;\; \text{or}$$

$$I=-\frac{1}{2}e^{-2x}\sin x-\frac{1}{4}e^{-2x}\cos x-\frac{1}{4}I \longrightarrow I+\frac{1}{4}I=-\frac{1}{2}e^{-2x}\sin x-\frac{1}{4}e^{-2x}\cos x$$

$$\frac{5}{4}I=-\frac{1}{4}e^{-2x}(2\sin x+\cos x) \longrightarrow 5I=-e^{-2x}(2\sin x+\cos x), \text{ thus}$$

$$I=\int e^{-2x}\sin x dx=-\frac{1}{5}e^{-2x}(2\sin x+\cos x)+c.$$

Example 14.32: Evaluate $I(n)=\int \sin^n ax dx$ (n: positive integers).

Solution

$$I(n)=\int \sin^n ax dx=\int \underbrace{(\sin^{n-1}ax)}_{v}\underbrace{(\sin ax dx)}_{du}$$

$$=\underbrace{(\sin^{n-1}ax)}_{v}\underbrace{(-\frac{1}{a}\cos ax)}_{u}-\int \underbrace{(-\frac{1}{a}\cos ax)}_{u}\underbrace{(n-1)\sin^{n-2}ax(\sin ax)'dx}_{dv}$$

$$=-\frac{1}{a}\cos ax\sin^{n-1}ax+\frac{n-1}{a}\int \cos ax\sin^{n-2}ax(a\cos ax)dx$$

$$=-\frac{1}{a}\cos ax\sin^{n-1}ax+(n-1)\int \cos^2 ax\sin^{n-2}ax dx$$

$$=-\frac{1}{a}\cos ax\sin^{n-1}ax+(n-1)\int (1-\sin^2 ax)\sin^{n-2}ax dx$$

$$=-\frac{1}{a}\cos ax\sin^{n-1}ax+(n-1)\int (\sin^{n-2}ax-\sin^n ax)dx$$

$$=-\frac{1}{a}\cos ax\sin^{n-1}ax+(n-1)\int \sin^{n-2}ax dx-(n-1)\underbrace{\int \sin^n ax dx}_{I(n)} \quad \text{or}$$

$$I(n)=-\frac{1}{a}\cos ax\sin^{n-1}ax+(n-1)\int \sin^{n-2}ax dx-(n-1)I(n)$$

$$I(n)+(n-1)I(n)=-\frac{1}{a}\cos ax\sin^{n-1}ax+(n-1)\int \sin^{n-2}ax dx$$

$$nI(n)=-\frac{1}{a}\cos ax\sin^{n-1}ax+(n-1)\int \sin^{n-2}ax dx \longrightarrow I(n)=-\frac{1}{na}\cos ax\sin^{n-1}ax+\frac{n-1}{n}\int \sin^{n-2}ax dx$$

$$\therefore I(n)=-\frac{1}{na}\cos ax\sin^{n-1}ax+\frac{n-1}{n}I(n-2) \longleftarrow I(n-2)=\int \sin^{n-2}ax dx.$$

This final expression is a reduction formula because the integral of a higher-order integrand (n) can be determined by that of a lower-order integrand ($n-2$). The following example shows how to use the reduction formula obtained in Example 14.32 to find the integrals for higher-order integrands.

Example 14.33: Given $\int dx = x$ and $\int \sin x dx = -\cos x$, find $\int \sin^2 x dx$, $\int \sin^3 x dx$, $\int \sin^4 x dx$, and $\int \sin^5 x dx$ respectively using the reduction formula obtained in Example 14.32. [Note $a = 1$]

Solution

The given integrals can be denoted as $I(0) = \int \sin^0 x dx = \int dx = x$ and $I(1) = \int \sin x dx = -\cos x$. Using the reduction formula obtained in Example 14.32, the required integrals can be determined as follows:

$$\int \sin^2 x dx = I(2) = -\frac{1}{2}\cos x \sin^{2-1} x + \frac{2-1}{2} I(2-2) = -\frac{1}{2}\cos x \sin x + \frac{1}{2} I(0) = -\frac{1}{2}\cos x \sin x + \frac{1}{2} x + c.$$

$$\int \sin^3 x dx = I(3) = -\frac{1}{3}\cos x \sin^{3-1} x + \frac{3-1}{3} I(3-2) = -\frac{1}{3}\cos x \sin^2 x + \frac{2}{3} I(1)$$

$$= -\frac{1}{3}\cos x \sin^2 x + \frac{2}{3}(-\cos x) + c = -\frac{1}{3}\cos x \sin^2 x - \frac{2}{3}\cos x + c.$$

$$\int \sin^4 x dx = I(4) = -\frac{1}{4}\cos x \sin^{4-1} x + \frac{4-1}{4} I(4-2) = -\frac{1}{4}\cos x \sin^3 x + \frac{3}{4} I(2)$$

$$= -\frac{1}{4}\cos x \sin^3 x + \frac{3}{4}(-\frac{1}{2}\cos x \sin x + \frac{1}{2} x) + c \longleftarrow I(2) = -\frac{1}{2}\cos x \sin x + \frac{1}{2} x$$

$$= -\frac{1}{4}\cos x \sin^3 x - \frac{3}{8}\cos x \sin x + \frac{3}{8} x + c.$$

$$\int \sin^5 x dx = I(5) = -\frac{1}{5}\cos x \sin^{5-1} x + \frac{5-1}{5} I(5-2) = -\frac{1}{5}\cos x \sin^4 x + \frac{4}{5} I(3)$$

$$= -\frac{1}{5}\cos x \sin^4 x + \frac{4}{5}(-\frac{1}{3}\cos x \sin^2 x - \frac{2}{3}\cos x) + c \longleftarrow I(3) = -\frac{1}{3}\cos x \sin^2 x - \frac{2}{3}\cos x$$

$$= -\frac{1}{5}\cos x \sin^4 x - \frac{4}{15}\cos x \sin^2 x - \frac{8}{15}\cos x + c.$$

Example 14.34: Evaluate $I(n) = \int x^n e^{ax} dx$ ($n > 0$). Use the obtained formula to find $I(1)$, $I(2)$ and $I(3)$ respectively, assuming $a = 1$.

Solution

$$I(n) = \int \underbrace{x^n}_{v} \underbrace{e^{ax} dx}_{du} = \underbrace{(x^n)}_{v} \underbrace{(\frac{1}{a} e^{ax})}_{u} - \int \underbrace{(\frac{1}{a} e^{ax})}_{u} \underbrace{(n x^{n-1} dx)}_{dv} = \frac{1}{a} x^n e^{ax} - \frac{n}{a} \int x^{n-1} e^{ax} dx$$

$$= \frac{1}{a} x^n e^{ax} - \frac{n}{a} I(n-1) \longleftarrow I(n-1) = \int x^{n-1} e^{ax} dx.$$

For $a = 1$, $I(n) = x^n e^x - n \int x^{n-1} e^x dx$. Hence

$$I(1)=x^1e^x-\int x^{1-1}e^x dx = xe^x-\int e^x dx = xe^x - e^x + c = e^x(x-1)+c;$$
$$I(2)=x^2e^x-2I(1)=x^2e^x-2e^x(x-1)+c=e^x(x^2-2x+2)+c;$$
$$I(3)=x^3e^x-3I(2)=x^3e^x-3e^x(x^2-2x+2)+c=e^x(x^3-3x^2+6x-6)+c.$$

14.2.3 Integration by complete differentials

The concept of integration by complete differentials

Integration by complete differentials is basically the inverse of quotient and logarithmic rules of differentiation. Assuming $u = f(x)$ and $v = g(x)$ are differentiable functions, from formulae (10.12) in Chapter 10, quotient rule of differentiation is

$$\frac{d(\frac{u}{v})}{dx}=\frac{\frac{du}{dx}v-u\frac{dv}{dx}}{v^2}.$$

Multiply dx to both sides of above equation to transfer it to its differential form:

$$\not{dx}\times\frac{d(\frac{u}{v})}{\not{dx}}=\frac{(\frac{du}{\not{dx}}v-u\frac{dv}{\not{dx}})\times\not{dx}}{v^2}\longrightarrow d(\frac{u}{v})=\frac{vdu-udv}{v^2}.$$

Apply integral to both sides of above differential equation and remember $F(x)=\int d[F(x)]$,

$$\int d(\frac{u}{v})=\int\frac{vdu-udv}{v^2}=\int\frac{du}{v}-\int\frac{udv}{v^2}\longrightarrow\frac{u}{v}=\int\frac{du}{v}-\int\frac{udv}{v^2}=\int\frac{vdu-udv}{v^2}\text{ or}$$

$$\boxed{\int\frac{vdu-udv}{v^2}=\int d(\frac{u}{v})=\frac{u}{v}}\qquad(14.6)$$

Formula (14.6) indicates that if a combinative integrand can be reorganised back to its original quotient differential, its integral is simply the quotient.

The logarithmic rule of formula (10.16) in Chapter 10 is

$$\frac{d[\ln f(x)]}{dx}=\frac{f'(x)}{f(x)}$$

Multiply dx to both sides of above equation to transfer it to its differential form:

$$\not{dx}\times\frac{d[\ln f(x)]}{\not{dx}}=\frac{f'(x)}{f(x)}\times dx\longrightarrow d[\ln f(x)]=\frac{f'(x)}{f(x)}dx.$$

Apply integral to both sides of above differential equation and remember $F(x) = \int d[F(x)]$,

$$\int d[\ln f(x)] = \int \frac{f'(x)}{f(x)}dx \longrightarrow \ln f(x) = \int \frac{f'(x)}{f(x)}dx \qquad \text{or}$$

$$\boxed{\int \frac{f'(x)}{f(x)}dx = \int d[\ln f(x)] = \ln f(x).} \qquad (14.7)$$

Formula (14.7) indicates that if a quotient integrand can be reorganised as the ratio of the derivative of a function to the function itself, its integral is simply the logarithm of that function.

Example 14.35: Evaluate $\int \frac{e^x}{x}dx - \int \frac{e^x}{x^2}dx$.

Solution

Try formula (14.6) as follows:

$$\int \frac{e^x}{x}dx - \int \frac{e^x}{x^2}dx = \int (\frac{e^x}{x} - \frac{e^x}{x^2})dx = \int (\frac{xe^x}{x^2} - \frac{e^x}{x^2})dx = \int \frac{xe^x - e^x}{x^2}dx = \int \frac{xe^x dx - e^x dx}{x^2}$$

$$= \int \frac{(x)d(e^x) - (e^x)d(x)}{x^2} = \int d(\frac{e^x}{x}) = \frac{e^x}{x} + c. \longleftarrow u = e^x,\ v = x$$

Example 14.36: Evaluate $\int \frac{x^3 \cos x}{\sin^2 x}dx - \int \frac{3x^2}{\sin x}dx$.

Solution

Try formula (14.6) as follows:

$$\int \frac{x^3 \cos x}{\sin^2 x}dx - \int \frac{3x^2}{\sin x}dx = \int (\frac{x^3 \cos x}{\sin^2 x} - \frac{3x^2}{\sin x})dx = \int (\frac{x^3 \cos x}{\sin^2 x} - \frac{3x^2 \sin x}{\sin^2 x})dx$$

$$= \int \frac{x^3 \cos x dx - 3x^2 \sin x dx}{\sin^2 x} = \int \frac{(x^3)d(\sin x) - (\sin x)d(x^3)}{\sin^2 x}$$

$$= -\int \frac{(\sin x)d(x^3) - (x^3)d(\sin x)}{\sin^2 x} \longleftarrow u = x^3,\ v = \sin x$$

$$= -\int d(\frac{x^3}{\sin x}) = -\frac{x^3}{\sin x} + c.$$

Example 14.37: Evaluate $\int \frac{3dx}{x\ln x}$.

Solution

Try formula (14.7) as follows:

$$\int \frac{3dx}{x\ln x} = 3\int \frac{1}{\ln x}(\frac{1}{x})dx = 3\int \frac{(\ln x)'}{\ln x}dx = 3\int d[\ln(\ln x)] = 3\ln(\ln x) + c \longleftarrow f(x) = \ln x,\ f'(x) = (\ln x)' = \frac{1}{x}$$

Example 14.38: Evaluate $\int \cot x dx$.

Solution

Try formula (14.7) as follows:

$$\int \cot x dx = \int \frac{\cos x}{\sin x}dx = \int \frac{(\sin x)'}{\sin x}dx = \int d[\ln \sin x] = \ln \sin x + c \longleftarrow f(x) = \sin x,\ f'(x) = (\sin x)' = \cos x$$

Example 14.39: Evaluate $\int \frac{2x-1}{x^2-x-2}dx$.

Solution

Try formula (14.7) as follows:

$$\int \frac{2x-1}{x^2-x-2}dx = \int \frac{(x^2)'-(x)'-(2)'}{x^2-x-2}dx = \int \frac{(x^2-x-2)'}{x^2-x-2}dx \longleftarrow f(x) = x^2-x-2,\ f'(x) = 2x-1$$

$$= \int d[\ln(x^2-x-2) = \ln(x^2-x-2) + c.$$

Example 14.40: Evaluate $\int \frac{3x-7}{x^2-5x+6}dx$.

Solution

Try formula (14.7) as follows:

$$\int \frac{3x-7}{x^2-5x+6}dx = \int \frac{2x-4+x-3}{(x-2)(x-3)}dx = \int \frac{2(x-2)+(x-3)}{(x-2)(x-3)}dx$$

$$= \int \frac{2(x-2)}{(x-2)(x-3)}dx + \int \frac{(x-3)}{(x-2)(x-3)}dx = 2\int \frac{1}{x-3}dx + \int \frac{1}{x-2}dx$$

$$= 2\int \frac{(x-3)'}{x-3}dx + \int \frac{(x-2)'}{x-2}dx \longleftarrow (x-3)' = 1,\ (x-2)' = 1$$

$$= 2\int d[\ln(x-3)] + \int d[\ln(x-2)] = 2\ln(x-3) + \ln(x-2) + c.$$

14.2.4 Integration by partial fractions

The concept of integration by partial fractions

A partial fraction is a quotient $g(x)/f(x)$, in which the order of the numerator $g(x)$ is lower than that of the denominator $f(x)$. Some partial fractions can be separated into the sum of a few lower-order fractions, and each of these lower-order fractions may be mapped to its integral in a relatively easier way. Two special types of quadratic partial fractions are introduced here. More techniques in dealing with partial fractions are introduced in advanced calculus courses.

Given a function in partial fraction $g(x)/f(x)$, assume the denominator $f(x)$ is a *quadratic function* and can be factorised to $f(x) = (x + a)(x + b)$; the numerator is $g(x) = cx + d$ where a, b, c and d are constants. The following relationships exist:

$$\boxed{\frac{1}{(x+a)(x+b)} = \frac{1}{b-a}\left(\frac{1}{x+a} - \frac{1}{x+b}\right)} \tag{14.8}$$

$$\boxed{\frac{cx+d}{(x+a)(x+b)} = \frac{d-ac}{b-a}\frac{1}{x+a} - \frac{d-bc}{b-a}\frac{1}{x+b}}. \tag{14.9}$$

Example 14.41: Redo Example 14.40 using formula (14.9) for $\int \frac{3x-7}{x^2-5x+6}dx = \int \frac{3x-7}{(x-2)(x-3)}dx$

.

Solution

For the given fractional function, $a = -2$, $b = -3$, $c = 3$, $d = -7$. The integral can be converted to two separate integrals using formula (14.9) as follows:

$$\int \frac{3x-7}{x^2-5x+6}dx = \int \frac{3x-7}{(x-2)(x-3)}dx$$
$$= \int \frac{-7-(-2)(3)}{-3-(-2)}\frac{1}{x+(-2)}dx - \int \frac{-7-(-3)(3)}{-3-(-2)}\frac{1}{x+(-3)}dx$$
$$= \int \frac{-7+6}{-3+2}\frac{1}{x-2}dx - \int \frac{-7+9}{-3+2}\frac{1}{x-3}dx = \int \frac{-1}{-1}\frac{1}{x-2}dx - \int \frac{2}{-1}\frac{1}{x-3}dx$$
$$= \int \frac{(x-2)'}{x-2}dx + 2\int \frac{(x-3)'}{x-3}dx = \int d[\ln(x-2)] + 2\int d[\ln(x-3)] = \ln(x-2) + 2\ln(x-3) + c.$$

Example 14.42: Evaluate $\int \frac{1}{x^2+x-6}dx$.

Solution

Since $\int \frac{1}{x^2+x-6}dx = \int \frac{1}{(x+3)(x-2)}dx$, $a = 3$ and $b = -2$. The integral can be converted to two separate integrals using formula (14.8) as follows:

$$\int \frac{1}{x^2+x-6}dx = \int \frac{1}{(x+3)(x-2)}dx = \int \frac{1}{-2-(3)}[\frac{1}{x+3}-\frac{1}{x-2}]dx$$
$$= \int \frac{1}{-5}[\frac{1}{x+3}-\frac{1}{x-2}]dx = -\frac{1}{5}\int \frac{1}{x+3}dx + \frac{1}{5}\int \frac{1}{x-2}dx = -\frac{1}{5}\int \frac{(x+3)'}{x+3}dx + \frac{1}{5}\int \frac{(x-2)'}{x-2}dx$$
$$= -\frac{1}{5}\int d[\ln(x+3)] + \frac{1}{5}\int d[\ln(x-2)] = -\frac{1}{5}\ln(x+3) + \frac{1}{5}\ln(x-2) + c$$
$$= \frac{1}{5}[\ln(x-2) - \ln(x+3)] + c = \frac{1}{5}\ln\frac{x-2}{x+3} + c.$$

Example 14.43: Evaluate $\int \frac{-x+4}{6x^2+x-2}dx$.

Solution

For $\frac{-x+4}{6x^2+x-2}$, neither formula above can be applied directly. We need going through a general process to separate it into two simpler fractions as follows:

$$\frac{-x+4}{6x^2+x-2} = \frac{-x+4}{(3x+2)(2x-1)} = \frac{A}{3x+2} + \frac{B}{2x-1} = \frac{A(2x-1)}{(3x+2)(2x-1)} + \frac{B(3x+2)}{(3x+2)(2x-1)}$$
$$= \frac{2Ax-A}{(3x+2)(2x-1)} + \frac{3Bx+2B}{(3x+2)(2x-1)} = \frac{2Ax-A+3Bx+2B}{(3x+2)(2x-1)} = \frac{(2A+3B)x+(-A+2B)}{(3x+2)(2x-1)} \text{ or}$$
$$\frac{-x+4}{(3x+2)(2x-1)} = \frac{(2A+3B)x+(-A+2B)}{(3x+2)(2x-1)}.$$

Equate the corresponding coefficients of the numerators on both sides:

$$\begin{cases} 2A+3B=-1 \\ -A+2B=4 \end{cases} \xrightarrow[(2)\times 2]{\to} \begin{cases} 2A+3B=-1 \\ -2A+4B=8 \end{cases} \xrightarrow{(1)+(2)} 7B=7 \longrightarrow \begin{cases} B=1 \\ A=2B-4=-2 \end{cases}. \text{ Thus}$$
$$\frac{-x+4}{6x^2+x-2} = \frac{A}{3x+2} + \frac{B}{2x-1} = \frac{-2}{3x+2} + \frac{1}{2x-1} = \frac{1}{2x-1} - \frac{2}{3x+2}.$$

$$\int \frac{-x+4}{6x^2+x-2}dx = \int \frac{1}{2x-1}dx - \int \frac{2}{3x+2}dx = \int \frac{1}{2(x-\frac{1}{2})}dx - \int \frac{2}{3(x+\frac{2}{3})}dx$$
$$= \frac{1}{2}\int \frac{(x-\frac{1}{2})'}{x-\frac{1}{2}}dx - \frac{2}{3}\int \frac{(x+\frac{2}{3})'}{x+\frac{2}{3}}dx = \frac{1}{2}\int d[\ln(x-\frac{1}{2})] - \frac{2}{3}\int d[\ln(x+\frac{2}{3})]$$
$$= \frac{1}{2}\ln(x-\frac{1}{2}) - \frac{2}{3}\ln(x+\frac{2}{3}) + c = \frac{1}{2}\ln(\frac{2x-1}{2}) - \frac{2}{3}\ln(\frac{3x+2}{3}) + c$$

$$=\frac{1}{2}[\ln(2x-1)-\ln 2]-\frac{2}{3}[\ln(3x+2)-\ln 3]+c=\frac{1}{2}\ln(2x-1)-\frac{2}{3}\ln(3x+2)+(c-\frac{\ln 2}{2}+\frac{2\ln 3}{3})$$
$$=\frac{1}{2}\ln(2x-1)-\frac{2}{3}\ln(3x+2)+C \longleftarrow C=c-\frac{\ln 2}{2}+\frac{2\ln 3}{3}.$$

By combining the above techniques (and other more advanced ones too) of integration, the integrals of more complicated functions could be found. Mappings of some combined or complicated functions are listed in Appendix as a quick reference.

Exercises 14.2

1. Find the integrals for the following integrands by substitution.

   a) $\int \frac{1}{(x-3)^2}dx$ and $\int [6(2x-1)^2+4x-2]dx$

   b) $\int e^{2x+3}dx$ and $\int \frac{3}{\sqrt{2-x}}dx$

   c) $\int \frac{2xdx}{4-x^2}$ and $\int 3x\sqrt{x^2-9}dx$

   d) $\int xe^{2x^2}dx$ and $\int 2x\sin x^2 dx$

   e) $\int \frac{1}{\sqrt{9-x^2}}dx$ and $\int \sqrt{100-x^2}dx$

   f) $\int \sin 2x \cos x dx$ and $\int (1+\cos^3 x \sin x)dx$

   g) $\int \frac{\cot x}{\sin x}dx$ and $\int \cos x \sin\frac{x}{2}dx$

2. Find the integrals for the following integrands by integration by parts.

   a) $\int x^2 e^x dx$ and $\int \frac{\ln x}{x^2} dx$
   b) $\int 4x^2 \cos 2x dx$ and $\int x^2 e^{-x} dx$
   c) $\int x^4 \ln x dx$ and $\int \frac{(\ln x)^2}{x^5} dx$
   d) $\int 3x\sqrt{x+5} dx$ and $\int 8x^3 e^{-2x} dx$
   e) $\int x^2 e^x dx$ and $\int \frac{\ln x}{x^2} dx$
   f) $\int \sin^4 x dx$ and $\int \sin^3 x \cos^4 x dx$
   g) $\int e^{-x} \cos 2x dx$
   h) $\int e^{ax} \sin bx dx$
   i) $\int 4x \arccos x dx$
   j) $\int x^2 \sinh x dx$
   k) $\int e^{2x} \cosh x dx$

3. Find the integrals for the following integrands by complete differentials.

   a) $\int \frac{4}{2x+5} dx$ and $\int \frac{12x+9}{2x^2+3x+1} dx$
   b) $\int \frac{dx}{x(\ln x)^2}$ and $\int 4\tan(2x-1) dx$
   c) $\int \frac{\cos x}{x} dx - \int \frac{\sin x}{x^2} dx$
   d) $\int \frac{\cos x - \sin x}{e^x} dx$
   e) $\int \frac{1-2\ln x}{x^3} dx$

4. Find the integrals for the following integrands by partial fractions.

   a) $\int \frac{3}{x^2-x-2} dx$ and $\int \frac{2x}{x^3+5x^2+6x} dx$
   b) $\int \frac{2-x}{x^2-3x+2} dx$ and $\int \frac{2x-5}{x^2-3x-4} dx$
   c) $\int \frac{3x+4}{x^2-7x-8} dx$

Chapter 14: Exercises Answers

Exercises 14.1

1. a) x^5+c and $\frac{3}{5}x^{\frac{5}{3}}+c,$ b) $-\frac{4}{x}+c$ and $4\sqrt{x}+c,$ c) $3e^{\frac{1}{3}x}+c$ and $-\frac{1}{2}e^{-2x}+c,$

   d) $x-\frac{1}{2}\sin 2x+c$ and $-3\cos\frac{x}{3}+c,$ e) $-\ln x+c$ and $\frac{1}{2}\ln x+c$

2. a) $\frac{1}{2}x^4-\frac{1}{3}x^3+2x^2-3x+c$ and $\frac{2}{3}\sqrt{x^3}-\sqrt{x}+c,$

   b) $2x^2+3\ln x-\frac{1}{x}-\frac{1}{x^2}+c$ and $\frac{1}{2}(e^{2x}-e^{-2x}-\ln x)+c,$

   c) $2e^{\frac{1}{2}x}-3e^{\frac{1}{3}x}+c$ and $-\cos 2x-\frac{1}{2}\sin 2x+c,$ d) $y=x^3-2x^2+5x-2,$

   e) $y=x-\ln x+\frac{1}{x}-4,$ f) $y=x+\frac{1}{2}\sin 2x+\frac{1}{2}\cos 2x+\frac{1}{2}=x+\frac{1}{2}(\sin 2x+\cos 2x+1),$

   g) $y=2x^{\frac{3}{2}}-2x^{\frac{1}{2}}+10-2\sqrt{2}$

Exercises 14.2

1. a) $y=-\frac{1}{x-3}+c$ and $y=\frac{3}{3}t^3+2x^2-2x+c=(2x-1)^3+2x^2-2x+c,$

   b) $y=\frac{1}{2}e^{2x+3}+c$ and $y=-6\sqrt{2-x}+c,$

   c) $y=-\ln(4-x^2)+c$ and $y=\sqrt{(x^2-9)^3}+c,$

   d) $y=\frac{1}{4}e^{2x^2}+c$ and $y=-\cos x^2+c,$

   e) $y=-\arccos\frac{x}{3}+c$ and $y=50\arcsin\frac{x}{10}+\frac{x}{2}\sqrt{100-x^2}+c,$

   f) $y=-\frac{2}{3}\cos^3 x+c$ and $y=x-\frac{1}{4}\cos^4 x+c,$

   g) $y=-\frac{1}{\sin x}+c$ and $y=2\cos\frac{x}{2}(1-\frac{2}{3}\cos^2\frac{x}{2})+c$

2. a) $y=e^x(x^2-2x+2)+c$ and $y=-\frac{\ln x+1}{x}+c,$

   b) $y=2x^2\sin 2x+2x\cos 2x-\sin 2x+c$ and $y=-e^{-x}(x^2+2x+2)+c,$

   c) $y=\frac{1}{25}x^5(5\ln x-1)+c$ and $y=-\frac{1}{32x^4}[8(\ln x)^2+4\ln x+1]+c,$

   d) $y=2x(x+5)^{\frac{3}{2}}-\frac{4}{5}(x+5)^{\frac{5}{2}}+c$ and $y=-e^{-2x}(4x^3+6x^2+6x+3)+c,$

   e) $y=e^x(x^2-2x+2)+c$ and $y=-\frac{1}{x}(\ln x+1)+c$

f) $y = -\sin^3 x \cos x + \frac{3}{8}(x - \frac{1}{4}\sin 4x) + c$ and $y = \frac{1}{7}\cos^7 x - \frac{1}{5}\cos^5 x + c,$

g) $I = \frac{1}{5}e^{-x}(2\sin 2x - \cos 2x) + c,$ h) $I = \frac{e^{ax}}{a^2 + b^2}(a\sin bx - b\cos bx) + c$

i) $y = 2x^2 \arccos x + \arcsin x - x\sqrt{1 - x^2} + c$

j) $y = x^2 \cosh x - 2x\sinh x + 2\cosh x + c$

k) $y = -\frac{1}{3}e^{2x}(\sinh x - 2\cosh x) + c = \frac{1}{3}e^{2x}(2\cosh x - \sinh x) + c$

3. a) $y = 2\ln(2x+5) + c$ and $y = 3\ln(2x^2 + 3x + 1) + c,$

b) $y = -\frac{1}{\ln x} + c$ and $y = -2\ln[\cos(2x-1)] + c,$

c) $y = \frac{\sin x}{x} + c,$ d) $y = e^{-x}\sin x + c,$ e) $y = \frac{\ln x}{x^2} + c$

4. a) $y = \ln\frac{x-2}{x+1} + c$ and $y = 2\ln\frac{x+2}{x+3} + c,$

b) $y = \ln|1-x| + c$ and $y = \frac{1}{5}[3\ln(x-4) + 7\ln(x+1)] + c,$

c) $y = \frac{1}{9}[28\ln(x-8) - \ln(x+1)] + c$

CHAPTER 15

15 Applications of Indefinite Integration

CHAPTER OBJECTIVES

- Use indefinite integrals to solve problems in science, engineering, and other fields.

Essential statements on applications of indefinite integration:

- Indefinite integration is widely used in solving differential equations encountered in science, engineering and other disciplines.
- An indefinite integral becomes the so-called 'initial problem' if some known conditions are provided for the differential equation.

Key topics:

- Differential equations
- Initial problem
- Solve simple differential equations

Flowchart of mathematical knowledge building

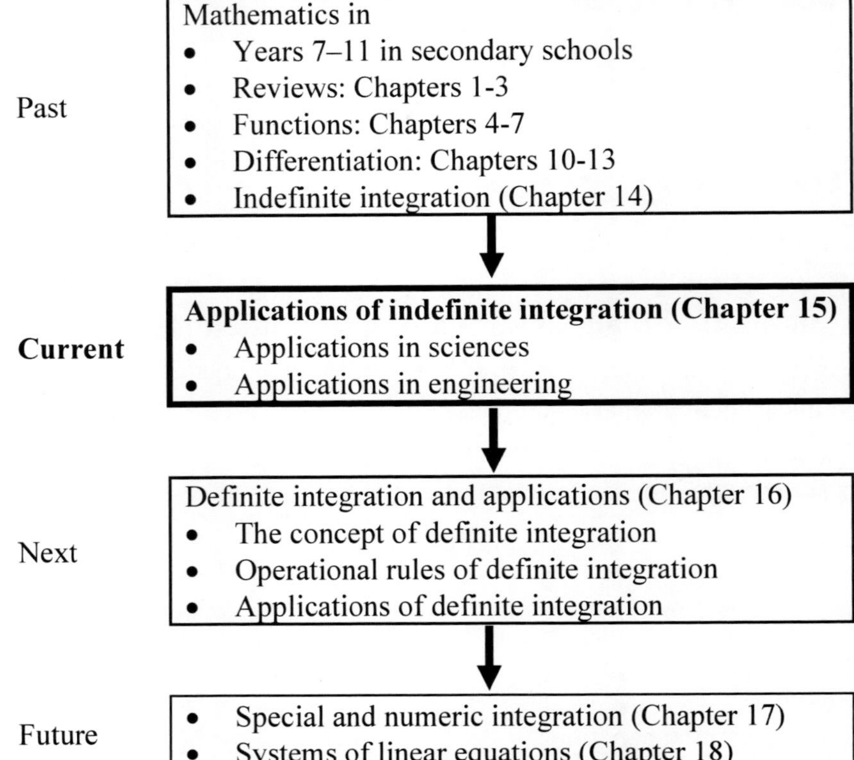

Chapter 15: Applications of Indefinite Integration

Indefinite integration is widely used in solving differential equations, in which one or more derivatives are involved. Differential equations can be resulted from mathematical description of a real-world scenario in science, engineering, technology, business and so forth. Some simple applications in various areas are presented in this chapter. More complicated applications should be studied in advanced mathematics courses.

15.1 Applications of Indefinite Integration in Sciences

Example 15.1: The gradient of a function is $\frac{dy}{dx} = 2x + \sin x - 3$. Find the function that goes through point (0, 1).

Solution

Direct integration should give the general formula of this function with one unknown constant.

$$\frac{dy}{dx} = 2x + \sin x - 3 \longrightarrow dy = (2x + \sin x - 3)dx$$

Integrate both sides simultaneously to get

$$\int dy = \int (2x + \sin x - 3)dx \longrightarrow y = \int (2x + \sin x - 3)dx = x^2 - \cos x - 3x + c.$$

As this function goes through point (0, 1), substitute $x = 0$ and $y = 1$ into above equation to determine constant c as follows:

$$1 = 0^2 - \cos 0 - 3 \times 0 + c \longrightarrow 1 = -1 + c \longrightarrow c = 2. \text{ Thus}$$
$$y = x^2 - \cos x - 3x + 2.$$

Example 15.2: A car travels on a straight road with a speed of 8 m/s. The driver stops acceleration and lets the car naturally slow down. Assuming the slowing-down (or the rate of change in speed) of the car equals 20% of its speed, how far will the car travel in 10 seconds after the driver stops acceleration?

Solution

The speed of the car is decreasing with time, and is also the rate of change in distance (s), i.e., $v = \frac{ds}{dt}$. As the car slows down in proportion to the speed with a constant –0.2 (slowing down), this means

$$\frac{dv}{dt} = -0.2v.$$

Rearrange this first-order differential equation by separating time and speed to two sides as

$$dt = \frac{dv}{-0.2v} = -5\frac{dv}{v}.$$

Integrate both sides simultaneously to get

$$\int dt = -5\int \frac{dv}{v} \longrightarrow t = -5\ln v + c.$$

At the time the driver stops acceleration ($t = 0$), the speed of the car is 8 m/s. Substitute these initial conditions into above equation to determine constant c:

$$0 = -5\ln 8 + c \longrightarrow c = 5\ln 8 \quad \text{thus}$$

$$t = -5\ln v + 5\ln 8 = 5\ln\frac{8}{v} \quad \text{or}$$

$$5\ln\frac{8}{v} = t \longrightarrow \ln\frac{8}{v} = \frac{t}{5} = 0.2t \longrightarrow \frac{8}{v} = e^{0.2t} \longrightarrow v = \frac{8}{e^{0.2t}} = 8e^{-0.2t}$$

As $v = \dfrac{ds}{dt}$, thus

$$\frac{ds}{dt} = 8e^{-0.2t} \longrightarrow ds = 8e^{-0.2t}dt \longrightarrow \int ds = 8\int e^{-0.2t}dt$$

$$s = 8\int e^{-0.2t}dt = \frac{8}{-0.2}e^{-0.2t} + c = -40e^{-0.2t} + c.$$

As the distance is counted after the driver stops acceleration, $s = 0$ when $t = 0$. Substitute these initial conditions into above equation to determine constant c as follows:

$$0 = -40e^{-0.2\times 0} + c \longrightarrow c = 40.$$

$$s = -40e^{-0.2t} + 40 = 40(1 - e^{-0.2t}).$$

Ten seconds ($t = 10$ s) after the driver stopped acceleration, the car would have run for

$$s = 40(1 - e^{0.2\times 10}) = 40(1 - e^{-2}) \approx 34.6 \text{ m}.$$

Example 15.3: Ignoring the resistance in the air, a free fall object has a constant acceleration g, known as the acceleration of Earth's gravity that always points to the centre of the Earth. Since acceleration a is the first-order derivative of velocity (v), $a = g$ can be expressed as $\frac{dv}{dt} = g$. At the start of the fall, if the object is placed at the initial point $y(0) = 0$, and $v(0) = 0$, determine the velocity and distance (y) of such free fall.

Solution

For $\frac{dv}{dt} = g$,

$$dv = gdt \longrightarrow \int dv = \int gdt \longrightarrow v = gt + c.$$

Substitute $v(0) = 0$ into above equation to determine constant c:

$$0 = g \times 0 + c \longrightarrow c = 0. \quad \longrightarrow v = gt.$$

$$\frac{dy}{dt} = v = gt \longrightarrow dy = gtdt \longrightarrow \int dy = \int gtdt \longrightarrow y = g\int tdt = \frac{1}{2}gt^2 + C.$$

Substitute $y(0) = 0$ into above equation to determine constant C:

$$0 = \frac{1}{2}g \times 0^2 + C \longrightarrow C = 0 \text{ m}. \quad \therefore y = \frac{1}{2}gt^2.$$

Example 15.4: The velocity of a parachute in the air is expressed as $m\frac{dv}{dt} = mg - kv$, where m = 50 kg, g = 10 m/s^2 and k = 10 kg/s are all constants. Solve this differential equation to find the velocity of the parachute at any time in the air, assuming $v(0) = 0$. If the parachute is released at an altitude of 9750 m above the ground, find the altitude (h) of the parachute in the air.

Solution

The differential equation can be rearranged as

$$m\frac{dv}{dt} = mg - kv \longrightarrow \frac{dv}{dt} = \frac{1}{m}(mg - kv) = \frac{-k}{m}v + g = \frac{-10}{50}v + 10 = -0.2v + 10$$

$$dv = (-0.2v + 10)dt = -0.2(v - \frac{10}{0.2})dt = -0.2(v - 50)dt$$

$$dv = -0.2(v - 50)dt \xrightarrow{\div(v-50)} \frac{dv}{v - 50} = -0.2dt.$$

Integrate both sides simultaneously,

$$\frac{dv}{v-50} = -0.2dt \longrightarrow \int \frac{(v-50)'}{v-50} dv = -0.2\int dt \longrightarrow \ln(v-50) = -0.2t + c$$
$$v - 50 = e^{-0.2t+c} = e^c e^{-0.2t} = Ce^{-0.2t} \longrightarrow v = Ce^{-0.2t} + 50.$$

Substitute $v(0) = 0$ into above equation to determine constant C:

$$0 = Ce^0 + 50 \longrightarrow 0 = C + 50 \longrightarrow C = -50. \quad \therefore v = -50e^{-0.2t} + 50 = 50(1 - e^{-0.2t}).$$

As the altitude h decreases while the parachute is falling, i.e., $\frac{dh}{dt} = -v$, hence

$$\frac{dh}{dt} = -v = -50(1 - e^{-0.2t}) \longrightarrow dh = -50(1 - e^{-0.2t})dt \longrightarrow \int dh = -50\int (1 - e^{-0.2t})dt$$
$$h = -50\int (1 - e^{-0.2t})dt = -50(t - \frac{1}{-0.2}e^{-0.2t}) + c = -50(t + 5e^{-0.2t}) + c.$$

Substitute $h(0) = 9750$ m into above equation to determine constant c:

$$9750 = -50(0 + 5e^0) + c \longrightarrow 9750 = -250 + c \longrightarrow c = 10000 \text{ m}.$$
$$\therefore h = -50(t + 5e^{-0.2t}) + 10000 = 10000 - 50t - 250e^{-0.2t}.$$

Example 15.5: The compound interest is the addition of interest to the principal sum of a loan or deposit. For a long-term account without withdrawing monies from it, it becomes a special case called as continuous compounding. The compound amount of an account of continuous compounding can be determined by solving the differential equation $\frac{dP}{dt} = rP$, where P is the amount of dollars in the account that attracts a yearly interest of r.

a) How much will be in the account after 3 years if $2000 initially in the account with a fixed interest 5% per year?
b) When will the account accumulate to $10,000?

Solution

Apply direct integration:

$$\frac{dP}{dt} = rP \longrightarrow dP = rPdt \longrightarrow rdt = \frac{dP}{P}. \text{ Thus}$$

$$\int rdt = \int \frac{dP}{P} \longrightarrow \ln P = rt + c \longrightarrow P = e^{rt+c} = e^c e^{rt} = Ce^{rt}.$$

a) Substitute $t = 0$ and $P = 2000$ into this general solution,

$$2000 = Ce^{0} \longrightarrow C = 2000.$$

Therefore, considering $r = 5\% = 0.05$, the solution is

$$P = 2000e^{0.05t}.$$

The following amount will be in the account after 3 years:

$$P = 2000e^{0.05\times 3} = 2000e^{0.15} = 2000 \times 1.1618 = \$2323.6.$$

b) Substitute $P = 10000$ into $P = 2000e^{0.05t}$,

$$10000 = 2000e^{0.05t} \longrightarrow e^{0.05t} = 10000/2000 = 5 \longrightarrow 0.05t = \ln 5 = 1.6094$$
$$t = 1.6094/0.05 = 32.2 \approx 33 \; years$$

This means it will need about 33 years to reach \$10000 if \$2000 initially in the account with a fixed interest 5% per year.

Example 15.6: A city currently has a population of 500,000 people with a natural growth rate of 2% annually. In the meantime, people also leave the city at a rate of $1000e^{0.03t}$ people annually (t is the number of years from today). Find a function that predicts the population of this city in 10 years.

Solution

The *population change* is *natural growth* × *population* – *number of people left*, i.e.,

$$\frac{dp}{dt} = 0.02p - 1000e^{0.03t} \longrightarrow dp = 0.02pdt - 1000e^{0.03t}dt$$

$$dp - 0.02pdt = -1000e^{0.03t}dt \longrightarrow e^{-0.02t}(dp - 0.02pdt) = e^{-0.02t}(-1000e^{0.03t}dt)$$

$$e^{-0.02t}dp - 0.02pe^{-0.02t}dt = -1000e^{0.03t-0.02t}dt$$

$$\underbrace{d(p)(e^{-0.02t})}_{du\cdot v} + \underbrace{(p)d(e^{-0.02t})}_{u\cdot dv} = -1000e^{0.01t}dt \longrightarrow d(pe^{-0.02t}) = -1000e^{0.01t}dt$$

$$\int d(pe^{-0.02t}) = -1000\int e^{0.01t}dt \longrightarrow pe^{-0.02t} = \frac{-1000}{0.01}e^{0.01t} + c$$

$$\therefore\ p = e^{0.02t}(-100000e^{0.01t} + c) = ce^{0.02t} - 100000e^{0.03t}.$$

Substitute $t = 0$ and $p = 500,000$ into above formula:

$$500000 = ce^{0} - 100000e^{0} \longrightarrow 500000 = c - 100000 \longrightarrow c = 600000$$
$$\therefore\ p = 600000e^{0.02t} - 100000e^{0.03t}.$$

In 10 years ($t = 10$), the population of this city would be:

$$p = 600000e^{0.02\times 10} - 100000e^{0.03\times 10} = 600000e^{0.2} - 100000e^{0.3} = 597856\ \text{(people)}.$$

Exercises 15.1

1. The gradient of a function is $\frac{dy}{dx} = 3x^2 - \frac{1}{\sqrt{x}}$. Find this function that crosses point (1, 1).

2. The bacteria in a certain culture increases according to $\frac{dN}{dt} = 0.125N$. Initially $N_0 = 500$; find N when $t = 10$ seconds.

3. A ball slides down a chute with acceleration 1.25 m/s^2. The chute is 25 meters long and the ball touches the bottom in 5 seconds. What was the initial velocity of the ball and how fast the ball moves when it is 10 meters from the bottom of the chute?

4. A car travels on a straight road with a speed of 8 m/s. The driver accelerates the car with a rate of 10% of the speed. How fast will the car travel after the driver keeps accelerating for 10 seconds?

5. Refer to Example 15.5. (a) How much will be in the account after 5 years if $5000 initially in the account with a fixed interest 6% per year? (b) When will the account accumulate $10,000?

6. Refer to Example 15.6. A city currently has a population of 900,000 people with a natural growth rate of 1.5% annually. In the meantime, people also leave the city at a rate of $1000e^{0.04t}$ people annually (t is the number of years from today). Find a function that predicts the population of this city in 5 and 10 years respectively.

15.2 Applications of Indefinite Integration in Engineering

Example 15.7: The bending moment M of a beam of L metres long is defined by $\frac{dM}{dx} = w(x-L)$, where w is a loading constant. Find moment M given $M = 0$ at $x = L$.

Solution

Direct integration should give the general formula of this moment with one unknown constant.

$$\frac{dM}{dx} = w(x-L) \longrightarrow dM = w(x-L)dx = (wx - wL)dx.$$

$$M = \int dM = \int (wx - wL)dx = \frac{1}{2}wx^2 - wLx + c.$$

Substitute $x = L$ and $M = 0$ into above equation to determine constant c as follows:

$$0 = \frac{1}{2}wL^2 - wLL + c \longrightarrow 0 = -\frac{1}{2}wL^2 + c \longrightarrow c = \frac{1}{2}wL^2.$$

$$\therefore M = \frac{1}{2}wx^2 - wLx + \frac{1}{2}wL^2 = \frac{1}{2}w(x^2 - 2Lx + L^2) = \frac{1}{2}w(x-L)^2.$$

Example 15.8: A metal bar cools down with a rate $\frac{dT}{dt} = -kT$, where k is a constant. Find the general formula between temperature T and time t.

Solution

Direct integration should lead to determining the general formula.

$$\frac{dT}{dt} = -kT \longrightarrow dT = -kTdt \xrightarrow{\div T} \frac{dT}{T} = -kdt$$

Integrate both sides simultaneously and remember $\int \frac{dx}{x} = \ln x$,

$$\int \frac{dT}{T} = -\int kdt \longrightarrow \ln T = -kt + c$$

$$\therefore T = e^{-kt+c} = e^c e^{-kt} = Ce^{-kt}. \longleftarrow C = e^c \text{ a new constant}$$

Example 15.9: The current i in an RL circuit with a constant input E is defined by $L\frac{di}{dt}+Ri=E$, where L is the inductance; R is the resistance; E is the voltage of the supply. Find the current given $i=0$ at $t=0$.

Solution

The differential equation can be rearranged as

$$L\frac{di}{dt}+Ri=E \longrightarrow \frac{di}{dt}=\frac{1}{L}(E-Ri)=\frac{-R}{L}(i-\frac{E}{R})$$

$$di=\frac{-R}{L}(i-\frac{E}{R})dt \xrightarrow{\div(i-E/R)} \frac{di}{i-\frac{E}{R}}=\frac{-R}{L}dt.$$

Integrate both sides simultaneously,

$$\int\frac{di}{i-\frac{E}{R}}=\int\frac{-R}{L}dt \longrightarrow \int\frac{(i-\frac{E}{R})'}{i-\frac{E}{R}}di=\frac{-R}{L}\int dt \longrightarrow \ln(i-\frac{E}{R})=\frac{-R}{L}t+c$$

$$i-\frac{E}{R}=e^{\frac{-R}{L}t+c}=e^{c}e^{\frac{-R}{L}t}=Ce^{\frac{-R}{L}t} \qquad \therefore i=Ce^{\frac{-R}{L}t}+\frac{E}{R}.$$

Substitute $t=0$ and $i=0$ into above equation to determine constant C as follows:

$$0=Ce^{0}+\frac{E}{R} \longrightarrow 0=C+\frac{E}{R} \longrightarrow C=-\frac{E}{R}. \qquad \therefore i=-\frac{E}{R}e^{\frac{-R}{L}t}+\frac{E}{R}=\frac{E}{R}(1-e^{\frac{-R}{L}t}).$$

Example 15.10: If a spring linked with a mass of m is placed in the static equilibrium position ($x=0$), the system will be static without movement. If the mass is pulled ($x>0$) or pushed ($x<0$) and then released, the spring will exert a resistant force against the movement of the mass (see the figure below). Without applying any other external force, establish the differential equation for this system and solve it given $m=1$ kg, $k=1$ N/m, $x(0)=1$ m and $v(0)=0$ m/s.

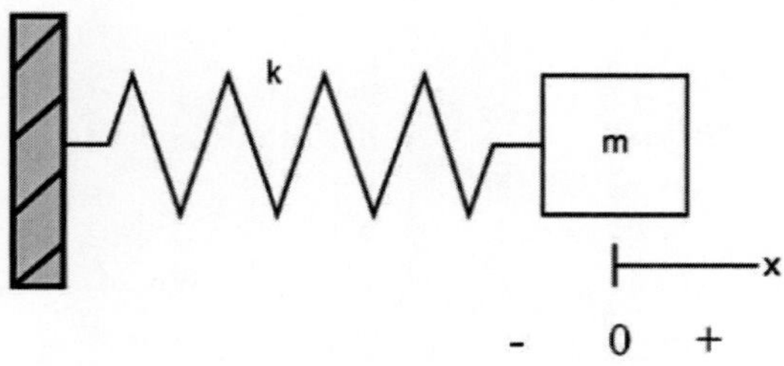

Solution

By Hooke's law, the resistance against any movement of the mass is negatively propositional to the distance of the mass moved away from the static equilibrium position, i.e., $F_r = -kx$ where k is the spring constant of the spring used in the system. By Newton's law, this force equals the product of the mass m and its acceleration $a = \frac{dv}{dt} = \frac{d^2x}{dt^2}$, or $m\frac{d^2x}{dt^2} = -kx$.

This is equivalent to

$$m\frac{d^2x}{dt^2} + kx = 0 \quad \text{or} \quad \frac{d^2x}{dt^2} + \frac{k}{m}x = 0.$$

Let $\omega^2 = \frac{k}{m}$; the equation becomes

$$\frac{d^2x}{dt^2} + \omega^2 x = 0.$$

This is a second-order differential equation of distance $x(t)$ with respect to time t. It describes the movement of free oscillations possessed by the spring-mass system or alike in scientific and engineering applications. Solving this equation requires advanced mathematical techniques. However, we can still solve this specific case using techniques learnt in this text as follows.

Since m = 1 kg and k = 1 N/m, $\omega^2 = \frac{k}{m} = 1$ or $\omega = 1$. Thus $\frac{d^2x}{dt^2} + x = 0$.

Since $\frac{dx^2}{dt^2} = \frac{d}{dt}(\underbrace{\frac{dx}{dt}}_{v}) = \frac{dv}{dt} = \frac{dv}{dx}\underbrace{\frac{dx}{dt}}_{v} = v\frac{dv}{dx}$, the above equation can be expressed as

$$v\frac{dv}{dx} + x = 0 \longrightarrow v\frac{dv}{dx} = -x \longrightarrow vdv = -xdx \longrightarrow \int vdv = -\int xdx$$

$$\frac{1}{2}v^2 = -\frac{1}{2}x^2 + c \longrightarrow v^2 = -x^2 + 2c.$$

In the beginning, $x(0)$ = 1 m and $v(0)$ = 0 m/s. Substitute these into above formula,

$$0^2 = -1^2 + 2c \longrightarrow 2c = 1. \longrightarrow v^2 = -x^2 + 1 = 1 - x^2 \longrightarrow v = \frac{dx}{dt} = \sqrt{1-x^2}$$

$$\frac{dx}{\sqrt{1-x^2}} = dt \longrightarrow \int \frac{dx}{\sqrt{1-x^2}} = \int dt$$

We obtained $\int \frac{1}{\sqrt{1-x^2}} dx = \arcsin x + c$ in Example 14.15. Thus

$$\arcsin x = t + c \longrightarrow \sin(\arcsin x) = \sin(t+c) \longleftarrow \text{apply sin() to both sides}$$
$$x = \sin(t+c).$$

Substitute $x(0) = 1$ m into above formula,

$$1 = \sin(0+c) \longrightarrow 1 = \sin c \longrightarrow c = \frac{\pi}{2} \qquad \therefore x = \sin(t + \frac{\pi}{2}) = \cos t.$$

This is a cosine function, which means the spring-mass system oscillates periodically around the equilibrium position.

Exercises 15.2

1. The bending moment M of a beam of L metres long is defined by $\frac{dM}{dx} = 2x(x-L)$. Find moment M given $x = \mathrm{L}$ and $M = 0$.

2. Assume the bending moment M of a beam of L metres long with a sine load is defined by $\frac{dM}{dx} = \sin x(x-L)$. Find moment M given $x = 0$ and $M = 0$.

3. A radioactive element decays with a rate $\frac{dm}{dt} = -km$, where k is a constant. Find the general formula between mass m and time t.

4. Refer to Example 15.10. Solve the spring-mass system given $m = 1$ kg, $k = 4$ N/m, $x(0) = 2$ m and $v(0) = 0$ m/s.

5. Are temperature decreases with altitude above the ground with a rate $\frac{dT}{dh} = \frac{-k}{h^2}$, where k is a constant. If the temperature 1 meter above the ground is 20° and that at 10,000 meters is –70°, find the formula between temperature T and altitude h.

Chapter 15: Exercises Answers

Exercises 15.1

1. $y = x^3 - 2\sqrt{x} + 2$
2. 1745
3. $v_0 = 1.875$ m/s, $v_{h=10} \approx 6.4$ m/s
4. $v = 21.746$ m/s
5. (a) $P = \$6749.29$ (b) $t = 11.55 \approx 12$ years
6. $p(5) \approx 964355$ (people) and $p(10) \approx 1032452$ (people)

Exercises 15.2

1. $M = \frac{2}{3}x^3 - Lx^2 + \frac{1}{3}L^3 = \frac{1}{3}(2x^3 - 3Lx^2 + L^3)$
2. $M = -x\cos x + \sin x + L\cos x - L$
3. $m = ce^{-kt}$
4. $x = 2\sin(t + \frac{\pi}{2}) = 2\cos t$
5. $T = \frac{90.009}{h} - 70.009$ (°)

CHAPTER 16

16 Definite Integration and Applications

CHAPTER OBJECTIVES

- Introduce the concept of definite integration and its geometric meaning
- Study general properties and operational rules of definite integration
- Utilise symmetric properties of functions to simplify definite integration over symmetric ranges
- Use definite integration to solve geometric, science and engineering problems.

Essential statements on definite integration:

- Definite integral is the indefinite integral of function $f(x)$ within range $[a, b]$.
- The definite integral of $f(x)$ within range $[a, b]$ is the area between $f(x)$ and the x-axis from a to b.
- The outcome of the definite integral of $f(x)$ within range $[a, b]$ is a certain number if both a and b are specified.
- The outcome of the definite integral of $f(x)$ within range $[a, b]$ is determined by the Newton-Leibniz formula.

Key topics:

- The concept of definite integral and geometric meaning
- Newton-Leibniz formula
- Limits (bounds) in definite integration
- Operational rules of definite integration
- Plane areas, volume of solids of revolution, length of arc, surface area of revolution
- Work, and plane mass centroids

Flowchart of mathematical knowledge building

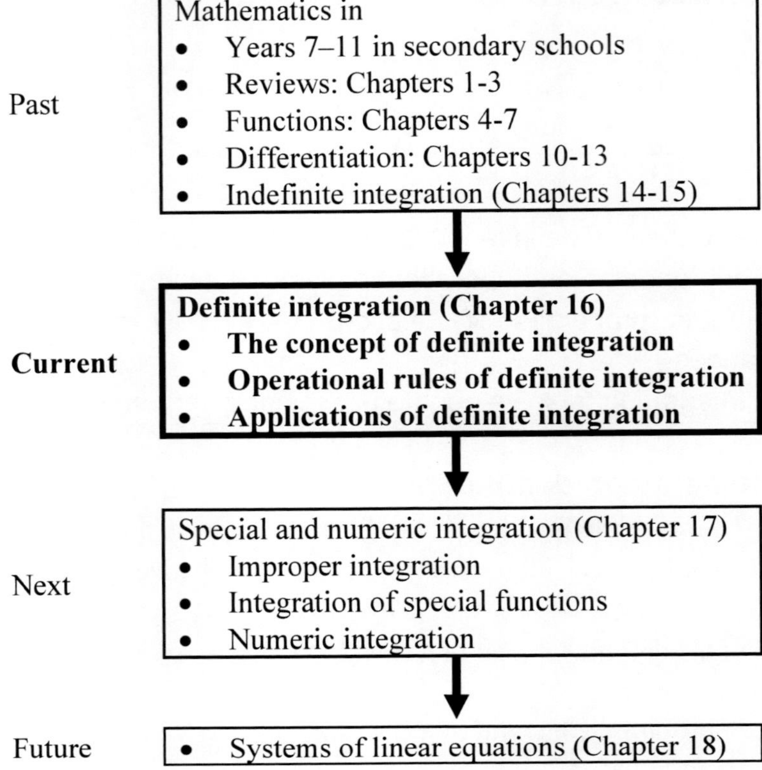

Chapter 16: Definite Integration and Applications

16.1 Essentials of Definite Integration

16.1.1 The concept and essential formulae of definite integrals

The concept of definite integrals

Let $f(x)$ be the derivative of function $F(x)$ or $f(x) = F'(x)$ and both be continuous functions defined in range $[a, b]$. The integral of integrand $f(x)$ within range $[a, b]$ is called the definite integral of function $f(x)$, denoted as:

$$\int_a^b f(x)dx \tag{16.1}$$

where a and b are called the lower and upper limits (bounds) respectively. Note that the outcome of the definite integral (16.1) is a certain number if both a and b are specified numbers.

Newton-Leibniz formula

If $F(x)$ is the indefinite integral of $f(x)$, the value of the definite integral of $f(x)$ within $[a, b]$ is

$$\int_a^b f(x)dx = F(x)\Big|_a^b = F(b) - F(a). \tag{16.2}$$

This is called the *Newton-Leibniz formula*.

Geometric explanation of definite integrals

The definite integral of $f(x)$ within range $[a, b]$ is the area between $f(x)$ and the x-axis from a to b as illustrated in Figure 16.1. This can be expressed by the limit of the following summation:

$$\int_a^b f(x)dx = \lim_{\delta x \to 0} \sum f(x)\delta x \xleftrightarrow{\delta x = \frac{b-a}{n}} \lim_{n \to \infty} \sum f(x)\delta x = \lim_{n \to \infty} \sum f(x)\frac{b-a}{n}. \tag{16.3}$$

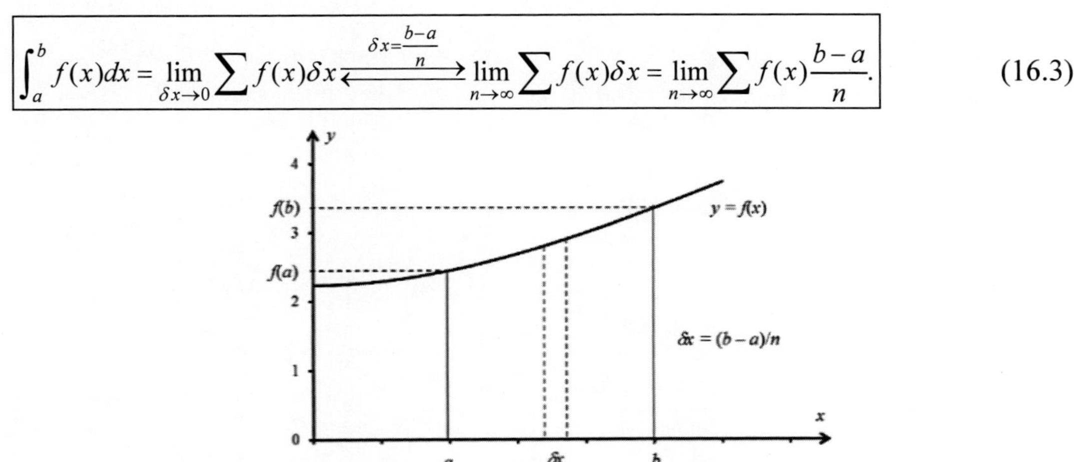

Figure 16.1: Geometric explanation of definite integral

It is easy to understand that a similar expression can be derived if given an integrand $x = g(y)$ within range $[c, d]$:

$$\int_c^d g(y)dy = \lim_{\delta y\to 0}\sum g(y)\delta y \xleftarrow{\delta y=\frac{d-c}{n}} \lim_{n\to\infty}\sum g(y)\delta y = \lim_{n\to\infty}\sum g(y)\frac{d-c}{n}. \quad (16.4)$$

Example 16.1: Given $f(x)=x+1$, find $\int_1^3 f(x)dx$ using Newton-Leibniz formula (16.2). Verify the result by using the limit of summation formula (16.3).

Solution

Given $f(x)=x+1$, its indefinite integral is

$$F(x)=\int f(x)dx=\int (x+1)dx=\frac{1}{2}x^2+x+c. \text{ Hence}$$

$$F(1)=\frac{1}{2}\times 1^2+1+c=\frac{1}{2}+1+c=1.5+c$$

$$F(3)=\frac{1}{2}\times 3^2+3+c=\frac{9}{2}+3+c=7.5+c.$$

By Newton-Leibniz formula (16.2),

$$\int_1^3 f(x)dx=F(x)|_1^3=F(3)-F(1)=(7.5+c)-(1.5+c)=7.5-1.5+c-c=6.$$

It can be seen the unknown constant c in $F(x)$ is naturally cancelled out by the difference between the upper and lower values. Therefore, *there is no need to retain unknown constant c in F(x) in definite integration*.

To verify the correctness of the outcome using the limit of summation formula (16.3), we divide range [1, 3] by n intervals of equal width δx, bounded at $x_0 = a$, $x_1 = x_0 + \delta x = a + \delta x$, $x_2 = a + 2\delta x$, …, $x_i = a + i\delta x$, …, $x_n = a + n\delta x = b$ (Figure 16.2). It is obvious that

$$\delta x=\frac{b-a}{n}=\frac{3-1}{n}=\frac{2}{n}, \text{ and } f(x_i)=x_i+1=a+i\delta x+1=2+\frac{2i}{n}.$$

Substitute δx and $f(x_i)$ into formula (16.3)

$$\int_1^3 f(x)dx=\lim_{n\to\infty}\sum f(x)\delta x=\lim_{n\to\infty}\sum(2+\frac{2i}{n})(\frac{2}{n})=\lim_{n\to\infty}\sum(\frac{4}{n}+\frac{4i}{n^2})$$

$$=\lim_{n\to\infty}\sum\frac{4}{n}+\lim_{n\to\infty}\sum\frac{4i}{n^2}=\lim_{n\to\infty}(\frac{4}{n}\sum_{i=1}^{n}1+\lim_{n\to\infty}(\frac{4}{n^2}\sum_{i=1}^{n}i)$$

$$= \lim_{n\to\infty}(\frac{4}{n}\times n) + \lim_{n\to\infty}[\frac{4}{n^2}\frac{n(n+1)}{2}] \longleftarrow \sum_{i=1}^{n} 1 = n \text{ and } \sum_{i=1}^{n} i = \frac{n(n+1)}{2} \longleftarrow \textit{see} \text{ Table 3.2.}$$

$$= \lim_{n\to\infty}(4) + 2\lim_{n\to\infty}\frac{n^2+n}{n^2} = 4 + 2\lim_{n\to\infty}(1+\frac{1}{n}) = 4 + 2(1+0) = 6.$$

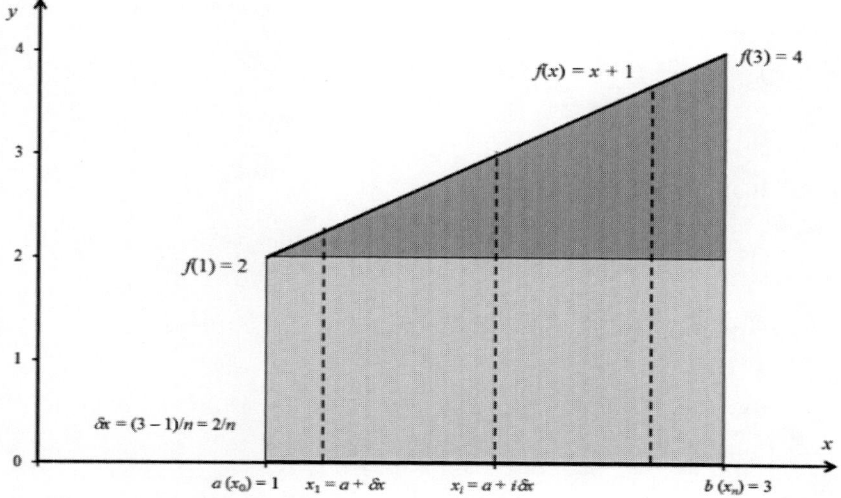

Figure 16.2: Division of the area between $f(x) = x + 1$ and the x-axis from 1 to 3

To check if the area between $f(x) = x + 1$ and the x-axis from 1 to 3 is 6, we regard the area as the combination of a rectangle and a right triangle shown in Figure 16.2. Thus

$$A = f(a)(b-a) + \frac{1}{2}(b-a)[f(b)-f(a)] = 2(3-1) + \frac{1}{2}(3-1)[4-2] = 4 + \frac{1}{2}\times 2\times 2 = 4 + 2 = 6.$$

These three approaches produced the same value.

The processes in a definite integral can be combined together without specifying $F(a)$ and $F(b)$ separately, which is shown in the following examples.

Example 16.2: Evaluate $\int_0^{\frac{\pi}{2}} \sin x dx$ and $\int_{-1}^{2} 3x^2 dx$.

Solution

By Newton-Leibniz formula (16.2),

$$\int_0^{\pi/2} \sin x dx = (-\cos x)\Big|_0^{\pi/2} = -(\cos\frac{\pi}{2} - \cos 0) = -(0-1) = 1,$$

$$\int_{-1}^{2} 3x^2 dx = \frac{3}{3}x^3\Big|_{-1}^{2} = 2^3 - (-1)^3 = 8 + 1 = 9.$$

16.1.2 Handling the limits in definite integration

Let $F(x)$ be the *indefinite* integral of $f(x)$ through substitution $t = t(x)$. There are two options to determine the *definite* integral of $f(x)$ in range $[a, b]$ through $F(x)$ with $x \in [a, b]$ or $G(t)$ with $t \in [c, d]$ where $c = t(a)$ and $d = t(b)$, mathematically denoted as:

$$\boxed{\int_a^b f(x)dx = F(x)\Big|_a^b \xrightarrow[c=t(a),d=t(b)]{t=t(x)} \int_c^d g(t)dt = G(t)\Big|_c^d.} \tag{16.5}$$

Example 16.3: Evaluate $\int_0^2 6(x+2)^2 dx$.

Solution

The corresponding indefinite integral was obtained in Example 14.11 in Chapter 14 using substitution $u = x + 2$ as

$$\int 6(x+2)^2 dx = \int 6u^2 du = \frac{6}{3}u^3 + c = 2u^3 + c = 2(x+2)^3 + c.$$

The definite integral can be directly obtained by substituting $a = 0$ and $b = 2$ as follows,

$$\int_0^2 6(x+2)^2 dx = 2(x+2)^3\Big|_0^2 = 2(2+2)^3 - 2(0+2)^3 = 2(4)^3 - 2(2)^3 = 128 - 16 = 112.$$

Alternatively, as $u = x + 2$, the new limits in u are $c = u(0) = 0 + 2 = 2$ and $d = u(2) = 2 + 2 = 4$. Thus,

$$\int_0^2 6(x+2)^2 dx = F(u)\Big|_c^d = 2u^3\Big|_2^4 = 2(4)^3 - 2(2)^3 = 128 - 16 = 112.$$

Both have the same result.

Example 16.4: Evaluate $\int_1^e \frac{3(\ln x)^2}{x} dx$.

Solution

Let $t = \ln x$; then $c = t(1) = \ln 1 = 0$, $d = t(e) = \ln e = 1$ and $dt = d(\ln x) = \frac{dx}{x}$. Substitute all these into the definite integral,

$$\int_1^e \frac{3(\ln x)^2}{x} dx = \int_1^e 3(\ln x)^2 \frac{dx}{x} = \int_0^1 3t^2 dt = \frac{3}{3}t^3\Big|_0^1 = 1^3 - 0^3 = 1.$$

Example 16.5: Evaluate $\int_0^3 \sqrt{9-x^2}\,dx$.

Solution

Let $x = 3\sin t \to dx = 3\cos t dt$; $c = t(0) = arc\sin\frac{0}{3} = 0$ and $d = t(3) = arc\sin\frac{3}{3} = \frac{\pi}{2}$. Substitute these into the definite integral:

$$\int_0^3 \sqrt{9-x^2}\,dx = \int_0^{\pi/2} \sqrt{3^3 - (3\sin t)^2}(3\cos t dt) = 9\int_0^{\pi/2} \sqrt{1-\sin^2 t}\cos t dt$$

$$= 9\int_0^{\pi/2} \sqrt{\cos^2 t}\cos t dt = 9\int_0^{\pi/2} \cos^2 t dt = 9\int_0^{\pi/2} \frac{\cos 2t + 1}{2} dt \longleftarrow \cos^2 t = \frac{\cos 2t + 1}{2}$$

$$= \frac{9}{2}\int_0^{\pi/2} (\cos 2t + 1)dt = \frac{9}{2}(\frac{1}{2}\sin 2t + t)\Big|_0^{\pi/2} = \frac{9}{2}[(\frac{1}{2}\sin\frac{2\pi}{2} + \frac{\pi}{2}) - (\frac{1}{2}\sin 0 + 0)]$$

$$= \frac{9}{2}[(\frac{1}{2}\sin\pi + \frac{\pi}{2}) - (\frac{1}{2}\times 0 + 0)] = \frac{9}{2}(\frac{1}{2}\times 0 + \frac{\pi}{2}) = \frac{9}{2}\times\frac{\pi}{2} = \frac{9\pi}{4}.$$

16.1.3 Operational rules for definite integration

General properties of definite integration

Since definite integral is the bounded case of the corresponding indefinite integral, many operational rules for indefinite integration are also applicable in definite integration. However, definite integration has some properties different from indefinite integration.

$$\boxed{\int_a^a f(x)dx = 0} \quad \text{(The area of a point is zero.)} \qquad (16.6)$$

$$\boxed{\int_a^b f(x)dx = -\int_b^a f(x)dx} \qquad (16.7)$$

$$\boxed{\int_a^b f(x)dx = \int_a^c f(x)dx + \int_c^b f(x)dx \quad a < c < b} \quad \text{(Sum of adjacent subareas)} \qquad (16.8)$$

$$\boxed{f(x) = \frac{d[F(x)]}{dx} = \frac{d}{dx}\int_a^x f(x)dx \quad (a < x)} \qquad (16.9)$$

$$\boxed{\begin{cases} f(x_m) = \frac{1}{b-a}\int_a^b f(x)dx \text{ or} \\ \int_a^b f(x)dx = (b-a)f(x_m) \end{cases} (a < x_m < b)} \quad \text{(The mean value theorem)} \qquad (16.10)$$

Properties (16.6) – (16.9) can be derived from the Newton-Leibniz formula.

Example 16.6: Derive properties (16.6) – (16.9) from the Newton-Leibniz formula.

Solution

By Newton-Leibniz formula $\int_a^b f(x)dx = F(b) - F(a)$,

$$\int_a^a f(x)dx = F(a) - F(a) = 0 \longleftarrow \text{Property (16.6)}$$

$$\int_b^a f(x)dx = F(a) - F(b) = -[F(b) - F(a)] = -\int_a^b f(x)dx \longleftarrow \text{Property (16.7)}$$

$$\int_a^c f(x)dx + \int_c^b f(x)dx = [F(c) - F(a)] + [F(b) - F(c)] = F(b) - F(a) + F(c) - F(c)$$

$$= F(b) - F(a) = \int_a^b f(x)dx \longleftarrow \text{Property (16.8)}$$

$$\int_a^x f(x)dx = F(x) - F(a) \longrightarrow \frac{d}{dx}\int_a^x f(x)dx = \frac{d[F(x)]}{dx} - \frac{d[F(a)]}{dx}$$

$$\therefore \frac{d}{dx}\int_a^x f(x)dx = f(x) \longleftarrow \frac{d[F(x)]}{dx} = f(x) \text{ and } \frac{d[F(a)]}{dx} = 0 \text{ or}$$

$$f(x) = \frac{d[F(x)]}{dx} = \frac{d}{dx}\int_a^x f(x)dx \longleftarrow \text{Property (16.9).}$$

Example 16.7: Use Example 16.1 to explain Property (16.10).

Solution

Since definite integral can be regarded as the area between $f(x)$ and the x-axis from a to b, Property (16.10) indicates that the area determined by the definite integral is equivalent to the area of a rectangle whose base length is $(b - a)$ and whose height is the value $f(x_m)$ on the curve $f(x)$ between $f(a)$ and $f(b)$. $f(x_m)$ is called the mean value of $f(x)$ between a and b and can be determined by

$$f(x_m) = \frac{1}{b-a}\int_a^b f(x)dx \quad a < x_m < b$$

In Example 16.1, the area is $A = \int_1^3 (x+1)dx = 6$, and $(b - a) = 2$. Thus the mean value is

$$f(x_m) = \frac{1}{3-1}\int_1^3 (x+1)dx = \frac{1}{2} \times 6 = 3.$$

$$\because f(x_m) = x_m + 1, \quad \therefore x_m + 1 = 3 \longrightarrow x_m = 2.$$

Hence the mean value is located at point (2, 3) in $f(x) = x + 1$.

Example 16.8: Find the mean value for $\int_0^{\pi} \sin x dx$.

Solution

By the mean value theorem (16.10), the mean value is

$$f(x_m) = \frac{1}{\pi - 0}\int_0^{\pi} \sin x dx = \frac{1}{\pi}(-\cos x)\Big|_0^{\pi} = -\frac{1}{\pi}(\cos\pi - \cos 0) = -\frac{1}{\pi}(-1-1) = \frac{2}{\pi}.$$

This mean point is located at

$$\sin x_m = \frac{2}{\pi} \longrightarrow x_m = \arcsin\frac{2}{\pi} \approx \arcsin 0.6366 \approx 39.54°.$$

Properties of definite integrals of symmetric functions over symmetric ranges

If $f(x)$ is a symmetric function in a symmetric range $[-a, a]$, its definite integral is

$$\boxed{\int_{-a}^{a} f(x)dx = 0} \quad \text{if } f(x) \text{ is an } \textit{odd function} \text{ in } [-a, a]; \tag{16.11}$$

$$\boxed{\int_{-a}^{a} f(x)dx = 2\int_0^{a} f(x)dx} \quad \text{if } f(x) \text{ is an } \textit{even function} \text{ in } [-a, a]. \tag{16.12}$$

Example 16.9: Evaluate $\int_{-3}^{3} x^2 dx$, $\int_{-3}^{3} x^3 dx$ and $\int_{-3}^{3} x(3x^2 + 12\cos 2x)dx$, respectively.

Solution

The two definite integrals can be regarded as the areas under x^2, and x^3 in the symmetric range [–3, 3] shown in Figure 16.3.

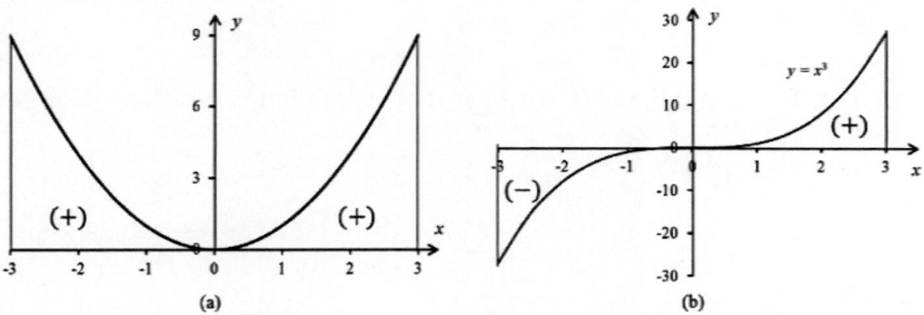

Figure 16.3: Areas of even function $f(x) = x^2$ (a) and odd function $f(x) = x^3$ in Example 16.9

As x^2 is an even function over [–3, 3], the two sub-areas on both sides of the y-axis have the same value with the *same sign* (+). Thus the total area under x^2 over [–3, 3] is the double of the sub-area on the right side (so true for the left side), this being the same meaning implied by formula (16.12).

$$\int_{-3}^{3} x^2 dx = 2\int_{0}^{3} x^2 dx = \frac{2}{3}x^3\Big|_0^3 = \frac{2}{3}(3^3 - 0^3) = 2\times 3^2 = 18.$$

As x^3 is an odd function over [–3, 3], the two sub-areas on both sides of the y-axis also have the same value but with *opposite signs*. Thus the total area of x^3 over [–3, 3] is zero as both sub-areas cancel with each other, this being the same meaning implied by formula (16.11), i.e.,

$\int_{-3}^{3} x^3 dx = 0.$ This can be verified by $\int_{-3}^{3} x^3 dx = \frac{1}{4}x^4\Big|_{-3}^{3} = \frac{1}{4}[3^4 - (-3)^4] = \frac{1}{4}(3^4 - 3^4) = 0.$

$$\int_{-3}^{3} x(3x^2 + 12\cos 2x)dx = \int_{-3}^{3} \underbrace{3x^3}_{odd} dx + \int_{-3}^{3} \underbrace{12x\cos 2x}_{odd\times even=odd} dx = 0 .$$

Example 16.10: Evaluate $\int_{-\frac{\pi}{2}}^{\frac{\pi}{2}} \sin x dx$ and $\int_{-\frac{\pi}{2}}^{\frac{\pi}{2}} \cos x dx$.

Solution

As sin x is an odd function and cos x is an even function over [–π/2, π/2], similar to the analysis in Example 16.9, the sub-areas under these functions are shown in Figure 16.4.

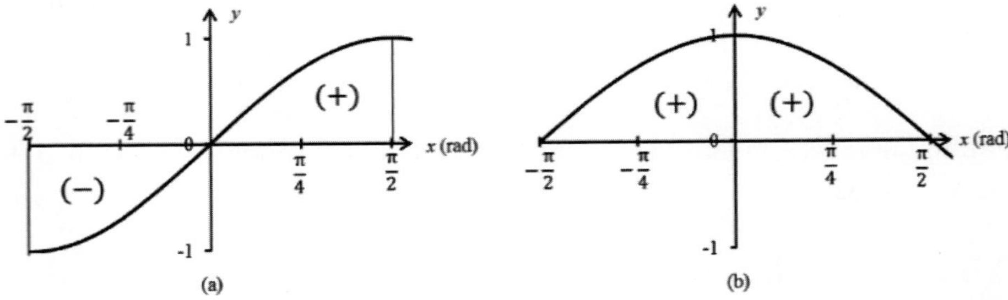

Figure 16.4: Areas of odd function sin x (a) and even function cos x in Example 16.10

By formulae (16.11) and (16.12),

$$\int_{-\frac{\pi}{2}}^{\frac{\pi}{2}} \sin x dx = 0,$$

$$\int_{-\frac{\pi}{2}}^{\frac{\pi}{2}} \cos x dx = 2\int_{0}^{\frac{\pi}{2}} \cos x dx = 2\sin x\Big|_0^{\frac{\pi}{2}} = 2(\sin\frac{\pi}{2} - \sin 0) = 2(1-0) = 2.$$

Exercises 16.1

1. Given $f(x)=2x^2$, find $\int_0^2 f(x)dx$ using Newton-Leibniz formula (16.2). Verify the result using the limit of summation formula (16.3). [note $\sum_{i=1}^{n} i^2 = \frac{n(n+1)(2n+1)}{6}$]

2. Evaluate

a) $\int_0^{\frac{\pi}{2}} \cos x \sin x dx$ and $\int_1^3 (4x^3 - 6x)dx$

b) $\int_0^{\frac{\pi}{2}} \sin^2 x dx$ and $\int_0^1 x^2 e^x dx$

c) $\int_0^3 3\sqrt{x+1}dx$ and $\int_0^2 8(x-2)^3 dx$

d) $\int_0^4 2x\sqrt{x^2+9}dx$ and $\int_0^5 \frac{1}{\sqrt{25-x^2}}dx$

e) $\int_0^{\frac{\pi}{2}} \cos^2 \frac{x}{2} dx$ and $\int_1^e x^2 \ln x dx$

3. Evaluate

a) $\int_{-\frac{\pi}{2}}^{\frac{\pi}{2}} \cos 2x \sin 2x dx$ and $\int_{-3}^3 (x^3 - 2x)dx$

b) $\int_{-\frac{\pi}{2}}^{\frac{\pi}{2}} 2\sin^2 x dx$ and $\int_{-1}^1 x^2(4x+3)dx$

c) $\int_{-3}^3 2x\sqrt{x^2+16}dx$ and $\int_{-4}^4 \frac{1}{\sqrt{16-x^2}}dx$

d) $\int_{-\pi}^{\pi} \cos\frac{x}{2} dx$ and $\int_{-1}^1 (e^x + e^{-x})dx$

4. Find the mean value of

a) $\int_0^3 3\sqrt{x+1}dx$ and $\int_0^2 8(x-2)^3 dx$

b) $\int_{-\frac{\pi}{2}}^{\frac{\pi}{2}} 2\sin^2 x dx$ and $\int_{-1}^1 x^2(4x+3)dx$

16.2 Applications of Definite Integration

16.2.1 Geometric applications

Physical plane areas

Because the geometric meaning of definite integral of $f(x)$ within range $[a, b]$ is the area between $f(x)$ and the x-axis from a to b, definite integration is naturally used for calculating the area enclosed by two or multiple functions in a given range. In real world, there is no '*negative area*'; thus any '*negative value*' from definite integration for a sub-area must be converted to positive value for the total area, which is often called the *physical area.*

To avoid adding 'a negative area' into the total area during integration, an effective approach is to use the '*difference of the two functions*' that form a sub-area or the entire area as a '*positive integrand*' of the definite integral in that sub-range or entire range by means of always *taking the lower curve, from the upper curve*, i.e.,

$$A=\int_a^c [f(x)-g(x)]dx+\int_c^d [g(x)-f(x)]dx \longleftarrow \begin{cases} f(x)>g(x) \text{ in } [a,c] \\ f(x)<g(x) \text{ in } [c,b] \end{cases} \tag{16.13}$$

As the x-axis is represented by function $y = 0$, formula (16.13) is the general method to determine the (physical) area enclosed by any two functions (including the x-axis as $y = 0$) in a specified range.

Example 16.11: Find the area of $\int_{-\frac{\pi}{2}}^{\frac{\pi}{2}} \sin x dx$ and $\int_{-3}^{3} x^3 dx$ respectively.

Solution

Referring to Figure 16.4a in Example 16.10, $\int_{-\frac{\pi}{2}}^{\frac{\pi}{2}} \sin x dx$ is an odd function with two sub-areas on both sides of the y-axis enclosed by $y_1 = \sin x$ and $y_2 = 0$ (the x-axis). On the left, $y_2 > y_1$ whereas on the right $y_2 < y_1$. By formula (16.13), the physical area is

$$\int_{-\frac{\pi}{2}}^{\frac{\pi}{2}} \sin x dx = \int_{-\frac{\pi}{2}}^{0} (0-\sin x)dx + \int_{0}^{\frac{\pi}{2}} (\sin x - 0)dx = \cos x\Big|_{-\frac{\pi}{2}}^{0} - \cos x\Big|_{0}^{\frac{\pi}{2}}$$

$$= [\cos 0 - \cos(-\frac{\pi}{2})] - (\cos\frac{\pi}{2} - \cos 0) = (1-0)-(0-1) = 1+1 = 2 \text{ (square units)}$$

Referring to Figure 16.3b, similarly the physical area of $\int_{-3}^{3} x^3 dx$ is

$$A = \int_{-3}^{3} x^3 dx = \int_{-3}^{0} (0-x^3)dx + \int_{0}^{3} (x^3-0)dx = -\frac{1}{4}x^4\Big|_{-3}^{0} + \frac{1}{4}x^4\Big|_{0}^{3}$$

$$= -\frac{1}{4}[(0^4-(-3)^4)] + \frac{1}{4}(3^4-0^4) = \frac{1}{4}\times 3^4 + \frac{1}{4}\times 3^4 = \frac{1}{2}\times 3^4 = \frac{81}{2} = 40.5 \text{ (square units)}$$

Since the default unit for areas is square units, without specifying a particular unit, the 'square units' are often omitted in the final result.

As we know $y = 0$ represents the x-axis, it does not need to be specified in the process of integration in the future unless necessary.

Example 16.12: The ellipse function is $\frac{x^2}{a^2}+\frac{y^2}{b^2}=1$, where a is the length of the semi-major axis and b is the length of the semi-minor axis (Figure 16.5). Use definite integral to find its area.

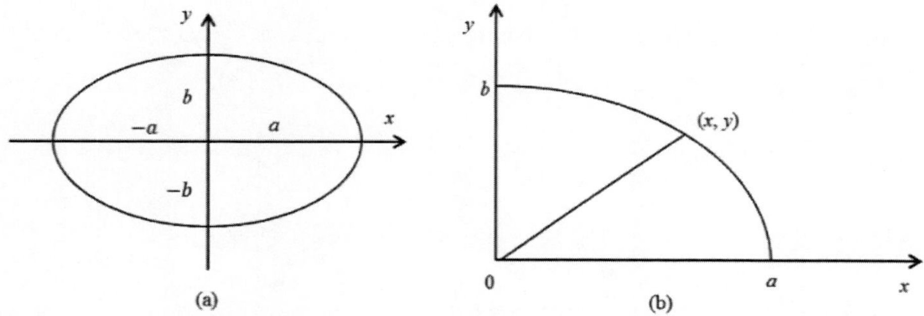

Figure 16.5: Ellipse (a) and the sketch of the first quadrant (b)

Solution

The ellipse function can be rearranged as follows,

$$\frac{x^2}{a^2}+\frac{y^2}{b^2}=1 \longrightarrow \frac{y^2}{b^2}=1-\frac{x^2}{a^2} \longrightarrow \frac{y}{b}=\sqrt{\frac{a^2-x^2}{a^2}} \longrightarrow y=f(x)=\frac{b}{a}\sqrt{a^2-x^2}.$$

Due to symmetry, the area of the ellipse is four times of the area of the first quadrant, i.e.,

$$A=4\int_0^a ydx=\frac{4b}{a}\int_0^a\sqrt{a^2-x^2}dx \longleftarrow x=a\sin t, dx=a\cos tdt$$

$$=\frac{4b}{a}\int_0^{\frac{\pi}{2}}\sqrt{a^2-a^2\sin^2 t}(a\cos t)dt \longleftarrow x=0\to t=0, x=a\to t=\frac{\pi}{2}$$

$$=4b\int_0^{\frac{\pi}{2}}a\sqrt{(1-\sin^2 t)}\cos tdt=4ab\int_0^{\frac{\pi}{2}}\sqrt{\cos^2 t}\cos tdt=4ab\int_0^{\frac{\pi}{2}}\cos^2 tdt$$

$$=4ab\int_0^{\frac{\pi}{2}}\frac{1+\cos 2t}{2}dt=2ab\int_0^{\frac{\pi}{2}}(1+\cos 2t)dt=2ab(t+\frac{\sin 2t}{2})\Big|_0^{\frac{\pi}{2}}$$

$$=2ab\{[\frac{\pi}{2}+\frac{1}{2}\sin(2\times\frac{\pi}{2})]-[(0+\frac{1}{2}\sin(2\times 0)]\}=2ab[(\frac{\pi}{2}+\frac{1}{2}\sin\pi)-0]$$

$$=2ab(\frac{\pi}{2}+0)=\pi ab.$$

Example 16.13: Find the area bounded by parabola $y=-x^2$ and line $y=x-2$.

Solution

We first determine the points where the two curves intersect with each other by solving the following system of equations:

$$\begin{cases} y=-x^2 \\ y=x-2 \end{cases} \longrightarrow x-2=-x^2 \longrightarrow x^2+x-2=0 \longrightarrow (x+2)(x-1)=0 \longrightarrow \begin{cases} x=-2, y=-4 \\ x=1, y=-1 \end{cases}.$$

Figure 16.6 shows these two points of intersection. The parabola $y=f(x)=-x^2$ is above the line $y=g(x)=x-2$ in the whole range [–2, 1]. Thus, by formula (16.13),

$$A=\int_{-2}^{1}[f(x)-g(x)]dx=\int_{-2}^{1}[-x^2-(x-2)]dx=-\frac{1}{3}x^3-\frac{1}{2}x^2+2x\Big|_{-2}^{1}$$

$$=[-\frac{1}{3}\times 1^3-\frac{1}{2}\times 1^2+2\times 1]-[-\frac{1}{3}\times(-2)^3-\frac{1}{2}\times(-2)^2+2\times(-2)]=-\frac{1}{3}-\frac{1}{2}+2-(\frac{8}{3}-2-4)$$

$$=-\frac{1}{3}-\frac{1}{2}+2-\frac{8}{3}+6=-\frac{9}{3}-\frac{1}{2}+8=-3-\frac{1}{2}+8=5-\frac{1}{2}=\frac{9}{2}=4.5.$$

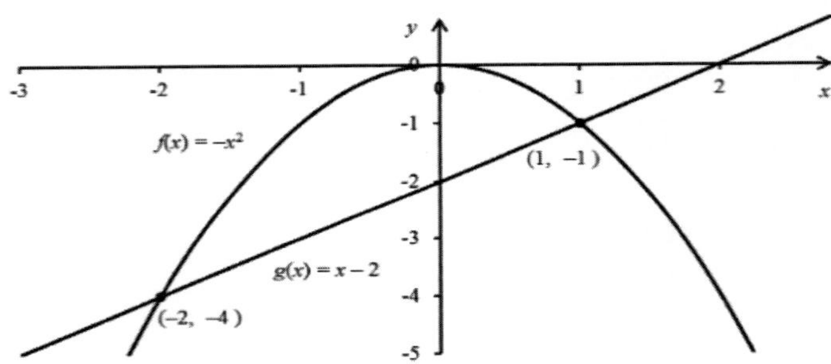

Figure 16.6: The sketch of $y=-x^2$ and $y=x-2$

Example 16.14: Find the area bounded by $y=\frac{1}{2}x^3+3$ and $y=2x+3$.

Solution

Solve the following system of equations to find the points where the two curves intersect:

$$\begin{cases} y=\frac{1}{2}x^3+3 \\ y=2x+3 \end{cases} \longrightarrow \frac{1}{2}x^3+3=2x+3 \longrightarrow x^3+6=4x+6 \longrightarrow x^3-4x=0 \longrightarrow x(x^2-4)=0$$

$$x(x+2)(x-2)=0 \longrightarrow \begin{cases} x=-2, y=-1 \\ x=0, y=3 \\ x=2, y=7 \end{cases}.$$

Figure 16.7 shows these three points of intersection, in which $f(x)=\frac{1}{2}x^3+3$ is above $g(x)=2x+3$ in range [–2, 0] whereas $g(x)=2x+3$ is above $f(x)=\frac{1}{2}x^3+3$ in range [0, 2]. By formula (16.13),

$$A=\int_{-2}^{0}[f(x)-g(x)]dx+\int_{0}^{2}[g(x)-f(x)]dx$$
$$=\int_{-2}^{0}[(\frac{1}{2}x^3+3)-(2x+3)]dx+\int_{0}^{2}[(2x+3)-(\frac{1}{2}x^3+3)]dx$$
$$=\int_{-2}^{0}(\frac{1}{2}x^3+3-2x-3)dx+\int_{0}^{2}(2x+3-\frac{1}{2}x^3-3)dx=\int_{-2}^{0}(\frac{1}{2}x^3-2x)dx+\int_{0}^{2}(2x-\frac{1}{2}x^3)dx$$
$$=(\frac{1}{8}x^4-x^2)\Big|_{-2}^{0}+(x^2-\frac{1}{8}x^4)\Big|_{0}^{2}=\{(\frac{0^4}{8}-0^2)-[\frac{(-2)^4}{8}-(-2)^2]\}+[(2^2-\frac{2^4}{8})-(0^2-\frac{0^4}{8})]$$
$$=[0-(\frac{16}{8}-4)]+[(4-\frac{16}{8})-0]=-(2-4)+(4-2)=4.$$

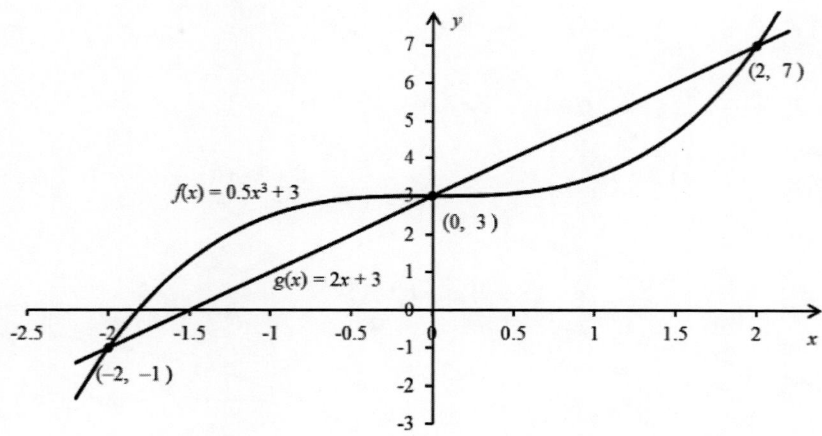

Figure 16.7: The sketch of $y=\frac{1}{2}x^3+3$ and $y=2x+3$

Example 16.15: Find the area bounded by $y^2=x$ and $y=x-2$.

Solution

Solve the following system of equations to find the points where the two curves intersect:

$$\begin{cases}y^2=x\\ y=x-2\end{cases}\longrightarrow(x-2)^2=x\longrightarrow x^2-4x+4=x\longrightarrow x^2-5x+4=0\longrightarrow(x-1)(x-4)=0$$

$$\begin{cases}x=1,y=-1\\ x=4,y=2\end{cases}.$$

Figure 16.8a shows these points of intersection. As $y^2 = x$ or $y = \pm\sqrt{x}$, $f(x) = \sqrt{x}$ is symmetric about the x-axis in range [0, 1], and above $g(x) = x - 2$ in range [1, 4], the enclosed area is,

$$A = 2\int_0^1 f(x)dx + \int_1^4 [f(x) - g(x)]dx = 2\int_0^1 \sqrt{x}dx + \int_1^4 [\sqrt{x} - (x-2)]dx$$

$$= 2\times\frac{2}{3}x^{\frac{3}{2}}\Big|_0^1 + (\frac{2}{3}x^{\frac{3}{2}} - \frac{1}{2}x^2 + 2x)\Big|_1^4 = \frac{4}{3}(1-0) + [(\frac{2}{3}\times 4^{\frac{3}{2}} - \frac{1}{2}\times 4^2 + 2\times 4) - (\frac{2}{3}\times 1^{\frac{3}{2}} - \frac{1}{2}\times 1^2 + 2\times 1)]$$

$$= \frac{4}{3} + [(\frac{2}{3}\times 2^3 - 8 + 8) - (\frac{2}{3} - \frac{1}{2} + 2)] = \frac{4}{3} + \frac{16}{3} - \frac{2}{3} + \frac{1}{2} - 2 = \frac{18}{3} + \frac{1}{2} - 2 = 6 + \frac{1}{2} - 2 = 4.5.$$

Alternatively, we can regard x as a function of y by exchanging x and y in formula (16.13). This sees $f(y) = y^2$ and $g(y) = y + 2$. The range of y is [–1, 2], in which $g(y)$ is above $f(y)$ (Figure 16.8b).

$$A = \int_{-1}^2 [g(y) - f(y)]dy = \int_{-1}^2 (y + 2 - y^2)dy = (\frac{1}{2}y^2 + 2y - \frac{1}{3}y^3)\Big|_{-1}^2$$

$$= (\frac{1}{2}\times 2^2 + 2\times 2 - \frac{1}{3}\times 2^3) - [\frac{1}{2}\times(-1)^2 + 2\times(-1) - \frac{1}{3}\times(-1)^3]$$

$$= 2 + 4 - \frac{8}{3} - (\frac{1}{2} - 2 + \frac{1}{3}) = 6 - \frac{8}{3} - \frac{1}{2} + 2 - \frac{1}{3} = 8 - \frac{9}{3} - \frac{1}{2} = 8 - 3 - \frac{1}{2} = 5 - \frac{1}{2} = 4.5.$$

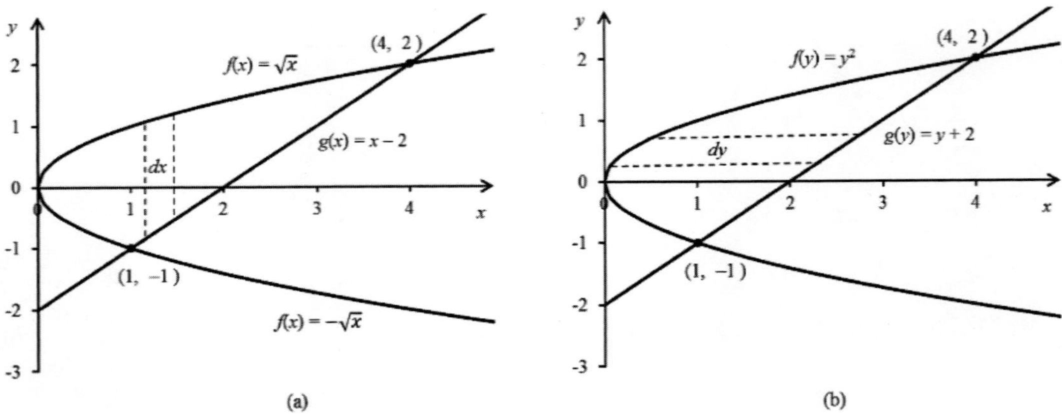

Figure 16.8: The sketch of $y^2 = x$ and $y = x - 2$

Volume of solids of revolution

A solid of revolution is generated by revolving a plane area about a line, often called the axis of revolution of the plane. The volume of a solid of revolution can be obtained by definite integration.

In mathematics, any line as the axis of revolution can be converted to either the x-axis or the y-axis. Therefore, assuming the axis of revolution is the x-axis and the solid of revolution is formed by rotating $y = f(x)$ defined in $[a, b]$ about the x-axis (Figure 16.9a), the volume of such solid of revolution can be obtained by

$$V = \pi\int_a^b y^2 dx = \pi\int_a^b [f(x)]^2 dx. \tag{16.14}$$

This is derived by the summation of the volumes of the vertical 'thin discs', each having a volume of $\delta v = \pi y^2 \delta x$ (Figure 16.9a), i.e.,

$$V = \lim_{\delta x \to 0} \sum \delta v = \lim_{\delta x \to 0} \sum \pi y^2 \delta x = \pi \lim_{\delta x \to 0} \sum y^2 \delta x = \pi\int_a^b y^2 dx.$$

Similarly the volume of the solid of revolution formed by rotating $x = g(y)$ defined in $[c, d]$ about the y-axis (Figure 16.9b) can be obtained by

$$V = \pi\int_c^d x^2 dy = \pi\int_c^d [g(y)]^2 dy. \tag{16.15}$$

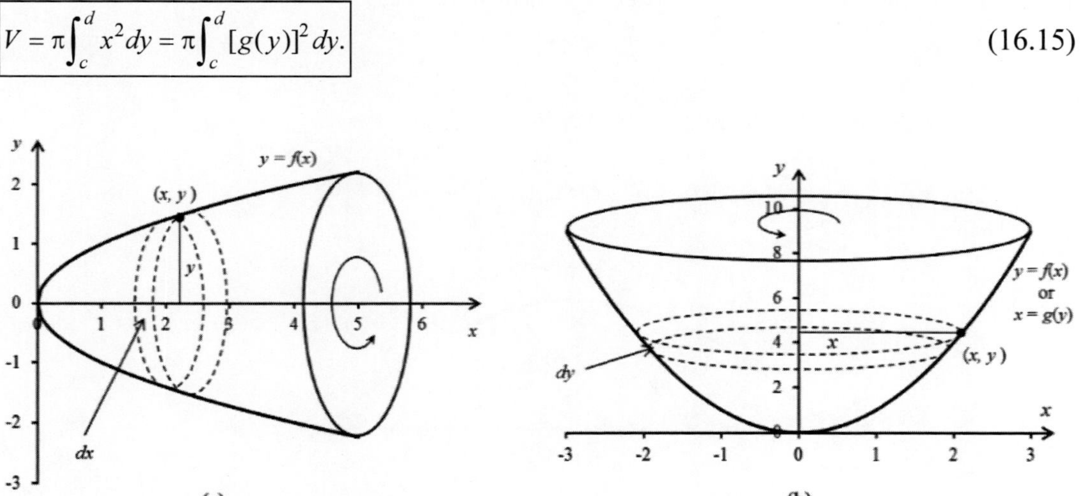

Figure 16.9: The sketch of solids of revolution about the x-axis (a) and the y-axis (b)

Example 16.16: In Figure 16.9a, if the solid of revolution is formed by rotating $y = \sqrt{x}$ about the x-axis in $[1, 3]$, find the volume of this solid of revolution.

<u>**Solution**</u>

By formula (16.14), the volume of this solid of revolution is

$$V = \pi\int_1^3 y^2 dx = \pi\int_1^3 (\sqrt{x})^2 dx = \pi\int_1^3 x dx = \frac{1}{2}\pi x^2\Big|_1^3 = \frac{1}{2}\pi(3^2 - 1^2) = \frac{8}{2}\pi = 4\pi.$$

Example 16.17: A solid of revolution is formed by rotating $y = x^2$ about the y-axis in [1, 3], find the volume of this solid of revolution.

Solution

By formula (16.15), the volume of this solid of revolution is

$$V = \pi\int_1^3 x^2 dy = \pi\int_1^3 y dy = \frac{1}{2}\pi y^2\Big|_1^3 = \frac{1}{2}\pi(3^2 - 1^2) = \frac{8}{2}\pi = 4\pi.$$

Example 16.18: If a solid of revolution is formed by rotating $y = \cos x$ about the x-axis in [−π/2, π/2], find the volume of this solid of revolution.

Solution

Considering symmetry of the solid about the y-axis in [−π/2, π/2] shown in Figure 16.10, by formula (16.14), the volume of this solid of revolution is

$$V = 2\pi\int_0^{\frac{\pi}{2}} \cos^2 x dx = 2\pi\int_0^{\frac{\pi}{2}} \frac{\cos 2x + 1}{2} dx = \pi(\frac{1}{2}\sin 2x + x)\Big|_0^{\frac{\pi}{2}}$$

$$= \pi[(\frac{1}{2}\sin\pi + \frac{1}{2}\pi) - (\frac{1}{2}\sin 0 + 0)] = \pi(0 + \frac{1}{2}\pi - 0) = \frac{\pi^2}{2}.$$

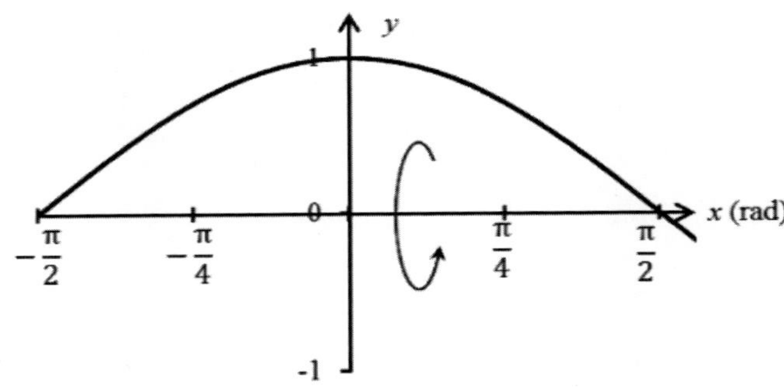

Figure 16.10: The sketch of solid of revolution about the x-axis by $y = \cos x$

Example 16.19: In Example 16.14, we obtained the area bounded by $y = \frac{1}{2}x^3 + 3$ and $y = 2x + 3$. If the bounded subarea ($x \geq 0$) is rotated around the x-axis, find the volume of the solid of revolution formed by such rotation (Figure 16.11).

Solution

Considering the right subarea, as $g(x)$ is above $f(x)$ in [0, 2] at any point of x, the volume formed by rotating a thin annulus (Figure 16.11) is $\delta v = \pi\{[g(x)]^2 - [f(x)]^2\}\delta x$. By formula (16.14), the volume of this solid of revolution is

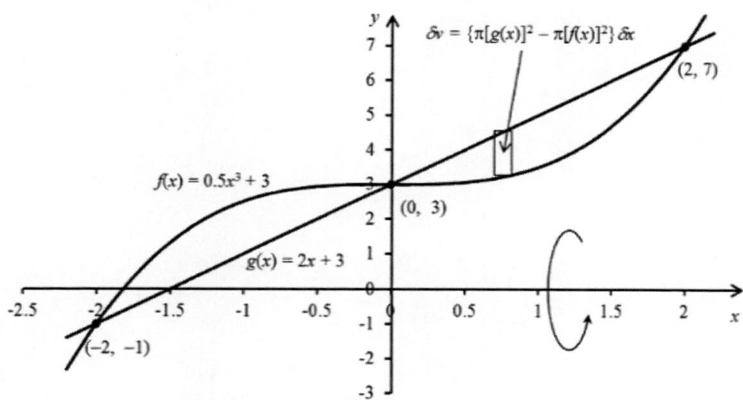

Figure 16.11: The sketch of solids of revolution of $f(x)$ and $g(x)$ in Example 16.19

$$V = \pi\int_0^2 [(2x+3)^2 - (\frac{1}{2}x^3+3)^2]dx$$

$$= \pi\int_0^2 [(4x^2+12x+9) - (\frac{1}{4}x^6+3x^3+9)]dx = \pi\int_0^2 (4x^2+12x-\frac{1}{4}x^6-3x^3)dx$$

$$= \pi(\frac{4}{3}x^3+6x^2-\frac{1}{28}x^7-\frac{3}{4}x^4)\Big|_0^2 = \pi[(\frac{4}{3}\times 2^3+6\times 2^2-\frac{1}{28}\times 2^7-\frac{3}{4}\times 2^4)-0]$$

$$= \pi(\frac{32}{3}+24-\frac{128}{28}-\frac{3\times 16}{4}) = \pi(\frac{32}{3}+24-\frac{32}{7}-12) = \pi(12+\frac{224}{21}-\frac{96}{21})$$

$$= \pi(12+\frac{128}{21}) = (\frac{252}{21}+\frac{128}{21})\pi = \frac{380}{21}\pi \approx 18.0952\pi \approx 56.85 \text{ (cubic units).}$$

Length of arc

Function $y = f(x)$ defined in $[a, b]$ can be regarded as a segment of arc (curve) within $x = a$ and $x = b$ (Figure 16.12). The length of this arc can be obtained by

$$\boxed{L = \int_a^b \sqrt{1+(\frac{dy}{dx})^2}\,dx.} \qquad (16.16)$$

This can be derived by the summation of the lengths of the 'hypotenuses of small right triangles' on top of the divided small vertical strips, each having a length of $\delta l = \sqrt{(\delta x)^2+(\delta y)^2}$, i.e.,

$$L = \lim_{\delta x\to 0}\sum \delta l = \lim_{\delta x\to 0}\sum\sqrt{(\delta x)^2+(\delta y)^2} = \lim_{\delta x\to 0}\sum\sqrt{(\delta x)^2[1+(\frac{\delta y}{\delta x})^2]}$$

$$= \lim_{\delta x\to 0}\sum\sqrt{1+(\frac{\delta y}{\delta x})^2}\,\delta x = \int_a^b\sqrt{1+(\frac{dy}{dx})^2}\,dx.$$

If the curve is defined by a set of parametric functions $x = x(t)$ and $y = y(t)$ within $t = a$ and $t = b$, the length of this arc can be obtained by

$$\boxed{L = \int_a^b \sqrt{(\frac{dx}{dt})^2 + (\frac{dy}{dt})^2}\, dt.} \qquad (16.17)$$

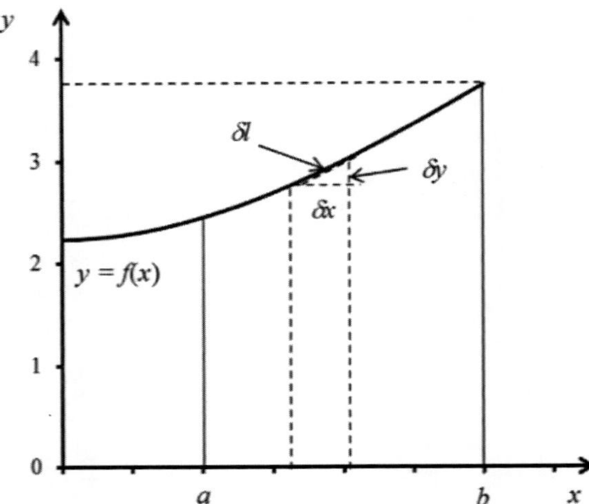

Figure 16.12: The sketch of arc formed by $f(x)$ between $[a, b]$

Example 16.20: Find the length of arc formed by $y^2 = x^3$ from $x = 0$ to $x = 4$.

Solution

The given function $y^2 = x^3$ can be expressed as

$$y = x^{\frac{3}{2}} \longrightarrow \frac{dy}{dx} = \frac{3}{2}x^{\frac{1}{2}}.$$

By formula (16.16), the length of the arc in [0, 4] is

$$L = \int_0^4 \sqrt{1 + (\frac{dy}{dx})^2}\, dx = \int_0^4 \sqrt{1 + (\frac{3}{2}x^{\frac{1}{2}})^2}\, dx = \int_0^4 \sqrt{1 + \frac{9}{4}x}dx \longleftarrow u = 1 + \frac{9}{4}x, du = \frac{9}{4}dx \leftrightarrow dx = \frac{4}{9}du$$

$$= \int_1^{10} (\sqrt{u})(\frac{4}{9}du) \longleftarrow x = 0 \to u = 1; x = 4 \to u = 10$$

$$= \frac{4}{9}\int_1^{10} \sqrt{u}du = \frac{4}{9}(\frac{2}{3}u^{\frac{3}{2}})\Bigg|_1^{10} = \frac{8}{27}(10^{\frac{3}{2}} - 1^{\frac{3}{2}}) = \frac{8}{27}(\sqrt{1000} - 1) \approx 9.0734.$$

Example 16.21: Find the length of the arc of $24xy = x^4 + 48$ from $x = 2$ to $x = 4$.

Solution

The given function $24xy = x^4 + 48$ can be expressed as

$$24xy = x^4 + 48 \longrightarrow y = \frac{x^4+48}{24x} = \frac{x^3}{24} + \frac{2}{x} \longrightarrow \frac{dy}{dx} = \frac{x^2}{8} - \frac{2}{x^2} = \frac{x^4-16}{8x^2}.$$

By formula (16.16), the length of the arc in [2, 4] is

$$L = \int_2^4 \sqrt{1+(\frac{dy}{dx})^2}\,dx = \int_2^4 \sqrt{1+(\frac{x^4-16}{8x^2})^2}\,dx = \int_2^4 \sqrt{\frac{(8x^2)^2}{(8x^2)^2} + \frac{(x^4-16)^2}{(8x^2)^2}}\,dx = \int_2^4 \frac{1}{8x^2}\sqrt{x^8+32x^4+16^2}\,dx$$

$$= \int_2^4 \frac{1}{8x^2}\sqrt{(x^4+16)^2}\,dx = \int_2^4 \frac{x^4+16}{8x^2}\,dx = \int_2^4 (\frac{x^2}{8} + \frac{2}{x^2})dx = \frac{x^3}{24} - \frac{2}{x}\Big|_2^4$$

$$= \frac{1}{24}(4^3-2^3) - (\frac{2}{4} - \frac{2}{2}) = \frac{1}{24}(64-8) + \frac{1}{2} = \frac{1}{24}\times 56 + \frac{1}{2} = \frac{7}{3} + \frac{1}{2} = \frac{14+3}{6} = \frac{17}{6} \approx 2.8333 \text{ units}$$

Example 16.22: Find the length of arc formed by $x = r\cos t$ and $y = r\sin t$ from $t = 0$ to $t = 2\pi$. Note r is a constant.

Solution

This is a set of parametric functions, whose derivatives are $\frac{dx}{dt} = -r\sin t$ and $\frac{dy}{dt} = r\cos t$.

By formula (16.17), the length of the arc from $t = 0$ to $t = 2\pi$ is

$$L = \int_0^{2\pi} \sqrt{(\frac{dx}{dt})^2 + (\frac{dy}{dt})^2}\,dt = \int_0^{2\pi} \sqrt{(-r\sin t)^2 + (r\cos t)^2}\,dt = \int_0^{2\pi} r\sqrt{\sin^2 t + \cos^2 t}\,dt = rt\Big|_0^{2\pi} = 2\pi r.$$

This set of parametric functions defines a circle of radius r. A full circle is $t = 2\pi$. No wonder the length is $2\pi r$.

Example 16.23: Find the length of arc formed by $x = e^t \sin t$ and $y = e^t \cos t$ from $t = 0$ to $t = \pi/2$.

Solution

This is a set of parametric functions, whose derivatives are

$$\frac{dx}{dt} = e^t(\sin t + \cos t) \text{ and } \frac{dy}{dt} = e^t(\cos t - \sin t).$$

By formula (16.17), the length of the arc from $t = 0$ to $t = \pi/2$ is

$$L=\int_0^{\pi/2}\sqrt{[e^t(\sin t+\cos t)\ ]^2+[e^t(\cos t-\sin t)]^2}\,dt=\int_0^{\pi/2}e^t\sqrt{(\sin t+\cos t)^2+(\cos t-\sin t)^2}\,dt$$

$$=\int_0^{\pi/2}e^t\sqrt{\sin^2 t+2\sin t\cos t+\cos^2 t+\cos^2 t-2\sin t\cos t+\sin^2 t}\,dt$$

$$=\int_0^{\pi/2}e^t\sqrt{2(\sin^2 t+\cos^2 t)}\,dt=\int_0^{\pi/2}\sqrt{2}e^t dt=\sqrt{2}\,e^t\Big|_0^{\pi/2}=\sqrt{2}(e^{\pi/2}-e^0)=\sqrt{2}(e^{\pi/2}-1)\approx 5.389.$$

Surface area of revolution

The solid of revolution formed by rotating $y=f(x)$ defined in $[a, b]$ about the x-axis has not only a volume but also surface area which can be calculated by

$$\boxed{S=2\pi\int_a^b y\sqrt{1+(\frac{dy}{dx})^2}\,dx.} \tag{16.18}$$

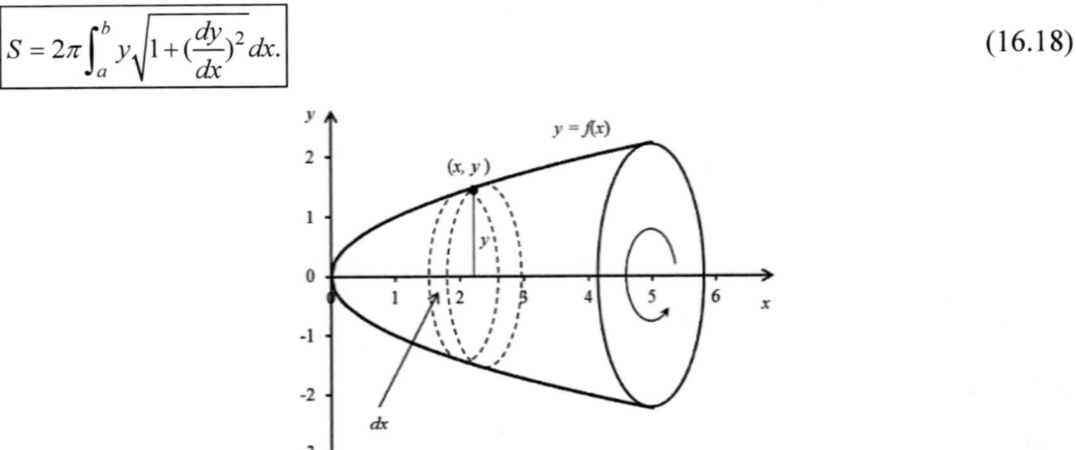

Figure 16.13: The sketch of surface area formed by rotating $f(x)$ about the x-axis between $[a, b]$

This can be derived by the summation of the surface areas of all the thin vertical disks (Figure 16.13). The circumference of a thin disk is $2\pi y$ and its width or height is the length of arc δl (see Figure 16.12). Therefore, the surface area can be derived by

$$S=\lim_{\delta x\to 0}\sum 2\pi y\delta l=2\pi\lim_{\delta x\to 0}\sum y\sqrt{(\delta x)^2+(\delta y)^2}=2\pi\lim_{\delta x\to 0}\sum y\sqrt{(\delta x)^2[1+(\frac{\delta y}{\delta x})^2]}$$

$$=2\pi\lim_{\delta x\to 0}\sum y\sqrt{1+(\frac{\delta y}{\delta x})^2}\,\delta x=2\pi\int_a^b y\sqrt{1+(\frac{dy}{dx})^2}\,dx.$$

This also applies to rotating an arc section about the x-axis, which forms a surface. If the curve is defined by a set of parametric functions $x = x(t)$ and $y = y(t)$ within $t = a$ and $t = b$, the length of this rotating arc can be obtained by

$$\boxed{S=2\pi\int_a^b y\sqrt{(\frac{dx}{dt})^2+(\frac{dy}{dt})^2}\,dt.} \tag{16.19}$$

Example 16.24: Find the surface area of revolution formed by rotating $y^2 = 12x$ about the x-axis from $x = 0$ to $x = 6$.

Solution

The given function $y^2 = 12x$ can be expressed as

$$y = \sqrt{12x} = 2\sqrt{3x} \longrightarrow \frac{dy}{dx} = 2\sqrt{3}(\frac{1}{2}x^{-\frac{1}{2}}) = \frac{\sqrt{3}}{\sqrt{x}} = \sqrt{\frac{3}{x}}$$

$$\sqrt{1+(\frac{dy}{dx})^2} = \sqrt{1+(\sqrt{\frac{3}{x}})^2} = \sqrt{1+\frac{3}{x}} = \sqrt{\frac{x+3}{x}}$$

By formula (16.18), the surface area in [0, 6] is

$$S = 2\pi\int_0^6 y\sqrt{1+(\frac{dy}{dx})^2}\,dx = 2\pi\int_0^6 2\sqrt{3x}\sqrt{\frac{x+3}{x}}\,dx = 2\pi\int_0^6 2\sqrt{3}\sqrt{x+3}\,d(x+3)$$

$$= 4\sqrt{3}\pi\int_0^6 (x+3)^{\frac{1}{2}}d(x+3) = 4\sqrt{3}\pi[\frac{2}{3}(x+3)^{\frac{3}{2}}]\Big|_0^6 = \frac{8\sqrt{3}\pi}{3}[(6+3)^{\frac{3}{2}} - 3^{\frac{3}{2}}] = \frac{8\sqrt{3}\pi}{3}(9^{\frac{3}{2}} - 3\sqrt{3})$$

$$= \frac{8\sqrt{3}\pi}{3}(3^3 - 3\sqrt{3}) = 8\sqrt{3}\pi(3^2 - \sqrt{3}) = 8\pi(9\sqrt{3} - 3) = 24\pi(3\sqrt{3} - 1) \approx 316.38 \text{ (square units)}$$

Example 16.25: Find the surface area of revolution formed by rotating $y^2 = 2\ln y - 4x$ about the y-axis from $y = 4$ to $y = 1$.

Solution

The given function $y^2 = 2\ln y - 4x$ can be expressed as

$$y^2 = 2\ln y - 4x \longrightarrow x = \frac{2\ln y - y^2}{4} = \frac{\ln y}{2} - \frac{y^2}{4} \longrightarrow \frac{dx}{dy} = \frac{1}{2y} - \frac{y}{2} = \frac{1-y^2}{2y}$$

$$\sqrt{1+(\frac{dx}{dy})^2} = \sqrt{1+(\frac{1-y^2}{2y})^2} = \sqrt{1+\frac{1-2y^2+y^4}{4y^2}} = \sqrt{\frac{4y^2+1-2y^2+y^4}{4y^2}}$$

$$= \frac{\sqrt{y^4+2y^2+1}}{2y} = \frac{\sqrt{(y^2+1)^2}}{2y} = \frac{y^2+1}{2y}$$

$$S = 2\pi\int_4^1 x\sqrt{1+(\frac{dx}{dy})^2}\,dy = 2\pi\int_4^1 (\frac{\ln y}{2} - \frac{y^2}{4})\frac{y^2+1}{2y}\,dy = 2\pi\int_4^1 [\frac{(y^2+1)\ln y}{4y} - \frac{y(y^2+1)}{8}]dy$$

$$= \frac{2\pi}{8}\int_4^1 [\frac{2(y^2+1)\ln y}{y} - (y^3+y)]dy = \frac{\pi}{4}\int_4^1 [2y\ln y + \frac{2\ln y}{y} - y^3 - y]dy$$

$$= \frac{\pi}{4}[(y^2\ln y - \frac{1}{2}y^2) + (\ln y)^2 - \frac{y^4}{4} - \frac{y^2}{2}]\Big|_4^1 = \frac{\pi}{16}[(4y^2\ln y - 2y^2) + 4(\ln y)^2 - y^4 - 2y^2]\Big|_4^1$$

$$=\frac{\pi}{16}\{[(4\ln 1-64\ln 4)-2(1-16)]+4[(\ln 1)^2-(\ln 4)^2]-(1^4-4^4)-2(1^2-4^2)\}$$

$$=\frac{\pi}{16}\{-64\ln 4+30-4(\ln 4)^2-(1-256)-2(1-16)\}=\frac{\pi}{16}[-64\ln 4+30-4(\ln 4)^2+255+30]$$

$$=\frac{\pi}{16}[315-64\ln 4-4(\ln 4)^2]\approx 42.92 \text{ (square units)}$$

Example 16.26: Find the surface area of revolution formed by $x=r\cos t$ and $y=r\sin t$ from $t=0$ to $t=\pi/2$ rotating about the x-axis. Note r is a constant.

Solution

This is a set of parametric functions, whose derivatives are $\frac{dx}{dt}=-r\sin t$ and $\frac{dy}{dt}=r\cos t$.

By formula (16.19), the surface area from $t=0$ to $t=\pi/2$ is

$$S=2\pi\int_0^{\pi/2} y\sqrt{(\frac{dx}{dt})^2+(\frac{dy}{dt})^2}\,dt=2\pi\int_0^{\pi/2} r\sin t\sqrt{(-r\sin t)^2+(r\cos t)^2}\,dt$$

$$=2\pi\int_0^{\pi/2} r^2\sin t\sqrt{\sin^2 t+\cos^2 t}\,dt=2\pi r^2\int_0^{\pi/2}\sin t dt=-2\pi r^2\cos t\Big|_0^{\pi/2}$$

$$=-2\pi r^2(\cos\frac{\pi}{2}-\cos 0)=-2\pi r^2(0-1)=2\pi r^2.$$

16.2.2 Applications in physics and engineering

Work

If a constant force ($\boldsymbol{F}$) exerted on an object moves the object alone a straight line for a distance $\boldsymbol{s}$, the *work* (W) of this force on the object is the product of the force and the distance moved, i.e., $W=\boldsymbol{F}\bullet\boldsymbol{s}$.

If the force varies at different locations of the straight line (e.g., the x-axis), the work done by the force $F(x)$ on the object over a small distance δx around point x can be approximated by $\delta W=F(x)\delta x$. the total work of the force on moving the object from $x=a$ to $x=b$ is defined as

$$\boxed{W=\lim_{\delta x\to 0}\sum\delta W=\lim_{\delta x\to 0}\sum F(x)\delta x=\int_a^b F(x)dx.} \qquad (16.20)$$

This formula only applied to situations where *the force and the distance are in the same direction*.

Example 16.27: In the spring and mass system shown in the figure below, if the mass is pulled out by x from the static equilibrium position, by Hooke's law, the resistance of the spring against such movement is $F_r = -kx$ where k is the spring constant. This force varies at different locations of x. Given $k = 2$ N/m, how much work must be done to pull the mass out by 1 metre from the static equilibrium position?

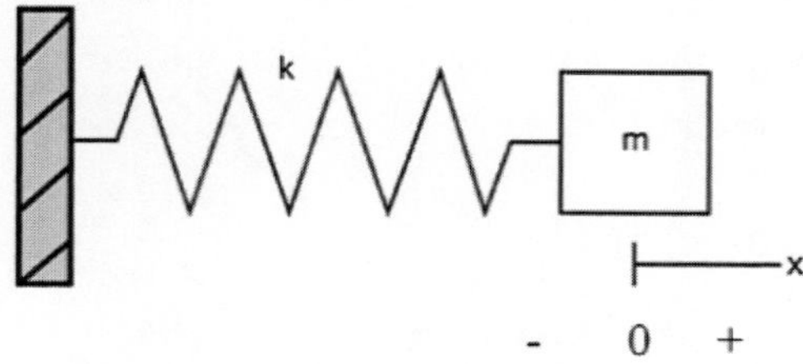

Solution
The force that pulls the mass out must equal the resistance of the spring but in the opposite direction, i.e., $F(x) = kx$. By formula (16.20), the required work is

$$W = \int_a^b F(x)dx = \int_0^1 2xdx = \frac{2}{2}x^2\Big|_0^1 = 1^2 - 0^2 = 1 \text{ (Nm)}.$$

Example 16.28: A horizontal force is controlled by time periodically as $F(x) = 4\cos t$ (N). This force is applied to an object so that it moves in the same direction horizontally with time as $x = \sin t$ (m). How much work has this force done on this object in a time period from 0 to 2π seconds?

Solution
Since both the force and movement are in the same direction, by formula (16.20), the work is

$$W = \int_a^b F(x)dx = \int_0^{2\pi} (4\cos t)d(\sin t) = \int_0^{2\pi} 4\cos t\cos t dt = 4\int_0^{2\pi} \cos^2 t dt = 4\int_0^{2\pi} \frac{1+\cos 2t}{2}dt$$

$$= 2(t + \frac{1}{2}\sin 2t)\Big|_0^{2\pi} = 2[(2\pi + \frac{1}{2}\sin 4\pi) - (0 + \frac{1}{2}\sin 0)] = 2[(2\pi + 0) - (0 + 0)] = 4\pi \text{ (Nm)}.$$

Plane mass centroids

The centre of mass for a plane is defined as the point (x_c, y_c) where the plane would still be balanced if it is underpinned right under the point (x_c, y_c) of the plane. In other words, the total mass m of the plane could be equivalently concentrated on the point (x_c, y_c). For a plane with a continuous density σ (kg/m^2) like abc in Figure 16.14, its centre of mass can be found by

$$x_c = \frac{\int_a^b xdm}{\int_a^b dm} \quad \text{and} \quad y_c = \frac{\frac{1}{2}\int_a^b ydm}{\int_a^b dm}, \tag{16.21}$$

where dm is the mass of the vertical (or horizontal) strips with respect to the y-axis (or the x-axis). Note that dm varies with both the density distribution and the geometric shape of a plane. For any symmetric plane with uniform density distribution, its centre of mass is located somewhere on the axis of symmetry.

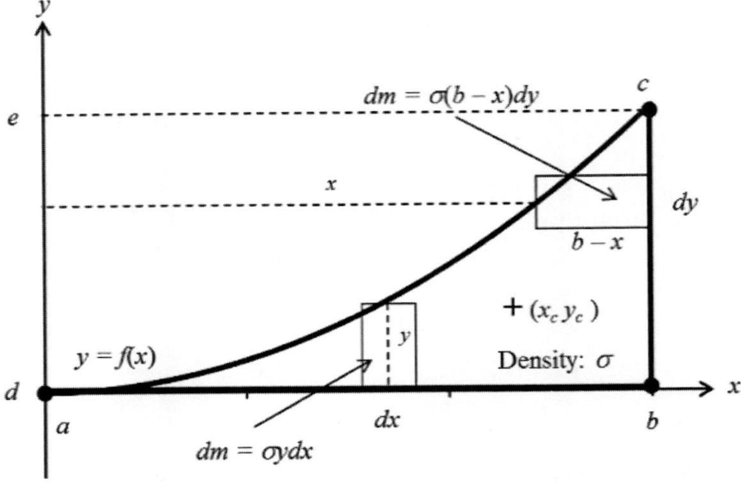

Figure 16.14: The sketch of plane mass centroids for $f(x)$ between $[a, b]$

Example 16.29: In Figure 16.14, assume the plane abc has a constant density σ and $y = f(x) = x^2$ defined in $x = 0$ and $x = 3$. Determine its centre of mass.

Solution

The denominator in formula (16.21) is the total mass of the plane. We can use $dm = \sigma ydx$ for the vertical strip to find the total mass of this plane.

$$M = \int_a^b dm = \int_0^3 \sigma ydx = \sigma\int_0^3 x^2dx = \frac{\sigma}{3}x^3\Big|_0^3 = \frac{\sigma}{3}\times 3^3 = 9\sigma.$$

The two numerators in formula (16.21) are called the moments with respect to the y-axis (M_y) and the x-axis (M_x) respectively.

$$M_y = \int_a^b xdm = \int_0^3 x\sigma ydx = \sigma\int_0^3 x^3dx = \frac{\sigma}{4}x^4\Big|_0^3 = \frac{\sigma}{4}\times 3^4 = \frac{81}{4}\sigma$$

$$M_x = \frac{1}{2}\int_a^b ydm = \frac{1}{2}\int_0^3 y\sigma ydx = \frac{1}{2}\sigma\int_0^3 x^4dx = \frac{\sigma}{10}x^5\Big|_0^3 = \frac{\sigma}{10}\times 3^5 = \frac{243}{10}\sigma.$$

Thus the centre of mass of this plane is located at

$$x_c = \frac{M_y}{M} = \frac{\frac{81}{4}\sigma}{9\sigma} = \frac{9}{4} \quad \text{and} \quad y_c = \frac{M_x}{M} = \frac{\frac{243}{10}\sigma}{9\sigma} = \frac{27}{10}.$$

Example 16.30: In the figure below, assume the plane is formed by $y = \cos x$ within [–π/2, π/2]. If the plane has a constant density σ, determine its centre of mass.

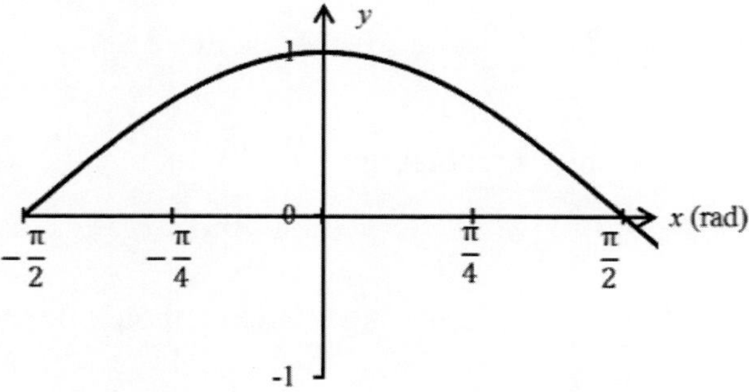

Solution

As the plane is symmetric about the y-axis, the mass centre must be located on the y-axis, or x_c = 0 (you may want to prove this by yourself!). For any vertical strip, $dm = \sigma y dx$ and considering the symmetry, the total mass of this plane is

$$M = \int_a^b dm = 2\int_0^{\pi/2} \sigma y dx = 2\sigma\int_0^{\pi/2} \cos x dx = 2\sigma \sin x\Big|_0^{\pi/2} = 2\sigma(1-0) = 2\sigma.$$

$$M_x = \frac{1}{2}\int_a^b y dm = 2\times\frac{1}{2}\int_0^{\pi/2} y\sigma y dx = \sigma\int_0^{\pi/2} y^2 dx = \sigma\int_0^{\pi/2} \cos^2 x dx$$

$$= \sigma\int_0^{\pi/2} \frac{1+\cos 2x}{2} dx = \frac{\sigma}{2}(x + \frac{1}{2}\sin 2x)\Big|_0^{\pi/2} = \frac{\sigma}{2}(\frac{\pi}{2} - 0) = \frac{\pi}{4}\sigma.$$

Thus the centre of mass of this plane is located at

$$x_c = 0 \quad \text{and} \quad y_c = \frac{M_x}{M} = \frac{\frac{\pi}{4}\sigma}{2\sigma} = \frac{\pi}{8}.$$

Exercises 16.2

1. Find the area of $\int_0^{\frac{\pi}{2}}(\sin x+\cos x)dx$ and $\int_{-3}^{3}(3x^2-1)dx$ respectively.

2. Find the area bounded by $x=9-y^2$ and the y-axis.

3. Find the area bounded by $y=4x-x^2$ and $y=x^2-6$.

4. Find the area bounded by $y=3x+1$, $y=x+3$ and $y=-x-3$.

5. A solid of revolution is formed by rotating $y=4\sqrt[3]{x}$ about the x-axis in [1, 8], find the volume of this solid of revolution.

6. A solid of revolution is formed by rotating $x=\ln y$ about the y-axis in [1, e], find the volume of this solid of revolution.

7. Find the length of arc formed by $x^2=y^3$ from $y=0$ to $y=4$.

8. Find the length of arc formed by $y=3x^{\frac{3}{2}}-1$ from $x=0$ to $x=1$.

9. Find the length of arc formed by $y=\frac{1}{20}x^4+\frac{5}{8x^2}$ from $x=1$ to $x=2$.

10. Find the length of arc formed by $x=t^3+3t^2$ and $y=t^3-3t^2$ from $t=0$ to $t=3$.

11. Find the length of arc formed by $x=\sin t-t\cos t$ and $y=\cos t+t\sin t$ from $t=0$ to $t=2\pi$.

12. Find the surface area of revolution formed by rotating $3y=x^3$ about the x-axis from $x=0$ to 3.

13. Find the surface area of revolution formed by rotating $y=e^x$ about the x-axis from $x=0$ to 2.

14. A horizontal force is controlled by time periodically as $F(x)=4t$ (N) in [0, 2] seconds. This force is applied to an object so that it moves in the same direction horizontally with time as $x=\sqrt{2t}$ (m). How much work has this force done on this object in one period?

15. Find the centre of mass of the area under the curve $x = 2\sin 2y$ from $y = 0$ to $y = \pi/2$. Assume the plane has a constant density σ.

16. Find the centre of mass of a semi-circular plane of radius r defined by $x = r\cos t$ and $y = r\sin t$ from $t = 0$ to $t = \pi$. Assume the plane has a constant density σ.

17. Find the centre of mass of the plane bounded by $y^2 = x$ and $x^2 = -8y$. Assume the plane has a constant density σ.

Chapter 16: Exercises Answers

Exercises 16.1

2. a) ½ and 56, b) $\pi/4$ and $e - 2$, c) 14 and –32, d) 196/3 and $\pi/2$, e) $(\pi + 2)/4$ and $(2e^3 + 1)/9$
3. a) 0 and 0, b) π and 2, c) 0 and π, d) 4 and $2(e^2 - 1)/e$
4. a) 14/3 and –16, b) 1 and 1

Exercises 16.2

1. 2 and 48
2. 36
3. 64/3
4. 8
5. $1488\pi/5$
6. $(e-2)\pi$
7. $L = \frac{8}{27}(\sqrt{1000} - 1)$
8. $L \approx 3.192$
9. $L = \frac{39}{32} = 1.21875$
10. $L \approx 54.9735$
11. $L = 2\pi^2$
12. $S = \frac{\pi}{9}[82^{\frac{3}{2}} - 1] \approx 82.3935\pi \approx 258.85$
13. $S \approx 55.5\pi \approx 174.35$
14. $W = \frac{16}{3}$ (Nm)
15. $x_c = \frac{\pi}{8}$, $y_c = \frac{\pi}{4}$
16. $x_c = 0$, $y_c = \frac{4r}{3\pi}$
17. $x_c = \frac{9}{5}$, $y_c = -\frac{9}{10}$

CHAPTER 17

17 Special and Numeric Integration

CHAPTER OBJECTIVES

- Introduce techniques to deal with improper integration
- Introduce integration of special functions
- Use Trapezium and Simpson's methods and Maclaurin series for numeric integration

Essential statements on improper and numeric integration:

- Limit operations play a key role in improper integration
- Trapezium and Simpson's methods are widely used in numeric integration.

Key topics:

- Improper integrals
- Integral of segmented functions
- Integral of rational sine/cosine functions
- Trapezium method and Simpson's method

Flowchart of mathematical knowledge building

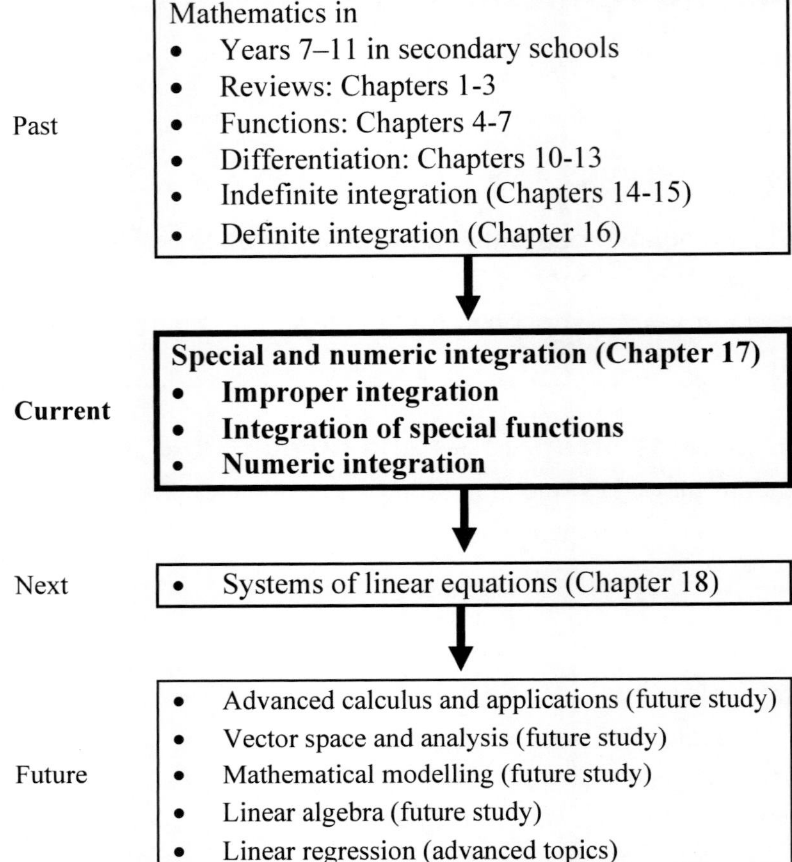

Chapter 17: Special and Numeric Integration

17.1 Improper Integration

Improper integration refers to definite integrals where either any limit (or boundary) of the range is towards infinity, or the integrand *f*(*x*) is undefined at one or more points within the range [*a*, *b*]. There is no guarantee that such integration comes with a finite value (or called convergence). It is a case-by-case assessment by means of limit operations.

17.1.1 Definite integration over infinite limit(s)

This refers to the situation where a definite integral is over the range with one or both boundaries towards infinity, i.e.,

$$\int_{-\infty}^{b} f(x)dx \text{ or } \int_{a}^{\infty} f(x)dx \text{ or } \int_{-\infty}^{\infty} f(x)dx . \qquad (17.1)$$

Such cases can be handled by the following steps

1. Using the usual process of *indefinite integration* to find the original function *F*(*x*);
2. Applying a limit operation towards the infinity to the original function *F*(*x*);
3. If each of all such limit operations produces a definite value, the improper integral converts to the value defined by the Newton-Leibniz formula. If any limit operation is undefined, the improper integral does not exist or is divergent.

Example 17.1: Evaluate $\int_{0}^{\infty} e^{-2x}dx$.

Solution

$$F(x) = \int e^{-2x}dx = -\frac{1}{2}e^{-2x} \text{ (ignore } c\text{)}$$

$$\because F(\infty) = \lim_{x\to\infty} F(x) = \lim_{x\to\infty} (-\frac{1}{2}e^{-2x}) = -\frac{1}{2}\lim_{x\to\infty}\frac{1}{e^{2x}} = -\frac{1}{2}\times 0 = 0. \text{ Hence}$$

$$\int_{0}^{\infty} e^{-2x}dx = F(\infty) - F(0) = 0 - (-\frac{1}{2}e^{0}) = \frac{1}{2}.$$

This improper integral converts to 1/2.

Example 17.2: Evaluate $\int_{-\infty}^{1} \frac{1}{2-x} dx$.

Solution

$$F(x) = \int \frac{1}{2-x} dx = \int \frac{-d(2-x)}{2-x} = -\ln(2-x) = \ln\frac{1}{2-x} \quad \text{(ignore } c\text{)}$$

$$\because F(-\infty) = \lim_{x\to-\infty} F(x) = \lim_{x\to-\infty} \ln\frac{1}{2-x} = \ln(\lim_{x\to-\infty} \frac{1}{2-x}) \to \ln(0). \text{ undefined}$$

Thus, this improper integral is divergent or does not exist.

Example 17.3: Evaluate $\int_{1}^{\infty} \frac{3}{x^2} dx$.

Solution

$$F(x) = \int \frac{3}{x^2} dx = -\frac{3}{x} \quad \text{(ignore } c\text{)}$$

$$\because F(\infty) = \lim_{x\to\infty} F(x) = \lim_{x\to\infty} (-\frac{3}{x}) = 0. \text{ Hence}$$

$$\int_{1}^{\infty} \frac{3}{x^2} dx = F(\infty) - F(1) = 0 - (-\frac{3}{1}) = 3.$$

This improper integral converts to 3.

Example 17.4: Evaluate $\int_{-\infty}^{\infty} xe^{-x} dx$.

Solution

$$F(x) = \int xe^{-x} dx = -xe^{-x} + \int e^{-x} dx = -xe^{-x} - e^{-x} = -e^{-x}(x+1) = -\frac{x+1}{e^x} \quad \text{(ignore } c\text{)}$$

$$\because F(\infty) = \lim_{x\to\infty} F(x) = \lim_{x\to\infty} (-\frac{x+1}{e^x}) = -\lim_{x\to\infty} \frac{(x+1)'}{(e^x)'} = -\lim_{x\to\infty} \frac{1}{e^x} = 0 \text{ and}$$

$$F(-\infty) = \lim_{x\to-\infty} F(x) = \lim_{x\to-\infty} (-\frac{x+1}{e^x}) \to \frac{\infty}{0}. \text{ undefined}$$

Thus, this improper integral is divergent or does not exist.

17.1.2 Definite integration of integrands with undefined points

This refers to the situation where an integrand has limited number of undefined (or discontinuous) points within the range of integration. For example, integrand $f(x)$ can be defined in $(a, b]$, or $[a, b)$, or (a, b), or $[a, c)$ and $(c, b]$ where c is between a and b. Limit operations play a vital role in dealing with such integration, with the following steps

1. Using the usual process of *indefinite integration* to find the original function $F(x)$;
2. If the undefined point is an endpoint of the whole range or a subrange, applying inner-side limit operation to the original function $F(x)$;
3. Repeating Step 2 to all undefined points in the whole range;
4. If each of all such limit operations produces a definite value, the improper integral converts to the value defined by the Newton-Leibniz formula over all subranges. If any limit operation is undefined, the improper integral does not exist or is divergent.

Example 17.5: Evaluate $\int_1^2 \frac{1}{x^2-1}dx$.

Solution

Note that $f(x)=\frac{1}{x^2-1}$ is not defined at $x = 1$.

$$F(x)=\int\frac{1}{x^2-1}dx=\int\frac{1}{(x-1)(x+1)}dx=\int\frac{1}{2}[\frac{1}{x-1}-\frac{1}{x+1}]dx=\frac{1}{2}\ln\frac{x-1}{x+1}\quad\text{(ignore } c\text{)}$$

$$\because F(1)=\lim_{x\to 1^+}F(x)=\lim_{x\to 1^+}(\frac{1}{2}\ln\frac{x-1}{x+1})=\frac{1}{2}\lim_{x\to 1^+}\ln\frac{x-1}{x+1}=\frac{1}{2}\times\ln\frac{0}{2}.\ \text{undefione}$$

Thus, this improper integral does not exist.

Example 17.6: Evaluate $\int_0^1 \frac{1}{\sqrt{x}}dx$.

Solution

Note that $f(x)=\frac{1}{\sqrt{x}}$ is not defined at $x = 0$.

$$F(x)=\int\frac{1}{\sqrt{x}}dx=2\sqrt{x}\quad\text{(ignore } c\text{)}$$

$$\because F(0)=\lim_{x\to 0^+}F(x)=\lim_{x\to 0^+}(2\sqrt{x})=0\ \text{ and }\ F(1)=2\sqrt{1}=2$$

$$\int_0^1\frac{1}{\sqrt{x}}dx=F(1)-F(0)=2-0=2.$$

Example 17.7: Evaluate $\int_{-1}^{1} \frac{1}{\sqrt[3]{x^2}} dx$.

Solution

$f(x) = \frac{1}{\sqrt[3]{x^2}} = x^{-\frac{2}{3}}$ is not defined at $x = 0$; two subranges [–1, 0) and (0, 1] are considered.

$$F(x) = \int \frac{1}{\sqrt[3]{x^2}} dx = 3\sqrt[3]{x} \text{ (ignore } c\text{)}$$

For $[-1, 0)$: $F(0) = \lim_{x \to 0^-} F(x) = \lim_{x \to 0^-} (3\sqrt[3]{x}) = 0$

For $(0, 1]$: $F(0) = \lim_{x \to 0^+} F(x) = \lim_{x \to 0^+} (3\sqrt[3]{x}) = 0$. Thus,

$$\int_{-1}^{1} \frac{1}{\sqrt[3]{x^2}} dx = \int_{-1}^{0} \frac{1}{\sqrt[3]{x^2}} dx + \int_{0}^{1} \frac{1}{\sqrt[3]{x^2}} dx = [F(0) - F(-1)] + [F(1) - F(0)] = [0 - 3\sqrt[3]{-1}] + [3\sqrt[3]{1} - 0]$$

$$= -3 \times (-1) + 3 \times 1 = 3 + 3 = 6.$$

Exercises 17.1

1. Evaluate

   a) $\int_{1}^{\infty} e^{-\sqrt{x}} dx$

   b) $\int_{e}^{\infty} \frac{1}{x \ln x} dx$

   c) $\int_{-\infty}^{\infty} \frac{1}{e^{-x} + e^{x}} dx$

   d) $\int_{-\infty}^{\infty} \frac{1}{x^2 + 2x + 2} dx$

   e) $\int_{-\infty}^{\infty} x e^{-x^2} dx$

2. Evaluate

   a) $\int_{1}^{2} \frac{3x}{\sqrt{x-1}} dx$

   b) $\int_{0}^{2} \frac{1}{(1-x)^2} dx$

   c) $\int_{0}^{4} \frac{2}{\sqrt[3]{x-1}} dx$

   d) $\int_{0}^{\pi/2} \frac{\cos x}{2\sqrt{1 - \sin x}} dx$

   e) $\int_{0}^{1} 3 \ln x dx$

17.2 Integration of Special Functions

17.2.1 Integration of segmented (piecewise) functions

The segmented function has discontinuous points at the connection point of two consecutive segments. Integration at these points can be handled by similar way used in dealing with improper integration. In most cases, if a segmented function takes limited values (i.e., not infinity) at all points, the integration of the segmented function can be handled as 'normal integration' over each of the segments without going through the 'inner-side limit operation'.

Example 17.8: Given $f(x)=\begin{cases}\dfrac{1}{x^2} & -\infty<x\le -1\\ 2x+5 & -1<x\le 0\\ e^{-x} & 0<x<\infty\end{cases}$, evaluate $\int_{-\infty}^{\infty} f(x)dx$.

Solution

$$\int_{-\infty}^{\infty} f(x)dx=\int_{-\infty}^{-1}\frac{1}{x^2}dx+\int_{-1}^{0}(2x+5)dx+\int_{0}^{\infty}e^{-x}dx$$

$$=-\frac{1}{x}\bigg|_{-\infty}^{-1}+(x^2+5x)\Big|_{-1}^{0}-e^{-x}\Big|_0^{\infty}=-(-1-0)+\{0-[(-1)^2+5\times(-1)]-[e^{-\infty}-e^0]$$

$$=1-1+5-0+1=6.$$

Example 17.9: Given $f(x)=\begin{cases}\sin x & -\pi<x\le 0\\ 2x+1 & 0<x\le\dfrac{\pi}{2}\\ \cos x & \dfrac{\pi}{2}<x<\pi\end{cases}$, evaluate $\int_{-\pi}^{\pi} f(x)dx$.

Solution

$$\int_{-\pi}^{\pi} f(x)dx=\int_{-\pi}^{0}\sin xdx+\int_{0}^{\pi/2}(2x+1)dx+\int_{\pi/2}^{\pi}\cos xdx$$

$$=-\cos x|_{-\pi}^{0}+(x^2+x)\Big|_0^{\pi/2}+\sin x|_{\pi/2}^{\pi}=-[\cos 0-\cos(-\pi)]+\{[(\frac{\pi}{2})^2+\frac{\pi}{2}]-0\}+[\sin\pi-\sin\frac{\pi}{2}]$$

$$=-[1-(-1)]+[\frac{\pi^2}{4}+\frac{\pi}{2}]+[0-1]=-2+\frac{\pi^2}{4}+\frac{\pi}{2}-1=\frac{\pi^2}{4}+\frac{\pi}{2}-3\approx 1.0382.$$

17.2.2 Integration of rational functions of sine or cosine

Integration of some rational functions of sin$x$ or cos$x$ that is part of the denominator in a fractional function can be resolved by a special substitution of $x = 2\arctan t$, which leads to

$$\begin{cases} x = 2\arctan t \rightleftarrows t = \tan\dfrac{x}{2} \\ \sin x = \dfrac{2t}{1+t^2}, \quad \cos x = \dfrac{1-t^2}{1+t^2}, \quad dx = \dfrac{2dt}{1+t^2}. \end{cases} \tag{17.2}$$

Such substitution is the Pythagorean theorem applied to the right triangle in Figure 17.1.

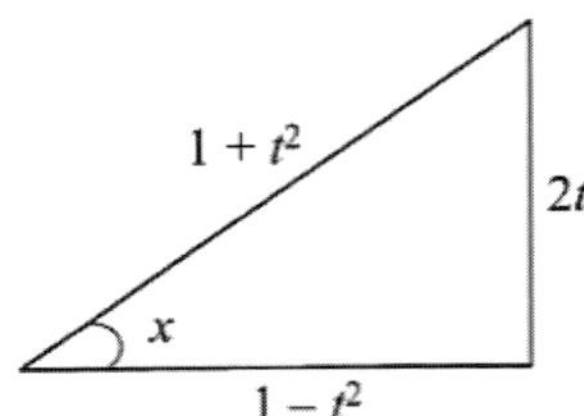

Figure 17.1: The right triangle for substitution

Example 17.10: Find $\int \dfrac{dx}{1+\sin x - \cos x}$.

Solution

Let $x = 2\arctan t$. By formulae (17.2), sin$x$, cos$x$, and dx in the integral are replaced

$$\int \frac{dx}{1+\sin x-\cos x} = \int \frac{\frac{2dt}{1+t^2}}{1+\frac{2t}{1+t^2}-\frac{1-t^2}{1+t^2}} = \int \frac{\frac{2dt}{1+t^2}}{\frac{1+t^2}{1+t^2}+\frac{2t-1+t^2}{1+t^2}} = \int \frac{\frac{2dt}{\cancel{1+t^2}}}{\frac{2t^2+2t}{\cancel{1+t^2}}} = \int \frac{2dt}{2t(t+1)} = \int \frac{dt}{t(t+1)}$$

$$= \int (\frac{1}{t}-\frac{1}{t+1})dt = \ln t - \ln(t+1) + c = \ln\frac{t}{t+1} + c = \ln\frac{\tan\frac{x}{2}}{1+\tan\frac{x}{2}} + c \quad \xleftarrow{(17.2)} t = \tan\frac{x}{2}$$

Example 17.11: Find $\int \dfrac{dx}{3+\cos x}$.

Solution

Let $x = 2\arctan t$. By formulae (17.2),

$$\int \frac{dx}{3+\cos x} = \int \frac{\frac{2dt}{1+t^2}}{3+\frac{1-t^2}{1+t^2}} = \int \frac{\frac{2dt}{1+t^2}}{\frac{3(1+t^2)}{1+t^2}+\frac{1-t^2}{1+t^2}} = \int \frac{\frac{2dt}{\cancel{1+t^2}}}{\frac{3+3t^2+1-t^2}{\cancel{1+t^2}}} = \int \frac{2dt}{2t^2+4} = \int \frac{dt}{t^2+2}$$

$$= \int \frac{dt}{t^2 + (\sqrt{2})^2} = \frac{1}{\sqrt{2}} \arctan \frac{t}{\sqrt{2}} + c = \frac{\sqrt{2}}{2} \arctan(\frac{\sqrt{2}}{2} t) + c \quad \xleftarrow{\#18, a=\sqrt{2}} \text{Table 14.1}$$

$$= \frac{\sqrt{2}}{2} \arctan(\frac{\sqrt{2}}{2} \tan \frac{x}{2}) + c$$

Example 17.12: Find $\int_0^{\pi/6} \frac{1+\sin 3x}{1+\cos 3x} dx$.

Solution

Let $u = 3x$ or $x = u/3$. Then $x = 0, u = 0; x = \pi/6, u = \pi/2$.

$$\int_0^{\pi/6} \frac{1+\sin 3x}{1+\cos 3x} dx = \int_0^{\pi/2} \frac{1+\sin u}{1+\cos u} d(\frac{u}{3}) = \frac{1}{3} \int_0^{\pi/2} \frac{1+\sin u}{1+\cos u} du$$

By formulae (17.2), $t = \tan\frac{u}{2}$; $u = 0, t = 0$; $u = \frac{\pi}{2}, t = \tan\frac{\pi}{4} = 1$.

$$\int_0^{\pi/6} \frac{1+\sin 3x}{1+\cos 3x} dx = \frac{1}{3} \int_0^{\pi/2} \frac{1+\sin u}{1+\cos u} du = \frac{1}{3} \int_0^1 \frac{1+\frac{2t}{1+t^2}}{1+\frac{1-t^2}{1+t^2}} \frac{2dt}{1+t^2} = \frac{1}{3} \int_0^1 \frac{\frac{1+t^2+2t}{1+t^2}}{\frac{1+t^2+1-t^2}{1+t^2}} \frac{2dt}{1+t^2} = \frac{1}{3} \int_0^1 \frac{1+t^2+2t}{2} \frac{2dt}{1+t^2}$$

$$= \frac{1}{3} \int_0^1 \frac{1+t^2+2t}{1+t^2} dt = \frac{1}{3} \int_0^1 [\frac{1+t^2}{1+t^2} + \frac{2t}{1+t^2}] dt = \frac{1}{3} \int_0^1 [1 + \frac{(1+t^2)'}{1+t^2}] dt = \frac{1}{3} \int_0^1 [dt + \frac{d(1+t^2)}{1+t^2}]$$

$$= \frac{1}{3} [t + \ln(1+t^2)] \Big|_0^1 = \frac{1}{3} [(1+\ln 2) - (0 + \ln 1)] = \frac{1+\ln 2}{3} \approx 0.5644$$

Exercises 17.2

1. Find the integral for the following functions

a) $f(x)=\begin{cases}\dfrac{1}{x^2} & elsewhere \\ x & -4\le x\le 4\end{cases}$

b) $f(x)=\begin{cases}2x & -10\le x<0 \\ x^2 & 0\le x\le 9\end{cases}$

c) $f(x)=\begin{cases}e^x & -\infty<x\le -1 \\ 4x^3 & -1<x\le 0 \\ 2e^{-x} & 0<x<\infty\end{cases}$

d) $f(x)=\begin{cases}\cos^2 x & -\pi<x\le -\dfrac{\pi}{2} \\ 2\sin x\cos x & -\dfrac{\pi}{2}<x\le\dfrac{\pi}{2} \\ \sin^2 x & \dfrac{\pi}{2}<x<\pi\end{cases}$

2. Find the integral for the following functions

a) $f(x)=\int\frac{4}{1-\sin x}dx$

b) $f(x)=\int\frac{2}{4\sin x-3\cos x}dx$

c) $f(x)=\int\frac{\tan x}{1+\cos x}dx$

d) $\int_0^{\pi/2}\frac{dx}{1+\sin x+\cos x}$

e) $\int_0^{\pi/2}\frac{dx}{4+5\sin x}$

17.3 Numeric Integration

17.3.1 Trapezium method

Some definite integrals cannot be solved exactly by any analytical techniques, for example $\int_0^1 \frac{\sin x}{x} dx$. Since the definite integral of $y = f(x)$ in $[a, b]$ is the area enclosed between $f(x)$ and the x-axis within $[a, b]$, one simple strategy to approximate this area is to divide the range $(b - a)$ into n vertical strips of equal width (in theory variable width works too); then calculate the subarea of each of these vertical strips using a known method; finally add all these subareas together as an approximate to the total area as the integral.

The trapezium method is to treat each vertical strip as a vertical trapezoid (Figure 17.2). If we turn the vertical strip as a horizontal strip, its subarea should be

$$A_i = \frac{1}{2}(y_{i-1} + y_i)h \quad \text{where} \quad h = \frac{b-a}{n}.$$

Adding the n subareas together, the integral sought is finally expressed as

$$\boxed{\int_a^b y dx = \frac{h}{2}[y_0 + 2(y_1 + y_2 + y_3 + ... + y_{n-1}) + y_n].} \quad (17.3)$$

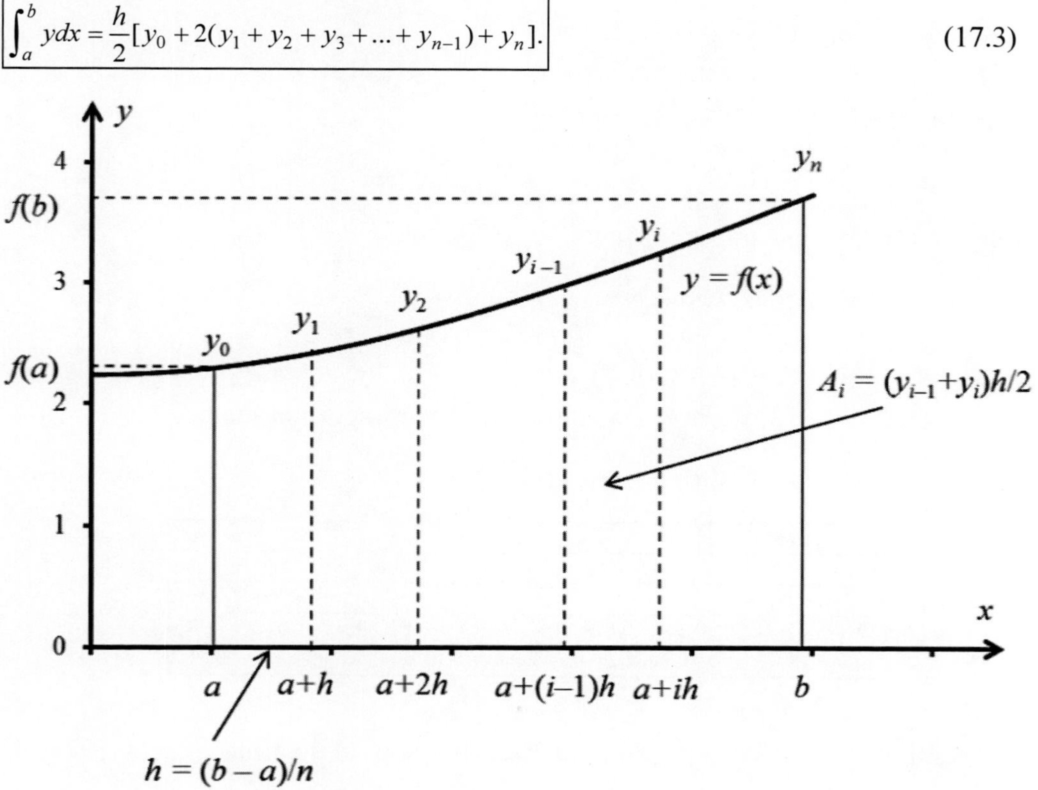

Figure 17.2: Sketch of the trapezium method

In numeric computation, if the final result demands an accuracy to n-decimal places, the intermediate results should be kept to at least n-decimal places, better to $(n+1)$-decimal places. This is to minimize the accumulation or propagation of errors during numeric computation.

Example 17.13: Use the trapezium method to approximate $\int_0^1 \frac{\sin x}{x}dx$ by dividing the range into ten strips of equal width.

Solution

The trapezium method (17.3) is often realised by tabulating the numbers of each strips as follows:

i	0	1	2	3	4	5	6	7	8	9	10	Sum
x_i	0	0.1	0.2	0.3	0.4	0.5	0.6	0.7	0.8	0.9	1	
y_0 or y_n	1										0.8415	*1.8415*
$2y_i$		1.9967	1.9867	1.9701	1.9471	1.9177	1.8821	1.8406	1.7934	1.7407		*17.0751*
Area	h×Sum/2 = 0.1×Sum/2 = **0.9458**											18.9166

Note that x is in radians; $h = \frac{1-0}{10} = 0.1$; $y_0 = \lim_{x\to 0}\frac{\sin x}{x} = 1$. The final result is $\int_0^1 \frac{\sin x}{x}dx \approx 0.9458$.

Example 17.14: Use the trapezium method to approximate $\int_1^{1.5} \frac{1}{1+x^2}dx$ by dividing the range into five strips of equal width.

Solution

Similar to Example 17.13, we use a table to realise the trapezium method (17.3). Note that $h = (1.5-1)/5 = 0.1$.

i	0	1	2	3	4	5	Sum
x_i	1	1.1	1.2	1.3	1.4	1.5	
y_0 or y_n	0.5					0.30769	*0.80769*
$2y_i$		0.90498	0.81967	0.74349	0.67568		*3.14382*
Area	h×Sum/2 = 0.1×Sum/2 = **0.1976**						3.95151

The final result is $\int_1^{1.5} \frac{1}{1+x^2}dx \approx 0.1976$, which is very close to the 'true value' 0.1974.

Example 17.15: Use the trapezium method to approximate the arc length of $y=\frac{1}{3}x^3$ between $x = 0$ and $x = 1$ by dividing the range into ten strips of equal width.

Solution

The arc length is defined by $L=\int_0^1\sqrt{1+(\frac{dy}{dx})^2}\,dx$.

$$\frac{dy}{dx}=(\frac{1}{3}x^3)'=x^2 \longrightarrow L=\int_0^1\sqrt{1+(x^2)^2}\,dx=\int_0^1\sqrt{1+x^4}\,dx$$

We use a table to realise the trapezium method (17.3). Note that $h=(1-0)/10=0.1$.

i	0	1	2	3	4	5	6	7	8	9	10	Sum
x_i	0	0.1	0.2	0.3	0.4	0.5	0.6	0.7	0.8	0.9	1	
y_0 or y_n	1										1.4142	*2.4142*
$2y_i$		2.0001	2.0016	2.0081	2.0254	2.0616	2.1257	2.2272	2.3745	2.5738		*19.3979*
Length		h×Sum/2 = 0.1×Sum/2 = **1.0906**										21.8122

The final result is $L=\int_0^1\sqrt{1+x^4}\,dx\approx 1.0906$.

17.3.2 Simpson's method

If dividing the range into an *even number* of vertical strips, a local quadratic interpolation can be created using the three known boundary points of any two adjunct strips (Figure 17.3) to replace $f(x)$ within the sub-range $[x_{i-1}, x_{i+1}]$. The area under the interpolation in $[x_{i-1}, x_{i+1}]$ is regarded as an approximate to the area under $f(x)$ in this sub-range. Sliding this window through the whole range should get the areas for all vertical strips with some been counted multiple times. The final area is adjusted as

$$\boxed{\int_a^b y\,dx=\frac{h}{3}[y_0+4(y_1+y_3+...)+2(y_2+y_4+...)+y_n].} \tag{17.4}$$

Note n must be an even number.

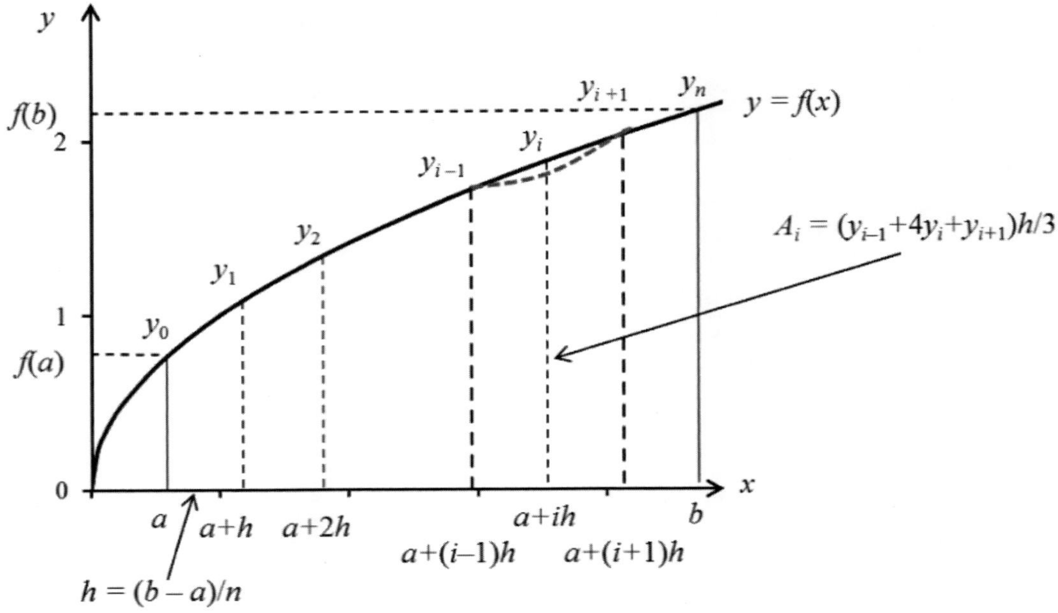

Figure 17.3: Sketch of the Simpson's method

Example 17.16: Use the Simpson's method to approximate $\int_0^1 \frac{\sin x}{x} dx$ by dividing the range into ten strips of equal width.

Solution

We use the following table to realise the Simpson's method (17.4). Note that x is in radians; $n = 10$ is an even number; $h = \frac{1-0}{10} = 0.1$ and $y_0 = \lim_{x \to 0} \frac{\sin x}{x} = 1$.

i	0	1	2	3	4	5	6	7	8	9	10	Sum
x_i	0	0.1	0.2	0.3	0.4	0.5	0.6	0.7	0.8	0.9	1	
y_0 or y_n	1										0.8415	*1.8415*
$4y_i$		3.9933		3.9403		3.8354		3.6812		3.4815		*18.9317*
$2y_i$			1.9867		1.9471		1.8821		1.7934			*7.6093*
Area		h×Sum/3 = 0.1×Sum/3 = **0.9461**										28.3826

The final result is $\int_0^1 \frac{\sin x}{x} dx \approx 0.9461$.

Example 17.17: Use the Simpson's method to approximate $\int_1^{1.5} \frac{1}{1+x^2} dx$ by dividing the range into four strips of equal width.

Solution

We use a table to realise the Simpson's method. Note that $n = 4$ and $h = (1.5-1)/4 = 0.125$.

i	0	1	2	3	4	Sum
x_i	1	1.125	1.25	1.375	1.5	
y_0 or y_n	0.5				0.30769	*0.80769*
$4y_i$		1.76552		1.38378		*3.14930*
$2y_i$			0.78049			*0.78049*
Area		h×Sum/3 = 0.125×Sum/3 = **0.1974**				4.73748

The final result is $\int_1^{1.5} \frac{1}{1+x^2} dx \approx 0.1974$, the same as the 'true value' even with only 4 strips, which is much more accurate than that from the trapezium method in Example 17.14.

Example 17.18: Use the Simpson's method to approximate the arc length of $y = \frac{1}{3}x^3$ between $x = 0$ and $x = 1$ by dividing the range into eight strips of equal width.

Solution

The arc length is defined by $L = \int_0^1 \sqrt{+(\frac{dy}{dx})^2} dx$.

$$\frac{dy}{dx} = (\frac{1}{3}x^3)' = x^2 \longrightarrow L = \int_0^1 \sqrt{1+(x^2)^2} dx = \int_0^1 \sqrt{1+x^4} dx$$

We use a table to realise the trapezium method (17.4). Note that $h = (1-0)/8 = 0.125$.

i	0	1	2	3	4	5	6	7	8	Sum
x_i	0	0.125	0.25	0.375	0.5	0.625	0.75	0.875	1	
y_0 or y_n	1								1.41421	*2.41421*
$4y_i$		4.00049		4.03936		4.29435		5.03775		*17.37194*
$2y_i$			2.00390		2.06155		2.29469			*6.36015*
length				h×Sum/3 = 0.125×Sum/3 = **1.0894**						26.14630

The final result is $L = \int_0^1 \sqrt{1+x^4} dx \approx 1.0894$.

Under the similar conditions, the Simpson's method often produces a more accurate numeric solution than the trapezium method.

17.3.3 Numeric integration by Maclaurin series

Maclaurin series can be used to obtain numeric solutions for some integrands that cannot be integrated by any integration techniques, such as $\int \frac{\sin x}{x} dx$. In such a case, the range of integration must be within the range of convergence of the Maclaurin series associated with the function. For example, the Maclaurin series for $\sin x$ is convergent in the entire domain; hence the Maclaurin series can be used to approximate $\int \frac{\sin x}{x} dx$ in any given range.

Example 17.19: Approximate $\int_0^1 \frac{\sin x}{x} dx$ within an error limit $\varepsilon = 0.0001$.

Solution

We use Maclaurin series for sin$x$ to approximate this integral as follows:

$$\frac{\sin x}{x} = \frac{x - \frac{x^3}{3!} + \frac{x^5}{5!} - \frac{x^7}{7!} \ldots}{x} = 1 - \frac{x^2}{3!} + \frac{x^4}{5!} - \frac{x^6}{7!} + \frac{x^8}{9!} \ldots$$

$$\int_0^1 \frac{\sin x}{x} dx = \int_0^1 (1 - \frac{x^2}{3!} + \frac{x^4}{5!} - \frac{x^6}{7!} + \frac{x^8}{9!} \ldots) dx = (x - \frac{x^3}{3\times 3!} + \frac{x^5}{5\times 5!} - \frac{x^7}{7\times 7!} + \frac{x^9}{9\times 9!} \ldots)\Big|_0^1$$

$$= (1 - \frac{1}{3\times 3!} + \frac{1}{5\times 5!} - \frac{1}{7\times 7!} + \frac{1}{9\times 9!} \ldots) - (0 - 0 + 0 - 0 + \ldots) = (1 - \frac{1}{3\times 3!} + \frac{1}{5\times 5!} - \frac{1}{7\times 7!} + \frac{1}{9\times 9!} \ldots)$$

Since $\frac{1}{7\times 7!} = 0.000028 < 0.0001$, the sum of the first three terms should make the error within the limit $\varepsilon = 0.0001$.

$$\int_0^1 \frac{\sin x}{x} dx \approx 1 - \frac{1}{3\times 3!} + \frac{1}{5\times 5!} \approx 0.9461.$$

This is the same as the result from using the Simpson's method in Example 17.16. Both are far more accurate than that from the trapezium method in Example 17.13.

Exercises 17.3

1. Use both the trapezium and Simpson's methods to approximate $\int_1^2 \frac{1}{x} dx$ by dividing the range into ten strips of equal width. Compare your results with the 'true value' to assess the error.

2. Use both the trapezium and Simpson's methods to approximate $\int_0^1 \sqrt{1+x^3} dx$ by dividing the range into eight strips of equal width.

3. Use both the trapezium and Simpson's methods to approximate $\int_0^1 e^{-x^2} dx$ by dividing the range into ten strips of equal width.

4. Use both the trapezium and Simpson's methods to approximate the arc length of $y = e^x$ from $x = 0$ to 1. Divide the range into ten strips of equal width.

5. Use both the trapezium and Simpson's methods to approximate the surface area of revolution formed by rotating $y = \ln x$ about the x-axis from $x = 1$ to 2. Divide the range into eight strips of equal width.

6. Use the Simpson's method to approximate $\int_1^{1.5} \frac{1}{\sqrt{1+x^2}} dx$ by dividing the range into ten strips of equal width.

7. Use Maclaurin series to approximate $\int_0^1 \sin\sqrt{x} dx$ within an error limit $\varepsilon = 0.0001$.

8. Use Maclaurin series to approximate $\int_0^1 \cos x^2 dx$ within an error limit $\varepsilon = 0.0001$.

9. Use Maclaurin series to approximate $\int_0^1 e^{-x^2} dx$ within an error limit $\varepsilon = 0.0001$.

Chapter 17: Exercises Answers

Exercises 17.1

1. a) 2; b) does not exist; c) π/2; d) π; e) 0
2. a) 8; b) does not exist; c) $3[\sqrt[3]{9}-1]$; d) 1; e) –3

Exercises 17.2

1. a) ½; b) 143; c) $1+e^{-1}$; d) π/2
2. a) $\dfrac{8\sin\frac{x}{2}}{\cos\frac{x}{2}-\sin\frac{x}{2}}+c$ b) $\dfrac{2}{5}\ln\left|\dfrac{3\sin\frac{x}{2}-\cos\frac{x}{2}}{\sin\frac{x}{2}+3\cos\frac{x}{2}}\right|+c$ c) $-\ln|1-\tan^2\frac{x}{2}|+c$

   d) $\ln 2 \approx 0.6931$ e) $\dfrac{1}{3}\ln 2 \approx 0.2310$

Exercises 17.3

1. 0.6938 (T), 0.6932 (S) 2. 1.1128 (T), 1.1114 (S)
3. 0.7462 (T), 0.7468 (S) 4. 2.0050 (T), 2.0035 (S) 5. 2.8580 (T), 2.8655 (S)
6. 0.3134 7. 0.6023 8. 0.9045 9. 0.7468

CHAPTER 18

18 Systems of Linear Equations

CHAPTER OBJECTIVES

- Introduce fundamentals of matrices, vectors, and operations
- Introduce determinants
- Express systems of linear equations in matrix format
- Solve systems of linear equations by Cramer's rule
- Solve systems of linear equations by Gauss elimination

Essential statements on systems of linear equations:

- Matrices are powerful tools in massive and numeric computation.
- The result of matrix operations is still a matrix whereas the result of operations on a determinant is a single value.
- Only square matrices have determinants.
- Cramer's rule is useful in solving 2 by 2 and 3 by 3 systems of linear equations.
- Gauss elimination is more useful in solving higher order systems of linear equations.

Key topics:

- Matrices, vectors, and operations
- Determinants
- Systems of linear equations in matrix format
- Cramer's rule
- Gauss elimination

Flowchart of mathematical knowledge building

Past	Mathematics in • Years 7–11 in secondary schools • Reviews: Chapters 1-3 • Functions: Chapters 4-7 • Differentiation: Chapters 10-13 • Integration: (Chapters 14-17)
	↓
Current	**Systems of linear equations (Chapter 18)** • **Matrices and vectors** • **Determinants** • **Systems of linear equations** • **Cramer's rule** • **Gauss elimination**
	↓
Future	• Advanced calculus and applications (future study) • Vector space and analysis (future study) • Mathematical modelling (future study) • Linear algebra (future study) • Linear regression (advanced topics)

Chapter 18: Systems of Linear Equations

18.1 Fundamentals of Matrices and Vectors

18.1.1 Matrices and vectors

Matrices

A matrix is a rectangular array of numbers or functions enclosed in brackets. The numbers or functions are called ***entries*** or ***elements*** of the matrix. If the array has m rows and n columns, it forms an $m \times n$ matrix **A**, written as

$$\mathbf{A} = [a_{ij}] = \begin{bmatrix} a_{11} & a_{12} & \cdots & a_{1n} \\ a_{21} & a_{22} & \cdots & a_{2n} \\ \cdots & \cdots & \cdots & \cdots \\ a_{m1} & a_{m2} & \cdots & a_{mn} \end{bmatrix}, \quad (i = 1, 2, \ldots, m; j = 1, 2, \ldots, n) \tag{18.1}$$

There are $m \times n$ entries in matrix **A** so the ***size*** of matrix **A** is $m \times n$. Each entry in (18.1) has two subscripts. The first refers to the row number and the second is the column number. Hence a_{ij} is the entry in Row i and Column j. Some examples of matrices are shown below:

$$\mathbf{X} = \begin{pmatrix} 3 & 1 & 2 \\ -2 & 2 & 3 \\ 0 & 5 & 3 \end{pmatrix}, \quad \mathbf{Y} = \begin{pmatrix} 9 & -2 \\ 0 & 1 \\ 5 & 7 \end{pmatrix}, \quad \mathbf{F} = \begin{bmatrix} \sin x & -\cos x \\ -\cos x & -\sin x \end{bmatrix}.$$

X is a 3×3 matrix with 9 entries of numbers whereas **Y** is a 3×2 matrix with 6 entries of numbers. **F** is a 2×2 matrix of functions. Entry x_{23} is 3; y_{31} is 5; f_{12} is $-\cos x$. If a matrix has the same number in rows and columns, it is called a ***square matrix***. Thus, **X** and **F** are square matrices.

Vectors

If a matrix has just one row or column, it becomes a vector. A one-row matrix is called a ***row vector*** whereas a one-column matrix is called a ***column vector***. Since vectors are one dimensional, it is normally denoted by ***lowercase*** boldface letters with only one subscript, i.e., **a** = $[a_j]$. Some examples of vectors are shown below:

$$\mathbf{a} = \begin{pmatrix} 3 \\ -1 \\ 0 \\ 2 \end{pmatrix}, \quad \mathbf{b} = \begin{bmatrix} 7 & 0.5 & -3 & 1 \end{bmatrix}, \quad \mathbf{c} = \begin{bmatrix} 1 \\ \sin x \\ \cos x \end{bmatrix}.$$

Two matrices (or vectors) are equal if and only if they have the same size and the corresponding entries are equal. This means if $\mathbf{A} = [a_{ij}]$ and $\mathbf{B} = [b_{ij}]$, $\mathbf{A} = \mathbf{B}$ if and only if $a_{ij} = b_{ij}$ for all i and j.

18.1.2 Special matrices

Triangular matrices

If a square matrix has nonzero entries only on and ***above*** the main diagonal (hence any entry below the diagonal being zero), the matrix is called an ***upper triangular matrix***. Similarly a ***lower triangular matrix*** has only nonzero entries on and ***below*** the main diagonal. In the following examples, **P** and **Q** are upper triangular matrices whereas **S** and **T** are lower triangular matrices.

$$\mathbf{P} = \begin{bmatrix} 3 & 2 & -1 \\ 0 & 5 & 1 \\ 0 & 0 & 2 \end{bmatrix}, \quad \mathbf{Q} = \begin{bmatrix} 1 & 3 & -1 & 2 \\ 0 & -1 & 0 & -4 \\ 0 & 0 & -2 & 1 \\ 0 & 0 & 0 & 3 \end{bmatrix},$$

$$\mathbf{S} = \begin{bmatrix} -6.7 & 0 & 0 & 0 \\ 6.3 & 4.7 & 0 & 0 \\ 1.3 & 2.2 & -2.0 & 0 \\ 5.2 & -3.1 & 2.7 & 3.3 \end{bmatrix}, \quad \mathbf{T} = \begin{bmatrix} \sin x & 0 \\ 1 & \cos x \end{bmatrix}.$$

Diagonal and unit matrices

If a square matrix has nonzero entries only on the main diagonal (hence any entry below and above the diagonal being zero), the matrix is called a ***diagonal matrix***. If all the entries on the main diagonal equal **1**, this special diagonal matrix is called a ***unit matrix*** (or an ***identity matrix***), especially denoted as **I**. In the following examples, **R** is a diagonal matrix and **I** is a unit matrix.

$$\mathbf{R} = \begin{bmatrix} 2 & 0 & 0 \\ 0 & 1 & 0 \\ 0 & 0 & -3 \end{bmatrix}, \quad \mathbf{I} = \begin{bmatrix} 1 & 0 & 0 & 0 \\ 0 & 1 & 0 & 0 \\ 0 & 0 & 1 & 0 \\ 0 & 0 & 0 & 1 \end{bmatrix}.$$

18.1.3 Addition and scalar multiplication of matrices and vectors

Addition of matrices and vectors

If **A**, **B**, and **C** are matrices with the same size, the following operations exist

$$\begin{cases} \mathbf{A}+\mathbf{B}=\mathbf{B}+\mathbf{A} \\ (\mathbf{A}+\mathbf{B})+\mathbf{C}=\mathbf{A}+(\mathbf{B}+\mathbf{C}) \\ \mathbf{A}+\mathbf{0}=\mathbf{A} \quad (\mathbf{0} \text{ denotes the zero matrix.}) \\ \mathbf{A}+(-\mathbf{A})=\mathbf{0} \end{cases} \tag{18.2}$$

Additions or subtractions are conducted on the corresponding entices in the matrices.

Example 18.1: Given **P** and **Q** as follows, find **P** + **Q** and **Q** – **P** respectively.

$$\mathbf{P}=\begin{bmatrix} 3 & -1 \\ 2 & 1 \end{bmatrix}, \quad \mathbf{Q}=\begin{bmatrix} -1 & 0 \\ 3 & 2 \end{bmatrix}.$$

Solution

$$\mathbf{P}+\mathbf{Q}=\begin{bmatrix} 3 & -1 \\ 2 & 1 \end{bmatrix}+\begin{bmatrix} -1 & 0 \\ 3 & 2 \end{bmatrix}=\begin{bmatrix} 3-1 & -1+0 \\ 2+3 & 1+2 \end{bmatrix}=\begin{bmatrix} 2 & -1 \\ 5 & 3 \end{bmatrix},$$

$$\mathbf{Q}-\mathbf{P}=\begin{bmatrix} -1 & 0 \\ 3 & 2 \end{bmatrix}-\begin{bmatrix} 3 & -1 \\ 2 & 1 \end{bmatrix}=\begin{bmatrix} -1-3 & 0-(-1) \\ 3-2 & 2-1 \end{bmatrix}=\begin{bmatrix} -4 & 1 \\ 1 & 1 \end{bmatrix}.$$

Example 18.2: Given **a**, **b** and **c** as follows, find **a** + **b** and **a** + **b** – **c**.

$$\mathbf{a}=\begin{bmatrix} a_1 \\ a_2 \\ a_3 \\ a_4 \end{bmatrix}, \quad \mathbf{b}=\begin{bmatrix} b_1 \\ b_2 \\ b_3 \\ b_4 \end{bmatrix}, \quad \mathbf{c}=\begin{bmatrix} c_1 \\ c_2 \\ c_3 \\ c_4 \end{bmatrix}.$$

Solution

$$\mathbf{a}+\mathbf{b}=\begin{bmatrix} a_1 \\ a_2 \\ a_3 \\ a_4 \end{bmatrix}+\begin{bmatrix} b_1 \\ b_2 \\ b_3 \\ b_4 \end{bmatrix}=\begin{bmatrix} a_1+b_1 \\ a_2+b_2 \\ a_3+b_3 \\ a_4+b_4 \end{bmatrix}, \quad \mathbf{a}+\mathbf{b}-\mathbf{c}=\begin{bmatrix} a_1+b_1 \\ a_2+b_2 \\ a_3+b_3 \\ a_4+b_4 \end{bmatrix}-\begin{bmatrix} c_1 \\ c_2 \\ c_3 \\ c_4 \end{bmatrix}=\begin{bmatrix} a_1+b_1-c_1 \\ a_2+b_2-c_2 \\ a_3+b_3-c_3 \\ a_4+b_4-c_4 \end{bmatrix}$$

Scalar multiplication (multiplication by a number)

If **A** and **B** are matrices with the same size and α and β are constants, then

$$\begin{cases} \alpha(\mathbf{A}+\mathbf{B}) = \alpha\mathbf{A}+\alpha\mathbf{B} \\ (\alpha+\beta)\mathbf{A} = \alpha\mathbf{A}+\beta\mathbf{A} \\ \alpha(\beta)\mathbf{A} = (\alpha\beta)\mathbf{A} \end{cases} \tag{18.3}$$

The product of matrix $\mathbf{A} = [a_{ij}]$ and any scalar β is a new matrix $\beta\mathbf{A} = [\beta a_{ij}]$, in which each entry is the product of the original entry in **A** by β.

Example 18.3: Given **P** as follows, find 2**P** and $-\frac{1}{3}\mathbf{P}$ respectively.

$$\mathbf{P} = \begin{bmatrix} 3 & -1 \\ 2 & 1 \end{bmatrix}.$$

Solution

$$2\mathbf{P} = 2\begin{bmatrix} 3 & -1 \\ 2 & 1 \end{bmatrix} = \begin{bmatrix} 2\times3 & 2\times(-1) \\ 2\times2 & 2\times1 \end{bmatrix} = \begin{bmatrix} 6 & -2 \\ 4 & 2 \end{bmatrix},$$

$$-\frac{1}{3}\mathbf{P} = -\frac{1}{3}\begin{bmatrix} 3 & -1 \\ 2 & 1 \end{bmatrix} = \begin{bmatrix} -\frac{3}{3} & -\frac{-1}{3} \\ -\frac{2}{3} & -\frac{1}{3} \end{bmatrix} = \begin{bmatrix} -1 & \frac{1}{3} \\ -\frac{2}{3} & -\frac{1}{3} \end{bmatrix}.$$

Example 18.4: Given **a** and **b** as follows, find 2**a** + **b**, 3(**a** + **b**) and (**a** – 2**b**) respectively.

$$\mathbf{a} = [a_1 \quad a_2 \quad a_3], \quad \mathbf{b} = [b_1 \quad b_2 \quad b_3].$$

Solution

$$\begin{aligned} 2\mathbf{a}+\mathbf{b} &= 2[a_1 \quad a_2 \quad a_3]+[b_1 \quad b_2 \quad b_3] = [2a_1 \quad 2a_2 \quad 2a_3]+[b_1 \quad b_2 \quad b_3] \\ &= [2a_1+b_1 \quad 2a_2+b_2 \quad 2a_3+b_3], \\ 3(\mathbf{a}+\mathbf{b}) &= 3[a_1+b_1 \quad a_2+b_2 \quad a_3+b_3] = [3(a_1+b_1) \quad 3(a_2+b_2) \quad 3(a_3+b_3)] = 3\mathbf{a}+3\mathbf{b}, \\ \mathbf{a}-2\mathbf{b} &= [a_1 \quad a_2 \quad a_3]-2[b_1 \quad b_2 \quad b_3] = [a_1 \quad a_2 \quad a_3]-[2b_1 \quad 2b_2 \quad 2b_3] \\ &= [a_1-2b_1 \quad a_2-2b_2 \quad a_3-2b_3] \end{aligned}$$

18.1.4 Matrix multiplications

Multiplication of two matrices

Let **A** be a $l\times m$ matrix and **B** is an $m\times n$ matrix, a case that the number of columns in matrix **A** equals the number of rows in matrix **B**, then the product of **A** by **B** is a new matrix **C** of size $l\times n$, i.e., the same number of rows in **A** and the same number of columns in **B**. Entries in **C** are determined by

$$c_{ij} = \sum_{k=1}^{m} a_{ik} b_{kj} = a_{i1}b_{1j} + a_{i2}b_{2j} + \ldots + a_{im}b_{mj}, \quad i = 1,2,\ldots,l;\ j = 1,2,\ldots,n \tag{18.4}$$

This is also called a ***multiplication of rows into columns***. It is easy to understand that the product of two $n\times n$ square matrices is also a new $n\times n$ square matrix.

Example 18.5: Given $\mathbf{A} = \begin{bmatrix} 2 & 1 & 4 & 0 \\ 1 & 1 & 3 & 4 \end{bmatrix}$ and $\mathbf{B} = \begin{bmatrix} 2 & 3 & 1 \\ 0 & 1 & 2 \\ 1 & 3 & 1 \\ 4 & 0 & 2 \end{bmatrix}$, calculate **AB**.

<u>**Solution**</u>

A is a 2×4 matrix and **B** is a 4×3 matrix. There are 4 columns in **A** and 4 rows in **B** (i.e., $m = 4$) so **AB** produces a new matrix **C** of size 2×3 (i.e., $l = 2$ and $n = 3$) as

$$\mathbf{C} = \begin{bmatrix} c_{11} & c_{12} & c_{13} \\ c_{21} & c_{22} & c_{23} \end{bmatrix}.$$

The entries are determined by formula (18.4) as follows:

$$c_{11} = a_{11}b_{11} + a_{12}b_{21} + a_{13}b_{31} + a_{14}b_{41} = 2\times2 + 1\times0 + 4\times1 + 0\times4 = 8 \quad \textit{(1st row into 1st column)}$$
$$c_{12} = a_{11}b_{12} + a_{12}b_{22} + a_{13}b_{32} + a_{14}b_{42} = 2\times3 + 1\times1 + 4\times3 + 0\times0 = 19 \quad \textit{(1st row into 2nd column)}$$
$$c_{13} = a_{11}b_{13} + a_{12}b_{23} + a_{13}b_{33} + a_{14}b_{43} = 2\times1 + 1\times2 + 4\times1 + 0\times2 = 8 \quad \textit{(1st row into 3rd column)}$$
$$c_{21} = a_{21}b_{11} + a_{22}b_{21} + a_{23}b_{31} + a_{24}b_{41} = 1\times2 + 1\times0 + 3\times1 + 4\times4 = 21 \quad \textit{(2nd row into 1st column)}$$
$$c_{22} = a_{21}b_{12} + a_{22}b_{22} + a_{23}b_{32} + a_{24}b_{42} = 1\times3 + 1\times1 + 3\times3 + 4\times0 = 13 \quad \textit{(2nd row into 2nd column)}$$
$$c_{23} = a_{21}b_{13} + a_{22}b_{23} + a_{23}b_{33} + a_{24}b_{43} = 1\times1 + 1\times2 + 3\times1 + 4\times2 = 14 \quad \textit{(2nd row into 3rd column)}$$

Therefore the new matrix **C** is

$$\mathbf{C} = \begin{bmatrix} 8 & 19 & 8 \\ 21 & 13 & 14 \end{bmatrix}.$$

Example 18.6: Given

$$\mathbf{A}=\begin{bmatrix}2 & 0 & 1\\ -2 & 3 & 2\\ 4 & 1 & 5\end{bmatrix} \text{ and } \mathbf{B}=\begin{bmatrix}-3 & 1 & 0\\ 2 & 0 & 1\\ 0 & -1 & 3\end{bmatrix}, \text{ calculate } \mathbf{AB}.$$

Solution

The product of two 3×3 matrices is a new 3×3 matrix **C**:

$$\mathbf{C}=\mathbf{AB}=\begin{bmatrix}c_{11} & c_{12} & c_{13}\\ c_{21} & c_{22} & c_{23}\\ c_{31} & c_{32} & c_{33}\end{bmatrix}, \text{ whose entries are determined by formula (18.4) as follows:}$$

$$\begin{aligned}
c_{11} &= a_{11}b_{11}+a_{12}b_{21}+a_{13}b_{31} = 2\times(-3)+0\times2+1\times0=-6 && \textit{(1st row into 1st column)}\\
c_{12} &= a_{11}b_{12}+a_{12}b_{22}+a_{13}b_{32} = 2\times1+0\times0+1\times(-1)=1 && \textit{(1st row into 2nd column)}\\
c_{13} &= a_{11}b_{13}+a_{12}b_{23}+a_{13}b_{33} = 2\times0+0\times1+1\times3=3 && \textit{(1st row into 3rd column)}\\
c_{21} &= a_{21}b_{11}+a_{22}b_{21}+a_{23}b_{31} = -2\times(-3)+3\times2+2\times0=12 && \textit{(2nd row into 1st column)}\\
c_{22} &= a_{21}b_{12}+a_{22}b_{22}+a_{23}b_{32} = -2\times1+3\times0+2\times(-1)=-4 && \textit{(2nd row into 2nd column)}\\
c_{23} &= a_{21}b_{13}+a_{22}b_{23}+a_{23}b_{33} = -2\times0+3\times1+2\times3=9 && \textit{(2nd row into 3rd column)}\\
c_{31} &= a_{31}b_{11}+a_{32}b_{21}+a_{33}b_{31} = 4\times(-3)+1\times2+5\times0=-10 && \textit{(3rd row into 1st column)}\\
c_{32} &= a_{31}b_{12}+a_{32}b_{22}+a_{33}b_{32} = 4\times1+1\times0+5\times(-1)=-1 && \textit{(3rd row into 2nd column)}\\
c_{33} &= a_{31}b_{13}+a_{32}b_{23}+a_{33}b_{33} = 4\times0+1\times1+5\times3=16 && \textit{(3rd row into 3rd column)}
\end{aligned}$$

Therefore the new matrix **C** is

$$\mathbf{C}=\begin{bmatrix}-6 & 1 & 3\\ 12 & -4 & 9\\ -10 & -1 & 16\end{bmatrix}.$$

Multiplication of two vectors of the same size

A row vector **a** of size n can be regarded as a $1\times n$ matrix whereas a column vector **b** of size n can be regarded as an $n\times1$ matrix. Thus the product of a row vector into a column vector of same size is a new matrix of size 1 (= 1×1) regardless of the number of entries n, i.e.,

$$\mathbf{c}=\mathbf{ab}=\begin{bmatrix}a_1 & a_2 & \dots & a_n\end{bmatrix}\begin{bmatrix}b_1\\ b_2\\ \vdots\\ b_n\end{bmatrix}=\begin{bmatrix}c_{11}\end{bmatrix}=\begin{bmatrix}\sum_{i=1}^{n}a_ib_i\end{bmatrix}. \tag{18.5}$$

On the other hand, the product of a column vector into a row vector of same size creates a new matrix of size $n \times n$, i.e.,

$$\boxed{\mathbf{D} = \mathbf{ba} = \begin{bmatrix} b_1 \\ b_2 \\ \vdots \\ b_n \end{bmatrix} \begin{bmatrix} a_1 & a_2 & \dots & a_n \end{bmatrix} = \begin{bmatrix} d_{11} & d_{12} & \dots & d_{1n} \\ d_{21} & d_{22} & \dots & d_{2n} \\ \dots & \dots & \dots & \dots \\ d_{n1} & d_{n2} & \dots & d_{nn} \end{bmatrix}}, \tag{18.6}$$

where the entries are determined by

$$d_{ij} = b_i a_j, \quad i = 1, 2, \dots, n; j = 1, 2, \dots, n \,. \tag{18.7}$$

These two special cases of multiplication involving row and column vectors indicate that matrix multiplication is ***not commutative*** in general, i.e., **AB** ≠ **BA** (or **ab** ≠ **ba**).

Example 18.7: Given

$$\mathbf{a} = \begin{bmatrix} 1 & 2 & 3 \end{bmatrix} \text{ and } \mathbf{b} = \begin{bmatrix} 3 \\ 2 \\ 1 \end{bmatrix}, \text{ calculate } \mathbf{ab} \text{ and } \mathbf{ba}.$$

<u>Solution</u>

a is a row vector and **b** is a column vector both with 3 entries. By formula (18.5), the product of **ab** is a single entry matrix as follows

$$\mathbf{c} = \mathbf{ab} = \begin{bmatrix} 1 & 2 & 3 \end{bmatrix} \begin{bmatrix} 3 \\ 2 \\ 1 \end{bmatrix} = \begin{bmatrix} 1 \times 3 + 2 \times 2 + 3 \times 1 \end{bmatrix} = \begin{bmatrix} 10 \end{bmatrix}.$$

By formula (18.6), the product of **ba** is a 3×3 matrix whose entries are determined by formula (18.7) as follows:

$$\begin{aligned} &d_{11} = b_1 a_1 = 3 \times 1 = 3, \quad d_{12} = b_1 a_2 = 3 \times 2 = 6, \quad d_{13} = b_1 a_3 = 3 \times 3 = 9, \\ &d_{21} = b_2 a_1 = 2 \times 1 = 2, \quad d_{22} = b_2 a_2 = 2 \times 2 = 4, \quad d_{23} = b_2 a_3 = 2 \times 3 = 6, \\ &d_{31} = b_3 a_1 = 1 \times 1 = 1, \quad d_{32} = b_3 a_2 = 1 \times 2 = 2, \quad d_{33} = b_3 a_3 = 1 \times 3 = 3. \end{aligned}$$

The new 3×3 matrix is

$$\mathbf{D} = \mathbf{ba} = \begin{bmatrix} 3 \\ 2 \\ 2 \end{bmatrix} \begin{bmatrix} 1 & 2 & 3 \end{bmatrix} = \begin{bmatrix} 3 & 6 & 9 \\ 2 & 4 & 6 \\ 1 & 2 & 3 \end{bmatrix}.$$

Some rules of matrix multiplication

If **A**, **B**, and **C** are matrices with the same size, the following rules of matrix multiplication exist

$$\begin{cases}(\mathbf{A}+\mathbf{B})\mathbf{C}=\mathbf{AC}+\mathbf{BC}\\ \mathbf{C}(\mathbf{A}+\mathbf{B})=\mathbf{CA}+\mathbf{CB}\\ \lambda(\mathbf{AB})=(\lambda\mathbf{A})\mathbf{B}=\mathbf{A}(\lambda\mathbf{B}) \quad (\lambda \text{ is a constant.})\\ \mathbf{A}(\mathbf{BC})=(\mathbf{AB})\mathbf{C}\\ \mathbf{AI}=\mathbf{IA}=\mathbf{A} \quad (\mathbf{I} \text{ is unit matrix.})\end{cases} \tag{18.8}$$

Example 18.8: Given $\mathbf{A}=\begin{bmatrix}3 & -1\\ 2 & 1\end{bmatrix}$, $\mathbf{B}=\begin{bmatrix}-1 & 0\\ 3 & 2\end{bmatrix}$, $\mathbf{C}=\begin{bmatrix}0 & 1\\ 4 & -2\end{bmatrix}$, prove $(\mathbf{A}+\mathbf{B})\mathbf{C}=\mathbf{AC}+\mathbf{BC}$, $\mathbf{C}(\mathbf{A}+\mathbf{B})=\mathbf{CA}+\mathbf{CB}$ and $(\mathbf{A}+\mathbf{B})\mathbf{C}\neq\mathbf{C}(\mathbf{A}+\mathbf{B})$.

Solution

$$\mathbf{A+B}=\begin{bmatrix}3 & -1\\ 2 & 1\end{bmatrix}+\begin{bmatrix}-1 & 0\\ 3 & 2\end{bmatrix}=\begin{bmatrix}3-1 & -1+0\\ 2+3 & 1+2\end{bmatrix}=\begin{bmatrix}2 & -1\\ 5 & 3\end{bmatrix}$$

$$\mathbf{(A+B)C}=\begin{bmatrix}2 & -1\\ 5 & 3\end{bmatrix}\begin{bmatrix}0 & 1\\ 4 & -2\end{bmatrix}=\begin{bmatrix}2\times0+(-1)\times4 & 2\times1+(-1)\times(-2)\\ 5\times0+3\times4 & 5\times1+3\times(-2)\end{bmatrix}=\begin{bmatrix}-4 & 4\\ 12 & -1\end{bmatrix}$$

$$\mathbf{AC}=\begin{bmatrix}3 & -1\\ 2 & 1\end{bmatrix}\begin{bmatrix}0 & 1\\ 4 & -2\end{bmatrix}=\begin{bmatrix}3\times0+(-1)\times4 & 3\times1+(-1)\times(-2)\\ 2\times0+1\times4 & 2\times1+1\times(-2)\end{bmatrix}=\begin{bmatrix}-4 & 5\\ 4 & 0\end{bmatrix}$$

$$\mathbf{BC}=\begin{bmatrix}-1 & 0\\ 3 & 2\end{bmatrix}\begin{bmatrix}0 & 1\\ 4 & -2\end{bmatrix}=\begin{bmatrix}(-1)\times0+0\times4 & (-1)\times1+0\times(-2)\\ 3\times0+2\times4 & 3\times1+2\times(-2)\end{bmatrix}=\begin{bmatrix}0 & -1\\ 8 & -1\end{bmatrix}$$

$$\mathbf{AC+BC}=\begin{bmatrix}-4 & 5\\ 4 & 0\end{bmatrix}+\begin{bmatrix}0 & -1\\ 8 & -1\end{bmatrix}=\begin{bmatrix}-4+0 & 5-1\\ 4+8 & 0-1\end{bmatrix}=\begin{bmatrix}-4 & 4\\ 12 & -1\end{bmatrix}=\mathbf{(A+B)C}$$

$$\mathbf{C(A+B)}=\begin{bmatrix}0 & 1\\ 4 & -2\end{bmatrix}\begin{bmatrix}2 & -1\\ 5 & 3\end{bmatrix}=\begin{bmatrix}0\times2+1\times5 & 0\times(-1)+1\times3\\ 4\times2+(-2)\times5 & 4\times(-1)+(-2)\times3\end{bmatrix}=\begin{bmatrix}5 & 3\\ -2 & -10\end{bmatrix}\neq\mathbf{(A+B)C}$$

Example 18.9: Prove $\mathbf{AI}=\mathbf{IA}=\mathbf{A}$ given $\mathbf{A}=\begin{bmatrix}2 & -2 & 4\\ 0 & 3 & 1\\ 1 & 2 & 5\end{bmatrix}$.

Solution

$$\mathbf{IA}=\begin{bmatrix}1 & 0 & 0\\ 0 & 1 & 0\\ 0 & 0 & 1\end{bmatrix}\begin{bmatrix}2 & -2 & 4\\ 0 & 3 & 1\\ 1 & 2 & 5\end{bmatrix}=\begin{bmatrix}1\times2+0\times0+0\times1 & 1\times(-2)+0\times3+0\times2 & 1\times4+0\times1+0\times5\\ 0\times2+1\times0+0\times1 & 0\times(-2)+1\times3+0\times2 & 0\times4+1\times1+0\times5\\ 0\times2+0\times0+1\times1 & 0\times(-2)+0\times3+1\times2 & 0\times4+0\times1+1\times5\end{bmatrix}$$

$$=\begin{bmatrix}2 & -2 & 4\\ 0 & 3 & 1\\ 1 & 2 & 5\end{bmatrix}=\mathbf{A}.$$

$$\mathbf{AI}=\begin{bmatrix}2&-2&4\\0&3&1\\1&2&5\end{bmatrix}\begin{bmatrix}1&0&0\\0&1&0\\0&0&1\end{bmatrix}=\begin{bmatrix}2\times1+(-2)\times0+4\times0&2\times0+(-2)\times1+4\times0&2\times0+(-2)\times0+4\times1\\0\times1+3\times0+1\times0&0\times0+3\times1+1\times0&0\times0+3\times0+1\times1\\1\times1+2\times0+5\times0&1\times0+2\times1+5\times0&1\times0+2\times0+5\times1\end{bmatrix}$$

$$=\begin{bmatrix}2&-2&4\\0&3&1\\1&2&5\end{bmatrix}=\mathbf{A}.$$

Therefore we proved **AI** = **IA** = **A**.

Exercises 18.1

1. Given $\mathbf{A}=\begin{bmatrix}4&1\\5&3\end{bmatrix}$ and $\mathbf{B}=\begin{bmatrix}2&3\\1&1\end{bmatrix}$, calculate **A** + **B**, and **B** – **A**.
2. Given $\mathbf{A}=\begin{bmatrix}1&2\\3&4\end{bmatrix}$ and $\mathbf{B}=\begin{bmatrix}1&1\\0&1\end{bmatrix}$, calculate **AB**, and **BA**.
3. Given $\mathbf{A}=\begin{bmatrix}2&0&1\\1&-4&-1\\-1&8&3\end{bmatrix}$, calculate **AA**.
4. Given $\mathbf{A}=\begin{bmatrix}1&1&1\\1&1&-1\\1&-1&1\end{bmatrix}$ and $\mathbf{B}=\begin{bmatrix}1&2&3\\-1&-2&4\\0&5&1\end{bmatrix}$, calculate **AB** and 3**AB** – 2**A**.
5. Given $\mathbf{A}=\begin{bmatrix}4&1&2&4\\1&2&0&2\\10&5&2&0\\0&1&1&7\end{bmatrix}$ and $\mathbf{B}=\begin{bmatrix}2&1&4&1\\3&-1&2&1\\1&2&3&2\\5&0&6&2\end{bmatrix}$, calculate **B** – 2**A**, **AB**, and **BA**.

18.2 Determinants

Determinants

A ***determinant of order*** n is a scalar associated with an $n \times n$ matrix $\mathbf{A} = [a_{ij}]$ denoted as

$$D = \det \mathbf{A} = |\mathbf{A}| = \begin{vmatrix} a_{11} & a_{12} & \cdots & a_{1n} \\ a_{21} & a_{22} & \cdots & a_{2n} \\ \cdots & \cdots & \cdots & \cdots \\ a_{n1} & a_{n2} & \cdots & a_{nn} \end{vmatrix}. \tag{18.9}$$

This definition means that determinants must be square. For $n = 1$, the determinant is $D = a_{11}$.

Determinants of 2×2 matrices

For $n = 2$, the determinant is

$$\boxed{D = \begin{vmatrix} a_{11} & a_{12} \\ a_{21} & a_{22} \end{vmatrix} = a_{11}C_{11} + a_{12}C_{12} = a_{11}a_{22} - a_{12}a_{21}\,.} \tag{18.10}$$

This is known as cross-multiplication for the second-order determinant that is the product of the two values on the main diagonal (*from the top left corner to the bottom right corner*) taken away the product of the two values on the minor diagonal (*from the top right to the bottom left corner*).

Determinants of 3×3 matrices

For $n = 3$, the determinant is

$$\boxed{\begin{aligned} D = \begin{vmatrix} a_{11} & a_{12} & a_{13} \\ a_{21} & a_{22} & a_{23} \\ a_{31} & a_{32} & a_{33} \end{vmatrix} &\longrightarrow \begin{vmatrix} a_{11} & a_{12} & a_{13} & | & a_{11} & a_{12} \\ a_{21} & a_{22} & a_{23} & | & a_{21} & a_{22} \\ a_{31} & a_{32} & a_{33} & | & a_{31} & a_{32} \end{vmatrix} \\ &= a_{11}a_{22}a_{33} + a_{12}a_{23}a_{31} + a_{13}a_{21}a_{32} - a_{11}a_{23}a_{32} - a_{12}a_{21}a_{33} - a_{13}a_{22}a_{31} \end{aligned}} \tag{18.11}$$

This is equivalent to the sum of the three multiplications along the three main diagonals in the expanded determinant (appending the first two columns to the right of the third column) by taking away the three multiplications along the minor diagonals in the expanded determinant.

Example 18.10: Given $\mathbf{A} = \begin{bmatrix} 2 & 0 & 1 \\ -2 & 3 & 2 \\ 4 & 1 & 5 \end{bmatrix}$, calculate its determinant.

Solution

$$D = \begin{vmatrix} 2 & 0 & 1 \\ -2 & 3 & 2 \\ 4 & 1 & 5 \end{vmatrix} \longrightarrow \begin{vmatrix} 2 & 0 & 1 & 2 & 0 \\ -2 & 3 & 2 & -2 & 3 \\ 4 & 1 & 5 & 4 & 1 \end{vmatrix}$$

$$= 2\times 3\times 5 + 0\times 2\times 4 + 1\times(-2)\times 1 - 2\times 2\times 1 - 0\times(-2)\times 5 - 1\times 3\times 4$$
$$= 30 + 0 - 2 - 4 - 0 - 12 = 30 - 18 = 12.$$

Determinants of triangular and diagonal matrices

The value of an ***nth-order upper or lower triangular determinant*** is the product of all values on the diagonal, i.e.,

$$D = \begin{vmatrix} a_{11} & a_{12} & \cdots & a_{1n} \\ & a_{22} & \cdots & a_{2n} \\ & & \cdots & \cdots \\ & & & a_{nn} \end{vmatrix} = \begin{vmatrix} a_{11} & & & \\ a_{21} & a_{22} & & \\ \cdots & \cdots & \cdots & \\ a_{n1} & a_{n2} & \cdots & a_{nn} \end{vmatrix} = a_{11}a_{22}...a_{nn}. \quad (18.12)$$

The value of any ***diagonal determinant*** is the product of all values on the diagonal:

$$D = \begin{vmatrix} \lambda_1 & & & \\ & \lambda_2 & & \\ & & \cdots & \\ & & & \lambda_n \end{vmatrix} = \lambda_1\lambda_2...\lambda_n. \quad (18.13)$$

Example 18.11: Determine the determinants of the matrices given below:

$$\mathbf{P} = \begin{bmatrix} 3 & 2 & -1 \\ 0 & 5 & 1 \\ 0 & 0 & 2 \end{bmatrix}, \quad \mathbf{Q} = \begin{bmatrix} -6.7 & 0 & 0 & 0 \\ 6.3 & 4.7 & 0 & 0 \\ 1.3 & 2.2 & -2.0 & 0 \\ 5.2 & -3.1 & 2.7 & 3.3 \end{bmatrix}, \quad \mathbf{R} = \begin{bmatrix} 2 & 0 & 0 \\ 0 & 1 & 0 \\ 0 & 0 & -3 \end{bmatrix}.$$

Solution

P is a lower triangular matrix whereas **Q** is an upper triangular matrix. By (18.12), the determinants of these two matrices can be obtained as follows:

$$\mathbf{P}=\begin{vmatrix}3 & 2 & -1\\ 0 & 5 & 1\\ 0 & 0 & 2\end{vmatrix}=3\times5\times2=30$$

$$\mathbf{Q}=\begin{vmatrix}-6.7 & 0 & 0 & 0\\ 6.3 & 4.7 & 0 & 0\\ 1.3 & 2.2 & -2.0 & 0\\ 5.2 & -3.1 & 2.7 & 3.3\end{vmatrix}=(-6.7)\times4.7\times(-2)\times3.3=207.834.$$

R is a diagonal matrix. By (18.13), its determinant is

$$\mathbf{R}=\begin{vmatrix}2 & 0 & 0\\ 0 & 1 & 0\\ 0 & 0 & -3\end{vmatrix}=2\times1\times(-3)=-6.$$

Exercises 18.2

1. Calculate the determinant of each of the following matrices:

a) $\begin{bmatrix}4 & 1\\ 5 & 3\end{bmatrix}$ b) $\begin{bmatrix}2 & 3\\ 1 & 1\end{bmatrix}$ c) $\begin{bmatrix}\cos x & -\sin x\\ \sin x & \cos x\end{bmatrix}$ d) $\begin{bmatrix}1 & 2 & -1\\ 3 & 4 & -2\\ 5 & -4 & 1\end{bmatrix}$

e) $\begin{bmatrix}2 & 0 & 1\\ 1 & -4 & -1\\ -1 & 8 & 3\end{bmatrix}$ f) $\begin{bmatrix}a & b & c\\ b & c & a\\ c & a & b\end{bmatrix}$

2. Calculate the determinant of each of the following matrices:

a) $\begin{bmatrix}3 & -2 & 0 & -1\\ 0 & 2 & 2 & 1\\ 1 & -2 & -3 & -2\\ 0 & 1 & 2 & 1\end{bmatrix}$ b) $\begin{bmatrix}1 & -2 & -3 & -2\\ 0 & 1 & 2 & 1\\ 0 & -2 & 0 & -1\\ 0 & 2 & 2 & 1\end{bmatrix}$ c) $\begin{bmatrix}2 & 1 & 4 & 1\\ 3 & -1 & 2 & 1\\ 1 & 2 & 3 & 2\\ 5 & 0 & 6 & 2\end{bmatrix}$

d) $\begin{bmatrix}1 & 3 & -1 & 2\\ 0 & -1 & 0 & -4\\ 0 & 0 & -2 & 1\\ 0 & 0 & 0 & 3\end{bmatrix}$ e) $\begin{bmatrix}\lambda_1 & & & 0\\ & \lambda_2 & & \\ & & \lambda_3 & \\ 0 & & & \lambda_4\end{bmatrix}$

18.3 Solving Systems of Linear Equations

18.3.1 Systems of linear equations

Systems of linear equations

A **system of *m* linear equations in *n* unknowns** $x_1, \ldots, x_n$ is a set of equations in the form of

$$\begin{array}{l} a_{11}x_1 + a_{12}x_2 + \ldots + a_{1n}x_n = b_1 \\ a_{21}x_1 + a_{22}x_2 + \ldots + a_{2n}x_n = b_2 \\ \ldots\ldots\ldots\ldots\ldots \\ a_{m1}x_1 + a_{m2}x_2 + \ldots + a_{mn}x_n = b_m \end{array}. \tag{18.14}$$

The coefficients a_{ij} on the left side and b_i on the right side are known values. If at least one b_i is nonzero, it is an ***inhomogeneous system***. If all values of b_i are zero, system (18.14) is called a ***homogeneous system***, i.e.,

$$\begin{array}{l} a_{11}x_1 + a_{12}x_2 + \ldots + a_{1n}x_n = 0 \\ a_{21}x_1 + a_{22}x_2 + \ldots + a_{2n}x_n = 0 \\ \ldots\ldots\ldots\ldots\ldots \\ a_{m1}x_1 + a_{m2}x_2 + \ldots + a_{mn}x_n = 0 \end{array}. \tag{18.15}$$

Matrix form for systems of linear equations

System (18.14) can be expressed in the matrix format as follows:

$$\mathbf{Ax} = \mathbf{b}, \tag{18.16}$$

where **A** is the coefficient matrix; **x** is the unknown vector and **b** is the known vector noted as

$$\mathbf{A} = \begin{bmatrix} a_{11} & a_{12} & \ldots & a_{1n} \\ a_{21} & a_{22} & \ldots & a_{2n} \\ \ldots & \ldots & \ldots & \ldots \\ a_{m1} & a_{m2} & \ldots & a_{mn} \end{bmatrix}, \ \mathbf{x} = \begin{bmatrix} x_1 \\ x_2 \\ \ldots \\ x_m \end{bmatrix} \text{ and } \mathbf{b} = \begin{bmatrix} b_1 \\ b_2 \\ \ldots \\ b_m \end{bmatrix}. \tag{18.17}$$

Augmented matrix

Since both **A** and **b** contain known values, both can be rewritten in one expanded matrix as

$$\tilde{\mathbf{A}} = \begin{bmatrix} a_{11} & \ldots & a_{1n} & | \, b_1 \\ . & \ldots & . & | \, . \\ . & \ldots & . & | \, . \\ a_{m1} & \ldots & a_{mn} & | \, b_m \end{bmatrix}. \tag{18.18}$$

This expanded matrix is called the ***augmented matrix*** of system (18.14).

A ***solution*** of system (18.14) is a set of values x_1, …, x_n that satisfies all the m equations. These values form a ***solution vector*** **x** for system (18.14).

Example 18.12: Given the following system of linear equations, find its coefficient matrix, known vector and unknown vector. Then express this system using its matrix form, construct its augmented matrix, and classify if it is a homogeneous system or an inhomogeneous system.

$$\begin{cases} 2x + z = -1 \\ -2x + 3y + 2z = 1. \\ 4x + y + 5z = 0 \end{cases}$$

Solution

By formula (18.17), the coefficient matrix, known vector and unknown vector can be found as

$$\mathbf{A} = \begin{bmatrix} 2 & 0 & 1 \\ -2 & 3 & 2 \\ 4 & 1 & 5 \end{bmatrix} \quad \mathbf{b} = \begin{bmatrix} -1 \\ 1 \\ 0 \end{bmatrix} \quad \mathbf{x} = \begin{bmatrix} x \\ y \\ z \end{bmatrix} \longrightarrow \begin{cases} 2x + z = -1 \\ -2x + 3y + 2z = 1 \\ 4x + y + 5z = 0 \end{cases} \Rightarrow \quad \begin{bmatrix} 2 & 0 & 1 \\ -2 & 3 & 2 \\ 4 & 1 & 5 \end{bmatrix} \begin{bmatrix} x \\ y \\ z \end{bmatrix} = \begin{bmatrix} -1 \\ 1 \\ 0 \end{bmatrix}$$

Its augmented matrix is

$$\widetilde{\mathbf{A}} = \left[\begin{array}{ccc|c} 2 & 0 & 1 & -1 \\ -2 & 3 & 2 & 1 \\ 4 & 1 & 5 & 0 \end{array}\right].$$

As the known vector **b** is nonzero, this is an inhomogeneous system.

Example 18.13: Verify $\mathbf{x} = \begin{bmatrix} -1 \\ -1 \\ 1 \end{bmatrix}$ is the solution to the system of equations in Example 18.12.

Solution

The solution vector means $x = -1$, $y = -1$, and $z = 1$. We can substitute these values into the system of equations to verify if all equations are still balanced on both sides after the substitutions.

$$2x + z = \underline{-1} \xrightarrow{x=-1, z=1} 2 \times (-1) + 1 = -2 + 1 = \underline{-1}$$
$$-2x + 3y + 2z = \underline{1} \xrightarrow{x=-1, y=-1, z=1} -2 \times (-1) + 3 \times (-1) + 2 \times 1 = 2 - 3 + 2 = \underline{1}$$
$$4x + y + 5z = \underline{0} \xrightarrow{x=-1, y=-1, z=1} 4 \times (-1) - 1 + 5 \times 1 = -4 - 1 + 5 = \underline{0}$$

This verifies that the given vector is the solution to the system of equations in Example 18.12.

18.3.2 Solving systems of linear equations by Cramer's rule

Cramer's rule

A **system of n linear equations in n unknowns** $x_1, \ldots, x_n$ can be expressed by matrix format as follows:

$$\begin{cases} a_{11}x_1 + a_{12}x_2 + \ldots + a_{1n}x_n = b_1 \\ a_{21}x_1 + a_{22}x_2 + \ldots + a_{2n}x_n = b_2 \\ \ldots\ldots\ldots\ldots\ldots \\ a_{n1}x_1 + a_{n2}x_2 + \ldots + a_{nn}x_n = b_n \end{cases} \Rightarrow \begin{pmatrix} a_{11} & a_{12} & \cdots & a_{1n} \\ a_{21} & a_{22} & \cdots & a_{2n} \\ \vdots & \vdots & \vdots & \vdots \\ a_{n1} & a_{n2} & \cdots & a_{nn} \end{pmatrix} \begin{pmatrix} x_1 \\ x_2 \\ \vdots \\ x_n \end{pmatrix} = \begin{pmatrix} b_1 \\ b_2 \\ \vdots \\ b_n \end{pmatrix} \quad \text{or} \qquad (18.19)$$

$$\mathbf{Ax} = \mathbf{b} \quad \text{where } \mathbf{A} = \begin{pmatrix} a_{11} & a_{12} & \cdots & a_{1n} \\ a_{21} & a_{22} & \cdots & a_{2n} \\ \vdots & \vdots & \vdots & \vdots \\ a_{n1} & a_{n2} & \cdots & a_{nn} \end{pmatrix}, \ \mathbf{x} = \begin{pmatrix} x_1 \\ x_2 \\ \vdots \\ x_n \end{pmatrix}, \ \mathbf{b} = \begin{pmatrix} b_1 \\ b_2 \\ \vdots \\ b_n \end{pmatrix}. \qquad (18.20)$$

Depending on conditions of the determinant of the coefficient matrix **A**, Cramer's rule states as follows:

1) If $D = \det\mathbf{A} \neq 0$, the system has precisely one set of solutions determined by

$$x_1 = \frac{D_1}{D}, \ x_2 = \frac{D_2}{D}, \ldots, \ x_n = \frac{D_n}{D} \qquad (18.21)$$

   where D_i is the determinant obtained from D by replacing the ith column in D by the column on the right, i.e., $b_1, b_2, \ldots, b_n$.
2) If $\mathbf{b} = \mathbf{0}$ in the system (18.20), known as homogeneous system, and $D \neq 0$, it has only the zero solution $x_1 = 0, \ldots, x_n = 0$.
3) If $D = 0$, the homogeneous system has nonzero solutions.
4) If $D = 0$, but any $D_i \neq 0$, the inhomogeneous system has no solution.

Example 18.14: Use Cramer's rule to solve the following system of equations:

$$\begin{cases} x + 2y = 5 \\ 4x - y = 2 \end{cases} \Rightarrow \begin{bmatrix} 1 & 2 \\ 4 & -1 \end{bmatrix} \begin{bmatrix} x \\ y \end{bmatrix} = \begin{bmatrix} 5 \\ 2 \end{bmatrix}$$

Solution

We first calculate the determinants for this 2×2 system using formula (18.10).

$$D = \begin{vmatrix} 1 & 2 \\ 4 & -1 \end{vmatrix} = 1 \times (-1) - 2 \times 4 = -1 - 8 = -9$$

$$D_x = \begin{vmatrix} 5 & 2 \\ 2 & -1 \end{vmatrix} = 5\times(-1) - 2\times 2 = -9 - 4 = -9, \quad D_y = \begin{vmatrix} 1 & 5 \\ 4 & 2 \end{vmatrix} = 1\times 2 - 5\times 4 = 2 - 20 = -18$$

$$x = \frac{D_x}{D} = \frac{-9}{-9} = 1, \quad y = \frac{D_y}{D} = \frac{-18}{-9} = 2$$

Example 18.15: Use Cramer's rule to solve the following system of equations:

$$\begin{cases} 2x + z = -1 \\ -2x + 3y + 2z = 1 \\ 4x + y + 5z = 0 \end{cases} \Rightarrow \begin{bmatrix} 2 & 0 & 1 \\ -2 & 3 & 2 \\ 4 & 1 & 5 \end{bmatrix} \begin{bmatrix} x \\ y \\ z \end{bmatrix} = \begin{bmatrix} -1 \\ 1 \\ 0 \end{bmatrix}$$

Solution

We used formula (18.11) to calculate the determinants for the system. The determinant of the coefficient matrix has been obtained in Example 18.10 as

$$D = \begin{vmatrix} 2 & 0 & 1 \\ -2 & 3 & 2 \\ 4 & 1 & 5 \end{vmatrix} = 12$$

$$D_x = \begin{vmatrix} -1 & 0 & 1 \\ 1 & 3 & 2 \\ 0 & 1 & 5 \end{vmatrix} = -1\times 3\times 5 + 0\times 2\times 0 + 1\times 1\times 1 - 1\times 3\times 0 - 0\times 1\times 5 - (-1)\times 2\times 1$$

$$= -15 + 0 + 1 - 0 - 0 + 2 = 3 - 15 = -12.$$

$$D_y = \begin{vmatrix} 2 & -1 & 1 \\ -2 & 1 & 2 \\ 4 & 0 & 5 \end{vmatrix} = 2\times 1\times 5 + (-1)\times 2\times 4 + 1\times(-2)\times 0 - 1\times 1\times 4 - (-1)\times(-2)\times 5 - 2\times 2\times 0$$

$$= 10 - 8 + 0 - 4 - 10 - 0 = -12.$$

$$D_z = \begin{vmatrix} 2 & 0 & -1 \\ -2 & 3 & 1 \\ 4 & 1 & 0 \end{vmatrix} = 2\times 3\times 0 + 0\times 1\times 4 + (-1)\times(-2)\times 1 - (-1)\times 3\times 4 - 0\times(-2)\times 0 - 2\times 1\times 1$$

$$= 0 + 0 + 2 + 12 - 0 - 2 = 12.$$

By formula (18.21), the solutions to the system are

$$x = \frac{D_x}{D} = \frac{-12}{12} = -1, \quad y = \frac{D_y}{D} = \frac{-12}{12} = -1, \quad z = \frac{D_z}{D} = \frac{12}{12} = 1$$

This is the same solution we verified in Example 18.13.

Example 18.16: A chemical engineer wishes to make 600 litres of 35% acid solution. He has solutions of 25% and 40% to mix up. How many litres of each solution should be mixed to make up the 600 litres of 35% solution? Use Cramer's rule to solve this equation.

Solution

Suppose that x litres of 25% solution and y litres of 40% solution are required. By the information provided, we can come up with the two equations as follows:

$$\begin{cases} 0.25x + 0.4y = 0.35 \times 600 \quad \text{(balance in total amount of acid)} \\ x + y = 600 \quad \text{(balance in total volume of solution)} \end{cases}$$

This translates to the following matrix equation

$$\begin{bmatrix} 0.25 & 0.4 \\ 1 & 1 \end{bmatrix} \begin{bmatrix} x \\ y \end{bmatrix} = \begin{bmatrix} 210 \\ 600 \end{bmatrix}$$

$$D = \begin{vmatrix} 0.25 & 0.4 \\ 1 & 1 \end{vmatrix} = 0.25 \times 1 - 0.4 \times 1 = -0.15$$

$$D_x = \begin{vmatrix} 210 & 0.4 \\ 600 & 1 \end{vmatrix} = 210 \times 1 - 0.4 \times 600 = 210 - 240 = -30$$

$$D_y = \begin{vmatrix} 0.25 & 210 \\ 1 & 600 \end{vmatrix} = 0.25 \times 600 - 210 \times 1 = 150 - 210 = -60$$

$$x = \frac{D_x}{D} = \frac{-30}{-0.15} = 200, \quad y = \frac{D_y}{D} = \frac{-60}{-0.15} = 400$$

Therefore, it requires 200 litres of 25% solution and 400 litres of 40% solution.

Example 18.17: An electrical network is shown in Figure 18.1. Solve this network using Cramer's rule.

Solution

We apply Kirchhoff's circuit laws to the given network to obtain the system of equations first. Kirchhoff's circuit laws state that

- At any node (junction) in an electrical circuit, the sum of currents flowing into that node is equal to the sum of currents flowing out of that node (Kirchhoff's current law).
- The algebraic sum of the products of the resistances of the conductors and the currents in them in a closed loop is equal to the total electrical potential of sources in that loop (Kirchhoff's voltage law).

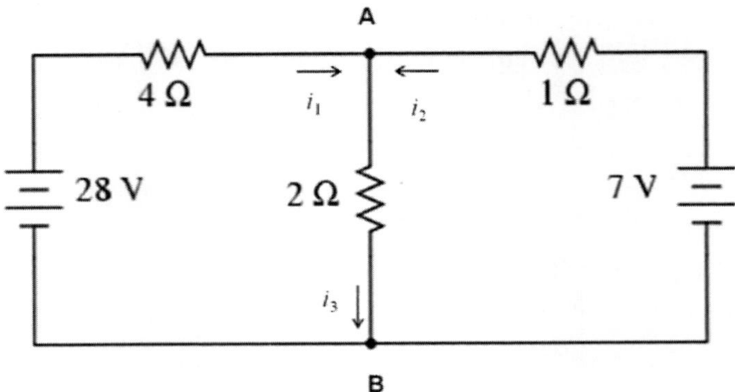

Figure 18.1: Electrical network in Example 18.17

At Node A, by Kirchhoff's current law, we have

$$i_1 + i_2 - i_3 = 0\,.$$

For the right loop, by Kirchhoff's voltage law, we have

$$i_2 + 2i_3 = 7\,.$$

For the left loop, by Kirchhoff's voltage law, we have

$$4i_1 + 2i_3 = 28\,.$$

Thus the system of equations is

$$\begin{aligned} i_1 + i_2 - i_3 &= 0 \\ i_2 + 2i_3 &= 7 \\ 4i_1 + 2i_3 &= 28 \end{aligned}\,.$$

In matrix format, this system is expressed as

$$\begin{bmatrix} 1 & 1 & -1 \\ 0 & 1 & 2 \\ 4 & 0 & 2 \end{bmatrix} \begin{bmatrix} i_1 \\ i_2 \\ i_3 \end{bmatrix} = \begin{bmatrix} 0 \\ 7 \\ 28 \end{bmatrix}.$$

We first calculate the determinants for the system.

$$D = \begin{vmatrix} 1 & 1 & -1 \\ 0 & 1 & 2 \\ 4 & 0 & 2 \end{vmatrix} = 14 \qquad D_1 = \begin{vmatrix} 0 & 1 & -1 \\ 7 & 1 & 2 \\ 28 & 0 & 2 \end{vmatrix} = 70$$

$$D_2 = \begin{vmatrix} 1 & 0 & -1 \\ 0 & 7 & 2 \\ 4 & 28 & 2 \end{vmatrix} = -14 \qquad D_3 = \begin{vmatrix} 1 & 1 & 0 \\ 0 & 1 & 7 \\ 4 & 0 & 28 \end{vmatrix} = 56$$

$$i_1 = \frac{D_1}{D} = \frac{70}{14} = 5, \quad i_2 = \frac{D_2}{D} = \frac{-14}{14} = -1, \quad i_3 = \frac{D_3}{D} = \frac{56}{14} = 4$$

Note that i_2 is negative, which means the current flows in the opposite direction as assumed in Figure 18.1.

Example 18.18: Use Cramer's rule to solve the following system of equations.

$$\begin{cases} x+y+5z=7 \\ x+2y+10z=13 \\ -5x+2y+10z=7 \end{cases} \quad \Rightarrow \quad \begin{bmatrix} 1 & 1 & 5 \\ 1 & 2 & 10 \\ -5 & 2 & 10 \end{bmatrix} \begin{bmatrix} x \\ y \\ z \end{bmatrix} = \begin{bmatrix} 7 \\ 13 \\ 7 \end{bmatrix}.$$

Solution

We need to calculate the determinant of matrix **A** first using formula (18.11).

$$D = \begin{vmatrix} 1 & 1 & 5 \\ 1 & 2 & 10 \\ -5 & 2 & 10 \end{vmatrix} = 1\times2\times10 + 1\times10\times(-5) + 5\times1\times2 - 5\times2\times(-5) - 1\times1\times10 - 1\times10\times2$$
$$= 20 - 50 + 10 + 50 - 10 - 20 = 0$$

According to Cramer's rule, if $D = 0$, the determinants of other replacement matrices must be zero so as to have nonzero solutions. Otherwise the system has no solution. We check if this condition is met.

$$D_x = \begin{vmatrix} 7 & 1 & 5 \\ 13 & 2 & 10 \\ 7 & 2 & 10 \end{vmatrix} = 7\times2\times10 + 1\times10\times7 + 5\times13\times2 - 5\times2\times7 - 1\times13\times10 - 7\times10\times2$$
$$= 140 + 70 + 130 - 70 - 130 - 140 = 0$$

$$D_y = \begin{vmatrix} 1 & 7 & 5 \\ 1 & 13 & 10 \\ -5 & 7 & 10 \end{vmatrix} = 1\times13\times10 + 7\times10\times(-5) + 5\times1\times7 - 5\times13\times(-5) - 7\times1\times10 - 1\times10\times7$$
$$= 130 - 350 + 35 + 325 - 70 - 70 = 490 - 490 = 0$$

$$D_z = \begin{vmatrix} 1 & 1 & 7 \\ 1 & 2 & 13 \\ -5 & 2 & 7 \end{vmatrix} = 1\times2\times7 + 1\times13\times(-5) + 7\times1\times2 - 7\times2\times(-5) - 1\times1\times7 - 1\times13\times2$$
$$= 14 - 65 + 14 + 70 - 7 - 26 = 98 - 98 = 0$$

Indeed, all determinants of this system are zero and thus it has nonzero solutions. This scenario indicates that the three equations are NOT independent from each other, i.e., one equation can be expressed by linear combinations of the other two equations. Further this means there are THREE unknowns but only TWO independent equations. This system has infinitely many solutions with one unknown taking any arbitrary value. We can choose and rearrange the first two equations in the system as

$$\begin{aligned} x+y&=7-5z \\ x+2y&=13-10z \end{aligned}.$$

We then apply Cramer's rule to this subsystem. The determinants for this system are

$$D=\begin{vmatrix} 1 & 1 \\ 1 & 2 \end{vmatrix}=2-1=1$$

$$D_x=\begin{vmatrix} 7-5z & 1 \\ 13-10z & 2 \end{vmatrix}=2(7-5z)-(13-10z)=1$$

$$D_y=\begin{vmatrix} 1 & 7-5z \\ 1 & 13-10z \end{vmatrix}=(13-10z)-(7-5z)=6-5z$$

$$x=\frac{D_x}{D}=\frac{1}{1}=1,\quad y=\frac{D_y}{D}=\frac{6-5z}{1}=6-5z\,.$$

In summary, the solution of this system of equations is

$$\begin{aligned} z&=c \\ y&=6-5z=6-5c\,, \\ x&=1 \end{aligned}$$

where c is any arbitrary value. Any value will satisfy z, which in turn satisfies y. Thus there are infinitely many solutions to the given system.

18.3.3 Elementary row operations and Gauss elimination

Elementary row operations with augmented matrix of a system of linear equations

We know that the matrix form of a system of linear equations is

$$\mathbf{Ax}=\mathbf{b}\quad\text{where }\mathbf{A}=\begin{pmatrix} a_{11} & a_{12} & \cdots & a_{1n} \\ a_{21} & a_{22} & \cdots & a_{2n} \\ \vdots & \vdots & \vdots & \vdots \\ a_{n1} & a_{n2} & \cdots & a_{nn} \end{pmatrix},\ \mathbf{x}=\begin{pmatrix} x_1 \\ x_2 \\ \vdots \\ x_n \end{pmatrix},\ \mathbf{b}=\begin{pmatrix} b_1 \\ b_2 \\ \vdots \\ b_n \end{pmatrix}.$$

If we ignore the **x** vector, matrix **A** and vector **b** are combined as the augmented matrix $\tilde{\mathbf{A}}$,

$$\tilde{\mathbf{A}} = \left(\begin{array}{cccc|c} a_{11} & a_{12} & \cdots & a_{1n} & b_1 \\ a_{21} & a_{22} & \cdots & a_{2n} & b_2 \\ \vdots & \vdots & \vdots & \vdots & \vdots \\ a_{n1} & a_{n2} & \cdots & a_{nn} & b_n \end{array}\right).$$

For any augmented matrix, the following ***elementary row operations*** are applicable:

1) Interchange of two rows does not change the final outcome;
2) Addition to or subtraction from a row of a multiple of another row does not change the final outcome;
3) Multiplication or division of a row by a constant factor does not change the final outcome.

Gauss elimination

The system of equations below forms an upper triangular matrix

$$\begin{aligned} 2x_1 + x_2 - 2x_3 &= 1 \\ 3x_2 + x_3 &= 4 \\ 3x_3 &= -6 \end{aligned} \quad \Rightarrow \quad \begin{bmatrix} 2 & 1 & -2 \\ 0 & 3 & 1 \\ 0 & 0 & 3 \end{bmatrix} \begin{bmatrix} x_1 \\ x_2 \\ x_3 \end{bmatrix} = \begin{bmatrix} 1 \\ 4 \\ -6 \end{bmatrix}.$$

Solution to this system can be worked out backwards from the last equation to the first in turn as follows:

$$\begin{aligned} 3x_3 &= -6 \\ 3x_2 + x_3 &= 4 \\ 2x_1 + x_2 - 2x_3 &= 1 \end{aligned} \quad \Rightarrow \quad \begin{aligned} x_3 &= \frac{-6}{3} = -2 \\ x_2 &= \frac{1}{3}(4 - x_3) = \frac{1}{3}(4+2) = 2 \\ x_1 &= \frac{1}{2}(1 - x_2 + 2x_3) = \frac{1}{2}(1-2-4) = \frac{-5}{2} \end{aligned}.$$

This backward process for solving systems of linear equations with an upper triangular matrix is called **back substitution**. This process leads to the idea of *Gauss elimination*, which is to use elementary row operations to reduce a coefficient matrix **A** to an upper triangular matrix. The back substitution is then used to obtain the solutions. Note that Gauss elimination works on the augmented matrix $\tilde{\mathbf{A}}$, rather than the coefficient matrix **A** alone.

Gauss elimination is very useful in solving higher-order systems of equations.

Example 18.19: Use Gauss elimination to solve the system of equations for the electrical network in Example 18.17.

Solution
We reduce the augmented matrix to an upper triangular matrix as follows

$$\begin{bmatrix} 1 & 1 & -1 | & 0 \\ 0 & 1 & 2 | & 7 \\ 4 & 0 & 2 | & 28 \end{bmatrix} \to \begin{matrix} \\ \\ R3-4R1 \end{matrix}\begin{bmatrix} 1 & 1 & -1 | & 0 \\ 0 & 1 & 2 | & 7 \\ 0 & -4 & 6 | & 28 \end{bmatrix} \to \begin{matrix} \\ \\ R3+4R2 \end{matrix}\begin{bmatrix} 1 & 1 & -1 | & 0 \\ 0 & 1 & 2 | & 7 \\ 0 & 0 & 14 | & 56 \end{bmatrix} \Rightarrow \begin{aligned} i_1 + i_2 - i_3 &= 0 \\ i_2 + 2i_3 &= 7 \\ 14i_3 &= 56 \end{aligned}$$

We then use back substitution to get the solutions as follows:

$$\begin{aligned} 14i_3 &= 56 \\ i_2 + 2i_3 &= 7 \\ i_1 + i_2 - i_3 &= 0 \end{aligned} \quad \Rightarrow \quad \begin{aligned} i_3 &= \frac{56}{14} = 4 \\ i_2 &= 7 - 2i_3 = 7 - 8 = -1 \\ i_1 &= -i_2 + i_3 = 1 + 4 = 5 \end{aligned}$$

This is the same as the solution obtained in Example 18.17 using Cramer's rule.

Example 18.20: The tensions, T_1, T_2, and T_3 in a framework, form a system of equations as follows. Use Gauss elimination to solve this system.

$$\begin{cases} T_1 + 2T_2 + 4T_3 = 15 \\ 2T_1 + T_2 + 3T_3 = 10 \\ 3T_1 + T_2 + T_3 = 9 \end{cases}$$

Solution

$$\begin{bmatrix} 1 & 2 & 4 | 15 \\ 2 & 1 & 3 | 10 \\ 3 & 1 & 1 | \ 9 \end{bmatrix} \to \begin{matrix} \\ R2-2R1 \\ R3-3R1 \end{matrix}\begin{bmatrix} 1 & 2 & 4 | & 15 \\ 0 & -3 & -5 | & -20 \\ 0 & -5 & -11 | & -36 \end{bmatrix} \to \begin{matrix} \\ -5R2 \\ -3R3 \end{matrix}\begin{bmatrix} 1 & 2 & 4 | & 15 \\ 0 & 15 & 25 | & 100 \\ 0 & 15 & 33 | & 108 \end{bmatrix}$$

$$\to \begin{matrix} \\ \\ R3-R2 \end{matrix}\begin{bmatrix} 1 & 2 & 4 | & 15 \\ 0 & 15 & 25 | & 100 \\ 0 & 0 & 8 | & 8 \end{bmatrix} \Rightarrow \begin{aligned} T_1 + 2T_2 + 4T_3 &= 15 \\ 15T_2 + 25T_3 &= 100 \\ 8T_3 &= 8 \end{aligned}$$

We then use back substitution to get the solutions as follows:

$$\begin{aligned} 8T_3 &= 8 \\ 15T_2 + 25T_3 &= 100 \\ T_1 + 2T_2 + 4T_3 &= 15 \end{aligned} \quad \Rightarrow \quad \begin{aligned} T_3 &= \frac{8}{8} = 1 \\ T_2 &= \frac{1}{15}(100 - 25T_3) = \frac{1}{15}(100 - 25 \times 1) = \frac{75}{15} = 5. \\ T_1 &= 15 - 2T_2 - 4T_3 = 15 - 10 - 4 = 1 \end{aligned}$$

Example 18.21: Use Gauss elimination to solve the following system of equations:

$$\begin{aligned} 4x_1 + 2x_2 - x_3 &= 2 \\ 3x_1 - x_2 + 2x_3 &= 10 \; . \\ 11x_1 + 3x_2 \qquad &= 8 \end{aligned}$$

Solution

We reduce the augmented matrix to an upper triangular matrix as follows

$$\begin{bmatrix} 4 & 2 & -1| & 2 \\ 3 & -1 & 2| & 10 \\ 11 & 3 & 0| & 8 \end{bmatrix} \to \begin{matrix} R1-R2 \\ \\ R3-3R2 \end{matrix} \begin{bmatrix} 1 & 3 & -3| & -8 \\ 3 & -1 & 2| & 10 \\ 2 & 6 & -6| & -22 \end{bmatrix} \to \begin{matrix} \\ R2-3R1 \\ R3-2R1 \end{matrix} \begin{bmatrix} 1 & \mathit{3} & \mathit{-3}| & \mathit{-8} \\ 0 & \mathit{-10} & \mathit{11}| & \mathit{34} \\ 0 & \mathit{0} & \mathit{0}| & \mathit{-6} \end{bmatrix}.$$

This system has no solution because $|\mathbf{A}| = 0$ but $|\tilde{\mathbf{A}}| \neq 0$ (the right 3×3 matrix with italic values). The last row indicates $0x_1 + 0x_2 + 0x_3 = -6$, which is impossible.

Example 18.22: Use Gauss elimination to solve the following system of equations:

$$\begin{cases} x_1 + x_2 + x_3 - x_4 = 2 \\ x_1 - x_2 + x_3 + x_4 = 6 \\ -x_1 + x_2 + x_3 - x_4 = 0 \\ -x_1 + x_2 - x_3 + 4x_4 = 14 \end{cases}$$

Solution

We reduce the augmented matrix to an upper triangular matrix as follows

$$\begin{bmatrix} 1 & 1 & 1 & -1| & 2 \\ 1 & -1 & 1 & 1| & 6 \\ -1 & 1 & 1 & -1| & 0 \\ -1 & 1 & -1 & 4| & 14 \end{bmatrix} \to \begin{matrix} \\ R2-R1 \\ R3+R1 \\ R4+R1 \end{matrix} \begin{bmatrix} 1 & 1 & 1 & -1| & 2 \\ 0 & -2 & 0 & 2| & 4 \\ 0 & 2 & 2 & -2| & 2 \\ 0 & 2 & 0 & 3| & 16 \end{bmatrix} \to \begin{matrix} \\ \\ R3+R2 \\ R4+R2 \end{matrix} \begin{bmatrix} 1 & 1 & 1 & -1| & 2 \\ 0 & -2 & 0 & 2| & 4 \\ 0 & 0 & 2 & 0| & 6 \\ 0 & 0 & 0 & 5| & 20 \end{bmatrix}$$

We then use back substitution to get the solutions as follows:

$$\begin{aligned} x_1 + x_2 + x_3 - x_4 &= 2 \\ -2x_2 \qquad + 2x_4 &= 4 \\ 2x_3 \qquad &= 6 \\ 5x_4 &= 20 \end{aligned} \Rightarrow \begin{aligned} 5x_4 &= 20 \\ 2x_3 \qquad &= 6 \\ -2x_2 \qquad + 2x_4 &= 4 \\ x_1 + x_2 + x_3 - x_4 &= 2 \end{aligned} \Rightarrow \begin{aligned} x_4 &= 4 \\ x_3 &= 3 \\ x_2 &= -2 + x_4 = -2 + 4 = 2 \\ x_1 &= 2 - x_2 - x_3 + x_4 = 2 - 2 - 3 + 4 = 1 \end{aligned}$$

Exercises 18.3

1. Solve the following systems of equations using Cramer's rule:

a) $\begin{cases} x_1 + 3x_2 + 2x_3 = -1 \\ 2x_1 + 3x_2 + 3x_3 = -2 \\ -2x_1 + 2x_2 - 3x_3 = 7 \end{cases}$ (b) $\begin{cases} 2x + y + 2z = 10 \\ 4x + 4y + 7z = 33 \\ 2x + 5y + 12z = 48 \end{cases}$

2. Solve the following systems of equations using Gauss elimination:

a) $\begin{cases} x_1 + x_2 + x_3 + x_4 = 5 \\ x_1 + 2x_2 - x_3 + 4x_4 = -2 \\ 2x_1 - 3x_2 - x_3 - 5x_4 = -2 \\ 3x_1 + x_2 + 2x_3 + 11x_4 = 0 \end{cases}$

b) $\begin{cases} 5x_1 + 6x_2 = 1 \\ x_1 + 5x_2 + 6x_3 = 0 \\ x_2 + 5x_3 + 6x_4 = 0 \\ x_3 + 5x_4 = 1 \end{cases}$

c) $\begin{cases} x_2 + x_3 + x_4 = 0 \\ 3x_1 + 3x_3 - 4x_4 = 7 \\ x_1 + x_2 + x_3 + 2x_4 = 6 \\ 2x_1 + 3x_2 + x_3 + 3x_4 = 6 \end{cases}$

d) $\begin{cases} 2x_1 + x_2 - x_3 + x_4 = 1 \\ 4x_1 + 2x_2 - 2x_3 + x_4 = 2 \\ 2x_1 + x_2 - x_3 - x_4 = 1 \end{cases}$

e) $\begin{cases} 2x_1 + 3x_2 + x_3 = 4 \\ x_1 - 2x_2 + 4x_3 = -5 \\ 3x_1 + 8x_2 - 2x_3 = 13 \\ 4x_1 - x_2 + 9x_3 = -6 \end{cases}$

Chapter 18: Exercises Answers

Exercises 18.1

1. $\mathbf{A}+\mathbf{B}=\begin{bmatrix}6 & 4\\6 & 4\end{bmatrix}$, $\mathbf{B}-\mathbf{A}=\begin{bmatrix}-2 & 2\\-4 & -2\end{bmatrix}$

2. $\mathbf{AB}=\begin{bmatrix}1 & 3\\3 & 7\end{bmatrix}$, $\mathbf{BA}=\begin{bmatrix}4 & 6\\3 & 4\end{bmatrix}$

3. $\mathbf{AA}=\begin{bmatrix}3 & 8 & 5\\-1 & 8 & 2\\3 & -8 & 0\end{bmatrix}$

4. $\mathbf{AB}=\begin{bmatrix}0 & 5 & 8\\0 & -5 & 6\\2 & 9 & 0\end{bmatrix}$, $3\mathbf{A}^{\mathrm{T}}\mathbf{B}-2\mathbf{A}=\begin{bmatrix}-2 & 13 & 22\\-2 & -17 & 20\\4 & 29 & -2\end{bmatrix}$

5. $\mathbf{B}-2\mathbf{A}=\begin{bmatrix}-6 & -1 & 0 & -7\\1 & -5 & 2 & -3\\-19 & -8 & -1 & 2\\5 & -2 & 4 & -12\end{bmatrix}$, $\mathbf{AB}=\begin{bmatrix}33 & 7 & 48 & 17\\18 & -1 & 20 & 7\\37 & 9 & 56 & 19\\39 & 1 & 47 & 17\end{bmatrix}$, $\mathbf{BA}=\begin{bmatrix}49 & 25 & 13 & 17\\31 & 12 & 11 & 17\\36 & 22 & 10 & 22\\80 & 37 & 24 & 34\end{bmatrix}$

Exercises 18.2

1. a) 7, b) –1, c) 1, d) 2, e) –4, f) $3abc-a^3-b^3-c^3$
2. a) 1, b) –2, c) 0, d) 6, e) $\lambda_1\lambda_2\lambda_3\lambda_4$

Exercises 18.3

1. a) $x_1=\dfrac{D_1}{D}=\dfrac{10}{5}=2,\ x_2=\dfrac{D_2}{D}=\dfrac{5}{5}=1,\ x_3=\dfrac{D_3}{D}=\dfrac{-15}{3}=-3$

   b) $x=\dfrac{D_x}{D}=\dfrac{16}{16}=1,\ y=\dfrac{D_y}{D}=\dfrac{32}{16}=2,\ z=\dfrac{D_z}{D}=\dfrac{48}{16}=3$

2. a) $x_1=1, x_2=2, x_3=3, x_4=-1$; b) $x_1=-\dfrac{151}{211}, x_2=\dfrac{161}{211}, x_3=-\dfrac{109}{211}, x_4=\dfrac{64}{211}$;

   c) $x_1=4, x_2=-3, x_3=1, x_4=2$

   d) $x_1=c_1$ (any arbitrary number), $x_2=c_2$ (any arbitrary number), $x_3=2x_1+x_2-1, x_4=0$,

   e) $x_1=-1-2x_3, x_2=2+x_3, x_3=c$ (any arbitrary number)

General References

[1] Catler, P.A., and Calter, M.A. Technology Mathematics, 4th Edition, Wiley, New York, 2000.

[2] Croft, A., Davison, R., Hargreaves, M., and Flint, J. Engineering Mathematics, 4th Edition, Pearson, England, 2013.

[3] Gillett, P. Calculus and Analytic Geometry, D.C. Heath and Company, USA, 1981.

[4] Guo, W.W. Essentials and Examples of Applied Mathematics, 1st Edition, Pearson, Sydney, 2018.

[5] Guo, W.W. Advanced Mathematics for Engineering and Applied Sciences, 3rd Edition, Pearson, Sydney, 2016.

[6] Guo, W.W. Magnetic petrophysics and density investigations of the Hamersley Province, Western Australia: implications for magnetic and gravity interpretation, The University of Western Australia, Perth, 1999.

[7] Guo, W.W. A novel application of neural networks for instant iron-ore grade estimation, *Expert Systems with Applications,* 37, 8729–8735, 2010.

[8] Haeussler, E.F., Paul, R.S., and Wood, R.J. Introductory Mathematical Analysis, 13th Edition, Pearson, Boston, 2011.

[9] Kasube, Herbert E. A Technique for Integration by Parts. *The American Mathematical Monthly*, 90(3), 210–211, 1983.

[10] Xuan, L.X. Advanced Mathematics, 1st Edition, Higher Education Press, Beijing, 1999.

Appendix: Formulae of Indefinite Integration

No	$f(x)$	$\int f(x)dx$
1	$\sin^2 ax$	$\frac{1}{4}(2x-\frac{1}{a}\sin 2ax)$
2	$\sin^n ax$ (*n*: positive integer)	$I(n)=-\frac{1}{na}\cos^{n-1}ax\sin ax+\frac{n-1}{n}I(n-2)$
3	$\frac{1}{\sin ax}$	$\frac{1}{a}\ln\tan\frac{ax}{2}=\frac{1}{a}\ln(\csc ax-\cot ax)$
4	$\frac{1}{\sin^2 ax}$	$-\frac{1}{a}\cot ax$
5	$\frac{1}{1+\sin ax}$	$-\frac{1}{a}\tan(\frac{\pi}{4}-\frac{ax}{2})$
6	$\frac{1}{1-\sin ax}$	$\frac{1}{a}\tan(\frac{\pi}{4}+\frac{ax}{2})$
7	$\cos^2 ax$	$\frac{1}{4}(2x+\frac{1}{a}\sin 2ax)$
8	$\cos^n ax$ (*n*: positive integer)	$I(n)=\frac{1}{na}\cos^{n-1}ax\sin ax+\frac{n-1}{n}I(n-2)$
9	$\frac{1}{\cos ax}$	$\frac{1}{a}\ln\tan(\frac{\pi}{4}+\frac{ax}{2})=\frac{1}{a}\ln(\tan ax+\sec ax)$
10	$\frac{1}{\cos^2 ax}$	$\frac{1}{a}\tan ax$
11	$\frac{1}{1+\cos ax}$	$\frac{1}{a}\tan\frac{ax}{2}$
12	$\frac{1}{1-\cos ax}$	$-\frac{1}{a}\cot\frac{ax}{2}$
13	$\sin ax\cos bx$ ($\lvert a\rvert\neq\lvert b\rvert$)	$-\frac{\cos(a-b)x}{2(a-b)}-\frac{\cos(a+b)x}{2(a+b)}$
14	$\frac{1}{\sin ax\cos ax}$	$\frac{1}{a}\ln\tan ax$
15	$\tan ax$	$-\frac{1}{a}\ln\cos ax$
16	$\tan^2 ax$	$\frac{1}{a}\tan ax-x$
17	$x\sin ax$	$\frac{1}{a^2}\sin ax-\frac{x}{a}\cos ax$
18	$x^n\sin ax$	$I(n)=-\frac{x^n}{a}\cos ax+\frac{n}{a^2}x^{n-1}\sin ax-\frac{n(n-1)}{a^2}I(n-2)$
19	$x\cos ax$	$\frac{1}{a^2}\cos ax+\frac{x}{a}\sin ax$
20	$x^n\cos ax$ ($n>0$)	$I(n)=\frac{x^n}{a}\sin ax+\frac{n}{a^2}x^{n-1}\cos ax-\frac{n(n-1)}{a^2}I(n-2)$

21	xe^{ax}	$\frac{e^{ax}}{a^2}(ax-1)$
22	$x^n e^{ax} \quad (n>0)$	$I(n)=\int x^n e^{ax}dx=\frac{x^n}{a}e^{ax}-\frac{n}{a}I(n-1)$
23	$e^{ax}\sin bx$	$\frac{e^{ax}}{a^2+b^2}(a\sin bx-b\cos bx)$
24	$e^{ax}\cos bx$	$\frac{e^{ax}}{a^2+b^2}(a\cos bx+b\sin bx)$
25	$e^{ax}\sin^n bx$	$I(n)=\frac{e^{ax}\sin^{n-1}bx(a\sin bx-nb\cos bx)}{a^2+(nb)^2}+\frac{n(n-1)b^2}{a^2+(nb)^2}I(n-2)$
26	$e^{ax}\cos^n bx$	$I(n)=\frac{e^{ax}\cos^{n-1}bx(a\cos bx+nb\sin bx)}{a^2+(nb)^2}+\frac{n(n-1)b^2}{a^2+(nb)^2}I(n-2)$
27	$\ln ax$	$x\ln ax-x$
28	$x\ln ax$	$\frac{x^2}{2}\ln ax-\frac{x^2}{4}$
29	$x^2\ln ax$	$\frac{x^3}{3}\ln ax-\frac{x^3}{9}$
30	$\frac{1}{a^2+x^2} \quad (a>0)$	$\frac{1}{a}\arctan\frac{x}{a}$
31	$\frac{1}{a^2-x^2} \quad (\|x\|<\|a\|)$	$\frac{1}{2a}\ln\frac{a+x}{a-x}$
32	$\frac{1}{x^2-a^2} \quad (\|x\|>\|a\|)$	$\frac{1}{2a}\ln\frac{x-a}{x+a}$
33	$\frac{1}{\sqrt{a^2-x^2}}$	$\arcsin\frac{x}{a}$
34	$\sqrt{a^2-x^2}$	$\frac{x}{2}\sqrt{a^2-x^2}+\frac{a^2}{2}\arcsin\frac{x}{a}$
35	$\frac{1}{\sqrt{a^2+x^2}}$	$\ln(x+\sqrt{a^2+x^2})$
36	$\sqrt{a^2+x^2}$	$\frac{x}{2}\sqrt{a^2+x^2}+\frac{a^2}{2}\ln(x+\sqrt{a^2+x^2})$